Springer Protocols Handbooks

More information about this series at http://www.springer.com/series/8623

Fluorescence In Situ Hybridization (FISH)

Application Guide

Editor

Thomas Liehr

Institut für Humangenetik, Jena, Germany

Second Edition

Springer

Editor
Thomas Liehr
Institut für Humangenetik
Jena, Germany

ISSN 1949-2448 ISSN 1949-2456 (electronic)
Springer Protocols Handbooks
ISBN 978-3-662-52957-7 ISBN 978-3-662-52959-1 (eBook)
DOI 10.1007/978-3-662-52959-1

Library of Congress Control Number: 2016954653

Printed on acid-free paper

This Springer imprint is published by Springer Nature
The registered company is Springer-Verlag GmbH Berlin Heidelberg

Foreword

Of all biological structures, chromosomes are the most fascinating. They embody the essence of life, containing the DNA blueprint of evolution and the source of variation in all organisms. Only by learning about chromosomes can one understand the life cycle of the cell and the mechanisms of heredity. A full appreciation of genetics and molecular biology requires knowledge of chromosome structure and function. There is beauty in the image of the chromosome, fixed, stained and observed with the compound microscope. The stains used in the discovery of bacteria in the nineteenth century led also to the serendipitous discovery of chromosomes. As methods improved, their detailed structure became more apparent. Chromosome variation in number and morphology were found to be species characteristics, and a pair associated with sex determination could be identified in many species. Each advance in chromosome biology followed an advance in chromosome technology. This is the history of cytogenetics.

My fascination with chromosomes makes it a great pleasure for me to read about modern methods for examining them. The second edition of this *Application Guide* to FISH contains the essential protocols required by anyone embarking on a study of molecular cytogenetics and its various applications. These protocols are laid out by an international group of experts with a track record of publications based on firsthand experience of the techniques they describe. The book contains all the well-tried recipes required by the beginner for making fluorescence-labelled probes used in the diagnosis of chromosome aberrations in patients with complex disorders including cancer, for the prenatal diagnosis of genetic disease, for gene mapping and for the elucidation of phylogenetic relationships through cross-species chromosome painting. The preparation of the three main types of DNA probes is described: whole chromosome painting probes made from chromosome-specific DNA labelled by PCR and obtained by flow sorting or microdissection; locus-specific probes isolated from DNA libraries and cloned in bacteria (BACs); and probes containing tandem repeats of DNA sequence, such as centromeric, telomeric or ribosomal DNA. Most of these probes are hybridized to metaphase chromosomes, but can be used also in interphase nuclei (to determine chromosome copy number) or in DNA fibre-FISH (to determine order of close sequences). Other cytogenetic methods are not forgotten, including several array-based comparative genome hybridization techniques useful for identifying small duplications and deletions and for assessing genomic variation at the molecular level.

The first edition was widely used, and this second edition now follows after seven years and is greatly enhanced by the addition of 17 chapters and the careful revision and/or new arrangement of all the original 36. The new additions include methods for breakpoint mapping, for the measurement of telomere length, for the analysis of micronuclei, for the study of epigenetic modifications and for much more. Also the aspect the exciting field of chromosome sequencing is included in the last chapter of this edition.

Thomas Liehr has not only gathered together a distinguished team of contributors but has been personally involved as coauthor in many of the chapters and has edited them all. His outstanding contributions to molecular cytogenetics are well known and serve to guarantee the scientific quality of the publication. This is more than a global cookbook of delightful chromosome fare and will surely lead to many enjoyable feasts to come!

Cambridge University Department of Veterinary Medicine
Cambridge, UK June 2016
Malcolm A. Ferguson-Smith

Preface to First Edition

This book is a unique source of information on the present state of the exciting field of molecular cytogenetics and how it can be applied in research and diagnostics. The basic techniques of fluorescence in situ hybridization and primed in situ hybridization (**PRINS**) are outlined, the multiple approaches and probe sets that are now available for these techniques are described, and applications of them are presented in 36 chapters by authors from ten different countries around the world. The book not only provides the reader with basic and background knowledge on the topic but also gives detailed protocols that show how molecular cytogenetics is currently performed by specialists in this field.

The *FISH Application Guide* initially provides an overview of the (historical) development of molecular cytogenetics, its basic procedures, the equipment required, and probe generation. The book then describes tips and tricks for making different tissues available for molecular cytogenetic studies. These are followed by chapters on various multicolor **FISH** probe sets, their availability, and their potential for use in combination with other approaches. The possible applications that are shown encompass the characterization of marker chromosomes, cryptic cytogenetic aberrations, and epigenetic changes in humans by interphase and metaphase cytogenetics, studies of nuclear architecture, as well as the application of molecular cytogenetics to zoology, botany, and microbiology. As comparative genomic hybridization (**CGH**), including array CGH, is currently indispensable for precisely characterizing minimal chromosomal aberrations, **CGH** and array-based chip techniques are reviewed, and protocols that describe how to perform them are also provided. Finally, an exclusive collection of Internet resources related to cytogenetics, molecular cytogenetics, and molecular genetics is given.

This up-to-date, comprehensive, and unique book is a valuable resource for lecturers and students, newcomers to the field of cytogenetics, as well as specialists in **FISH** techniques. Apart from cytogeneticists, molecular cytogeneticists, and human and clinical geneticists, this book is also of the greatest relevance to those working in the fields of reproduction medicine, oncology, hematology, pathology, cell biology, botany, zoology, evolutionary biology, and microbiology.

Jena, Germany — *Thomas Liehr*
April 2008

Contents

About the Editor

Thomas Liehr, a biologist (diploma, PhD, and postdoctoral degree in human genetics), has been working in molecular cytogenetics since 1991. He has been head of the molecular cytogenetic group at the Institute of Human Genetics, Jena, Germany, since 1998; before he was a PhD fellow of the Herbert Quandt Stiftung der VARTA-AG at the Institute of Human Genetics, Erlangen, Germany. Working intensively in diagnostics, since 2002, he has been a "specialist in human genetics diagnostics" (FHG = Fachhumangenetiker), a title awarded by the German Society of Human Genetics (GfH), and a European registered clinical laboratory geneticist (ErCLG) since 2015. Dr. Liehr's research fields include clinical genetics, leukemia cytogenetics, and the (3D) structures of the interphase and the human chromosomes, as well as breakpoint characterization. The results of his research have been published in 5 books, >50 book chapters, >550 peer-reviewed papers, and ~800 abstracts. His particular expertise is molecular cytogenetics including chromosomal heteromorphisms, small supernumerary marker chromosomes (sSMC), and uniparental disomy, fields for which he has collected all available literature and made it freely available at http://ssmc-tl.com/Start.html. Further, Dr. Liehr has multiple long-standing international collaborations with many scientists around the world including researchers in, e.g., Armenia, Croatia, India, Morocco, Russia, Serbia, Thailand, and Brazil.

Background

Thomas Liehr and Anja Weise

Abstract

The concept of molecular cytogenetics, its history, and perspectives are introduced here. FISH applications in clinical and tumor genetic diagnostics, including diagnostic guidelines and quality control, are reviewed. The impact of molecular cytogenetics in nowadays' research is discussed, and finally a unique collection of internet pages is provided dealing with cytogenetics, molecular cytogenetics, as well as closely related fields.

Keywords Concept of molecular cytogenetics, History and perspectives of molecular cytogenetics, Clinical genetic diagnostics, Tumor genetic diagnostics, Diagnostic guidelines, Quality control, Molecular cytogenetic research, Interphase architecture

1 Molecular Cytogenetics: History and Perspectives

The history of human cytogenetics is marked mainly by different technical developments and can be divided into three major time periods: the pre-banding era (1879–1970), the pure chromosomal banding era (1970–1986), and the molecular cytogenetic era (since 1986). The pre-banding era performed classical cytogenetics only and is characterized by the first visualization of human chromosomes in 1879 [1], the creation of the word "chromosome" (from chroma = color, and soma = body) in 1888 [2], the determination of the correct modal human chromosome number in 1956 [3], and the identification of the first inherited chromosomal abnormality, trisomy 21 in Down syndrome, in 1959 [4]. Interestingly, classical cytogenetics started originally in plants and practically all nonmedical human cytogeneticists founding the field were originally botanists.

The banding era started with the invention of the Q-banding method by Dr. Lore Zech (Uppsala) in 1968 [5, 6]. Many more chromosomal abnormalities, like translocations, inversions, deletions, and insertions, could be detected from then on (for review, see [7]). The GTG banding approach (G-bands by trypsin using Giemsa [8]) is still the gold standard of all cytogenetic techniques

Thomas Liehr (ed.), *Fluorescence In Situ Hybridization (FISH)*, Springer Protocols Handbooks,
DOI 10.1007/978-3-662-52959-1_1,

(chapters by Anja Weise and Thomas Liehr "Pre- and Postnatal Diagnostics and Research on Peripheral Blood, Bone Marrow, Chorion, Amniocytes, and Fibroblasts" and Thomas Liehr and Monika Ziegler "Application of FISH to Previously GTG-Banded and/or Embedded Cytogenetic Slides").

However, the pure banding era ended in 1986 with the first molecular cytogenetic experiment on human chromosomes [9], which was also the starting point for the youngest discipline in human genetics. The major techniques used in molecular cytogenetics are fluorescence in situ hybridization (FISH; Fig. 1) and primed in situ hybridization (PRINS; Fig. 2).

In situ hybridization is an approach that allows nucleic acid sequences to be examined inside cells or on chromosomes and was described first in 1969 as a radioactive variant [10]. Nonradioactive probe labeling, such as biotin detected by avidin coupled to a fluorochrome, was invented in 1981 [11]. In 1989, the primed in situ (PRINS) labeling technique was introduced alternative to conventional FISH for in situ chromosomal detection [12]. PRINS is based on the principles of the polymerase chain reaction (PCR); it uses oligonucleotide primers and a Taq DNA polymerase for the in situ detection of target DNA sequences. The approach has proven its efficiency in numerous types of cells, but its current utilization is limited to the detection of repeat sequences [13]. PRINS labeling combines the high sensitivity of the PCR with the cytological localization of specific DNA sequences. The key to the PRINS reaction is the use of short, unlabeled, and specific oligonucleotide primers. The primers are annealed in situ to their denatured complementary DNA target sequences and then extended by a Taq DNA polymerase in the presence of free nucleotides. The visualization of the generated fragments is enabled by the incorporation of one labeled nucleotide-type. Also, two-color PRINS was reported [14]. As PRINS is applied only in exceptional cases, no corresponding protocols are included here, but may be found elsewhere [13–15].

FISH has been continuously developed and improved, and it is now the most widely used technique for in situ localization, as illustrated by the great variety of applications that use it in research and diagnostics. FISH, like other DNA-based approaches, takes advantage of the ability of nucleic acids to de- and renature. Here, the most relevant feature of nucleic acids is that, in single-stranded DNA, homologous sequences find each other and build a double helix again. In a regular FISH experiment, the formation of DNA-DNA hybrids is intended normally. In other words, the target DNA is fixed on a slide, and the probe DNA (chapter by Fengtang Yang et al. "Generation of Paint Probes from Flow-Sorted and Micro dissected Chromosomes"; chapter by Nadezda Kosyakova et al. "FISH-Microdissection"; chapter by Thomas Liehr "Homemade Locus-Specific FISH Probes: Bacterial Artificial Chromosomes") is labeled; both of these DNAs are denatured to single-stranded DNAs

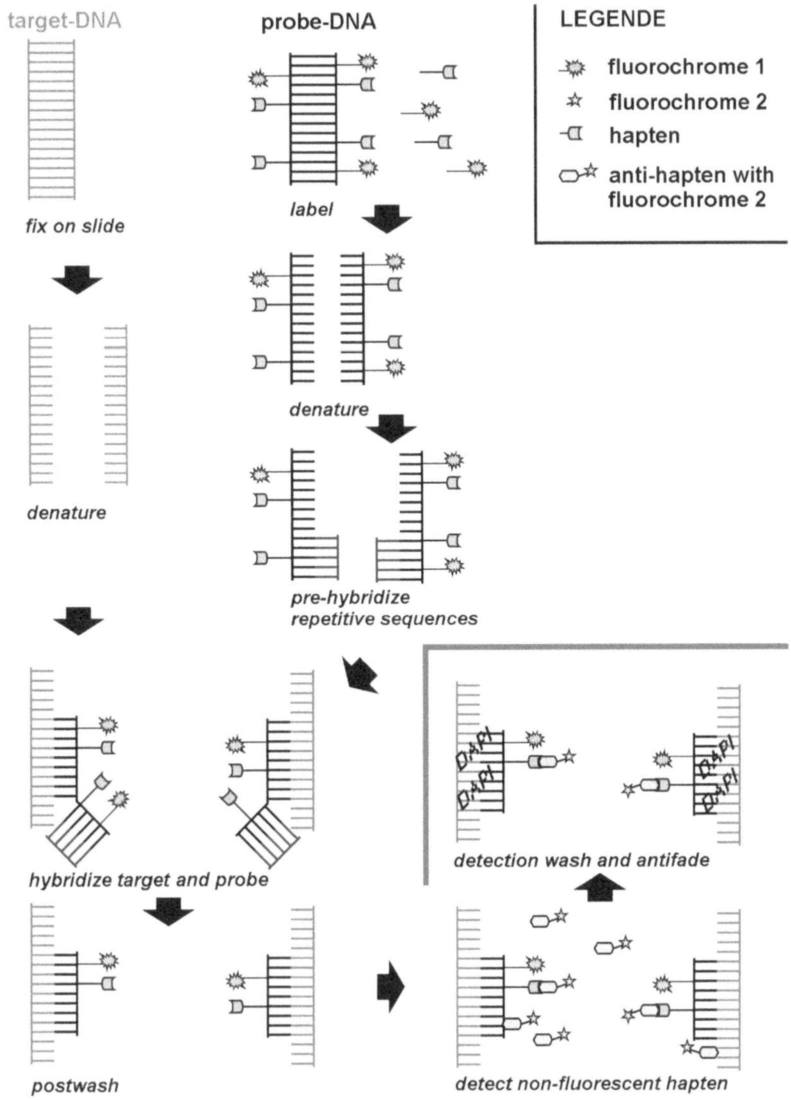

Fig. 1 Flow diagram for fluorescence in situ hybridization (FISH). The target DNA (*light gray*) is fixed onto the slide surface and denatured. In parallel, the probe DNA (*black*) must be labeled by a fluorochrome and/or a nonfluorescent hapten, denatured and pre-hybridized with unlabeled repetitive DNA (*dark gray*). Then probe DNA is brought together with target DNA and hybridized. Postwashing procedures remove unbound single-stranded DNA as well as nonspecifically bound DNA. When a nonfluorescent hapten is used, this must be detected by a fluorescence-coupled anti-hapten. After detection, washing, and application of an antifade solution with DAPI (4,6-diamidino-2-phenylindol.2HCl l), FISH is finished and the slide is ready for inspection under the microscope

and unified in a hybridization mixture for reaction. There are, however, exceptions where PNA or RNA is used as the probe and/or target (chapters by Nikolay Rubtsov and Natalya Zhdanova "The Replicative Detargeting FISH (ReD-FISH) Technique in Studies of Telomere Replication"; by Gordana Joksic et al. "Telomere Length

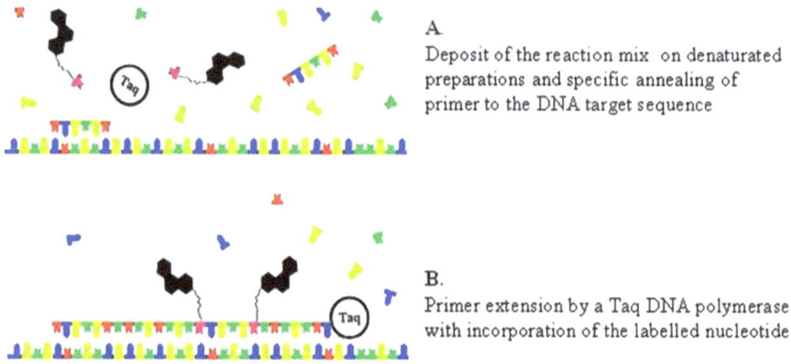

Fig. 2 Principle of the standard PRINS reaction. (**a**) Deposition of the reaction mix onto denatured preparations (involving the specific primer, the dNTP mix with one labeled nucleotide, the Taq DNA polymerase and its buffer) and specific annealing of the primer to the DNA target sequence. (**b**) Primer extension by Taq DNA polymerase with incorporation of the labeled nucleotide, leading to the in situ generation of fluorescent fragments that are directly detectable by fluorescence microscopy

Measurement by FISH"; by Thomas Liehr "Classification of FISH Probes"; chapter by Bin Ma and Naoko Tanese "RNA-Directed FISH and Immunostaining").

The principle of DNA-DNA FISH is as follows (Fig. 1):

- Fix the target DNA onto a slide surface. The target DNA can be cells, nuclei, fixed tissue sections, metaphase chromosomes, or pure DNA.
- Label the probe DNA. Labeling can be direct or indirect. Direct labeling means that the fluorochrome(s) that are to be detected in the microscope are directly bound to the probe DNA. An indirect label refers to the incorporation of a hapten that is not visible under a fluorescence microscope into the probe DNA. However, the hapten can be detected immunohistochemically by a fluorophore-tagged antibody against the hapten: biotin and digoxigenin are the most frequently used haptens for FISH.
- Denature the target and probe DNA.
- In most cases an excess of unlabeled repetitive COT1-DNA is added to the labeled probe DNA and a pre-hybridization is allowed in order to block repetitive elements.
- Renature the target and probe DNA together.
- Perform post-hybridization washes.
- When applying indirectly labeled probes, fluorophore-tagged antibodies should now be used for detection.
- Perform detection washes.
- Add the counterstain, antifade, and coverslip to finish the procedure.

However, variants of this basic protocol are widespread and diverse—some examples may be found in this volume (chapters by Anja Weise and Thomas Liehr "Microwave Treatment for Better FISH Results in a Shorter Time"; Vladimir Trifonov et al. "FISH with and Without COT1 DNA"; chapter by Emanuela Volpi "Form amide-Free Fluorescence In Situ Hybridization (FISH)"; Gábor Méhes et al. "One Day Quick-FISH"; Bin Ma and Naoko Tanese "RNA Imaging in Living Cells"; Sandra Louzada et al. "Fluores cence In Situ Hybridization onto DNA Fibers Generated Using Molecular Combing"; chapter by Galina Hovhannisyan and Rouben Aroutiounian "Comet-FISH"; chapter by Galina Hovhannisyan et al. "Micronucleus FISH"; chapter by Hannes Schmidt, Thilo Eickhorst "Gold-FISH: In Situ Hybridization of Microbial Cells for Combined Fluorescence and Scanning Electron Microscopy").

Molecular cytogenetics is still a rapidly growing field in terms of applications as well as newly developed probe sets (chapter by Thomas Liehr et al. "Two- to Three-Color FISH") and combinations with immunohistochemistry (chapter by Anna Pendina et al. "Immunofluorescent Staining for Cytosine Modifications like 5-Methylcytosine and Its Oxidative Derivatives and FISH"; chapter by Elisabeth Klein and Thomas Liehr "CENP-Antibodies Used Additionally to FISH"; chapter by Tiphaine Aguirre-Lavin and Nathalie Beaujean "Three-Dimensional Immuno-Fluorescence In Situ Hybridization in Preimplantation Mouse"). Besides being applied in human diagnostics (Parts III to V), it is also more and more applied in animals, plants, and microbiology (Part VI). For new approaches we just can refer to the multiple new probe sets being established during the last decade, e.g., parental origin determination FISH ([16], chapter by Anja Weise and Thomas Liehr "Parental Origin Determination FISH: pod-FISH") or pericentromeric-region-oriented probe sets ([17], chapter by Thomas Liehr et al. "Bar-Coding Is Back"). Also array-comparative genomic hybridization (aCGH; (chapter by Eftychia Dimitriadou and Joris Vermeesch et al. "Array CGH")) is a legal child of molecular cytogenetic field, as it is a higher-resolution variant of chromosome-based comparative genomic hybridization (CGH; ([18], chapter by Thomas Liehr et al. "Comparative Genomic Hybridization (CGH) and Microdissection-Based CGH (micro-CGH)")).

2 Applications in Clinical Diagnostics

Molecular cytogenetic clinical diagnostics includes pre- and postnatal questions showing up in connection with human genetic counseling. The families or individuals may go there due to infertility or previous abortions; also genetic counselling and subsequent (molecular) cytogenetic diagnostics can be indicated due to affected own children or close relatives and/or own clinical problems.

Preimplantation diagnostics and studies on gametes by FISH are only performed by rather specialized laboratories and are mostly interphase oriented or done by (single cell based) aCGH ([19, 20], chapter by Maria Bonet Oliver "Sperms, Spermatocytes and Oocytes"; chapter by Eftychia Dimitriadou and Joris Vermeesch et al. "Array CGH"; chapter by Maria Isabel Melaragno and Mariana Moysés-Oliveira "Breakpoint Mapping of Balanced Chromo somal Rearrangements Using Array-CGH of Microdissection Derived FISH-Probes"; chapter by Jiří Štika and Oldřich Mazal "Sequencing of Microdissection Derived FISH-Probes").

Prenatal diagnosis can be done by FISH from amnion, chorion, or umbilical cord blood (chapter by Anja Weise et al. "Blood, Bone Marrow, Chorion, Amniocytes and Fibroblasts"). It may be metaphase or interphase FISH oriented and the latter can be done from native or cultivated material (chapters by Anja Weise and Thomas Liehr "Pre- and Postnatal Diagnostics and Research on Peripheral Blood, Bone Marrow, Chorion, Amniocytes, and Fibroblasts"; Thomas Liehr and Nadezda Kosyakova "Characterization of Mosai cism in Different Easy-to-Acquire Body Tissues such as Buccal Smears, Skin Abrasions, Hair Root Cells or Urine"; and Anja Weise et al. "FISH in Uncultivated Amniocytes"). Also postnatal diagnostics can be clinically or tumor oriented. Probes applied are mainly commercial (chapter by Thomas Liehr et al. "Commercial FISH Probes") and, e.g., centromere, whole chromosome, or locus specific (chapter by Thomas Liehr "Classification of FISH Probes") to study numerical or structural aberrations like trisomies [20], translocations [20], and/or microdeletion syndromes [21].

Another field of FISH application in clinical genetics is the verification of aCGH results ([17], chapter by Eftychia Dimitriadou and Joris Vermeesch et al. "Array CGH") or other submicroscopic aberrations found with molecular techniques (e.g., multiplex ligation-dependent probe amplification) [21]. As these techniques can detect genomic imbalances (either only on several loci or genome-wide), the physical location and the true nature of the chromosomal aberration behind are not resolvable without FISH and might be important for genetic counseling, recurrence risk calculation, or detecting causative balanced rearrangements in relatives.

3 Applications in Tumor Diagnostics

As recently stated in [22]: "it goes without saying that in neoplasia the identification of cytogenetic markers is of high clinical significance for diagnostics, follow-up studies and prognosis. In the first years after introduction of molecular cytogenetics into cancer diagnostics, FISH was most often considered as a tool to continue and refine previous cytogenetic studies. This way to choose and apply corresponding FISH-probes represents still a major part of molecular cytogenetic diagnostics. Besides, molecular cytogenetics is more and more performed independently from banding cytogenetic analyses in all kinds of tumors, too. This development was, among others, supported by the fact that every cytogenetic analysis is in need of dividing cells to produce metaphase spreads. In other words, time-consuming cell culture is necessary (and might produce culture induced artefacts). Thus, interphase directed FISH analyses on tumor cell smear, touch preparations or tissue sections are more and more in use with the goal to achieve a quick ("in vivo") result. FISH approaches are especially suited to characterize chromosomal and subchromosomal copy number changes and gene fusions due to translocations or other rearrangements. All these features are characteristically found acquired aberrations in cancer [22]".

Centromeric region-directed probes in tumor diagnostics are mainly applied for determination and/or confirmation of clones with mono-, tri-, or tetrasomies of single chromosomes; also they are suited for follow-up of sex-mismatched bone marrow transplantation; to detect and characterize tumor-specific translocations whole chromosome paints or locus-specific probes may be applied. The latter can also be used to screen for tumor suppressor gene loss or oncogene amplification ([22], chapter by Eyad Alhourani et al. "Tumorcytogenetic Diagnostics and Research on Blood and Bone Marrow Smears or Effusions"; chapter by Thomas Liehr "Charac terization of Archived Formalin-Fixed/Paraffin-Embedded or Cryofixed Tissue, Including Nucleus Extraction").

Another feature of FISH and cytogenetics is the analysis on single cell level, which is, in case of mosaic aberrations, as to be expected in tumor diagnostics, very useful to detect even small cell clones; if tissue sections are studied, FISH gives additional information on tumor morphology/heterogeneity ([17], chapter by Ivan Iourov et al. "Interphase FISH for Detection of Chromosomal Mosaicism"; chapter by Thomas Liehr and Sven Hauke "Interphase FISH in Diagnostics").

4 Diagnostic Guidelines and Quality Control in FISH

Even though molecular cytogenetics is one of the youngest disciplines of cytogenetics, diagnostic guidelines are available already in Europe [23–25] and North America (e.g., [26, 27]), as well as on selected national levels (e.g., [28]). Also internal and external quality controls are obligatory for diagnostic FISH labs in many countries (http://www.eurogentest.org). In the following, we refer to [29] where we outlined "our thoughts on important points to be taken into account to perform high quality molecular cytogenetic diagnostics and research," as based on [23].

In molecular cytogenetic diagnostics, a profound knowledge on human chromosomes is a necessary prerequisite, i.e., one needs well-trained personnel. The staff has to be able to identify the 24 different human chromosomes on a 300–550 (or higher) haploid band level and recognize alterations. Also for correct interpretation of interphase FISH-based results, knowledge in cytogenetics and heteromorphisms and underlying principles of signal splitting and flaring are necessary [23].

For laboratory equipment it is imperative to have every apparatus in double, starting from backup-storage place (in case a refrigerator breaks down to store probes and samples in the right way) and stopping at fluorescence microscopes. Also the latter must be maintained according to manufacturers' instructions and especially the quality of fluorescence filters and lamps must be monitored in an appropriate way ([23], chapter by Ivan Iourov "Microscopy and Imaging Systems"; chapter by Michael Sommerauer et al. "Optical Filters and Light Sources for FISH").

As most probes for FISH are sold as "not suited for diagnostic purposes" (chapter by Thomas Liehr "Commercial FISH Probes"), quality tests of each probe and each batch need to be done to confirm their identity. This is especially essential if the probes are only applied in interphase FISH. Furthermore, it is important to define the number of cells to be evaluated per test to be performed and set cut-off levels [30, 31].

Also, handling of samples has to be regulated in a way which avoids a mix-up, e.g., that always two people are present at critical steps; documentation and also archiving should be done according to the national guidelines. For quality, it means a certain turnover of diagnostic cases has to be done; if a test is only performed for few cases, routine handling is difficult. Additionally, it may then not be evaluated and finished in a reasonable, diagnostically relevant time frame. Molecular cytogenetic results should be communicated according to the actual version of the international system of cytogenetic nomenclature (ISCN). Even though it is a matter of discussion for some fields of cytogenetics [32–34], overall ISCN-based nomenclature is well established. Still what we wrote for ish-nomenclature previously is valid: it "is at present not really well-engineered and reason for discussions. To achieve the main goal of a molecular cytogenetic report, i.e. to be understandable for the 'customer' (= MD or patient), we tend to avoid the ish nomenclature. We give a standard karyotype formula, and add clear information which probes where applied to come to that result. Also a clear explanation of the obtained results is given in the result interpretation part" [23].

Finally, nowadays, one important criterion for good laboratory quality is also the environmental aspect and by that also the protection of staff, i.e., toxic substances have to be discarded properly according to national safety rules. Overall, "quality is nothing which is achieved once and then present for all future, but it is hard, continuous and daily work to keep and always enhance an achieved standard“[23].

5 Applications in Research

Molecular cytogenetic research is a field, which cannot be reviewed comprehensively here; an impression on its width may be gained by visiting the web site denominated as the “Basics and literature on multicolor fluorescence in situ hybridization application” [35]. Clinical and tumor molecular cytogenetics performed for diagnostic reasons may easily develop to a “research case” and lead to a publication. More complexly composed multicolor-FISH probe sets (chapter by Thomas Liehr and Nadezda Kosyakova “Multiplex FISH and Spectral Karyotyping”; chapter by Thomas Liehr et al. “FISH Banding Techniques”; chapter by Thomas Liehr et al. “cenM-FISH Approaches”; chapter by Thomas Liehr et al. “Het erochromatin Directed M-FISH”; chapter by Anja Weise and Thomas Liehr “Subtelomeric and/or Subcentromeric Probe Sets”; chapter by Thomas Liehr et al. “Bar-Coding Is Back”; chapter by Anja Weise and Thomas Liehr “Parental Origin Determina tion FISH: pod-FISH”; chapter by Christine Ye et al. “Simultaneous Fluorescence Immunostaining and FISH”) are only applied in few specialized laboratories as routine application and may be used for basic research in tumor genetics (e.g., [36]), clinical genetics (e.g., [37]), karyotype evolution (e.g., [38], chapter by Fengtang Yang et al. “Animal Probes and ZOO-FISH”), interphase-architecture studies (e.g., [39], chapter by Thomas Liehr et al. “Three-Dimensional Interphase Analysis Enabled by Suspension FISH”), and many others (Part IV), like specific approaches where FISH is combined with electron microscopy (chapter by Hannes Schmidt, Thilo Eickhorst “Gold-FISH: In Situ Hybridization of Microbial Cells for Combined Fluorescence and Scanning Electron Microscopy”). However, also, the application of simple FISH approaches in complex questions may lead to high-ranked papers [40, 41].

6 FISH and Chips on the Internet

In this last subchapter of this introduction part, we provide a unique collection of websites related to cytogenetics, FISH and FISH probes, array techniques, genome browsers, cytogenetic associations, (molecular) cytogenetic courses, medical literature, genetics and biology, genetics and education, and diagnostic addresses. They are summarized in Table 1. We can obviously only give a subjective selection of pages here. Readers are therefore

Table 1
Web pages in connection with the topics of the present book

Cytogenetics	
Atlas of Genetics and Cytogenetics in Oncology and Haematology	http://atlasgeneticsoncology.org/
cancer genome data @ progenetix.org	http://www.progenetix.de/progenetix/index.html
Chromosome Anomaly Collection	http://www.ngrl.org.uk/wessex/collection/
Chromosome empiric risk calculations	http://www.thegeneticscenter.com/transrsk.htm
Chromosome size	https://en.wikipedia.org/wiki/Human_genome
Chromosomal Variation in Man	http://www.wiley.com/legacy/products/subject/life/borgaonkar/access.html
Cytogenetics Gallery	http://www.pathology.washington.edu/Cytogallery/
Cytogenetics in plants	http://bot.biologia.unipi.it/
DACRO—Disease-Associated Chromosomal Rearrangements Online	https://www1.hgu.mrc.ac.uk/Softdata/Translocation/
DECIPHER—database of unbalanced chromosome aberrations	http://www.sanger.ac.uk/science/tools/decipher-mapping-clinical-genome
Drawing Derivative Chromosomes Online	http://www.cydas.org/OnlineAnalysis/WebExample2.aspx
ECARUCA—database of unbalanced chromosome aberrations	http://www.ecaruca.net
Mitelman Database of Chromosome Aberrations in Cancer	http://cgap.nci.nih.gov/Chromosomes/Mitelman
Small supernumerary marker chromosomes	http://ssmc-tl.com/sSMC.html
FISH	
BAC clones, FISH-mapped	http://cgap.nci.nih.gov/Chromosomes/CCAP_BAC_Clones
BACPAC Resources Center (BPRC)	http://bacpac.chori.org/
e-FISH, an in silico FISH simulation tool	http://projects.tcag.ca/cgi-bin/efish/index.cgi
European Nucleotide Archive	http://www.ebi.ac.uk/ena
Genomic clone database	http://projects.tcag.ca/gcd/
IHC World	http://www.ihcworld.com/in-situ-hybridization.htm
ISCA, International Standards for Cytogenomic Arrays	http://www.ncbi.nlm.nih.gov/projects/gap/cgi-bin/study.cgi?study_id=phs000205.v2.p1
Multicolor FISH database	http://ssmc-tl.com/mfish.html
Research Genetics	http://www.resgen.com/resources/index.php3
SKY/M-FISH and CGH database	http://www.ncbi.nlm.nih.gov/projects/sky/

(continued)

Table 1
(continued)

The Sanger Center	http://www.sanger.ac.uk/Teams/Team63/CloneRequest
Array-CGH and genome browsers for human	
Brown lab guide protocols	http://brownlab.stanford.edu/Pat_Brown_Lab_Home_Page/Resources.html
Browser ENSEMBL	http://www.ensembl.org/index.html
Browser NCBI	http://www.ncbi.nlm.nih.gov/genome/guide/human/
Browser UCSC	http://genome.ucsc.edu/cgi-bin/hgGateway
HUSAR bioinformatics lab	http://genius.embnet.dkfz-heidelberg.de/menu/
Genomic variation database	http://dgv.tcag.ca/dgv/app/home
Segmental duplication database	http://humanparalogy.gs.washington.edu/
THE I.M.A.G.E. Consortium	http://www.imageconsortium.org/
Cytogenetic associations	
Association for Clinical Cytogenetics (ACC)	http://www.cytogenetics.org.uk/
Association des Techniciens en Cytogénétique	http://www.techniciens-cytogenetique.com/#sthash.cUQyQVwa.dpbs
European Cytogenetic Association (ECA)	http://www.e-c-a.eu/en/
l'Association des Cytogénéticiens de Langue Française:/	http://www.eaclf.org
(Molecular) cytogenetic courses announced by	
ECA	http://www.e-c-a.eu/en/COURSES.html
EMBG	https://www.eshg.org/669.0.html
ESHG	https://www.eshg.org/courses.0.html
Medical background	
Human genome project	http://web.ornl.gov/sci/techresources/Human_Genome/
Information for genetic professionals	http://www.kumc.edu/gec/geneinfo.html
NLM catalogue of Multiple Congenital Anomaly/Mental Retardation Syndromes	https://www.nlm.nih.gov/archive/20061212/mesh/jablonski/syndrome_toc/toc_w.html
OMIM—Online Mendelian Inheritance in Man	http://www.ncbi.nlm.nih.gov/omim
Orphanet—portal for rare diseases, syndrome search	http://www.orpha.net/consor/cgi-bin/Disease_DiagnosisAssistance_Simple.php?lng=EN
PubMed	http://www.ncbi.nlm.nih.gov/pubmed
Unique—understanding chromosome disorders	http://www.rarechromo.org/html/home.asp

(continued)

Table 1 (continued)

Genetics background	
Cell and Molecular Biology Online	http://www.cellbio.com/
DNA from the Beginning	http://www.dnaftb.org/
Genetics Home Reference	http://ghr.nlm.nih.gov/ghr/chromosomes
Genetics Made Easy	http://geneticsmadeeasy.com/
History of Genetics	http://homepage.smc.edu/hgp/history.htm
NIH image Gallery	https://www.genome.gov/dmd/
Uniparental Disomy	http://upd-tl.com/upd.html
Diagnostic lists and quality control	
BVDH Qualitycontrol	http://www.bvdh-ringversuche.de/
CEQAS	http://www.ceqas.org/
EuroGeneTest	http://www.eurogentest.org/web/qa/basic.xhtml
GeneTests	https://www.genetests.org/
Human Genetics Quality Network database by BVDH	http://www.hgqn.org/index.php?lang=en

also encouraged to use the multiple links provided on many of the pages listed below to find other fascinating pages. All of the links given below were active in May 2016.

7 Conclusion

It is still valid what Serakinci and Koelvraa stated in 2009 [42]: "FISH techniques were originally developed as extra tools in attempts to map genes and a number of advances were achieved with this new technique. However, it soon became apparent that the FISH concept offered promising possibilities also in a number of other areas in biology and its use spread into new areas of research and also into the area of clinical diagnosis. In very general terms the virtues of FISH are in two areas of biology, namely genome characterization and cellular organization, function and diversity. (...) To what extend FISH technology will be further developed and applied in new areas of research in the future remains to be seen, but the following chapters in the book will give numerous examples of possible future developments" [42].

References

1. Arnold J (1879) Über feinere Strukturen der Zelle unter normalen und pathologischen Bedingungen. Virchows Arch Pathol Anat 77:181–206
2. Waldeyer W (1888) Über Karyogenese und ihre Beziehung zu den Befruchtungsvorgängen. Arch Mikrosc Anat 32:1–22
3. Tijo JH, Levan A (1956) The chromosome number in man. Hereditas 42:1–6
4. Lejeune J, Turpin R, Gautier M (1959) Chromosomic diagnosis of mongolism. Arch Fr Pediatr 16:962–963 (in French)
5. Schlegelberger B (2013) In memoriam: Prof. Dr. rer. nat. Dr. med. h.c. Lore Zech; 24.9.1923 - 13.3.2013: Honorary member of the European Society of Human Genetics, Honorary member of the German Society of Human Genetics, Doctor laureate, the University of Kiel, Germany. Mol Cytogenet 6:20
6. Caspersson T, Farber S, Foley GE et al (1968) Chemical differentiation along metaphase chromosomes. Exp Cell Res 49:219–222
7. Pathak S (1979) Cytogenetic research techniques in humans and laboratory animals that can be applied most profitably to livestock. J Dairy Sci 62:836–843
8. Drets ME, Shaw MW (1971) Specific banding patterns of human chromosomes. Proc Natl Acad Sci USA 68:2073–2077
9. Pinkel D, Straume T, Gray JW (1986) Cytogenetic analysis using quantitative, high sensitivity, fluorescence hybridization. Proc Natl Acad Sci USA 83:2934–2938
10. Gall G, Pardue ML (1969) Formation and detection of RNA-DNA hybrid molecules in cytogenetical preparations. Proc Natl Acad Sci USA 63:378–381
11. Langer PR, Waldrop AA, Ward DC (1981) Enzymatic synthesis of biotin-labeled polynucleotides: novel nucleic acid affinity probes. Proc Natl Acad Sci USA 78:6633–6637
12. Koch JE, Kolvraa S, Petersen KB et al (1989) Oligonucleotide-priming methods for the chromosome-specific labelling of alpha satelliteDNA in situ. Chromosoma 98:259–265
13. Pellestor F (2006) Development and adaptation of the PRINS technology: an overview. In: Pellestor F (ed) PRINS and in situ PCR protocols. Humana, Totowa, NJ, pp 211–220
14. Pellestor F, Girardet A, Lefort G et al (1995) Selection of chromosome specific primers and their use in simple and double PRINS technique for rapid identification of human chromosomes. Cytogenet Cell Genet 70:138–142
15. Liehr T (ed) Fluorescence in situ hybridization (FISH) Application guide, 2009. ISBN-10: 3540705805; ISBN-13: 978-3540705802
16. Weise A, Gross M, Mrasek K et al (2008) Parental-origin-determination fluorescence in situ hybridization distinguishes homologous human chromosomes on a single-cell level. Int J Mol Med 21:189–200
17. Hamid AB, Fan X, Kosyakova N et al (2015) New BAC probe set to narrow down chromosomal breakpoints in small and large derivative chromosomes, especially suited for mosaic conditions. Methods Mol Biol 1227:279–287
18. Kallioniemi A, Kallioniemi OP, Sudar D et al (1992) Comparative genomic hybridization for molecular cytogenetic analysis of solid tumours. Science 258:818–821
19. Fernández SF, Toro E, Colomar A et al (2015) A 24-chromosome FISH technique in preimplantation genetic diagnosis: validation of the method. Syst Biol Reprod Med 61:171–177
20. Liss J, Kiewisz J, Zabielska J et al (2015) Application of FISH method for preimplantation genetic diagnostics of reciprocal and Robertsonian translocations. Folia Histochem Cytobiol 53:162–168
21. Weise A, Mrasek K, Klein E et al (2012) Microdeletion and microduplication syndromes. J Histochem Cytochem 60:346–358
22. Liehr T, Othman MA, Rittscher K et al (2015) The current state of molecular cytogenetics in cancer diagnosis. Expert Rev Mol Diagn 15:517–526
23. E.C.A. (2012) Cytogenetic guidelines and quality assurance, Guidelines Version 2.0. http://www.e-c-a.eu/files/downloads/Guidelines/E.C.A._General_Guidelines_Version-2.0.pdf
24. Hastings RJ, Cavani S, Bricarelli FD et al (2007) Cytogenetic Guidelines and Quality Assurance: a common European framework for quality assessment for constitutional and acquired cytogenetic investigations. Eur J Hum Genet 15:525–527
25. Hastings RJ, Bown N, Tibiletti MG et al (2014) Guidelines for cytogenetic investigations in tumours. E C A Newsletter 34:7–18
26. ACC (2007) Association for Clinical Cytogenetics, Professional Guidelines for Clinical Cytogenetics, Postnatal Best Practice Guidelines v1.01. http://www.cytogenetics.org.uk/prof_standards/acc_postnatal_bp_mar2007_1.01.pdf
27. American College of Medical Genetics (2008) ACMG standards and guidelines for clinical

genetic laboratories. http://www.acmg.net/AM/Template.cfm?Section=Laboratory_Standards_and_Guidelines&Template=/CM/HTMLDisplay.cfm&ContentID=6044

28. Stuhrmann-Spangenberg M, Engels H, Fritz B et al (2011) S2-Leitlinie Humangenetische Diagnostik. Medgen 23:281–323
29. Liehr T (2010) Fluorescence in situ hybridization (FISH) - quality issues in molecular cytogenetics. In: Kristoffersson U, Schmidtke J, Cassiman JJ (eds) Quality issues in clinical genetic services. Springer, Berlin, pp 315–320
30. Liehr T, Ziegler M (2005) Rapid prenatal diagnostics in the interphase nucleus: procedure and cut-off rates. J Histochem Cytochem 53:289–291
31. Gebhart E, Trautmann U, Reichardt S et al (1993) Chromosomal heterogeneity of aneuploid leukemic cell populations detected by conventional karyotyping and by fluorescence in situ hybridization (FISH). Anticancer Res 13:1857–1862
32. Mitelman F, Rowley JD (2007) ISCN (2005) is not acceptable for describing clonal evolution in cancer. Genes Chromosomes Cancer 2007(46):213–214
33. Liehr T (2009) Small supernumerary marker chromosomes (sSMCs): a spotlight on some nomenclature problems. J Histochem Cytochem 57:991–993
34. Kosyakova N, Weise A, Mrasek K et al (2009) The hierarchically organized splitting of chromosomal bands for all human chromosomes. Mol Cytogenet 2:4
35. Liehr T (2016) Basics and literature on multicolor fluorescence in situ hybridization application. http://fish-tl.com/mfish.html
36. Othman MA, Melo JB, Carreira IM et al (2015) High rates of submicroscopic aberrations in karyotypically normal acute lymphoblastic leukemia. Mol Cytogenet 8:45
37. Liehr T (2016) Cytogenetically visible copy number variations (CG-CNVs) in banding and molecular cytogenetics of human; about heteromorphisms and euchromatic variants. Mol Cytogenet 9:5
38. Weise A, Kosyakova N, Voigt M et al (2015) Comprehensive analyses of white-handed gibbon chromosomes enables access to 92 evolutionary conserved breakpoints compared to the human genome. Cytogenet Genome Res 145:42–49
39. Manvelyan M, Kempf P, Weise A et al (2009) Preferred co-localization of chromosome 8 and 21 in myeloid bone marrow cells detected by three dimensional molecular cytogenetics. Int J Mol Med 24:335–341
40. Reichwald K, Petzold A, Koch P et al (2015) Insights into sex chromosome evolution and aging from the genome of a short-lived fish. Cell 163(6):1527–1538
41. Maass PG, Aydin A, Luft FC et al (2015) PDE3A mutations cause autosomal dominant hypertension with brachydactyly. Nat Genet 47:647–653
42. Serakinci N, Koelvraa S (2009) Molecular cytogenetic applications in diagnostics and research – an overview. In: Liehr T (ed) Fluorescence in situ hybridization (FISH) – application guide. Springer, Berlin, pp 3–21. ISBN 9-78-3-540-70580-2

Part I

Required Equipment and Probes for FISH Procedures

Microscopy and Imaging

Ivan Y. Iourov

Abstract

Microscopy is an integral part of fluorescence in situ hybridization (FISH) and related techniques. In combination with imaging technologies, microscopy has become a basis for a variety of FISH-based approaches to qualitative and quantitative molecular cytogenetic analysis of nucleic acids in situ. Here, these basic components of FISH—microscopy and imaging—are discussed. To avoid reproducing numerous textbooks dedicated to the fundamentals of microscopy (including the previous edition of this chapter in 2009) and manufacturer's brochures about commercially available imaging systems, the basic aspects of microscopy and imaging are reviewed in the light of the latest FISH-based molecular cytogenetic developments.

Keywords Microscopy, Imaging, Filters, Autofluorescence, Resolution, Fluorochromes, Digital image analyses, Quantification of FISH results, Multicolor FISH

1 Introduction

Cytogenetics (classical, banding and molecular) is unimaginable without the microscope. Regardless of numerous technologies (i.e., array/on-chip technologies) providing alternative ways of studying the genome at chromosomal and subchromosomal level, microscopy remains essential for almost all diagnostic and basic research of chromosome abnormalities, structure, and behavior (chapter by Thomas Liehr and Anja Weise "Background"). Currently, microscopy is the basis for studying chromatin and chromosomes at molecular/supramolecular levels in cell and structural biology, genome and chromosome research, oncology, reproductive medicine, medical genetics [1–10], and many others (Part IV). Imaging is a technology that has become indispensable for several molecular cytogenetic technological platforms based on fluorescence in situ hybridization (FISH) (i.e., multiplex/multicolor/multiprobe FISH (MFISH) and spectral karyotyping (SKY)) (chapter by Thomas Liehr and Nadezda Kosyakova "Multiplex FISH and Spectral Karyotyping"), FISH banding (chapter by Thomas Liehr

Thomas Liehr (ed.), *Fluorescence In Situ Hybridization (FISH)*, Springer Protocols Handbooks,
DOI 10.1007/978-3-662-52959-1_2, © Springer-Verlag Berlin Heidelberg 2017

et al. "FISH Banding Techniques"), and metaphase comparative genomic hybridization (CGH) (chapter by Thomas Liehr et al. "Comparative Genomic Hybridization (CGH) and Microdissection-Based CGH (micro-CGH)"). Furthermore, machine learning allows extending computational imaging approaches to automation in FISH-based (molecular) cytogenetic studies [11, 12]. Generally, basic concepts of fluorescence microscopy and imaging have not been changed significantly over the last decade, in contrast to the dynamically developing field of molecular cytogenetics [4–7, 9, 10]. Therefore, in the context of molecular cytogenetic technologies based on microscopic visualization, it appears to be more pertinent to focus on microscopy and imaging in the light of specific research tasks that can be solved by different FISH-based techniques.

2 Fluorescence Microscopy

2.1 Basic Concepts

The resolution of visualization in molecular cytogenetic techniques (i.e., FISH-based techniques) is determined by the smallest visible molecular target in the field of microscopic vision. From a microscopic point of view, the analysis is generally referred to as the registration of small point-like (roundish, globular, or spherical) objects, which can be blurred due to diffraction and can impede distinguishing of closely positioned small fluorescing objects. Thus, the closest distinguishable microscopic objects are determined by the resolution. The latter depends on the wavelength (λ), the numerical aperture (NA), the magnification, and detection device resolution. The smallest diameter of a fluorescing roundish/globular object (i.e., FISH signal) to be visualized is defined by the following ratio: $d = 1.22\ \lambda/\mathrm{NA}$. Close-positioned fluorescing objects are still distinguished when the distance between them exceeds half a diameter ($d/2$) [13–15].

Background autofluorescence affects almost all assays that use fluorescence. Indeed, the solution of this problem refers to a decrease of the effective contrast of the adjacent fluorescing objects impeding distinguishing between the signals. The resolution thereby is closely related to the contrast or, in other words, reducing the image contrast results in resolution reduction. Consequently, the background is to be removed using imaging (imaging software). In addition, a treatment of microscopic slides prior to FISH or after the hybridization/detection to increase the signal fluorescence and to remove tissue-specific background autofluorescence can be performed [9, 16–19].

Microscopy resolution of FISH-based techniques can be increased through the adaptation of detection system parameters. Most commonly, the device for detecting FISH results is a digital camera. If so, the resolution can be changed by increasing the quantum efficiency of the detection system, automatic or manual, reducing the noise and optical NA of the microscope and camera (increasing NA decreases *d*) [13, 15]. Furthermore, the resolution is affected by the dynamic range of the acquired data or the maximum number of distinguishable intensity levels. The latter is defined by the technical parameters of the detection system and exposure time is adjusted manually or automatically. Alternatively, digitalization and quantitative FISH analysis is the way to differ between signal fluorescence and background autofluorescence. These manipulations can be applied to evaluate the shape and nature of FISH signals [20–22]. Actually, digitalization and quantitative analysis represent a basis for quantitative FISH (QFISH; chapter by Gordana Joksic et al. "Telomere Length Measurement by FISH").

2.2 Fluorescence Microscope

A fluorescence microscope is basically composed of a light source or an epifluorescence lamp and fluorescence filters, which include three components: (1) excitation filter, (2) dichroic mirror, and (3) emission filter. Although light source position is not mandatory for a microscope used for molecular cytogenetic analyses, the latter are usually performed by epifluorescence microscopes. A high-pressure mercury (Hg) lamp is considered to be the best for cytogenetic purposes giving high-energy excitations at specific wavelengths suitable for the overwhelming majority of FISH-based assays. Such lamps are rarely working more than 200 burning hours (for more details, see chapter by Michael Sommerauer et al. "Optical Filters and Light Sources for FISH"). Producing a considerable amount of heat, lamp cooling is mandatory. Practically, it is better to switch off the lamp when a pause for over 30 min is made (never turn on a lamp which is still hot, an explosion is possible!). Most fluorescence microscopes are equipped with a light shutter to block the exposure of specimen to the light. Exposure to the light is the cause of the decrease of the FISH signal intensity. The fluorescence filters are required for (1) selection of an optimal wavelength to excite the fluorochromes by excitation filter and (2) suppression of the excess excitation light through filtering out the emission wavelengths of the fluorochrome by emission filter. Dichroic mirror and excitation filter produce specific excitation wavelengths and direct the light onto the specimen through an objective lens to produce emission wavelengths seen by a cytogeneticist or registered by an imaging system [13, 14]. Each fluorochrome should contrast sufficiently with the other fluorochromes. The contrast is achieved through proper filter combinations. Single emission filters (one fluorochrome is visible) and multiple (dual or triple) emission filters (where multiple fluorochromes are visible) are available (for more details, see chapter by Michael Sommerauer et al. "Optical Filters and Light Sources for FISH"). FISH can even be applied in electron microscopy (chapter by Hannes Schmidt,

Thilo Eickhorst "Gold-FISH: In Situ Hybridization of Microbial Cells for Combined Fluorescence and Scanning Electron Microscopy").

3 FISH Imaging

FISH imaging system should possess at least these components: a fluorescence microscope, camera (the image sensor, i.e., a charge-coupled device (CCD) camera is a device of choice in FISH-based techniques), and hardware with software for digitizing and processing acquired microscopic images. FISH imaging is generally performed via three steps: (1) acquisition, (2) image pre-processing, and (3) digital analysis.

1. Acquisition of microscopic (FISH) images is naturally influenced by focusing process. To bias the focus, a live image (video) has to be provided by the imaging system. Autofocus option offered by a number of imaging systems does not allow skipping empirical testing of the system, especially in a more sophisticated FISH methodology. Automatic systems are unable to differ between intense autofluorescence particles and FISH signals. On the other hand, automatic protocols are useful for volumetric FISH analysis by producing numerous images (or stacks) based on the specimen [11, 12]. Using additional software simulating three-dimensional nuclei via pseudo images or via videos using similar approaches, are a technological basis for nuclear organization studies [5, 7–9].
2. Pre-processing of microscopic (FISH) images is needed inasmuch as raw images are usually difficult to interpret. The main aim of this manipulation is to remove fluorescence background by thresholding and contrast normalization using the software (interactive or manual). This is especially needed for FISH-based methods for multicolor in situ detection of multiple DNA/RNA targets or multicolor chromosomal banding (MCB) ([2, 6, 10, 21–23], part IV).
3. A digital analysis of microscopic (FISH) images is always specific for each FISH-based approach. The latter determines the choice of imaging systems according to the options of software, which are feasible to perform a molecular cytogenetic analysis. Still, imaging system abilities are secondary to basic FISH procedures (i.e., tissue preparation, FISH conditions, etc.) in an attempt to increase the potential of a FISH-based technique [2, 6, 9, 15, 19].

4 Digital Image Analyses for FISH

4.1 Standard FISH Analysis

Basic FISH imaging using digitalization of a microscopic image only requires a camera, hardware, and software. Software should be a hybrid between a raster graphics and image database management system. Acquiring of FISH microscopic images is needed for molecular cytogenetic preimplantation and pre- and postnatal and oncocytogenetic diagnosis. For instance, far infrared and near ultraviolet fluorochromes undetectable by simple visual analysis are commonly used in diagnostic and research FISH-based assays. In basic research, it is useful for interphase FISH assays, inasmuch as it simplifies the precise scoring of nuclei with chromosome imbalances [2–4, 6, 17]. Since structural chromosome abnormalities are also detectable in interphase by an analysis of juxtapositions of differently colored signals, the digital image registration of nuclei is required to perform this kind of FISH assays ([6, 23], chapter by Ivan Iourov et al. “Interphase FISH for Detection of Chromosomal Mosaicism”).

Digital analysis of FISH microscopic images is suggested to be used for solving problems caused by differences in hybridization efficiency between DNA probe types and chromosomal heteromorphisms (benign loss or gain of pericentromeric alpha satellite DNA) ([24], chapter by Thomas Liehr et al. “Heterochromatin Directed M-FISH (HCM-FISH)”). Similar problems during FISH interpretation can result from nonspecific cross hybridization.

4.2 Q-FISH

Interactive quantification of FISH results is the basis for several multicolor FISH approaches (chapter by Anja Weise and Thomas Liehr “Parental Origin Determination FISH: pod-FISH”; chapter by Galina Hovhannisyan and Rouben Aroutiounian “Comet-FISH”; chapter by Gordana Joksic et al. “Telomere Length Mea surement by FISH”; chapter by Benedetta Bottari et al. “FISHing for Food Microorganisms”). FISH signal quantification per se can be used for detecting DNA content variations within the genomic loci of a signal using different types of DNA probes or genome/chromosome organization in interphase nuclei or on metaphase chromosomes [20, 22]. Thus, Q-FISH is used for differentiating between chromosome abnormalities and variations in interphase intranuclear DNA organization and for the definition of the parental origin of homologous chromosomes [20, 24]. Measuring relative intensities by integrating the signal intensity profiles allow a differentiation between associated/overlapped signals and chromosome loss. The reproducibility is hardly achieved in Q-FISH analyses because of the variability in the intensity of signals and background autofluorescence (chapter by Thomas Liehr et al. “Telomere Length Measurement by FISH”). Therefore, the ratio of relative intensities is the best way for defining homologous chromosome parental origins. Since the ratio is relatively stable because the DNA content in a chromosomal region remains unchanged, variation in signal intensity between different slides can be ignored [20]. Q-FISH is able to help in reducing the background fluorescence to precisely define signal (chromosome region) borderlines through the construction of surface plots depicting intensity variations within the nuclear area or a genomic locus (three-dimensional intensity profiles). For more details, see [20, 22].

4.3 Multicolor FISH and CGH

Using specific imaging technologies (specific imaging systems or imaging software), digital analysis of color combinations or pseudocolors can be put into use for developing multicolor FISH approaches ([3, 10], Part IV). Pseudocolors are obtained by ratio labeling or combinatorial labeling. The former procedure is a result of a highly reproducible labeling. The latter is a combination of pseudocolors according to the formula $N = 2n - 1$, where N is the number of pseudocolors and n is the number of fluorochromes. Thus, the application of 4–7 color combinations and different labeling schemes can produce up to about 100 pseudocolors. Psuedocolor numbers are certainly enough to solve the majority of molecular cytogenetic tasks via FISH-based assays. Moreover, to obtain multicolor bands comparable to paint-specific genomic/chromosomal loci, the alignment of different FISH signals along the chromosomal axis overlapping with each other is used to produce a sequence of color combinations based on creating pseudocolors according to mixing fluorescence intensity ratios of MCB probes (chapter by Thomas Liehr et al. "FISH Banding Techniques").

These approaches require specific imaging and microscopy technologies. To succeed, an imaging system has to comprise a microscope equipped by a corresponding filter set, camera, hardware, and software. The latter is applied for (1) classifying (separating) objects on a microscopic image (metaphase chromosomes, interphase FISH signals, fluorescing objects creating autofluorescence/background fluorescence), (2) interactive intensity measuring to determine variability between signal intensities for virtual intermixing of colors according to relative DNA content (i.e., classical/metaphase CGH analysis), and (3) creating pseudocolors according to input color schemes, intensity ratio combinations, and/or variations. In-house and commercially available software packages for multicolor FISH are available and include all the aforementioned options. Despite developments in high-resolution genome scan by array-CGH, molecular karyotyping (chapter by Eftychia Dimitriadou and Joris Vermeesch "Array CGH"; chapter by Maria Isabel Melaragno and Mariana Moysés-Oliveira "Breakpoint Mapping of Balanced Chromosomal Rearrangements Using Array-CGH of Microdissection Derived FISH-Probes";

chapter by Jiří Štika and Oldřich Mazal "Sequencing of Microdis section Derived FISH-Probes"), and related molecular cytogenetic techniques, multicolor-FISH-based methods are still widely used and are successfully applied in basic and diagnostic research.

In addition to FISH-based approaches for visualization of nucleic acids in situ at single-cell level, there exists another molecular cytogenetic approach roughly based on FISH and related microscopy and unique imaging technology known as CGH. Based on comparison of differentially labeled signals ("whole genome paints") of patient genomic DNA and normal reference DNA hybridization, CGH is performed on metaphase plates of a karyotypically normal male (chapter by Thomas Liehr et al. "Com parative Genomic Hybridization (CGH) and Microdissection-Based CGH (micro-CGH)"). This technique has become a platform for array-based/on-chip high-resolution genomic screening of chromosome imbalances (molecular karyotyping). Due to numerous drawbacks and poor reproducibility, classical or metaphase CGH is almost completely excluded from "technological arsenal" of molecular cytogenetic laboratories. The application of this technique is rare and is not associated with valuable genomic or cytogenetic data.

5 Conclusions

The first imaging systems were applicable for a slight increase in efficiency of FISH-based approaches and molecular cytogenetic analysis [2, 13, 16]. During the last two decades, imaging has become a basis of a variety of FISH-based molecular cytogenetic techniques (i.e., multicolor FISH, MCB, and CGH) ([6, 10], Part IV). Additionally, Q-FISH is a method of choice for solution of technical problems that are difficult or impossible to process by a visual microscopic analysis.

As it is repeatedly emphasized, the results of a FISH experiment are determined by tissue preparation and correctly selected denaturation and hybridization conditions suggesting that the role of imaging technologies is limited to processing the raw microscopic image data. Basic FISH procedures also define the resolution and image quality. Despite the value of microscopy and imaging for FISH, these are always modified following the lab-based elaboration of basic procedures for a new approach. To this end, bearing in mind that imaging and microscopy are fundamental for visualization molecular cytogenetic techniques, these should be still considered as secondary components for the developments of new FISH-based approaches.

Acknowledgments

The chapter is dedicated to Ilia V Soloviev. I would like to express my gratitude to Prof. Svetlana G. Vorsanova (Moscow, Russia) and Prof. Yuri B. Yurov (Moscow, Russia) for helping in preparation of this chapter. Author is supported by a grant from the Russian Science Foundation (project #14-15-00411).

References

1. Claussen U, Michel S, Mühlig P et al (2002) Demystifying chromosome preparation and the implications for the concept of chromosome condensation during mitosis. Cytogenet Genome Res 98:136–146
2. Iourov IY, Vorsanova SG, Yurov YB (2008) Recent patents on molecular cytogenetics. Recent Pat DNA Gene Seq 2:6–15
3. Iourov IY, Vorsanova SG, Yurov YB (2012) Single cell genomics of the brain: focus on neuronal diversity and neuropsychiatric diseases. Curr Genomics 13:477–488
4. Iourov IY, Vorsanova SG, Yurov YB (2016) Detection of nuclear DNA by interphase fluorescence in situ hybridization. In: Encyclopedia of analytical chemistry. Wiley, New York, pp 1–12
5. Liehr T (ed) (2009) Fluorescence *in situ* hybridization (FISH) – application guide, Springer protocols. Springer, Heidelberg
6. Vorsanova SG, Yurov YB, Iourov IY (2010) Human interphase chromosomes: a review of available molecular cytogenetic technologies. Mol Cytogenet 3:1
7. Iourov IY (2012) To see an interphase chromosome or: how a disease can be associated with specific nuclear genome organization. BioDiscovery 4:5
8. Rapkin LM, Anchel DR, Li R et al (2012) A view of the chromatin landscape. Micron 43:150–158
9. Yurov YB, Vorsanova SG, Iourov IY (eds) (2013) Human interphase chromosomes: biomedical aspects. Springer Science + Business Media, New York
10. Bakker B, van den Bos H, Lansdorp PM et al (2015) How to count chromosomes in a cell: an overview of current and novel technologies. Bioessays 37:570–577
11. Pajor G, Kajtár B, Pajor L et al (2012) State-of-the-art FISHing: automated analysis of cytogenetic aberrations in interphase nuclei. Cytometry A 81:649–663
12. Hamey FK, Shavit Y, Maciulyte V, et al (2015) Automated detection of fluorescent probes in molecular imaging. In: Computational intelligence methods for bioinformatics and biostatistics. Springer International Publishing, pp 68–75
13. Tanke HJ, Florin RJ, Wiegant J et al (1995) CCD microscopy and image analysis of cells and chromosomes stained by fluorescence in situ hybridization. Histochem J 27:4–14
14. McAleer C (2005) The fundamentals of microscopy. In: Gersen SL, Keagle MB (eds) The principles of clinical cytogenetics, 2nd edn. Humana, Totowa, NJ, pp 81–91
15. Garini Y, Young IT, McNamara G (2006) Spectral imaging: principles and applications. Cytometry A 69:735–747
16. Soloviev IV, Yurov YB, Vorsanova SG, et al (1994) Application of digital imaging microscopy for the fluorescence detection of signals of in situ hybridization with directly cyanine-labeled DNA probes. In: CA-AMCA: Microscope instrumentation, Instituto di Biofisica del CNR, Pisa, Italy, pp 81–82
17. Yurov YB, Soloviev IV, Vorsanova SG et al (1996) High resolution fluorescence in situ hybridization using cyanine and fluorescein dyes: ultra-rapid chromosome detection by directly fluorescently labeled alphoid DNA probes. Hum Genet 97:390–398
18. Henegariu O, Heerema NA, Lowe Wright L et al (2001) Improvements in cytogenetic slide preparation: controlled chromosome spreading, chemical aging and gradual denaturing. Cytometry 43:101–109
19. Iourov IY, Vorsanova SG, Pellestor F et al (2006) Brain tissue preparations for chromosomal PRINS labeling. Methods Mol Biol 334:123–132
20. Iourov IY, Soloviev IV, Vorsanova SG et al (2005) An approach for quantitative assessment of fluorescence in situ hybridization (FISH) signals for applied human molecular cytogenetics. J Histochem Cytochem 53:401–408
21. Iourov IY, Liehr T, Vorsanova SG et al (2007) Interphase chromosome-specific multicolor banding (ICS-MCB): a new tool for analysis

of interphase chromosomes in their integrity. Biomol Eng 24:415–417

22. Iourov IY, Vorsanova SG, Yurov YB (2008) Fluorescence intensity profiles of in situ hybridization signals depict genome architecture within human interphase nuclei. Tsitol Genet 42:3–8
23. Iourov IY, Vorsanova SG, Liehr T et al (2009) Increased chromosome instability dramatically disrupts neural genome integrity and mediates cerebellar degeneration in the ataxia-telangiectasia brain. Hum Mol Genet 18:2656–2669
24. Vorsanova SG, Iourov IY, Beresheva AK et al (2005) Non-disjunction of chromosome 21, alphoid DNA variation, and sociogenetic features of Down syndrome. Tsitol Genet 39:30–36

Optical Filters and Light Sources for FISH

Michael Sommerauer, Ingrid Feuerbacher, and Alexander Krause

Abstract

Brilliant fluorescence signals with almost no background and cross talk are the aim of FISH analysis in imaging systems. The precise selection of hardware components like optical filters and light sources plays a major role. Considering fluorescent dye characteristics is the base of configuring perfectly matched multicolor-FISH (mFISH) filters, which allow the simultaneous application of up to seven dyes. The spectral characteristics of filters are here explained with respect to microscope setups. Spectral cross talk, pixel-shift effects, and stable energy output will be the main issues in daily work. Specific hard-coated single-band filters with small bandwidth but maximum transmission avoid cross talk to a high degree; the use of multiband filters allows simultaneous imaging of up to four dyes; multiband systems with separate exciters and emitters in filter wheels can be controlled by software. The comparison of mercury and metal halide light sources to new light-emitting diode (LED)-based light sources is shown, which reveals the chance of stable and long-term light output of the new LED light sources. White-light LED sources are nowadays a perfect choice for replacing mercury-based lamps. The combination of hard-coated mFISH filters with stable LED light sources is a very helpful tool in daily work.

Keywords Light-emitting diode (LED) light sources, Comparative genomic hybridization (CGH) filter sets, Multicolor fluorescence in situ hybridization (mFISH) filter sets, Filter handling, Fluorophores, Mercury lamp, Exciters, Emitters, Beam splitter, Mounting of filters

1 Introduction

1.1 Fluorescence, Fluorescence Spectra, and Fluorescent Dyes

Fluorescence is a material intrinsic property which is common to all materials. It is the ability to absorb energy—here light ($E = h^*\nu$)—and to emit light of lower energy than the absorbed. The absorption is strongly depending on the wavelength. Absorption only occurs if a photon has enough energy to push an electron from the ground state S^0 into an excited state S^1. The extinction coefficient is a degree of the efficiency of the absorption [1].

Electrons in excited states are relaxing into the ground state in nanosecond scale. Hereby several pathways are possible. One way of releasing energy is the emission of light which can be detected as fluorescence; another way is relaxation without radiation. A degree of this emission is the quantum yield. In Fig. 1a the absorption and

Thomas Liehr (ed.), *Fluorescence In Situ Hybridization (FISH)*, Springer Protocols Handbooks,
DOI 10.1007/978-3-662-52959-1_3,

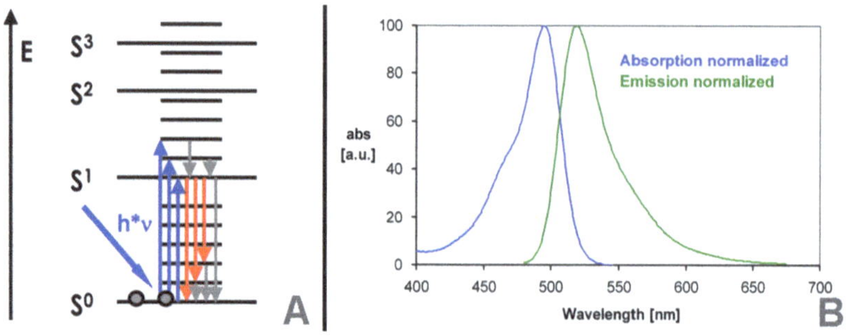

Fig. 1 (**a**) Jablonski term scheme. (**b**) Spectrum of FITC: The difference between the excitation maximum and the emission maximum is called Stokes shift

emission of a photon is illustrated. Further the term scheme shows that the electronic levels are divided into sublevels which belong to the vibronic states of substances. An electron can change these vibronic sublevels without any emission of light. This explains the energy difference between the light which is used for pushing the electron into an excited state and the light which is released by the electron during relaxation. The term scheme also shows an energy distribution for absorption and emission. The energy (E) can be transformed into a wavelength (λ) by the equitation $\lambda \sim 1/E$. Figure 1b shows the wavelength distribution or spectrum for the common fluorescence dye FITC [2]. The extinction coefficient and the quantum yield are characterizing each substance. The higher both parameters are, the better a substance (molecule or nanocrystal) can be used as fluorescence marker. In most cases the fluorescence markers are organic molecules based on aromatic ring systems. They are soluble in common solvents and can be chemically bound to proteins, DNA or RNA. This gives the user the possibility to label samples specifically [1].

1.2 Filter Characteristics

Fluorescence can be detected by using spectrometers or fluorescence microscopes. These instruments have light sources which are used for exciting the fluorescent dyes. In most cases such with a more or less continuous emission spectrum is applied, e.g., mercury, xenon, metal halide, or light-emitting diode (LED) light sources. Optical filters with a dye-specific transmission band are placed between these light sources and the labeled sample. For FITC (green) only a part of the blue emission of such a lamp is used. Every other wavelength is blocked by this so-called excitation filter or exciter. The detection of the fluorescence can be done with CCD cameras or with the eyes. In comparison with the light which is used for excitation, the emitted fluorescence is more than 10,000 times less intense. This means another filter which is called emission filter or emitter has to block the light which was used for excitation. Exciter and emitter have to be blocked against each other;

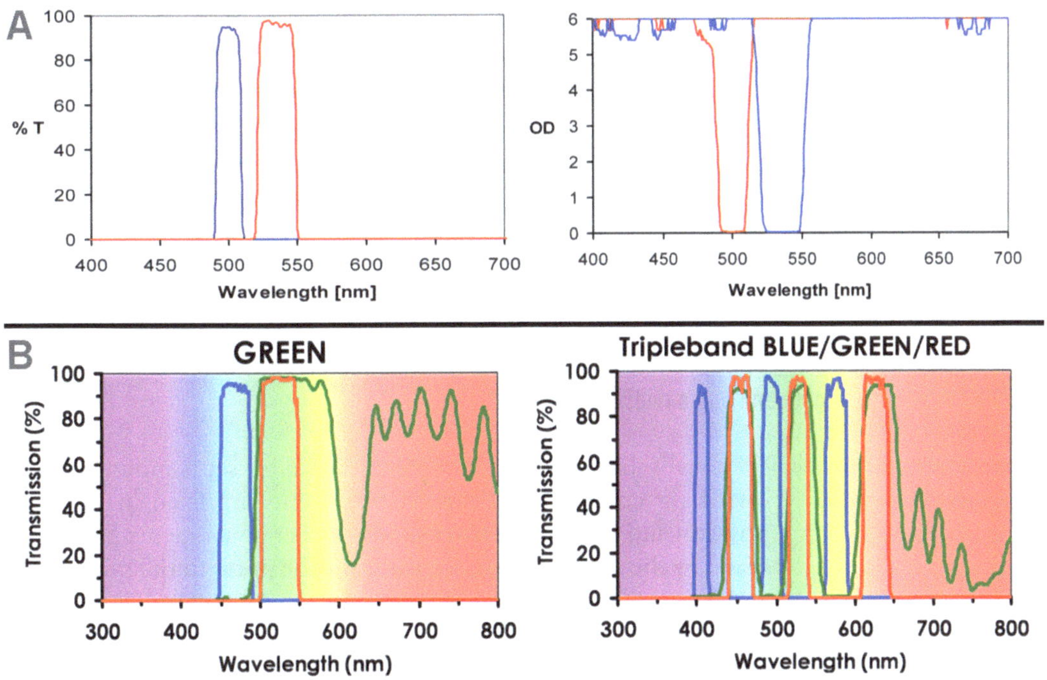

Fig. 2 (**a**) Spectral data of filters in linear and logarithmic scaling. (**b**) Single-band filter set for green (e.g., Sp. Green) and triple-band filter set for blue, green, and red dyes (e.g., DAPI, Sp. Green, Sp. Red). Exciter blue line, dichroic green line, and emitter red line

otherwise the fluorescence signals cannot be detected. Due to the big energy difference between excitation light and fluorescence, a filter must suppress undesired light very efficiently. The transmission (T) in these blocking areas should not exceed 0.0001 %. Instead of using the transmission, the blocking of a filter is determined by the optical density OD = $-\log(T)$. Figure 2a shows a pair of exciter and emitter.

Normally the fluorescence is measured perpendicular to the excitation. Therefore another optical component called "beam splitter" is used to reflect the light which is used for excitation and to transmit the fluorescence light of the sample. According to its function, the beam splitter has to be mounted under 45° into the light path. This beam splitter only supports the blocking of the excitation and emission filters against each other, but its own blocking properties are not sufficient [3]. For application of FISH in electron microscopy, see chapter by Hannes Schmidt and Thilo Eickhorst "Gold-FISH: In Situ Hybridization of Microbial Cells for Combined Fluorescence and Scanning Electron Microscopy".

1.3 Filters in a Microscope

In a fluorescence microscope, perfectly matched filter sets are used. A set consists of an exciter, an emitter, and a beam splitter as described (chapter by Ivan Iourov "Microscopy and Imaging

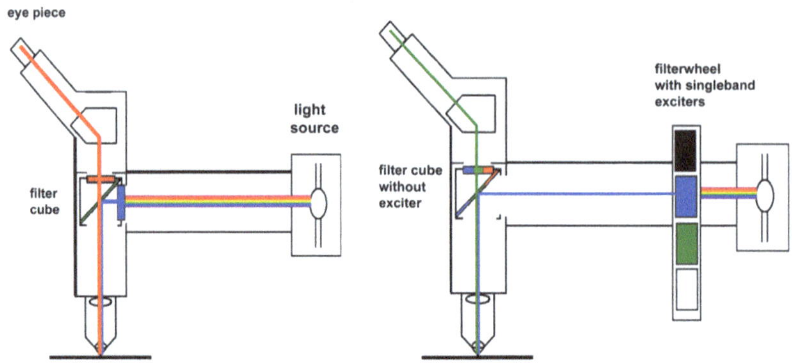

Fig. 3 Imaging with single-band and a multiband filter set

Systems"). In current microscopes each filter set is mounted in its own filter cube [4]. All three components are fixed in the right position and the correct angles. No further adjustments have to be made by the user. Such a cube can easily be put into the microscope and removed from it. For every dye a special filter set mounted in a cube has to be chosen (Fig. 2b). Nevertheless it is possible to make filter sets with two or three transmission bands, so that three different dyes can be detected simultaneously. These filter sets are called full dual- or triple-band filter sets as shown in Fig. 2b.

If the microscope has an additional filter wheel between the lamp housing and the microscope stand, further possibilities for placing filters are given. The dual-, triple-, and even quad-band filter sets—shortly described as multiband filter sets—can be divided into two sections. The beam splitter and emitter stay in the filter cube and the exciters are placed in the additional filter wheel. A filter set for a single color has one exciter. For each color channel of these multiband filter sets, one exciter (called single-band exciter, Fig. 3) can be placed in the described filter wheel. This allows a sequential imaging of the different colors, which can easily be automated. This avoids pixel shift (see below), because the same beam splitter and emitter stays in the light path.

Also the corresponding dual-band or triple-band exciters can be placed in these filter wheels for simultaneous imaging of the sample. Some FISH kits (e.g., UroVysion kit [5]) need this dual-band excitation (chapter by Thomas Liehr "Commercial FISH Probes"). A multiband filter set (Fig. 3) with a dual-band exciter or a full dual-band filter set itself can be used. The given example is using green and red dyes which have been imaged simultaneously. The green and red color can be easily detected in such a dual-band filter set. If both colors are co-localized, the addition of the two dyes can be detected as yellow (same principle as in every TV). A full triple-band filter set or the triple-band excitation of a multiband filter set will show the same results. Mostly the counterstaining (DAPI) is used as the third channel (blue channel). If the concentration of DAPI is too high, the blue channel is overbalanced and complicates the detection of the other channels.

A few comments to the abovementioned pixel shift [3]. Pixel shift can occur by switching from one filter set to the other. The reason is the difference in the beam deviation which is caused by the wedge angle of every component in the light path. If both surfaces of an optical filter or beam splitter are exactly coplanar, the beam deviation is zero. A small deviation of the parallelism cannot be detected with cameras, but if the deviation at the camera chip is more than 6.7 μm, pixel shift is detected. The identical signal of the sample in different color channels is not registered at the same place.

To avoid this pixel shift, so-called zero pixel shift certified filter sets can be used. These sets have nearly coplanar substrate surfaces (wedge angle $\leq$ 10 arc sec). Small deviations can be adjusted with an autocollimator, when the filters (emitter and beam splitter) are mounted into the filter cube.

1.4 Choosing Filters

According to the possibilities of the microscope setup, filter sets can be chosen. Most essential is to know the spectral behavior of the dyes used. Each filter set has to be chosen according to the dye spectrum. That means the exciter should include the absorption maximum of the dye. Also the emission maximum should be included in the emission filter. Only in this case a good signal/noise ratio can be achieved. That means the smaller the Stokes shift is, the steeper the filters. Figure 4 shows a filter set for FITC which fulfills the abovementioned criteria.

By choosing dyes and filters, a closer look to the microscope-related equipment like the camera and the excitation source should be done. Dyes in the spectral region between DAPI (UV excitation and blue detection) and Texas Red (yellow excitation and red detection) can easily be determined by the eyes or any camera system (color or black & white). If Far Red dyes like Cy5 (red excitation and NIR detection) are used, it is essential to use a black and white camera system, because the eyes are insensitive in

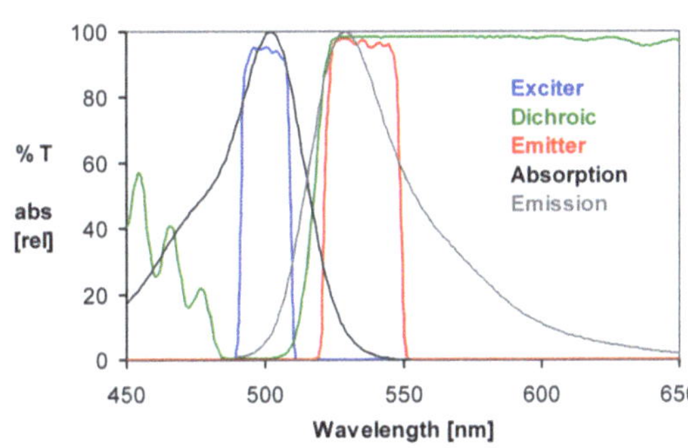

Fig. 4 Spectrum and filter set for FITC

these spectral regions. Color cameras will also cut a part of the emission spectra of these dyes. The excitation can be done with any kind of excitation sources like white-light LED, mercury, xenon, or metal halide lamps. If dyes like Cy5.5 (NIR) and Cy7 (IR) are used, black and white cameras and special LED or xenon excitation sources are essential. The output of mercury and metal halide lamps in the dark red and NIR are not powerful enough to excite these dyes properly. Every IR blocking filter or heat protection filter has to be removed from the microscope as well as from the camera.

1.5 Choosing the Right Light Sources

In the past mainly mercury or metal halide light sources, so-called arc lamps, were used. The principle of generating light is for all these sources the same. It is based on electrical discharge in a gas generated by high voltage. This leads to a broad emission of light which lasts from UV to the visible spectrum. In all those light sources, a major amount of mercury in the range of 20 mg is present. This can be seen at the characteristic peaks at 365, 405, 436, 546, 577, and 579 nm. Besides this emission, which is seen as white light, the lamps produce a lot of heat; only a few percent of their intake power is converted into light.

During the lifetime of these bulbs, the electrodes wear off. This strongly impacts the light output of these light sources (Table 1). LED light sources have generally a 10–100 times longer lifetime compared to arc lamps. They do not produce much heat, because the generation of light is different compared to arc lamps. Electrons and holes in a semiconductor material recombine and emit light of a defined wavelength range when current (DC) is switched on [6]. Depending on the semiconductor material, different emission bands from UV/NIR can be produced. Special high-power LEDs can be used for fluorescence excitation, but a single LED only covers a small range of the spectrum which is used for fluorescence excitation. A LED light source which can be used like a mercury lamp contains always several high-power LEDs of different wavelengths. In white-light LED light sources, up to six different LEDs

Table 1
Approximate lifetimes of different light sources of fluorescence microscopes

Light source	Lifetime (h)	Output at the end of lifetime (%)
Mercury 100 W	400	35 %
Mercury 50 W	200	40 %
Metal halide 120 W	1500–2000	Approx. 50 %
LED light source	20,000	>70 %

with bandwidths from 10 to 85 nm are combined. The result is a white-light spectrum between 350 and 680 nm. This allows exciting nearly all commonly used dyes even into the NIR range. In FISH samples DAPI is the dye with the lowest and Far Red or Cy5.5 is the one with the highest excitation wavelength. Every dye including the above mentioned ones can be properly excited.

One big advantage of LED light sources is they can be switched on and off instantaneously without compromising the lifetime of the LEDs. They reach their maximum output level in less than 2 s. No warm-up or cooling-down time has to be considered. This means that a LED light source has only to be switched on when it is needed. This reduces their "on" time up to 95 % compared to mercury or metal halide light sources. Arc lamps normally burn the whole day because they have to warm up before and cool down after use. Considered a 40 h week, an arc lamp will be switched on 40 h, even if it is used only for 20 h a week. This means the mercury arc lamp has to be replaced after 10 weeks (lifetime 400 h). A LED light source will be only switched on for the time in use. In this example it is 20 h a week. After a bit more than 19 years, the 20,000 h lifetime is reached. During their lifetime LED light sources are maintenance-free. An easy integration into automated slide scanners is possible without having any mechanically moving items (e.g., filter wheels or shutters). One last benefit is that LED light sources are completely mercury-free, which helps to protect the environment.

1.6 LED Light Sources for FISH

The question is if arc lamps generally can be replaced by LED light sources when using the microscopes for comparative genomic hybridization (CGH; chapter by Thomas Liehr et al. "Comparative Genomic Hybridization (CGH) and Microdissection-Based CGH (micro-CGH)"), FISH (chapter by Thomas Liehr et al. "Two- to three-color FISH"), or multicolor FISH (mFISH; chapters by Thomas Liehr and Nadezda Kosyakova "Multiplex FISH and Spec tral Karyotyping"; Thomas Liehr et al. "FISH Banding Tec hniques"; Thomas Liehr et al. "cenM-FISH Approaches"; Thomas Liehr et al. "Heterochromatin Directed M-FISH (HCM-FISH)"; Anja Weise and Thomas Liehr "Subtelomeric and/or Subcentromeric Probe Sets"; Thomas Liehr et al. "Bar-Coding Is Back"). Here the white-light LED light sources are the best choice, but not every LED light source will show necessarily a good result.

Due to technical reasons, yellow and green LEDs are at present much less powerful compared to mercury or metal halide lamps in this specific region. This means that orange and red dyes might be not bright enough to guarantee evaluable results. New technologies however, like the "light pipe technology," allow to overcome this disadvantage of common LEDs [7] resulting in a comparable output power to a newly installed mercury/metal halide lamp (Fig. 5a). This means all LED light sources with a "boosted"

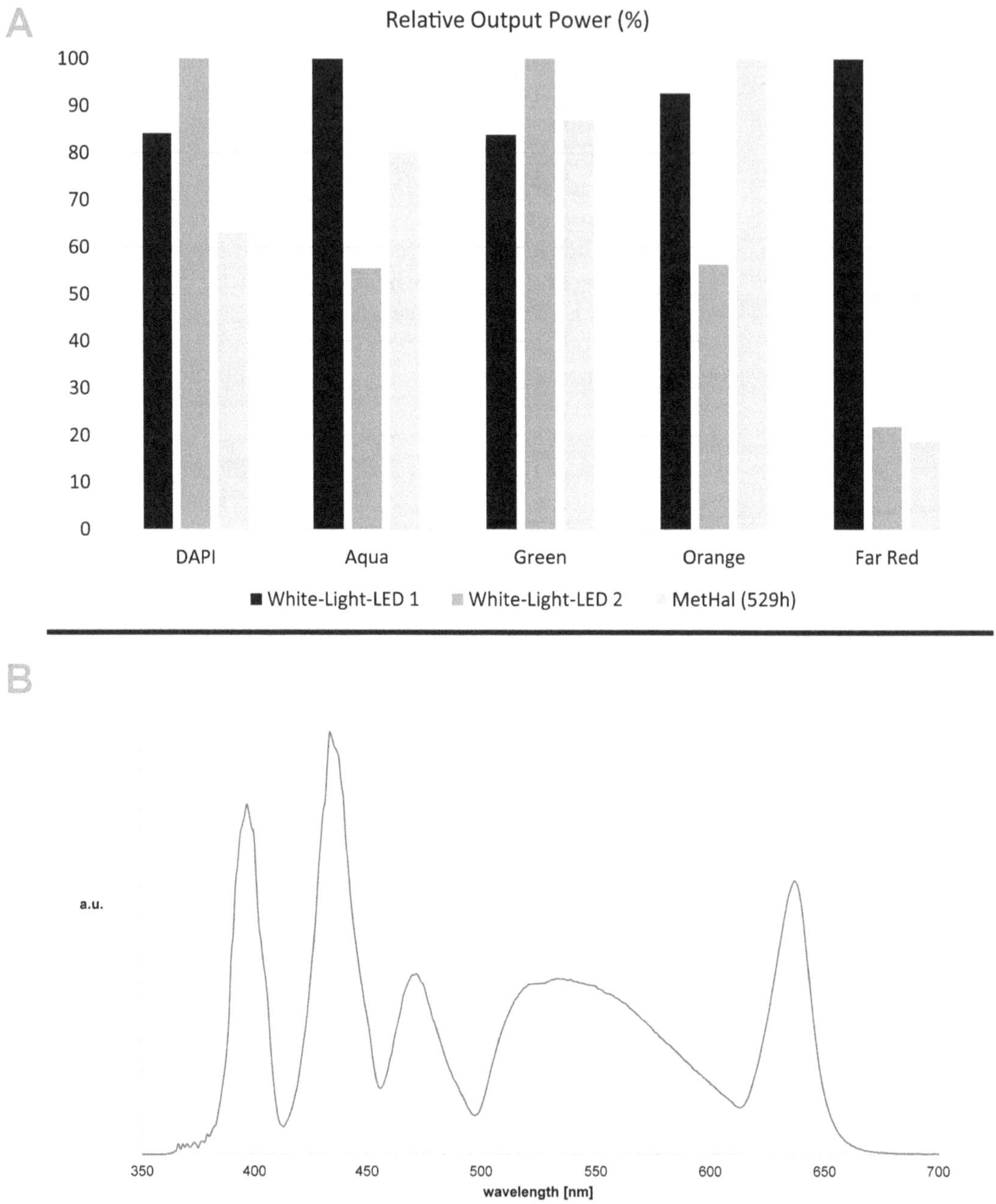

Fig. 5 (**a**) Comparison of the output power of metal halide and boosted LED light sources. (**b**) Spectrum of an LED light source with boosted yellow-green emission and red LED

yellow-green range can be used for FISH techniques. If mFISH is made, it is highly recommended that the LED light source has a red LED as shown in Fig. 5b.

If Cy7 is used as an IR dye, an additional NIR LED must be combined with the LED light source. This is technically no problem and can be easily integrated into any kind of fluorescence microscope.

2 Filter Sets for FISH

2.1 Filter Sets for FISH or CGH

The selection of the correct and suitable filter setup cannot be linked only with the type of dye which will be used in a typical FISH (chapter by Thomas Liehr et al. “Two- to Three-Color FISH”) or CGH experiment (chapter by Thomas Liehr et al. “The Standard FISH Procedure”; chapter by Thomas Liehr et al. “Comparative Genomic Hybridization (CGH) and Microdissection-Based CGH (micro-CGH)”; chapter by Thomas Liehr “Commercial FISH Probes”). The use of only two or three dyes with wide spectral distance allows the use of completely different filters in comparison with multicolor applications with six to seven dyes and small spectral distance (see Sect. 1.4). Typically users start with blue, green, and orange signal combinations. This combination allows the use of filter sets with “standard specifications” that means filter sets which are typically used also in immunofluorescence applications. Older microscopes are often equipped with corresponding filter sets, e.g., blue or DAPI basic and green and red filter sets. A typical filter set is shown in Fig. 6a.

Fluorescence images taken with this type of filters will be always “bright,” showing background signals from all components present in the sample which can be excited in the range from 450 to 490 nm. The emission window is “open.” In some cases people complain that their signals will be not as specific as they want to have, e.g., red signals will be seen in the green filter set. To avoid this drawback, band-pass filters are used, there is only a specific “optic window,” and background signals are blocked.

The signals will appear with more contrast and increased signal/noise ratio due to the black background. The most simple trick will be to exchange only the emitter by a suitable band-pass filter which fits to the exciter and dichroic in the setup.

If signals with very low intensity have to be detected, the excitation must be as efficient as possible in combination with very effective signal detection. This can be achieved by the new series of hard-coated band-pass filter sets which show maximum of transmission in the excitation and emitter band passes (Fig. 6b).

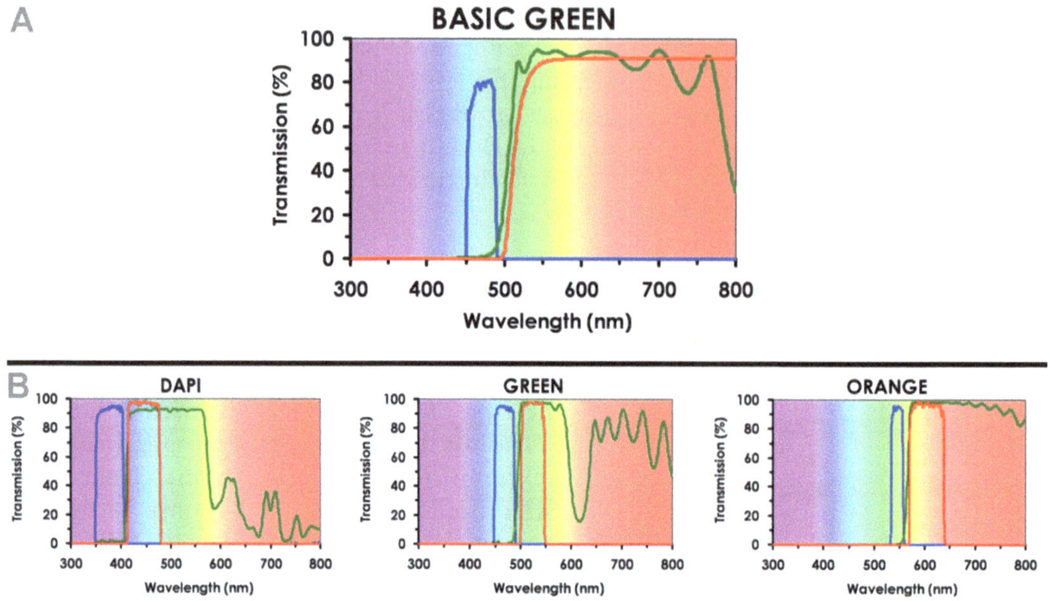

Fig. 6 (**a**) Basic green or FITC filter set with long-pass emitter. (**b**) Optimized spectra of three hard-coated filter sets used for DAPI, green, and orange dyes in FISH or CGH applications

The use of these high-efficient filter sets with >95 % transmission will reduce the exposure time in the range of 30–50 % in comparison with filter sets with around 70–80 % transmission. Additionally the pixel shift of these hard-coated filters will be very close to zero due to their perfect surface flatness (see Sect. 1.3).

2.2 Filter Sets for Multicolor FISH (mFISH)

The use of six to seven dyes in an mFISH experiment (chapters by Thomas Liehr and Nadezda Kosyakova "Multiplex FISH and Spec tral Karyotyping"; Thomas Liehr et al. "FISH Banding Tec hniques"; Thomas Liehr et al. "cenM-FISH Approaches"; Thomas Liehr et al. "Heterochromatin directed M-FISH (HCM-FISH)"; Anja Weise and Thomas Liehr "Subtelomeric and/or Subcentromeric Probe Sets"; Thomas Liehr et al. "Bar-Coding Is Back") requires very well-matched filter combinations. Even best-blocked filters cannot completely avoid spectral interferences due to the spectral overlapping of the dyes. As a rule of thumb, the spectral distance of neighbored dyes should be about 50–60 nm, e.g., the difference between Sp. Gold and Sp. Orange is about 30 nm, and between Sp. Gold and Sp. Red, the distance is about 55 nm. Sp. Gold and Sp. Orange can't be selectively separated by using filters. Sp. Gold and Sp. Red in the same sample can be detected selectively. The bleed through between these two dyes is only a few percent, if specific band-pass filters are used (Table 2). At present series of mFISH filters are available from different companies. As example we will present a complete series of hard-coated filter sets, which show maximum transmission for each filter set matched to the mFISH dyes (Fig. 7). Due to their small spectral

Table 2
Calculated spectral overlapping of the fluorophores DAPI, Sp. Aqua, Sp. Green, Sp. Gold, Sp. Orange, Sp. Red, Cy5/Far Red, Cy5.5, and Cy7 into the specific mFISH filter sets

Filter set	DAPI	Sp. Aqua	Sp. Green	Sp. Gold	Sp. Orange	Sp. Red	Cy 5/Far Red	Cy 5.5	Cy 7
DAPI	100 %	30 %	0 %						
Sp. Aqua	0 %	100 %	1 %						
Sp. Green		0 %	100 %	3 %	0 %				
Sp. Gold			2 %	100 %	49 %	1 %			
Sp. Orange			0 %	36 %	100 %	11 %			
Sp. Red				0 %	15 %	100 %	1 %		
Cy 5						12 %	100 %	53 %	1 %
Cy 5.5						0 %	53 %	100 %	6 %
Cy 7							0 %	12 %	100 %

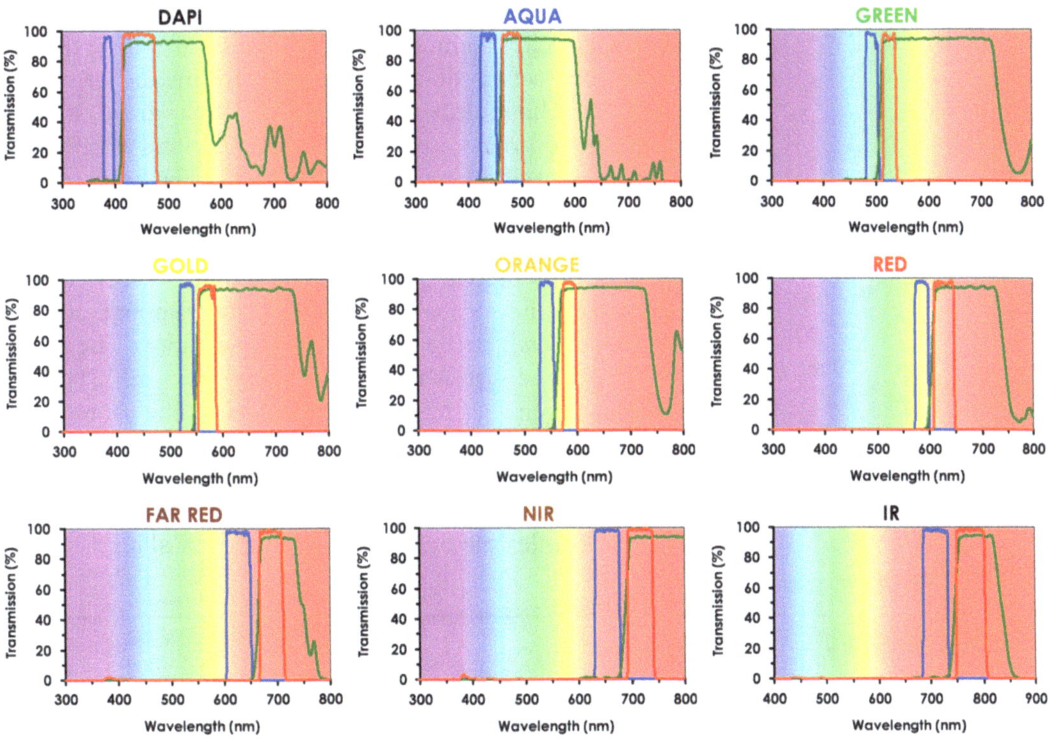

Fig. 7 Filter spectra of the series of specific narrowband hard-coated mFISH filter sets designed for mFISH dyes (list of dyes see Table 2)

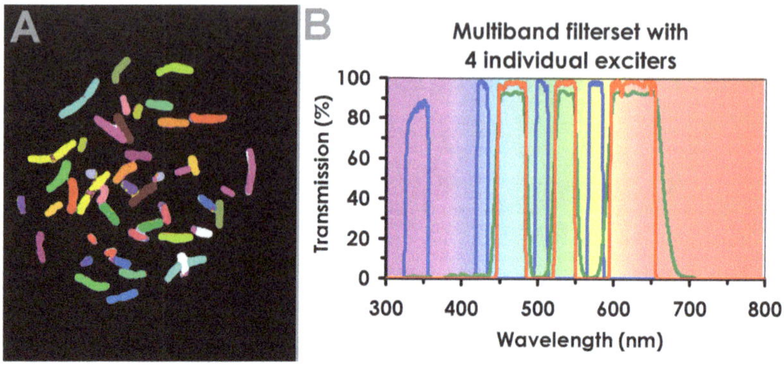

Fig. 8 (**a**) Metaphase spread labeled with DAPI, DEAC, FITC, Sp. Orange, Texas Red, and Cy5 taken with mFISH filter series, pseudo colors (MetaSystems, Altlussheim, Germany). (**b**) Spectrum of a hard-coated multiband filter set with four single-band exciters (blue lines), polychroic (green line), and multiband emitter (red line). Exciters will be mounted in separate filter wheel

bandwidth, the analysis will be mainly done with CCD cameras in combination with software programs. The theoretical overlap between the different channels is given in Table 2 [8]. Figure 8a shows the result of an mFISH experiment taken with this corresponding mFISH filter series.

A very elegant filter technique in mFISH analysis will be the use of multiband filter sets as described in Sect. 1.3 instead of single-band-pass filters. The multiband set for blue/aqua/green/orange (Fig. 8b) consists of four separate single-band exciters which must be mounted in an exciter filter wheel. The polychroic beam splitter and polychroic emitter will be mounted in the filter cube. Additionally available, precisely matched dual- or triple-band exciters allow the detection of two to four dyes simultaneously, a very helpful tool for quick visual detection of overlapping signals. Changing of exciters can be controlled by software. The background signal will be not as dark as in single-band mFISH filter sets, as long as the emitter has three band passes (like three "windows") instead of one specific. Microscope setups with a filter wheel in the emission path allow the use of specific band-pass emitters, which can be controlled as well by software. This configuration will be most flexible but affords precise controlling.

3 Filter Handling

3.1 Cleaning

Coated substrates should only be touched at the edges. Handling of exposed coatings with bare fingers has to be avoided.

3.1.1 Exciters and Emitters

Gently cleaning should be done only if necessary. Loose particles should be removed with a bulb puffer or filtered pressurized air cleaner. If necessary, the surfaces should be gently wiped using alcohol (ethanol, isopropanol, or methanol) and a lint-free towel. A new surface of the towel should be used with each wipe. Touching or wiping of A/R (antireflective) coated surfaces should be avoided. Fingerprints on the surface of the excitation filter will burn and might shorten the lifetime of the exciter.

3.1.2 Beam Splitters

Loose particles should be removed with a bulb puffer or filtered pressurized air cleaner. Touching or wiping of A/R coated surfaces should be avoided. If filters or beam splitters need a special cleaning, they should be sent back to the manufacturer. Exciters are exposed to the light source. Exciters in additional sliders or filter wheels close to a light source must be protected with appropriate heat protection filters. A heat protection filter has to be mounted in a filter wheel or in the lamp housing of the light source. The filters and beam splitters in the microscope should be checked from time to time.

3.2 Mounting

Most of the filter cubes allow mounting and demounting all filters and beaming splitters. The fixation of the filters is done with screws or special filter rings; if necessary special tools are delivered with the microscope or the filter cube. Some microscope manufacturers glue the filters into the filter cubes. In this case the filters have to be changed by specialists. To obtain the most optimal performance of a filter set, the filters and the beam splitter should be orientated in a filter cube.

Exciters and emitters are mostly labeled with arrows on the side of the filter ring. Often the arrow(s) point into the direction of propagation of the light, but this is not a general rule. Follow the instructions of the manufacturer carefully. Beam splitters (dichroics, polychroics, mirrors) have to be mounted with the coated side toward the light source (Fig. 9a). A dot, arrow, small scratch, or bevel edge on the beam splitters indicates the coated side.

If the beam splitter is not labeled, it can be illuminated with any light source. When viewing the beam splitter with the reflecting side up, only a predominantly single reflection of the light source can be seen. The thickness of the beam splitter at the far edge is not visible. When viewing the beam splitter with the reflecting side down, a double reflection of the light source occurs. The thickness of the beam splitter at the far edge is visible (Fig. 9b).

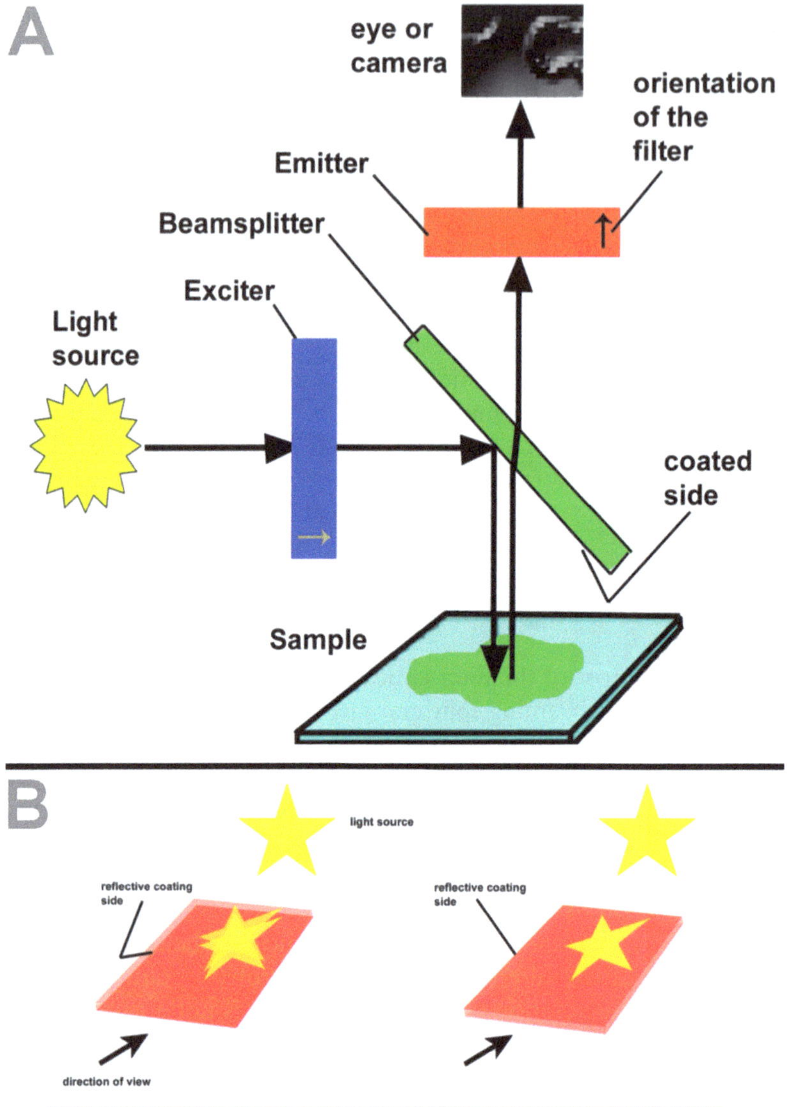

Fig. 9 (**a**) Orientation of filter sets in a filter cube. (**b**) Determination of the reflecting surface of a beam splitter

4 Troubleshooting

4.1 Uneven Illumination of the Sample

If every filter set shows the same uneven illumination, align the excitation source. If the light source is connected by a light guide, make sure that the light guide is mounted properly on both ends. If only one of the filter sets shows uneven illumination, check if the exciter is burned. You will see brown or black spots, which can't be removed by cleaning the filter. Exchange the excitation filter with a hard-coated filter. Hard-coated filters won't burn and age anymore.

4.2 A Lot of Background and Low Signal Intensity

Make sure that you are using the right filter set for the dye(s) in the sample. Maybe you are using a long-pass filter set. Use a specific band-pass filter set. If you are using a specific band-pass filter set and the described problems occur, try a new filter set. The used filter set might be aged (older than 10 years or daily used over years).

4.3 I See Nothing in My Blue, Green, Yellow, Orange, or Red Filter Set(s)

1. Turn on the lamp, open the shutter, or switch your filter wheel into the right position.
2. Make sure that the light is guided to the camera or the eyepiece of the microscope.

If only one filter set shows the problem, please check if the filters are mounted in the right way into the cube.

4.4 I Only See Very Dim Signals in All Color Channels

1. Make sure that the arc lamp didn't reach the end of life.
2. Check the intensity control of the light source, if the light source has an internal attenuation.
3. If your metal halide light source is coupled via liquid light guide to the microscope stand, check if the light guide is aged due to heat and UV of the light source. Change the liquid light guide after two to three lamp changes. This won't be an issue with LED light sources.

4.5 I See Nothing in My Dark Red, NIR, or IR Filter Set (e.g., Far Red, Cy5, Alexa 647, Cy5.5, Cy7)

1. Turn on the lamp, open the shutter, or switch your filter wheel into the right position.
2. Make sure that you have an IR sensitive camera. None of the abovementioned dyes can be seen with the eyes.
3. Remove every IR blocking glass from the light path. Check if older cameras have an IR blocking filter as protection glass in front of the CCD chip. Your camera supplier will help you.
4. If you are working with dyes like Cy5.5 or Cy7, use a LED, xenon, or metal halide light source. Mercury lamps are unsuitable for these dyes.

4.6 I See Only a Very Bright Even Illumination (White Light or a Specific Color)

1. Check if the exciter and emitter fit to each other. The labels of the filter rings often show the transmission band (e.g., 450–490, 470 ± 20, or 470/40). Compare the labeling with the provided spectral data. For further help, call your filter supplier.
2. If you are using a multiband filter set, make sure that your filter wheel is in the right position. If no exciter is in the light path, you will see scattered light of your excitation source. Be careful; bright unfiltered light can harm your eyes.

References

1. Hermann B (1998) Fluorescence microscopy (2nd edn). Springer, New York, Berlin, Heidelberg; Bios Scientific Publishers, Oxford
2. ThermoFisher (2016) Fluorescence spectra viewer. ThermoFisher Corporation. http://www.thermofisher.com/de/de/home/life-science/cell-analysis/labeling-chemistry/fluorescence-spectraviewer.html cited 21 Mar 2016
3. Reichman J (2000) Handbook of optical filters for fluorescence microscopy. Chroma Technology Corporation, Brattleboro, VT
4. Davidson M (2015) Florida State University. micro.magnet.fsu.edu/primer/techniques/fluorescence/fluorhome.html, cited 22 Mar 2016
5. Halling KC, King W, Sokolova IA et al (2000) A comparison of cytology and fluorescence in situ hybridization for the detection of urothelial carcinoma. J Urol 164:1768–1775
6. Dohlus R (2010) Photonik. Oldenbourg Wissenschaftsverlag München
7. Jaffe C (2010) Solid state light engines for bioanalytical instruments and biomedical devices. Lumencor Inc., Beaverton, OR http://lumencor.bydesign-pdx.com/wp-content/uploads/sites/11/2014/09/Lumencor-White-Paper-II.pdf cited 21 Mar 2016
8. Erdogan T (2007) BrightLine fluorescence filters for FISH. Semrock Inc., Rochester, NY

Classification of FISH Probes

Thomas Liehr

Abstract

Besides basic equipment, among consumables necessary for molecular cytogenetics, the choice of probes is the most critical point for a successful FISH experiment. Here the available FISH probes are reviewed and classified in different groups, i.e., according to their chemical properties, labeling, or target size.

Keywords DNA, RNA, PNA, LNA, Cloning, Microdissection, Flow sorting, Indirect labeling, Direct labeling, Chemical properties, Labeling, Target size

As stated in [1], "Humans like classifying. We often classify objects without even noticing. But there is normally a good reason for it." Thus, here we review how the probes used for FISH can be classified.

1 Probes for Molecular Cytogenetics According to Their Chemical Properties

Originally there was only one option for probes to be used in situ hybridization experiments: in the radioactive as well as in the non-radioactive variant, probes consisting from DNA were applied [2]. Another possible way was to create cDNA from RNA and use this as probe to target either chromosomal DNA ([3], e.g., Parts III–IV) or RNA ([4], chapter by Tiphaine Aguirre-Lavin and Nathalie Beaujean "Three-Dimensional Immuno-fluorescence In Situ Hybridization in Preimplantation Mouse"). Besides DNA cloned into bacterial artificial chromosomes or other plasmid-like constructs ([5], chapter by Thomas Liehr et al. "Bar-Coding Is Back"; chapter by Anja Weise and Thomas Liehr "Subtelomeric and/or Subcentromeric Probe Sets"; chapter by Anja Weise and Thomas Liehr "Parental Origin Determination FISH: pod-FISH"), derived from microdissection ([6], chapter by Fengtang Yang et al. "Generation of paint probes from flow-sorted and microdissected chromosomes") or flow sorting may be used ([6, 7], chapter by

Thomas Liehr (ed.), *Fluorescence In Situ Hybridization (FISH)*, Springer Protocols Handbooks,
DOI 10.1007/978-3-662-52959-1_4, © Springer-Verlag Berlin Heidelberg 2017

Fengtang Yang et al "Generation of Paint Probes from Flow-Sorted and Microdissected Chromosomes"). Those can either be PCR amplified and labeled ([6], chapter by Nadezda Kosyakova et al. "FISH-Microdissection"), or labeled by nick translation ([8], chapter by Nadezda Kosyakova et al. "FISH-Microdissection"). Nowadays most DNA probes for human are commercially available (chapter by Thomas Liehr "Commercial FISH Probes") as already labeled probes (see Sect. 2; chapter by Thomas Liehr "Commercial FISH Probes"). Another possibility is to use in vitro synthesized short DNA oligonucleotides (~20–50 nucleotides in length) to detect either repeats ([9], chapter by Cassia Yano et al. "Fish-FISH: molecular cytogenetics in fish species"), or if different oligonucleotides are pooled, they may span a region of several hundred nucleotides and serve as locus-specific, RNA-directed probes [10].

Besides DNA-based oligonucleotides, also peptide nucleic acid (PNA) and locked nucleic acid (LNA) probes were introduced. As outlined in the chapter by Nikolay Rubtsov and Natalya Zhdanova ("The Replicative Detargeting FISH (ReD-FISH) Technique in Studies of Telomere Replication"), "in PNA, the sugar-phosphate backbone is replaced by a synthetic peptide formed from *N*-(2-amino-ethyl)-glycine monomers. PNA oligonucleotides form hybrid duplexes with single stranded DNA which show higher thermal stabilities than the duplex of normal DNA, as PNA does not contain negatively charged phosphate groups. In LNA, the ribose ring is 'locked' by a methylene bridge connecting the 2′-O-atom with the 4′-C-atom. Like PNA, LNA can harbor the identical four bases being present in DNA and RNA and thus are able to hybridize as PNA/DNA or LNA/DNA, however, with increased thermal stability" and thus shorter hybridization time.

2 Probes for Molecular Cytogenetics According to Their Labeling

When FISH era started, there was only one possibility to label the applied probes, i.e., incorporate a biotin-labeled nucleotide analogue into the probe DNA and detect it after hybridization by avidin or streptavidin coupled to a fluorescence molecule—at that time most often FITC was applied in one-color FISH approaches [11, 12]. This labeling method was called "indirect labeling," as the fluorochrome was not directly attached to the probe DNA. Yet, routinely applied as "indirectly labeled probes" are only biotin and digoxigenin, which can be detected by antibodies specifically tagged with different fluorochromes; besides also dinitrophenol may be used besides biotin or digoxigenin as hapten.

Nowadays there are dozens of fluorochromes available on the market which can be purchased as already attached to a nucleotide analogue as well as detected by suited filters (chapter by Michael Sommerauer et al. "Optical filters and light sources for FISH").

Such kinds of probes are called "directly labeled probes." Directly labeled probes have the advantage to lead to quicker results, as no detection step is necessary (chapter by Thomas Liehr "Commercial FISH Probes"). However, especially in case of small probes, signal intensity of indirectly labeled probes can be enhanced by signal amplification by building up a "sandwich" as follows:

- First layer: biotin detected by avidin-FITC
- Second layer: anti-avidin in mouse
- Third layer: anti-mouse-FITC (chapter by Thomas Liehr and Anja Weise "Background")

Besides there are many other possibilities of signal amplification (e.g., [13])—however, most of these protocols amplify, besides the desired specific signal, also the background on the sample.

3 Probes for Molecular Cytogenetics According to Their Target Size

The most common classification of probes used in FISH is that putting them into the following four main groups (Fig. 1):

(a) Whole chromosome paints
(b) Partial chromosome paints
(c) Locus-specific probes
(d) Repetitive probes

Probes from groups (a) and (b) are in routine FISH only applicable on metaphases, while most probes from group c and d can be informative both in meta- and interphase. Group (a)–(c) probes need normally to be applied together with COT1 DNA and pre-hybridization (chapter by Vladimir Trifonov et al. "FISH with and without COT1 DNA"), while this may be omitted for group (d) probes. Normally group (a)–(c) probes are expected to result in specific FISH signals on a certain chromosome or chromosome

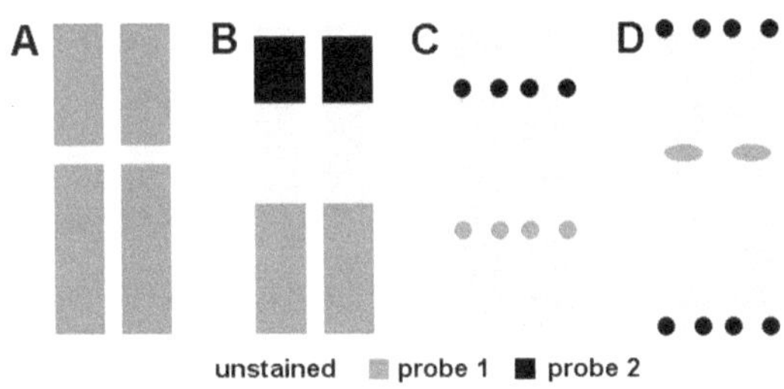

Fig. 1 Schematic drawings of the four different kinds of fluorescence in situ hybridization probes according to their target size. (**a**) Whole chromosome painting probe. (**b**) Two partial chromosome painting probes. (**c**) Two locus-specific single-copy probes. (**d**) Two probes specific for repetitive sequences like telomeric (probe 1) and centromeric regions (probe 2)

pair, even though exceptions are possible due to peculiarities of the studied genome [14]. Also repetitive probes from group (d) may only be detectable on one certain chromosome (e.g., DYZ1 on the Y chromosome) or chromosome pair (e.g., D7Z1 on chromosome 7, centromeric region). However, already among the so-called satellite probes specific for human centromeres, there are several ones which have complementary sequences on more than one chromosome (e.g., D14/22Z1 on chromosomes 14 and 22) [15]. Besides there are heterochromatic regions which may share homologous repetitive sequences like the ten human acrocentric short arms [15]. Also the telomeric sequences consist of repeats and thus belong here to group (d). They are present at each terminus of every chromatid in eukaryotes (chapter by Nikolay Rubtsov and Natalya Zhdanova "The Replicative Detargeting FISH (ReD-FISH) Technique in Studies of Telomere Replication") and may be targeted by FISH as well (chapter by Gordana Joksic et al. "Telomere Length Measurement by FISH").

FISH probes from group (d) may consist from repeats of different length. Telomeres have six base pair repeats, and centromeres have repeats of 171 base pairs. Besides, all genomes harbor different kinds of repeats which may be targeted in studies aimed to provide preliminary characterization of yet not molecular cytogenetically studied species ([9], chapter by Cassia Yano et al. "Fish-FISH: molecular cytogenetics in fish species"; chapter by Ana Paula Alves-Silva et al. "General Protocol of FISH for Insects"; chapter by Ekaterina Badaeva et al. "In Situ Hybridization to Plant Chromosomes").

4 Conclusion

Classification of FISH probes into whole chromosome paints, partial chromosome paints, locus-specific probes, and repetitive probes is the most commonly used in molecular cytogenetics. However, decision on which kind of probe to be used for a specific question needs to include all kinds of other aspects discussed above, like choice of labeling, fluorochrome used for visualizing the applied probe, and if to use a DNA, PNA, or LNA probe.

References

1. Bradley I, Gale P, Winterbottom M (2001) The Heinemann science scheme. Heinemann Educational Publishers, Oxford, p 48
2. Gall G, Pardue ML (1969) Formation and detection of RNA-DNA hybrid molecules in cytogenetical preparations. Proc Natl Acad Sci USA 63:378–381
3. Tillmann H, Stein S, Liehr T et al (2000) Structure and chromosomal localization of the human and mouse muscle fructose-1,6-bisphosphatase genes. Gene 247:241–253
4. Baechner D, Liehr T, Hameister H et al (1995) Widespread expression of the peripheral myelin protein-22 gene (PMP22) in neural and non-

neural tissues during murine development. J Neurosci Res 42:733–741

5. Hamid AB, Fan X, Kosyakova N et al (2015) New BAC probe set to narrow down chromosomal breakpoints in small and large derivative chromosomes, especially suited for mosaic conditions. Methods Mol Biol 1227:279–287
6. Liehr T, Heller A, Starke H et al (2002) Microdissection based high resolution multicolor banding for all 24 human chromosomes. Int J Mol Med 9:335–339
7. Romanenko SA, Perelman PL, Trifonov VA et al (2015) A First Generation comparative chromosome map between guinea pig (Cavia porcellus) and humans. PLoS One 10: e0127937
8. Rigby PW, Dieckmann M, Rhodes C et al (1977) Labeling deoxyribonucleic acid to high specific activity in vitro by nick translation with DNA polymerase I. J Mol Biol 113:237–251
9. Supiwong W, Liehr T, Cioffi MB et al (2013) Karyotype and cytogenetic mapping of 9 classes of repetitive DNAs in the genome of the naked catfish *Mystus bocourti* (Siluriformes, Bagridae). Mol Cytogenet 6:51
10. Orjalo AV Jr, Johansson HE (2016) Stellaris (®) RNA fluorescence in situ hybridization for the simultaneous detection of immature and mature long noncoding RNAs in adherent cells. Methods Mol Biol 1402:119–134
11. Pinkel D, Straume T, Gray JW (1986) Cytogenetic analysis using quantitative, high sensitivity, fluorescence hybridization. Proc Natl Acad Sci USA 83:2934–2938
12. Liehr T, Pfeiffer RA, Trautmann U (1992) Typical and partial cat eye syndrome: identification of the marker chromosome by FISH. Clin Genet 42:91–96
13. Pérez R, de Bustos A, Jouve N et al (2009) Localization of Rad50, a single-copy gene, on group 5 chromosomes of wheat, using a FISH protocol employing tyramide for signal amplification (Tyr-FISH). Cytogenet Genome Res 125:321–328
14. Weise A, Starke H, Mrasek K et al (2005) New insights into the evolution of chromosome 1. Cytogenet Genome Res 108:217–222
15. Liehr T (2014) Benign and pathological chromosomal imbalances, 1st edition Microscopic and submicroscopic copy number variations (CNVs) in genetics and counseling. Academic Press

Commercial FISH Probes

Thomas Liehr

Abstract

Sources for commercially available FISH probes, labeled or unlabeled ones, are urgently necessary prerequisites of molecular cytogenetic field. Here some basics on commercially available probes are provided, and most commonly by such probes tracked loci in prenatal, postnatal, tumor, and pathology molecular cytogenetics are provided.

Keywords Commercially available probes, Probe labeling, DNA, PNA, LNA, Indirect labeling, Direct labeling, Prenatal diagnostics, Postnatal diagnostics, Clinical genetics, Tumor genetics, Pathology

1 Commercial Probes in General

In the literature one can find repeatedly the statement that cytogenetics and molecular cytogenetics are outdated since decades [1]. However, the fact that especially molecular cytogenetics still is and will also in future be an indispensable tool for diagnostics could not be highlighted better by anything, than by the facts listed in the following. In our present world, all is about business and earning money [2]; and surprisingly there are numerous smaller and larger successful companies around, which are exclusively based on the concept to sell FISH probes. Thus, there must be a market for this kind of products. Furthermore, as still new companies are established and existing companies still invest money in enlarging their portfolios of probes, the molecular cytogenetic field must be far from being endangered or being outdated; it seems to be in contrary a growing field of interest. Still, most commercially available probes are human ones. There are only few companies also selling murine probes.

Thomas Liehr (ed.), *Fluorescence In Situ Hybridization (FISH)*, Springer Protocols Handbooks,
DOI 10.1007/978-3-662-52959-1_5, © Springer-Verlag Berlin Heidelberg 2017

2 Sources of Commercially Available Probes

Besides unlabeled locus-specific probes available from different sources like BACPAC Resources Center (http://bacpac.chori.org/) or the Wellcome Trust Sanger Institute (http://www.sanger.ac.uk/Teams/Team63/CloneRequest), labeled probes can be purchased from different sources. According to our subjective assessment, most relevant providers of FISH probes are in alphabetical order:

- Abbott Molecular/Vysis (https://www.abbottmolecular.com/vysis-fish-chromosome-search.html)
- Applied Spectral Imaging (http://www.spectral-imaging.com/)
- Cytocell (http://www.cytocell.com/)
- Dako/Agilent (http://www.dako.com/de/index.htm)
- Leica/Kreatech (http://www.leicabiosystems.com/ihc-ish-fish/kreatech-fish-probes/?no_cache=1)
- MetaSystems (http://www.metasystems-international.com/xcyting-dna-probes)
- Obio (http://www.obio.science)
- ZytoVision (www.zytovision.com)

There are also other providers which we cannot mention here in a comprehensive way.

3 Applications

Commercial probes are available for single, dual, and multicolor applications. Each kind of probe as specified in the chapter by Thomas Liehr "Classification of FISH Probes" is offered by aforementioned providers. The majority of those commercial probes are provided as directly labeled and DNA based (chapter by Thomas Liehr "Classification of FISH Probes"). Oligonucleotide-based probes can be purchased, but they still represent a minority compared to molecular clone-, microdissection-, or flow-sorting-derived probes.

Almost every company offering FISH probes provides whole chromosome paints and repetitive centromere-specific and locus-specific probes. Partial chromosome paints are available rather seldom (chapter by Thomas Liehr "Classification of FISH Probes"). All those probes can be provided alone, together with a control probe, in one-, two-, three-, or multicolor-FISH probe sets (Part IV) according to the market needs.

Commercially available FISH probes are, besides for basic research, directed toward the following fields of medicine: diagnostics in preimplantation medicine ([3, 4], chapter by Maria Bonet Oliver "Sperms, Spermatocytes and Oocytes"); in prenatal ([5], chapter by Anja Weise and Thomas Liehr "Pre- and Postna tal Diagnostics and Research on Peripheral Blood, Bone Marrow, Chorion, Amniocytes, and Fibroblasts"; chapter by Anja Weise et al. "FISH in Uncultivated Amniocytes") and postnatal clinical field ([6], chapter by Anja Weise and Thomas Liehr "Pre- and Postnatal Diagnostics and Research on Peripheral Blood, Bone Marrow, Chorion, Amniocytes, and Fibroblasts"); in solid tumors, lymphoma, and leukemia; and in human genetics, pathology, and oncology ([7], chapter by Eyad Alhourani et al. "Tumorcytogenetic Diagnostics and Research on Blood and Bone Marrow Smears or Effusions").

Especially for studying malignancies, sophisticated locus-specific probes and probe sets were introduced. Those are needed for detection of tumor suppressor gene loss, oncogene amplification, and characterization of translocations. Especially the latter ones can be tracked by so-called two-color break apart [7] or two-color fusion probe sets [7], some of them even being able to detect routinely deletions within the breakpoint region [8]. In Tables 1, 2, and 3, to the best of our knowledge, the most frequently studied targets of commercially available FISH probes in clinical genetics (including pre- and postnatal FISH—Tables 1 and 2), tumor genetics, and pathology (Table 3) are summarized. Interestingly, most commercially available diagnostic probes are labelled by the companies as "not being meant for diagnostic purposes".

Table 1
List of most important targets of commercially available FISH probes for prenatal diagnostics in clinical genetics

Syndrome to be tested	Cytoband/locus to be tested	Sonographic sign being indicative for syndrome
1p36 microdeletion syndrome	1p36/e.g., subtelomere 1pter	Heart defect, Ebstein's anomaly
Beckwith–Wiedemann syndrome	11p15/e.g., subtelomere 11pter	Omphalocele
DiGeorge syndrome	22q11.2/e.g., *HIRA*	Heart defect, not specified; pulmonary stenosis, tetralogy of Fallot, hygroma colli, omphalocele
Down syndrome	n.a./n.a. Any locus-specific probe for #21	Heart defect, tetralogy of Fallot, hygroma colli
Edwards syndrome	n.a./n.a. Any locus-specific probe or centromeric probe for #18	Heart defect, tetralogy of Fallot, hygroma colli
Jacobsen syndrome	11qter/e.g., subtelomere 11qter	Diaphragmic hernia, heart defect, pulmonary stenosis
Miller–Diecker syndrome	17p13.3/*LIS1*	Lissencephaly
Pätau syndrome	n.a./n.a. Any locus-specific probe for #13	Heart defect, tetralogy of Fallot, hygroma colli
Pallister–Killian syndrome	n.a./n.a. Any locus-specific probe or centromeric probe for #12	Omphalocele
Turner syndrome	n.a./n.a. Any locus-specific probe or centromeric probe for X	Hydrops fetalis
Williams–Beuren syndrome	7q11.23/*ELN*	Heart defect, pulmonary stenosis
Wolf–Hirschhorn syndrome	4p16.3/*WHSC1*	Diaphragmic hernia

Table 2
List of most important targets of commercially available FISH probes for postnatal diagnostics in clinical genetics

Syndrome to be tested	Cytoband/Locus to be tested	OMIM
1p36 microdeletion syndrome	1p36/e.g., subtelomere 1pter	607872
17q21.31 microdeletion syndrome	17q21.31/*KANSL1*	610443
Angelman syndrome	15q11 ~ 13/*UBE3A*	105830
Beckwith-Wiedemann syndrome	11p15/e.g., subtelomere 11pter	130650
Cri du Chat syndrome	5p15.2/*CTNND*	123450
DiGeorge syndrome	10p14 ~ 13/*NEBL* 22q11.2/e.g., *HIRA*	601362 188400
Down syndrome critical region	21q22.11/*SOD*	190685
Infertility SRY deletion	Yp11.3/*SRY*	480000
Jacobsen syndrome	11qter/e.g., subtelomere 11qter	147791
Kallmann syndrome	Xp22.3/*KAL1*	308700
Miller–Diecker syndrome	17p13.3/*LIS1*	247200
Neurofibromatosis type 1	7q11.2/*NF1*	162200
Prader–Willi syndrome	15q11 ~ 13/*SNRPN*	176270
Turner syndrome *SHOX* gene deletion	Xp22/*SHOX* any locus-specific probe or centromeric probe for X and Y	312865
Smith–Magenis syndrome	17p11.2/*FLI1*	182290
Sotos syndrome	5q35/*NSD1*	117550
Steroid sulfatase deficiency	Xp22.32/*STS*	300747
Terminal regions of all chromosomes	Subtelomeric region specific probes	e.g., #610253
Williams–Beuren syndrome	7q11.23/*ELN*	194050
Wolf–Hirschhorn syndrome	4p16.3/*WHSC1*	194190
XIST deletion	Xq13.2/*XIST*	314670

Table 3
List of most important targets of commercially available FISH probes for diagnostics of leukemia, lymphoma, and solid tumors

Tumor type	Target region(s)	Gene
Leukemia		
Myelodysplastic syndrome (MDS)	– 3q26 – 4q24 – 5q31.2 – 6p22 and 9q34 – 7q22 and 7q31 – 11q21 – 16p13 and 16q22 – 20q12 and 20q13.12	– *EVI1* – *TET2* – *EGR1* – *DEK/NUP214* – *RELN/TES* – *MAML2* – *MYH11/CBFB* – *PTPRT/MYBL2*
Chronic myeloid leukemia (CML)	– 4q12 – 5q32 ~ 33 – 9p24 – 9q34 and 22q11 – 11q22 – 17p13	– *FIP1L1/CHIC2/PDGFRα* – *PDGFRB* – *JAK2* – *BCR/ABL* – *ATM* – *P53*
Acute myeloid leukemia (AML)	– 3q26 – 4q12 – 5q31.2 – 5q32 – 5q35 – 6p22 and 9q34 – 6q23 – 6q27 – 7q22 and 7q31 – 9p24 – 9p21.3 – 11p15 – 11q23 – 15q24 and 17q21.2 – 16p13 and 16q22 – 20q12 and 20q13.12 – 21q22 – 22q22 and 8q21	– *EVI1* – *KIT* – *EGR1* – *CSF1R* – *NPM1* – *DEK/NUP214* – *MYB* – *MLLT4* – *RELN/TES* – *JAK2* – *MLLT3* – *NUP98* – *MLL* – *PML/RARα* – *MYH11/CBFB* – *PTPRT/MYBL2* – *ERG* – *RUNX1/RUNX1T1*
Chronic lymphocytic leukemia (CLL)	– 3q26 – 5q32 – 6q21 – 6q23 – 11q22 – 11q13 – 11q22 and 18q21 – 12q13 – 13q14.3 – 14q32 and 11q13 – 17p13 – 19q13	– *TERC* – *CD74* – *SEC63* – *MYB* – *ATM* – *Cyclin D1* – *BIRC3/MALT1* – *GLI* – *DLEU2* or *D13S25* – *IGH/CCND1* – *P53* – *BCL3*

(continued)

Table 3 (continued)

Tumor type	Target region(s)	Gene
Acute lymphocytic leukemia (ALL)	– Xp22.3 – Xp22.3 – 1p32 – 1q23 and 19p13.3 – 4q21 and 11q23 – 5q35 – 6q23 – 7q34 – 8q24 – 9p21 – 9p13 – 9q34 and 22q11 – 10q23 – 10q24.3 – 11q23 – 12p13 and 22q22 – 14q11 – 14q32.13 – 14q32.3 – 19p13 – 22q22 and 8q21	– *CRFL2* – *P2RY8* – *SIL/TAL1* – *PBX1/TCF3* – *MLL/AFF1* – *TLX3* – *MYB* – *TCRB* – *C-MYC* – *P16* or *CDKN2A* – *PAX5* – *BCR/ABL* – *PTEN* – *TLX1* – *MLL* – *TEL/AML1* – *TCR A/D* – *TCL1* – *IGH* – *E2A* – *RUNX1/RUNX1T1*
Lymphoma		
Anaplastic large-cell lymphoma	– 2p23 – 5q35	– *ALK* – *NPM1*
Burkitt lymphoma	– 2p11 – 8q24 – 14q32.3 – 17p13 – 21q11	– *IGK* – *C-MYC* – *IGH* – *P53* – *IGL*
Diffuse large B-cell lymphoma	– 2p16 – 2p11 – 3q27 – 8q24 – 9p21 – 14q32 and 18q21.33 – 17p13 – 19q13 – 21q11	– *REL* – *IGK* – *BCL6* – *C-MYC* – *P16* or *CDKN2A* – *IGH/BCL2* – *P53* – *BCL3* – *IGL*
Follicular lymphoma	– 3q27 – 6q23 – 9p21 – 14q32 and 18q21.33 – 17p13	– *BCL6* – *MYB* – *P16* or *CDKN2A* – *IGH/BCL2* – *P53*

(continued)

Table 3
(continued)

Tumor type	Target region(s)	Gene
Mantel cell lymphoma	– 5q32 – 9p21 – 11q22 and 18q21 – 13q14.3 – 14q32 and 11q13 – 17p13 – 19q13	– *CD74* – *P16* or *CDKN2A* – *BIRC3/MALT1* – *DLEU2* – *IGH/CCND1* – *P53* – *BCL3*
Other lymphoma	– 2p23 – 3q12 – 3q27 – 5q35 – 6q23 – 10p11.2 – 11q21 and 18q21 – 11q22 – 13q14.3 – 14q32 and 18q21.33 – 17p13	– *ALK* – *TFG* – *BCL6* – *NPM1* – *MYB* – *KIF5B* – *API/MALT1* – *ATM* – *DLEU2* – *IGH/BCL2* – *P53*
Multiple myeloma	– 1q21 and 1p36 – 1q21 and 8p21 – 4p16.3 – 5q32 – 6q23 – 11q22 – 13q14 – 14q32 and 4p16 – 14q32 and 11q13 – 14q32 and 16q23 – 14q32 and 20q12 – 15q22 and 9q34 – 17p13	– *c-MAF/SRD* – *c-MAF*/n.a. – *FGFR3* – *CD74* – *MYB* – *ATM* – *DLEU2* – *IGH/FGFR3* – *IGH/CCND1* – *IGH/MAF* – *IGH/MAFB* – n.a. → detection of hyperdiploidy – *P53*
Solid tumors		
Bladder	– 9p21 – 17p13	– *P16* or *CDKN2A* – *P53*
Bone and soft tissue	– 1p36.2 and 3q25 – 1p36 – 2q33 – 2q36 – 3q12 – 6p21 – 7p21 – 9q22 – 11p15.5 – 11p13 – 11q24 and 22q12	– *CAMTA1/WWTR1* – *PAX7* – *CREB1* – *PAX3* – *TFG* – *PHF1* – *ETV1* – *NR4A3* – *CARS* – *WT1* – *FLI1/EWSR1*

(continued)

Table 3
(continued)

Tumor type	Target region(s)	Gene
	– 12q13 – 12q13 ~ q14 – 12q14 – 12q15 – 13q14 – 16p11 – 17q21 and 22q13 – 18q11.2 – 21q22 – 22q12	– *DDIT3* – *CDK4* – *HMGA2* – *MDM2* – *FOXO1* – *FUS* – *COL1A1/PDGFB* – *SS18* – *ERG* – *EWSR1*
Breast	– 1q32 – 1q41 – 3q26 – 5q31.2 – 6q23 – 6q25 – 7p12 – 8p11.2 – 8q24 – 10q23 – 10q26 – 11q13 – 11q22.3 – 12p12 – 12q14 – 15q25 – 17p13.1 – 17q11.2 ~ 12 – 17q21 ~ 22 – 20q13	– *MDM4* – *CENPF* – *SOX2* – *EGR1* – *MYB* – *ESR1* – *EGFR* – *FGFR1* – *C-MYC* – *PTEN* – *FGFR2* – *CCND1* – *ATM* – *KRAS* – *HMGA2* – *NTRK3* – *P53* – *HER2/NEU1/ERBB2* – *TOP2A* – *ZNF217*
Central nervous system	– 1p36.2 and 3q25 – 1p36 – 1q25 – 1q41 – 2p24 – 3p25 – 3q26 – 6q22 – 7p11.2 – 9p21 – 10q23 – 12q13 ~ q14 – 15q25 – 17p13 – 19p13 – 19q13	– *CAMTA1/WWTR1* – *MEGF6* – *ABL2* – *CENPF* – *NMYC* – *VHL* – *SOX2* – *ROS1* – *EGFR* – *CDNK2A* – *PTEN* – *CDK4* – *NTRK3* – *P53* – *ZNF44/ZNF* – *CRX*

(continued)

Table 3
(continued)

Tumor type	Target region(s)	Gene
Colorectal	– 3q26 – 6q23 – 6q24.3 – 7q34 – 10q23 – 12p12 – 17p13.1 – 18p11.32	– *SOX2* – *MYB* – *RREB1* – *BRAF* – *PTEN* – *KRAS* – *P53* – *TYMS*
Esophagus	– 8q24 – 9p21 – 17p13.1 – 17q11.2 ~ 12 – 18p11.32 – 20q13	– *C-MYC* – *P16* or *CDKN2A* – *P53* – *HER2/NEU1/ERBB2* – *TYMS* – *ZNF217*
Eye	– 1q32 – 13q14	– *MDM4* – *RB1*
Head and neck	– 1q41 – 3p25 – 5q32 – 11q21 – 12p13.3 – 19p13.2	– *CENPF* – *VHL* – *CD74* – *MAML2* – *FOXM1* – *BRD4*
Kidney	– Xp11.23 – 3p25 – 3p14 – 6p21 – 7q31 – 10q23 – 17p13	– *TFE3* – *VHL* – *FHIT* – *TFEB* – *MET* – *PTEN* – *YWHAE*
Liver	– 4q12 – 8q24 – 9p21 – 11q13.3 – 12p12 – 17p13.1 – 18q21	– *KIT* – *CMYC* – *P16* – *FGF3,4,19* – *KRAS* – *P53* – *BCL2*
Lung	– 1q32 – 2p23 and 2p21 – 3p14 – 3q12 – 3q26 – 4q12 – 5q32 – 6q22 – 7p12	– *MDM4* – *ALK/EML4* – *FHIT* – *TFG* – *SOX2* – *PDGFRA* – *CD74* – *ROS1* – *EGFR*

(continued)

Table 3
(continued)

Tumor type	Target region(s)	Gene
	– 7q34 – 10p11.2 – 10q26	– *BRAF* – *KIF5B* – *FGFR2*
Skin (melanoma)	– 6q23 – 6p25 – 7p21 – 7q34 – 9p21 – 10q23 – 11q13 – 22q12	– *MYB* – *RREB1* – *ETV1* – *BRAF* – *P16* – *PTEN* – *CCND1* – *EWSR1*
Stomach	– 3q26 – 4q12 – 4q12 – 7q31 – 8q24 – 10q23 – 10q26 – 11q22 and 18q21 – 17p13.1 – 17q21 – 18p11.32	– *SOX2* – *KIT* – *PDGFRA* – *MET* – *CMYC* – *PTEN* – *FGFR2* – *BIRC3/MALT1* – *TP53* – *ERBB2* – *TYMS*
Ovary	– 3q26 – 8q24 – 9p21 – 10q26 – 11q13 – 12p12 – 17p13.1 – 19q13 – 20q13	– *PIK3CA* – *CMYC* – *P16* – *FGFR2* – *CCND1* – *KRAS* – *P53* – *CRX* – *NCOA3(AIB1)*
Pancreas	– 5q32 – 6q24.3 – 7q34 – 9p21 – 10q23 – 11q22.3 – 12p12 – 17q13	– *CD74* – *RREB1* – *BRAF* – *P16* – *PTEN* – *ATM* – *KRAS* – *P53*
Prostate	– Xq12 – 3p14 – 3q27 – 7p21 – 8q24 – 9p21	– *AR* – *FHIT* – *ETV5* – *ETV1* – *C-MYC* – *P16*

(continued)

Table 3
(continued)

Tumor type	Target region(s)	Gene
	– 10q23 – 12p13.3 – 12q13q14 – 17p13.1 – 21q22	– *PTEN* – *FOXM1* – *CDK4* – *P53* – *ERG*
Thyroid gland	– 1q22 ~ q23 – 2q13 – 3q12 – 7q34 – 10q11.2 – 10q23	– *NTRK1* – *PAX8* – *TFG* – *BRAF* – *RET* – *PTEN*
Uterus	– 3q26 – 5q32 – 6p21.3 – 7p15 – 8q24 – 9p21 – 10q23 – 10q26 – 12p12 – 17p13 – 17p13.1 – 17q12	– *PIK3CA* – *CSF1R* – *PHF1* – *JAZF1* – *CMYC* – *P16* – *PTEN* – *FGFR2* – *KRAS* – *YWHAE* – *P53* – *HER2/NEU1/ERBB2*
Other solid tumors	– 1p36 – 1p32 and 1q21 – 3p14 – 3q26 – 5p15 – 6q22 – 7q31 – 12p13.3	– *SRD* – *CKS1B/CDKN2C* – *FHIT* – *TERC* – *TERT* – *MET* – *ROS1* – *FOXM1*

4 Conclusion

Overall, the sources of commercially available FISH probes, labeled or unlabeled ones, are urgently necessary prerequisites of the molecular cytogenetic field. However, it is important to select these probes carefully according to the (clinical) question to be studied. Information on probe localization provided by the suppliers need to be checked carefully, as even probes with similar names from different providers may target completely different loci!

References

1. Grati FR, Malvestiti F, Grimi B et al (2013) QF-PCR as a substitute for karyotyping of cytotrophoblast for the analysis of chorionic villi: advantages and limitations from a cytogenetic retrospective audit of 44,727 first-trimester prenatal diagnoses. Prenat Diagn 33:502–508
2. Vanderbyl S, Kobelak S (2007) Critical success factors for biotechnology industry in Canada. J Commerc Biotechnol 13:68–77
3. Enciso M, Alfarawati S, Wells D (2013) Increased numbers of DNA-damaged spermatozoa in samples presenting an elevated rate of numerical chromosome abnormalities. Hum Reprod 28:1707–1715
4. Chiamchanya C, Visutakul P, Gumnarai N et al (2008) Preimplantation genetic screening (PGS) in infertile female age > or = 35 years by fluorescence in situ hybridization of chromosome 13, 18, 21, X and Y. J Med Assoc Thai 91:1644–1650
5. Liehr T, Ziegler M (2005) Rapid prenatal diagnostics in the interphase nucleus: procedure and cut-off rates. J Histochem Cytochem 53:289–291
6. Dufke A, Seidel J, Schöning M et al (2000) Microdeletion 4p16.3 in three unrelated patients with Wolf-Hirschhorn syndrome. Cytogenet Cell Genet 91:81–84
7. Othman MA, Melo JB, Carreira IM et al (2015) High rates of submicroscopic aberrations in karyotypically normal acute lymphoblastic leukemia. Mol Cytogenet 8:45
8. Loncarevic IF, Römer J, Starke H et al (2002) Heterogenic molecular basis for loss of ABL1-BCR transcription: deletions in der(9)t(9;22) and variants of standard t(9;22) in BCR-ABL1-positive chronic myeloid leukemia. Genes Chromosomes Cancer 34:193–200

Generation of Paint Probes from Flow-Sorted and Microdissected Chromosomes

Fengtang Yang, Vladimir Trifonov, Bee Ling Ng, Nadezda Kosyakova, and Nigel P. Carter

Abstract

FISH with whole chromosome or region-specific painting probes made from either flow-sorted or microdissected chromosomes has revolutionized cytogenetics. Generation of paints from flow-sorted chromosomes relies on the use of an expensive and sophisticated fluorescence-activated cell sorter and suspensions of freshly prepared chromosomes. Preparation of paints from microdissected materials requires an inverted microscope with appropriate micromanipulators and metaphase chromosome spreads on coverslips. Painting probes made from flow-sorted chromosomes generally have better chromosomal coverage and can be used in a wide range of applications between distantly related species, while region-specific probes from microdissection enable higher-resolution analyses restricted to comparative painting between closely related species. Here we provide detailed protocols on generation probes from both flow-sorted and microdissected chromosomes.

Keywords Probe for FISH, Chromosome painting, Flow sorting, Microdissection, Human chromosome-specific paints, Animal chromosome-specific paints

1 Introduction

Whole chromosome or region-specific paint probes ("paints") are collections of labeled DNA sequences derived from a specific type of chromosome or chromosomal segment. When the technique of chromosome painting was first developed, paints were made from cloned DNA libraries established using flow-sorted chromosomes as the source of DNA, by pooling plasmid clones from chromosome-specific DNA libraries and subsequent labeling by nick translation ([1], chapter by Thomas Liehr "Homemade Locus-Specific FISH Probes: Bacterial Artificial Chromosomes"; chapter by Thomas Liehr et al. "Two- to Three-Color FISH"). Such a process was labor intensive and time-consuming. From 1992 onward, most paints have been made by direct PCR amplification, using a degenerate universal primer (DOP-PCR) [2], of either microdissected

Thomas Liehr (ed.), *Fluorescence In Situ Hybridization (FISH)*, Springer Protocols Handbooks,
DOI 10.1007/978-3-662-52959-1_6,

([3], chapter by Nadezda Kosyakova et al. "FISH-Microdissection") or flow-sorted chromosomes [2, 4]. The introduction of DOP-PCR has greatly simplified the process of paint production, and the paints derived from DOP-PCR produce superior signal intensity. For these reasons, DOP-PCR using flow-sorted and microdissected chromosomes as template DNA has become the method of choice for the production of paint probes. In addition to the widely used chromosome-specific paints from human [1] and mouse [5] available commercially, paints from more than 100 vertebrate species have been generated, including representative species from the major branches of mammalian phylogenetic tree (http://www.vet.cam.ac.uk/genomics/; chapter by Fengtang Yang et al. "Animal Probes and ZOO-FISH").

Chromosome paints have become indispensable tools for clinical, cancer, and comparative cytogenetics. In particular, these reagents have enabled cross-species chromosome painting on a genome-wide scale between both closely and distantly related species [6–10]. Chromosome painting, in conjunction with the conventional chromosomal banding, has revitalized the cytogenetic analyses of both pathological and evolutionary chromosomal rearrangements and has made a great contribution to the success of molecular cytogenetics during the past two decades. Most animal chromosome-specific paints developed over the past two decades by the Ferguson-Smith group in the University of Cambridge are available to the research community via the Cambridge Resource Centre for Comparative Genomics (http://www.vet.cam.ac.uk/genomics/).

Chromosome microdissection is another useful technique for isolation of chromosomes or chromosomal regions of interest directly from metaphase plates using micromanipulation. The method was first applied to polytene chromosomes of *Drosophila* ([11], chapter by Amanda Larracuente "FISH in Drosophila"). The fragments of DNA isolated were cloned in lambda phage, and chromosome specificity was shown by in situ hybridization. However, early work on the microdissection of human chromosomes demonstrated the poor efficiency and reproducibility of this specific approach. Further improvement of the technique was accomplished by Senger and colleagues [12], who employed an inverted microscope (×1,250 magnification), which increased the resolution of the technique and enabled regions to be cut very precisely (about one GTG-band resolution). In addition, the use of rotating plate and dissection in air made microdissection more easily accomplishable. Following dissection, chromosomal fragments were deproteinated; the DNA was digested using restriction enzymes and cloned into plasmids, followed by DNA amplification with Klenow fragment DNA polymerase. These developments allowed the number of starting copies to be reduced from 100–200 to 20–40. The procedures for a rapid generation of region-specific libraries by microdissection followed by DOP-PCR were established independently [3, 13], which eliminated the time-consuming micro-cloning step. Subsequent major technological improvements included the addition of a second left-hand micromanipulator with a pipette, containing collection drop [14]. Chromosome-specific libraries for multicolor FISH and multicolor banding constructed by microdissection (chapter by Thomas Liehr et al. "FISH Banding Techniques") have found wide and successful application in clinical genetics (chapter by Anja Weise and Thomas Liehr "Pre- and Post natal Diagnostics and Research on Peripheral Blood, Bone Marrow, Chorion, Amniocytes, and Fibroblasts"; chapter by Thomas Liehr and Monika Ziegler "Application of FISH to Previously GTG-Banded and/or Embedded Cytogenetic Slides"; chapter by Anja Weise et al. "FISH in Uncultivated Amniocytes"), tumor genetics (chapter by Eyad Alhourani et al. "Tumorcytogenetic Diagnostics and Research on Blood and Bone Marrow Smears or Effusions"), and comparative interspecies studies and interphase cytogenetics ([15, 16], chapter by Fengtang Yang et al. "Animal Probes and ZOO-FISH", chapter by Thomas Liehr and Nadezda Kosyakova "Three-Dimensional Interphase Analysis Enabled by Suspension FISH"). Here we provide the detailed protocols for generating paint probes by DOP-PCR from flow-sorted and microdissected chromosomes or chromosomal segments.

2 Materials

2.1 Flow Sorting

2.1.1 Instruments

- Fluorescence microscope
- Flow cytometer with sorting capability [e.g., MoFlo® (Beckman Coulter, Fullerton, CA, USA), equipped with two water-cooled Innova 300 series lasers (Coherent, Santa Clara, CA, USA), or FACStar Plus (Becton-Dickinson, Rutherford, NJ, USA)]
- Sterilin™ 30ml Polypropylene universal container (Thermo Scientific, Waltham, MA, USA)

2.1.2 Tissue Culture

- Demecolcine solution 10 μg ml^{-1} (Sigma-Aldrich, St. Louis, MO, USA)
- DMEM supplemented with 15 % fetal bovine serum and antibiotics (penicillin-streptomycin)
- Histopaque-1083 (Sigma-Aldrich)
- Lipopolysaccharide, LPS (Sigma-Aldrich)
- 1 × PBS
- RPMI 1640 medium supplemented with 15 % fetal bovine serum and antibiotics (penicillin-streptomycin)
- 0.25 % Trypsin-EDTA (Sigma-Aldrich)

2.1.3 Chromosome Isolation Buffers

- Hypotonic solution: 75 mM KCl, 0.2 mM spermine, 0.5 mM spermidine, 10 mM $MgSO_4{\cdot}7H_2O$. Top up with HPLC water to 50 ml and then pH to 8.0 by adding 20 μl of 0.25 M NaOH). Store on ice. Prepare fresh on the day of chromosome isolation.
- Polyamine isolation buffer (PAB): 15 mM Tris, 2 mM EDTA, 0.5 mM EGTA, 80 mM KCl, 3 mM dithiothreitol, 0.25 % Triton X-100, 0.2 mM spermine, 0.5 mM spermidine. Top up with HPLC water to 50 ml. Mix well on a rotator for 30 min, and then adjust pH to 7.5 (add 20 μl of 0.5M NaOH). Filter through a 0.22 μm filter before storage at 4 °C. The solution is stable for a month.
- Propidium iodide: 1 mg ml^{-1} in sterile distilled water. Store in dark at 4 °C.
- Turk's stain: 0.01 % gentian violet in 1 % glacial acetic acid. Store at 4 °C.

2.1.4 Chromosome Staining Solutions

- Chromomycin A3: 10 mg ml^{-1} in absolute ethanol. Store in dark at −20 °C.
- Hoechst 33258: 1 mg ml^{-1} in sterile distilled water. Store in dark at 4 °C.
- Sodium citrate: 1M in HPLC water. Store at −20 °C.
- Sodium sulfite: 500 mM in HPLC water. Store at −20 °C.

2.1.5 Sheath Buffer

- 1 mM EDTA
- 0.05 % Na azide
- 100 mM NaCl
- 10 mM Tris-HCl, pH 7.4

Autoclave and store at room temperature (RT).

2.1.6 Generation of Paint Probes by DOP-PCR

- 300–1000 copies of flow-sorted chromosomes in a PCR tube that contains 27.5 μl sterile dH_2O
- 5 × PCR optimized buffer D (Invitrogen, Carlsbad, CA, USA)
- 10 × PCR buffer (Bioline, London, UK)
- 5 U μl^{-1} BIOTAQ DNA Polymerase (Bioline)
- 50 mM $MgCl_2$ (Bioline)
- 2.5 mM dNTP mix
- 10 mM individual dATP, dCTP, dGTP, and dTTP (Bioline or Jena BioScience, Jena, Germany)
- Biotin-16-dUTP (Jena BioScience)
- Aminoallyl-dUTP-ATTO-425 (Jena BioScience)

- Fluorescein-12-dUTP (Thermo Fisher Scientific, Waltham, MA, USA)
- Aminoallyl-dUTP-XX-ATTO-488 (Jena BioScience)
- Dig-11-dUTP (Jena BioScience)
- Aminoallyl-dUTP-Texas Red (Jena BioScience)
- Aminoallyl-dUTP-Cy3 (Jena BioScience)
- Aminoallyl-dUTP-Cy5 (Jena BioScience)
- 10X ½ dTTP dNTP solution (with 2 mM dATP/dCTP/dGTP, 1 mM dTTP)
- 1 mM dTTP solution
- PCR water
- 1 % Polyethylene glycol ether W-1 (Sigma-Aldrich)
- 20 μM 6-MW (DOP) primer: 5′ CCG ACT CGA GNN NNN NAT GTG G 3′

2.2 Microdissection

2.2.1 Instruments

- Inverted microscope Axiovert 10 or 135 (Zeiss, Jena, Germany) or the IX Series (Olympus, Shinjuku, Tokyo, Japan)
- Right-handed and left-handed micromanipulators (Zeiss, Germany, or Narishige, Tokyo, Japan)
- Pipette puller (Narishige, Japan)
- Glass rods, 2 mm diameter (Schott Glass, Mainz, Germany)
- Pasteur pipettes, 230 mm (Assistent, Sondheim, Germany)
- Aspirator with a trap flask (Grant Instruments, Royston, UK)

2.2.2 Reagents

- Glycerol (Sigma-Aldrich)
- Proteinase K from *Tritirachium album* (Sigma-Aldrich)
- Trypsin (Sigma-Aldrich)
- 0.025 M phosphate buffer pH 6.88 (Merck)
- Giemsa (Merck)
- Carbon tetrachloride (Sigma-Aldrich)
- Dimethyldichlorosilane (Sigma-Aldrich)
- DOP primer (5′-CCG ACT CGA GNN NNN NAT GTG G-3′)
- T7 DNA Polymerase 10 U μl^{-1} (Thermo Fischer Scientific)
- T7 DNA Polymerase Buffer (Thermo Fischer Scientific)
- AmpliTaq Gold DNA Polymerase (Thermo Fisher Scientific)
- 10× AmpliTaq Gold DNA Polymerase Buffer (Thermo Fisher Scientific)
- Biotin-16-dUTP (Roche, Basel, Switzerland)

2.2.3 Solutions

- Collection drop solution: 30 % glycerol, 10 mM Tris-HCl pH 7.5, 10 mM NaCl, 0.1 % SDS, 1 mM EDTA, 0.1 % Triton X-100, 1.44 mg ml^{-1} Proteinase K

3 Methods

3.1 Generation of Chromosome-Specific Paints by DOP-PCR Using Flow-Sorted Chromosomes

3.1.1 Cell Culture

Fibroblast Cell Lines

1. Culture cells in DMEM (Gibco) supplemented with 15 % fetal bovine serum (FBS, Gibco) and antibiotics (penicillin and streptomycin, Sigma-Aldrich) using 150 cm^2 flasks.
2. Subculture near confluent cells at a ratio of 1:4 or 1:3 in eight flasks depending on the rate of cell growth in order to obtain ~2 ml of chromosome suspension.
3. After 24 h growth, add demecolcine to the cell culture to a final concentration of 0.1 μg ml^{-1}; incubate for 6–16 h (*see* **Note 1**).
4. Collect the supernatant from eight flasks (150 cm^2) after mitotic shake-off using 50 ml falcon tubes, and centrifuge at 289× *g* for 5 min.
5. Discard the supernatant without disturbing the pellet, and place the tube upside down on an absorbent paper to drain off most of the medium.
6. Gently resuspend the cell pellets in hypotonic solution (~2 ml per falcon tube) with a plastic pipette. Pool the cell suspension together into one tube and incubate at RT for 15 min.
7. Proceed to Sect. 3.1.2.

Mouse Spleen Culture

1. Squash a mouse spleen into a cell suspension via a cell strainer using a plunger from a 10 ml syringe. Wash cells off from the cell strainer in between rounds of squashing with 10 ml of PBS.
2. Load the cell suspension onto Histopaque-1083 (Sigma-Aldrich) in a 30 ml universal container, and centrifuge at 800× *g* for 20 min.
3. Carefully remove the middle white layer (which contains lymphocytes) using a pipette and transfer to a new universal container.
4. Add an equal volume of RPMI 1640 to the cells, and centrifuge at 400× *g* for 10 min.

5. Discard the supernatant, and resuspend the cell pellet in RPMI 1640 to a final concentration of 10^6 cells ml^{-1}.
6. Add LPS to a final concentration of 50 μg ml^{-1}, and transfer cells to a 75 cm^2 tissue culture flask, and place the flask inside a 37 °C CO_2 incubator.
7. After 44–48 h of incubation, add demecolcine to a final concentration of 0.1 μg ml^{-1}, and incubate for 3.5 h.
8. Harvest the cell culture and centrifuge at 289× *g* for 5 min.
9. Discard the supernatant and gently resuspend the cell pellet in 5 ml of hypotonic solution, and incubate at RT for 15 min.
10. Proceed to Sect. 3.1.2.

3.1.2 Preparation of Chromosome Suspension Using Polyamine Isolation Buffer

1. Monitor the swelling of cells by mixing 5 μl of cell suspension with 5 μl of Turk's stain on slide. View under the light microscope. If cells are not swollen, leave for another 5–10 min. Continue to monitor the swelling of cells and incubate for an increased time as necessary.
2. After ~15 min in hypotonic solution at RT, centrifuge the swollen cells at 289× *g* for 5 min.
3. Discard the supernatant, drain tube briefly on absorbent paper, resuspend the cell pellet gently in 3 ml of ice-cold polyamine isolation buffer (PAB), and incubate on ice for 10 min.
4. Vortex the suspension vigorously for 10–20 s.
5. Monitor the chromosome suspension under the fluorescence microscope by mixing 5 μl of suspension with 5 μl of propidium iodide on a microscope slide and cover with a coverslip. Check for "singly floating chromosomes" in suspension (*see* **Note 2**).
6. Briefly centrifuge the chromosome suspension at 201× *g* for 2 min. Filter supernatant through 20 μm mesh filter (CellTrics, Partec).
7. Stain chromosomes overnight with 5 μg ml^{-1} of Hoechst (Sigma-Aldrich), 40 μg ml^{-1} of Chromomycin A3 (Sigma-Aldrich), and 10 mM $MgSO_4$.
8. To the stained chromosome suspension, add 10 mM of sodium citrate and 25 mM of sodium sulfite at least an hour before flow analysis and sorting (*see* **Note 3**).

3.1.3 Flow Analysis and Sorting of Chromosomes on a MoFlo®

1. Setting up the lasers and optics:

   The stained chromosome suspensions are analyzed on a flow cytometer equipped with two lasers spatially separated at the flow chamber. The first laser is tuned to emit multiline UV (330–360 nm) to efficiently excite Hoechst. The second laser is tuned to emit light at 457.9 nm to excite chromomycin A3.

The powers of both lasers are set to 300 mW and kept constant using light control feedback. Fluorescence emitted from Hoechst was collected using a 400 nm long-pass filter combined with a 480 nm short-pass filter. Chromomycin fluorescence is collected using a 490 nm long-pass filter. The optical light path of the flow cytometer is aligned before chromosome analysis using 3 μm beads (Sphero™ Rainbow Fluorescent Particles, Spherotech) with minimum peak coefficient of variance for both fluorescence channels (*see* **Note 4**).

2. Setting up the flow sorter:

   The instrument is configured for four-way sorting on a high-purity sort option of single mode per single drop envelope, with a data rate ranging from 2,000 to 12,000 events per second and an optimal setting of the sheath pressure to ~60 psi, using sheath buffer as the fluidics media and a drop drive frequency to ~95 KHz with a 70 μm Cytonozzle tip. The chromosomes are flow sorted into sterile 500 μl Eppendorf tubes containing 33 μl of UV-treated, sterile, distilled water (*see* **Note 5**).

3. Data analysis and gating:
   Data for forward scatter, pulse width, Hoechst fluorescence, and chromomycin fluorescence were collected using Hoechst fluorescence as the trigger signal. Flow karyotypes for all cell lines were displayed as a bivariate flow karyograms of Hoechst versus chromomycin fluorescence (Fig. 1b–d) after gating on forward scatter (FSC) versus pulse width to exclude doublets, clumps, and debris (Fig. 1a, region R1). A total of 100,000 events were acquired for each cell line at a rate of 1000 events s^{-1}.

3.1.4 Generation of Paint Probes by DOP-PCR

Primary DOP-PCR Using (5 ×) PCR Buffer D (Invitrogen) (50 μl Reaction)

Starting materials: 300–1,000 chromosomes flow sorted into a 0.5 ml PCR tube containing 27.5 μl of UV-treated PCR water (*see* **Note 6**).

1. Perform the whole procedure in a clean DNA-free hood.
2. Before setting up the reaction, place the pipettes, pipette tips, tubes for master mix, PCR buffer, and W1 solution into the hood, and turn on the UV light for 30 min to eliminate any trace of contamination of unwanted DNA (*see* **Note 7**).
3. If several tubes of flow-sorted chromosomes are to be amplified, prepare a master mix. Mix thoroughly before aliquoting.

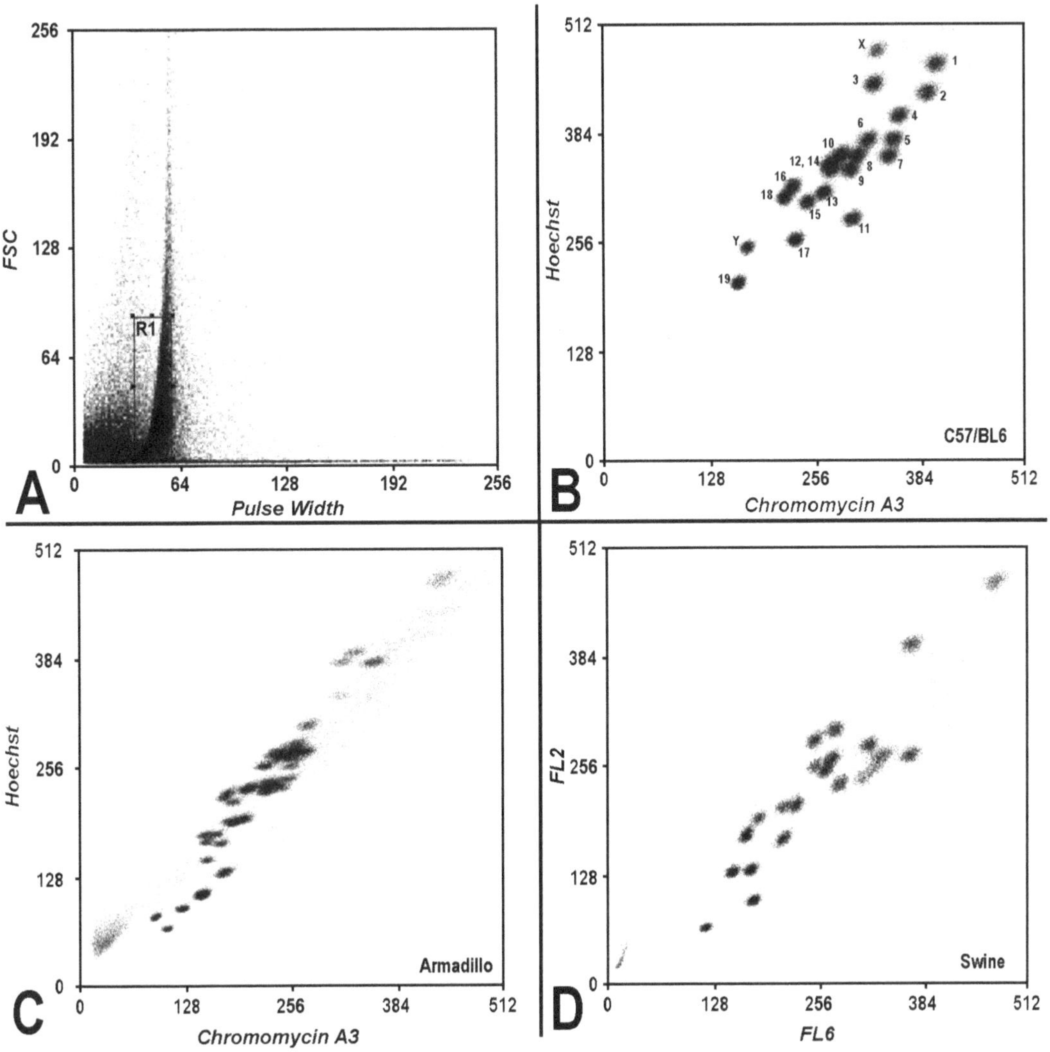

Fig. 1 Flow karyotypes. (**a**) A region (R1) is created on the plot of forward scatter (FSC) versus pulse width to exclude doublets, clumps, and debris. Bivariate plots of Hoechst versus chromomycin fluorescence are gated on this region [see figure parts (**b**) to (**d**)]. (**b**) Bivariate flow karyotype of chromosomes from c57/BL6 LPS stimulated B lymphocyte mouse cell culture. (**c**) Bivariate flow karyotype of chromosomes from an Armadillo fibroblast cell line. (**d**) Bivariate flow karyotype of chromosomes from a swine fibroblast cell line

	Master mix (per sample)
5× PCR optimized buffer D	10 μl
20 μM DOP primer	5 μl
2.5 mM dNTP mixture	4 μl
1 % polyoxyethylene ether W1	2.5 μl
BIOTAQ DNA Polymerase (5 U μl^{-1})	1 μl
Total:	22.5 μl

4. Aliquot 22.5 μl of master mix into each tube containing flow-sorted chromosomes. Vortex and spin briefly to collect all chromosomes into the bottom of PCR tubes. Always include a negative control (i.e., a PCR tube with 27.5 PCR water only).
5. Overlay with 30 μl of mineral oil if a thermal cycler with non-heated lid is used. Place the PCR in a thermocycler and start the following program:
   (a) 94 °C, 9 min
   (b) 94 °C, 1 min
   (c) 30 °C, 2.5 min (ramp at 0.1 °C s^{-1} to 72 °C)
   (d) 72 °C, 3 min
   Repeat 2–4 for nine cycles.
   (e) 94 °C, 1 min
   (f) 62 °C, 1.5 min
   (g) 72 °C, 2.5 min
   Repeat 5–7 for 25–30 cycles.
   (h) 72 °C, 8 min
   (i) 4 °C, hold
6. Run 3 μl of the PCR products on a 1 % agarose gel (100 V, 10–15 min). The products should be a clearly visible smear, with an average size of 0.2–2 kb. There should be no amplification products in the negative control.

Secondary DOP-PCR Labeling

1. Set up the secondary DOP-PCR labeling on the laboratory bench.
2. Calculate the number of probe tubes to be labeled, make a master mix, and then make aliquots. Use 1–2 μl of primary DOP-PCR product for each 25 μl reaction. Do not carry out large scale of probe labeling until the quality of each paint probe has been checked using FISH.
3. For each 25 μl reaction with Atto 488-, biotin-, Cy3-, Cy5-, Dig-, and FITC-conjugated dUTPs, combine the following solutions; for labeling with Texas Red-conjugated dUTP, adjust the final ratio of dTTP/dUTP in the PCR mixture from 140 μM/60 μM to 180 M/20 μM.

25 μl reaction	Master mix (per sample)
dH_2O	11.4 μl
10 × NH4 PCR buffer (Bioline)	2.5 μl
20 μM DOP primer	2.5 μl
½ dT dNTP mixture	2.5 μl
[d(A,C,G)TP, 2 mM each, dTTP 1 mM]	

1 mM labeled dUTP	1.5 μl
1 mM dTTP	1.0 μl
50 mM $MgCl_2$	1.3 μl
1 % W1	1.0 μl
BIOTAQ DNA Polymerase (5 U μl^{-1})	0.3 μl
Subtotal	24.0 μl
Primary DOP-PCR product:	1.0 μl

4. Overlay with 20 μl of mineral oil if a thermocycler with non-heated lid is used.
5. Place the PCR in a thermocycler and start the following program:
   (a) 94 °C, 3 min
   (b) 94 °C, 1 min
   (c) 62 °C, 1 min 30 s
   (d) 72 °C, 2 min 45 s
   Repeat 2–5 for 25–30 cycles.
   (e) 72 °C, 10 min
   (f) 4 °C hold
6. Run 3 μl of PCR products on a 1 % agarose gel (100 V, 10–15 min). The products should be a clearly visible smear with an average size of 0.2–2 kb.

Reamplification of Primary DOP-PCR Products (Optional)

1. In the situation where only a limited amount of primary DOP-PCR products (instead of flow-sorted chromosomes) are available, it may be necessary to reamplify the primary DOP-PCR products before labeling the probes in a secondary round of DOP-PCR amplification. For a 100 μl reaction, combine the following solutions:

	Master mix (per sample)
dH_2O	56 μl
10 x NH_4 PCR buffer (Bioline)	10 μl
20 μM DOP primer	10 μl
dNTP mix [d(A, T, C, G)TP, 2.5 mM each]	8 μl
50 mM $MgCl_2$	5 μl
1 % W1	5 μl
BIOTAQ DNA Polymerase (5 U μl^{-1})	1 μl
Subtotal	95 μl
Primary DOP-PCR product	5 μl

2. Mix thoroughly and overlay with 20 μl of mineral oil, if a thermal cycler with non-heated lid is used. Place the PCR in a thermocycler and start the following program:
   (a) 94 °C, 3 min
   (b) 94 °C, 1 min
   (c) 62 °C, 1 min 30 s
   (d) 72 °C, 2 min
   Repeat 2–4 for 25 cycles.
   (e) 72 °C, 10 min
   (f) 4 °C hold
3. Run 3 μl of the PCR products on a 1 % agarose gel (100 V, 10–15 min). The products should be a clearly visible smear with an average size of 0.2–2 kb.

3.2 Generation of Region-Specific Paint by Microdissection and DOP-PCR

An alternative protocol of microdissection to the one outlined below is presented by Nadezda Kosyakova et al. in chapter "FISH Microdissection".

3.2.1 Microneedle and Micropipette Preparation

1. Prepare microneedles from 2 mm glass rods on a vertical two-step puller, to get very sharp but not very long edges.
2. Expose the needles to UV light for at least 30 min.
3. Prepare micropipettes on the same puller from Pasteur pipettes.
4. Break the micropipette tips carefully to obtain a small round opening.
5. Siliconize micropipettes by immersing into 1 % dimethyldichlorosilane in carbon tetrachloride using an aspirator with a trap flask (Grant).
6. Air-dry and wash in 1mM EDTA (pH 7.5).
7. Incubate micropipettes for 30 min at 100 °C.
8. Keep micropipettes and microneedles in closed boxes prior to use (*see* **Note 8**).

3.2.2 Coverslip Preparation

1. Incubate 24 × 60 mm coverslips (Menzel-Gläser, Braunschweig, Germany) in 10 % SDS solution at RT for 1–60 days.
2. Rinse thoroughly with DNA-free distilled water (i.e., PCR water).
3. While the coverslip is still wet, drop 10 μl of 3:1 methanol/acetic acid fixative followed by 10 μl of metaphase preparation in 3:1 methanol/acetic acid fixative onto the coverslip.
4. Dry the coverslip in air, and then incubate the coverslip in 0.025M phosphate buffer for approximately 1 min.

5. Incubate the coverslip in trypsin solution (100 μl of 5 % trypsin mix in 35 ml 0.025M phosphate buffer) for 20–60 s (the required time will vary depending on chromosome preparation).
6. Rinse in 0.025 M phosphate buffer.
7. Incubate in Giemsa solution (35 ml phosphate buffer and 3.5 ml Giemsa) for 2–4 min (the time will vary depending on chromosome preparation).
8. Rinse in sterile water.
9. Dry the coverslips in air.
10. Assess the quality of the spreading and banding achieved under the microscope.
11. Keep the coverslips refrigerated until the microdissection step (see Sect. 3.2.3).

3.2.3 Microdissection

1. Prepare 20 μl of collection drop solution (see Sect. 2.2.3).
2. Use a 10× or 20× objective.
3. Move the objective down far enough to allow the needles in.
4. Load a needle into the holder, and then load onto the micromanipulator, so that the tip of the needle is close to the center of visual field of the objective.
5. Find the needle in the visual field and center it; then move to 40× objective, and ensure that the tip of the needle can be seen with the 40× objective.
6. Put the slide on the specimen stage and find a suitable metaphase spread. Turn the chromosome to be cut at a right angle to the needle.
7. Bring the needle down carefully until it is just above the chromosome. Forward movement of the tip will lead to the excision of the chromosome material.
8. Touch the excised fragment with the tip of the needle several times until it is retained on the needle. Carefully elevate the needle.
9. Change to a 10× or 20× objective.
10. Remove the coverslip.
11. Take a siliconized pipette; gently touch the surface of the collection drop solution using the siliconized pipette, so that a tiny amount of collection drop solution will be sucked up by capillary force.
12. Suspend the pipette above the objective.
13. Move needle to same height as pipette tip using micromanipulator. Transfer the chromosome fragment by touching inside of the pipette with the needle, and then withdraw the needle.

Large fragments will remain visible in the collection drop for several seconds.

14. Cut 10–20 more fragments (or chromosomes), and transfer them into the same collection drop.
15. Put the pipette in a humidified tray at 60 °C for 1–2 h.

3.2.4 Generation of Paint Probe by DOP-PCR

1. Prepare the T7 DNA Polymerase mix under a hood as follows: mix 0.25 μl T7 DNA Polymerase 10 U μl^{-1} (Thermo Fischer Scientific) with 1.75 μl T7 DNA Polymerase Dilution Buffer (Thermo Fischer Scientific) per sample.
2. Prepare the PCR master mix for low-temperature cycles in a hood: mix 0.6 μl 5×T7 DNA Polymerase Reaction Buffer (Thermo Fischer Scientific), 0.4 μl 0.2 mM dNTPs, 0.6 μl 40 μM DOP primer, and 3.4 μl of PCR water per sample.
3. Aliquot 5 μl from the PCR master mix into 0.5 ml tube.
4. Transfer the collection drop containing the microdissected chromosomes or fragments from the Pasteur pipette into the tube containing the PCR mix by breaking off the pipette tip into the tube.
5. Briefly spin tube, and keep on ice until ready to PCR (or store at −20 °C for longer periods).
6. Start low-temperature cycle (LTC) PCR program:

   (a) 92 °C, 5 min

   (b) 25 °C, 2 min 20 s

   During Step 2 add 0.2 μl of T7 DNA Polymerase mix to each sample.

   (c) 34 °C, 2 min

   (d) 90 °C, 1 min

   Repeat 2–4 for 7 more cycles.
7. Add 45 μl of PCR mix (below) for high-temperature cycles (HTC) into each sample:

Water	28.0 μl
10x AmpliTaq Gold DNA Polymerase Buffer	5.0 μl
2 mM dNTP	5.0 μl
25 mM $MgCl_2$	5.0 μl
20 μM DOP primer	1.4 μl
5 U μl^{-1} AmpliTaq Gold DNA Polymerase	0.6 μl
Total	45 μl

8. Place the PCR in a thermocycler and start the following HTC PCR program:
   (a) 92 °C, 1 min
   (b) 56 °C, 2 min
   (c) 72 °C, 2 min
   Repeat 1–3 for 32 more cycles.
   (d) 72 °C, 5 min
   (e) 4 °C, hold
9. Run 3 µl of PCR products on a 1 % agarose gel. The products should be a smear with an average size of 0.2–0.8 kb.
10. For a 20 µl DOP-PCR labeling, prepare the following reaction mix (prepare a master mix if several tubes of microdissected chromosomes are to be amplified):

Water	10.9 µl
10× AmpliTaq Gold DNA Polymerase Buffer	2.0 µl
20 µM DOP primer	1.0 µl
2 mM d(A,C,G)TP, 1 mM dTTP	2.0 µl
25 mM $MgCl_2$	2.0 µl
1 mM Biotin-16-dUTP	2.0 µl
5 u μl^{-1} AmpliTaq Gold DNA Polymerase	0.1 µl
Total	20 µl

11. Place the PCR in a thermocycler and start the following program:
   (a) 95 °C, 3 min
   (b) 94 °C, 1 min
   (c) 56 °C, 1 min
   (d) 72 °C, 2 min
   Repeat 2–4 for 19 more cycles.
   (e) 72 °C, 5 min
   (f) 4 °C, hold
12. Run 3 µl of PCR products on a 1 % agarose gel. The products should appear as a smear with an average size of 0.2–0.8 kb.
13. For FISH use, precipitate the PCR product in ethanol, and resuspend in 30–40 µl hybridization buffer.

4 Notes

1. Incubation time for demecolcine varies dependently on rate of cell growth. A high percentage of "rounded cells" (i.e., cells in M phase) should be observed when cells approach metaphase arrest.
2. If a large number of chromosome clusters are observed (chromosomes which appear to be stuck to each other in clumps), vortex for another 20 s. If this treatment does not improve the numbers of free single chromosomes, syringe the suspension twice through a 22.5 gauge needle using a 5 ml syringe.
3. The effect of sodium citrate and sodium sulfite on the resolution of the flow karyotype is observed to vary among cell lines. It is recommended to check the flow karyotype of the suspension with or without the addition of either reagent. (b) Improvement in the resolution of flow karyotypes has been observed for most lymphoblastoid cell lines with overnight incubation with both reagents.
4. In this protocol, the data are primarily collected using MoFlo® (Beckman Coulter, Fullerton, CA, USA) equipped with two water-cooled Innova 300 series lasers (Coherent). For cytometers using "jet in air" analysis such as MoFlo®, at sheath pressure of 60 psi, the lasers are inter-delayed from each pinhole by 1.6 μs. This time delay can be easily determined by using pulse monitor oscilloscope. For other flow cytometers, the signals are delayed and processed in different ways, using dual-beam delay compensation circuitry in the case of a Becton-Dickinson FACStar Plus or through a gated amplifier as in Beckman Coulter EPICS Elite ESP.
5. A lower differential pressure of 0.1–0.2 psi is set to regulate a data rate of ~2,000 event(s) to sort chromosomes when making paints. A higher differential pressure of 0.4–0.6 psi is set for high-speed sorting at a rate of >10,000 event(s).
6. Not all brands of Taq polymerase and PCR buffer will work with flow-sorted chromosomes as the templates. The BIOTAQ (Bioline) DNA Polymerase and 5 × PCR optimized buffer D (Invitrogen) have given consistent results at our hands. The pH of the PCR buffer and the magnesium ion concentration are the two most critical parameters for the success of DOP-PCR amplification of flow-sorted chromosomes.
7. Contamination is one of the major problems when doing sequence-independent DNA amplification starting from low amounts of DNA. We recommend the use of separate micropipettes for pre- and postprimary DOP-PCR procedures. UV treatment of the microdissection room and instruments as well as DNA-EX (Genaxis) cleaning of working surfaces is highly recommended.

8. Caution: Dimethyldichlorosilane and carbon tetrachloride are highly toxic—all manipulations should be done under the draft hood. Two milliliters of siliconization solution is enough to siliconize more than 30 micropipettes.

Acknowledgments

While writing this protocol, VT was supported by Budget Projects 0310-2014-0003, 0310-2014-0008, 0310-2014-0009, and DFG. FY, BLN, and NPC are supported by the Wellcome Trust (Grant number WT098051).

References

1. Pinkel D, Straume T, Gray JW (1986) Cytogenetic analysis using quantitative, high-sensivity, fluorescence hybridization. Proc Natl Acad Sci USA 83:2934–2938
2. Telenius H, Carter NP, Bebb CE et al (1992) Degenerate oligonucleotide-primed PCR: general amplification of target DNA by a single degenerate primer. Genomics 13:718–725
3. Meltzer PS, Guan XY, Burgess A et al (1992) Rapid generation of region specific probes by chromosome microdissection and their application. Nat Genet 1:24–28
4. Carter NP, Ferguson-Smith MA, Perryman MT et al (1992) Reverse chromosome painting: a method for the rapid analysis of aberrant chromosomes in clinical cytogenetics. J Med Genet 29:299–307
5. Rabbitts P, Impey H, Heppell-Parton A et al (1995) Chromosome specific paints from a high resolution flow karyotype of the mouse. Nat Genet 9:369–375
6. Wienberg J, Jauch A, Stanyon R et al (1990) Molecular cytotaxonomy of primates by chromosomal in situ suppression hybridization. Genomics 8:347–350
7. Jauch A, Wienberg J, Stanyon R et al (1992) Reconstruction of genomic rearrangements in great apes and gibbons by chromosome painting. Proc Natl Acad Sci USA 89:8611–8615
8. Scherthan H, Cremer T, Arnason U et al (1994) Comparative chromosome painting discloses homologous segments in distantly related mammals. Nat Genet 6:342–347
9. Yang F, Carter NP, Shi L et al (1995) A comparative study of karyotypes of muntjacs by chromosome painting. Chromosoma 103:642–652
10. Yang F, O'Brien P, Milne B et al (1999) A complete comparative chromosome map for the dog, red fox, and human and its integration with canine genetic maps. Genomics 62:189–202
11. Scalenghe F, Turco E, Edstrom JE et al (1981) Microdissection and cloning of DNA from a specific region of Drosophila melanogaster polytene chromosomes. Chromosoma 82:205–216
12. Senger G, Lüdecke HJ, Horsthemke B et al (1990) Microdissection of banded human chromosomes. Hum Genet 84:507–511
13. Guan XY, Meltzer PS, Cao J et al (1992) Rapid generation of region-specific genomic clones by chromosome microdissection: isolation of DNA from a region frequently deleted in malignant melanoma. Genomics 14:680–684
14. Weimer J, Kiechle M, Senger G et al (1999) An easy and reliable procedure of microdissection technique for the analysis of chromosomal breakpoints and marker chromosomes. Chromosome Res 7:355–362
15. Liehr T, Heller A, Starke H et al (2002) Microdissection based high resolution multicolor banding for all 24 human chromosomes. Int J Mol Med 9:335–339
16. Liehr T, Starke H, Heller A et al (2006) Multicolor fluorescence in situ hybridization (FISH) applied to FISH-banding. Cytogenet Genome Res 114:240–244

FISH-Microdissection

Nadezda Kosyakova, Thomas Liehr, and Ahmed B. Hamid Al-Rikabi

Abstract

FISH-microdissection (FISH-MD) is an approach combining FISH technique and chromosome microdissection in one experiment. This method enables reliable and straightforward identification of target chromosomes or chromosomal regions by FISH with specific probes and immediate microdissection of the chromosomal region of interest. FISH-MD can be applied when chromosome identification by trypsin-Giemsa staining/banding is complicated or not possible due to chromosomal morphology.

Keywords FISH-MD, FISH-microdissection, Chromosome microdissection, DOP-PCR, DNA library, Array CGH, Next-generation sequencing

1 Introduction

Chromosome microdissection is commonly used to prepare DNA libraries (chapter by Fengtang Yang et al. "Generation of Paint Probes from Flow-Sorted and Microdissected Chromosomes") and painting probes to identify breakpoints and origin of marker chromosomes. Microdissected material can be successfully used for further array CGH ([1], chapter by Maria Isabel Melaragno and Mariana Moysés-Oliveira "Breakpoint Mapping of Balanced Chromosomal Rearrangements Using Array CGH of Microdissection Derived FISH-Probes") or for next-generation sequencing applications ([2], chapter by Jiří Štika and Oldřich Mazal "Sequencing of Microdissection Derived FISH-Probes"). Whole-chromosome probes and partial-chromosome probes as well as probes covering tiny chromosomal regions of 2–4 Mb can be established by chromosome microdissection as it was recently shown in the experiments with microdissection from chicken lampbrush chromosomes ([3], chapter by Anna Zlotina and Alla Krasikova "FISH in Lampbrush-Chromosomes").

However, one is required to unambiguously identify the target chromosome/chromosomal region to perform chromosome microdissection. Chromosomal suspensions with high mitotic index and good chromosome spreading are required for successful

Thomas Liehr (ed.), *Fluorescence In Situ Hybridization (FISH)*, Springer Protocols Handbooks,
DOI 10.1007/978-3-662-52959-1_7, © Springer-Verlag Berlin Heidelberg 2017

microdissection. Though human chromosomes from peripheral blood preparation can be identified by trypsin-Giemsa staining/ banding in most of the cases (chapter by Anja Weise and Thomas Liehr "Pre- and Postnatal Diagnostics and Research on Peripheral Blood, Bone Marrow, Chorion, Amniocytes, and Fibroblasts"), chromosome suspensions prepared from bone marrow often show short chromosomes and very poor differential Giemsa staining (chapter by Eyad Alhourani et al. "Tumorcytogenetic Diagnostics and Research on Blood and Bone Marrow Smears or Effusions"). Chromosomes involved in a rearrangement might lose their typical p/q arm ratio and/or banding pattern and therefore mimic other chromosomes. In addition, chromosomes of other species (e.g., *Mus musculus*) are almost impossible to distinguish based on conventional staining due to their similar morphology (chapter by Thomas Liehr et al. "FISH Banding Techniques"; chapter by Fengtang Yang et al. "Animal Probes and ZOO-FISH").

FISH-microdissection (FISH-MD), an approach combining FISH technique and microdissection, was originally developed by Jörg Weimer and colleagues in 2000 [4]. Aberrant/target chromosomes were identified by FISH and then directly microdissected. Later, the FISH-MD technique was successfully used to characterize chromosomal aberrations and breakpoints in prenatal and postnatal cases and in leukemia patients and to establish a full set of murine multicolor banding probes [5–8].

Different types of probes can be used to identify a target chromosome for dissection: whole-chromosome paints and partial-chromosome paints (chapter by Fengtang Yang et al. "Gen eration of Paint Probes from Flow-Sorted and Microdissected Chromosomes"), pooled BAC probes, or centromeric probes (chapter by Thomas Liehr "Homemade Locus-Specific FISH Probes: Bacterial Artificial Chromosomes"). The choice of the probe to be used is dependent on the target chromosome/chromosomal region. However, the use of centromeric probes is beneficial as they do not require prehybridization with Cot-1 DNA and cover relatively small chromosomal segment but result in a strong FISH signal. In case homemade probes (whole- or partial-chromosomal paints, locus-specific probes) are applied, they should be precipitated with the sufficient amount of Cot DNA, and the concentration of DNA carrier (e.g., tRNA) in the precipitation mix can be reduced twice or more. All the probes should be tested prior to FISH-MD on test slides. It is advantageous to use probes that can help to identify the target chromosomes but do not cover the region of interest, especially if it is planned to use the dissected and amplified DNA for sequencing (chapter by Jiří Štika and Oldřich Mazal "Sequencing of Microdissection Derived FISH-Probes").

2 Materials

Standard laboratory equipment and reagents like mini-centrifuges, autoclaves, heating plates, water baths, the Milli-Q® water purification system, laboratory glassware, ethanol, etc. are not listed below. It is extremely important that microdissection and pre-PCR/PCR procedures are separated spatially from the post-PCR steps including probe precipitation, FISH, and post-hybridization wash. Ideally, a separate room with a UV lamp should be used for chromosome microdissection. The microdissection room, the PCR hood, and the working surfaces should be regularly cleaned. The equipment and instruments should not be moved from one room to another; for example, the microdissection room should have its own water bath for proteinase K incubation.

2.1 Instruments and Equipment

- Inverted microscope Axiovert 135 (Zeiss, Germany) equipped both for fluorescence and light microscopy with phase contrast is used by the authors. Other inverted microscopes can be used as well if they have a round rotatable table and can be supplemented with a fluorescence block and micromanipulators. Microscope should be equipped with a CCD camera connected to dedicated FISH image analysis software. In case that no software is available, FISH-MD is still possible; however, only very bright probes can be used for chromosome identification. Microscope should have at least two objectives with magnification of 10× and 40× and an additional 2.5× lens usually located in the microscope's base.
- One three-axis high-magnification motor-drive micromanipulator with joystick for dissection and one three-axis manual coarse manipulator for the micropipette with pipette holders (both should be suitable to use with the available inverted microscope).
- Pipette puller PB-7 or PC-10 (Narishige, Japan). Other pipette pullers that are designed both for single- and two-stage pulling can be used.
- Sterile hood for PCR.
- Fume hood with faucet aspirator vacuum pump.
- Thermal cycler for 0.5 ml tubes with heated lid.
- DURAN® glass rods, 2 ± 0.15 mm in diameter (SCHOTT AG, Germany).
- Pasteur pipettes, ISO 7712, total length 230 mm (Cat N 567/2, Hecht-Assistent, Germany).
- 1.5 ml Safe-Lock Tubes Eppendorf BIOPUR® (Eppendorf AG, Germany, Cat N 0030 121.589).
- 0.5 ml Safe-Lock Tubes Eppendorf BIOPUR® (Eppendorf AG, Germany, Cat N 0030 121.570).

- 50 ml sterile tubes free of DNase and RNase (Greiner Bio-One International GmbH, Cat N 227261).
- Two rust-free cover glass forceps (Karl Hammacher GmbH, Germany, Cat N 513-10): for FISH room and for microdissection room.
- Three or four stainless steel instrument trays, with lids, long enough to store micropipettes (Carl Roth GmbH, Germany, Cat N C850.1).
- Several Nunc 4-well ×22 ml cell culture rectangular dishes with lid (Thermo Fisher Scientific, USA, Cat N 267061).
- Filtropur S 0.2 syringe filter for sterile filtration (Sarstedt AG & Co, Germany, Cat N 83.1826.001).
- Transfer tube: take a PVC tube with a small diameter (e.g., from medical infusion set) and insert a self-made cotton wool filter in one of its openings. Put the transfer tube in a plastic bag and sterilize under UV light.

2.2 Reagents

- Phosphate buffer pH 6.88 (VWR International LLC, USA, Cat N 83601.290 or equivalent).
- Dichlorodimethylsilane (Sigma-Aldrich, USA, Cat N 440272).
- Carbon tetrachloride (Sigma-Aldrich, USA, Cat N 319961-500ML-D).
- Proteinase K, recombinant, PCR grade, 18 ± 4 mg ml^{-1} (Roche, Switzerland, Cat N 03115887001).
- Glycerol for molecular biology (Sigma-Aldrich, USA, Cat N G5516-100ML).
- Water, PCR grade (Roche, Switzerland, Cat N 03315932001 or 03315959001).
- Deoxynucleotide Set, 10 mM, individual dNTPs for routine PCR (Sigma-Aldrich, USA, Cat N DNTP10-1KT or equivalent).
- DOP primer (degenerate oligonucleotide primer): 6MW primer 5′-CCG ACT CGA GNN NNN NAT GTG G-3′ [9] or anti-6MW primer 5′-CCG TGA GCT CNN NNN NTA CAC C-3′ [10]. 6MW primer was successfully used for amplification of chromosomes of most species; anti-6MW primer was proposed for the amplification of murine chromosomes.
- Invitrogen Platinum® *Tfi* Exo(-) DNA Polymerase (Thermo Fisher Scientific, USA, Cat N 60684050)
- Applied Biosystems™ AmpliTaq™ DNA Polymerase with Buffer II (Thermo Fisher Scientific, USA, Cat N N8080172).
- Sequenase Version 2.0 DNA Polymerase (Affymetrix Inc., USA, Cat N 70775Z).

- TWEEN® 20 (Sigma-Aldrich, USA, Cat N P2287-100ML).
- Invitrogen UltraPure™ SSC, 20× (Thermo Fisher Scientific, USA, Cat N 15557-044).
- NP-40 (Abbott Molecular, USA, Cat N 07J05-001).
- Fixogum rubber cement (Marabu GmbH, Germany, Cat N 2901 10 000).

2.3 Other Solutions

- 2 % paraformaldehyde solution in PBS without Ca++ and Mg++ (use sterile PBS and sterile tubes for preparing the solution). Prepare under the fume hood; wear gloves and safety glasses. Store at +4 °C for 2–3 months.
- 1M $MgCl_2$. Prepare from magnesium chloride hexahydrate; sterilize by filtration. Store at +4 °C for 4–5 months.
- Postfix solution: mix 500 μl of 2 % paraformaldehyde solution, 450 μl of PBS, and 50 μl of 1M $MgCl_2$. Store +4 °C for 1–2 weeks.
- 10 % SDS solution. Prepare with autoclaved Milli-Q® water. Store in a bottle with a lid at room temperature (RT) for several months.
- 100 mM NaCl, 100 mM tris-HCl pH 7.5, 1 % triton X-100, 1 % SDS, 100 mM NaEDTA, 10 mM NaEDTA. Eppendorf BIO-PUR® tubes and water for PCR should be used for preparation of these solutions (very small volumes of about 1 ml are required). Solutions should be filter sterilized in sterile conditions and aliquoted, and the tubes should be put under UV light for at least 2 h. After that, the tubes should be kept for several days at +4 °C, and then they can be frozen and stored in the freezer for several years. Alternatively, some of these solutions can be purchased—in this case, they all should be "molecular biology grade."

3 Methods

Microdissection and FISH should be separated spatially; there should be two separate rooms dedicated to pre- and post-PCR processing, respectively (*see* **Note 1**). The person who works/worked the same day in FISH room is not allowed to enter the microdissection room. The microdissection room should have the set of all required equipment, lab coats (which should be changed regularly), and a supply of gloves. All the solutions required for pipette siliconization, collection drop, and PCR Master Mix preparation should be stored in a separate drawer in a refrigerator/freezer that is *not* placed in the FISH room.

3.1 Preparation of Microinstruments

Microinstruments (microneedles and micropipettes) can be prepared ahead of time and stored for a period of a couple of months. They should be prepared in the microdissection room that was previously cleaned and UV irradiated. Gloves should be worn.

3.1.1 Preparation of Microneedles

Microneedles are prepared from glass rods using the pipette puller and two-stage pulling method. The following conditions were established for the Narishige pipette puller PB-7. If another pipette puller is used, the conditions may require adjustment.

1. Use 33 g of weight; install the heating value of No. 1 heater to maximum and of No. 2 heater to 7.0–7.5.
2. Place the glass rod in the puller and press the start switch (so that heater No. 1 is active). As the glass rod melts, the weight pulls the melting glass down; heating stops as the weight drops down on a special plate.
3. Move the spiral heating element down to the middle of the pulled area. During the second-stage pulling, heater No. 2 is active, and the glass rod is pulled into two microneedles.
4. Put the microneedles in a rectangular cell culture dish with lid and a special "bedding" made of Plasticine/modeling clay in the middle (so the microneedle can be safely "fixed" in the dish with its middle part, leaving the tip of the needle free).
5. The microneedles must be treated with UV light prior to use (while being in the closed plastic dish).

3.1.2 Preparation of Micropipettes

Micropipette Pulling

Micropipettes are prepared from Pasteur pipettes using the pipette puller and one-stage pulling. The following conditions were established for the Narishige pipette puller PB-7. If another pipette puller is used, the conditions might need some adjustment.

1. Use 200 g of weight; install the No. 2 heater to 7.5–8.0.
2. Place the Pasteur pipette in the puller and press the start switch (heater No. 2 is active).
3. The resulting micropipette is "closed" due to the melted glass—break the tip of the micropipette by touching the precleaned metal surface.
4. Evaluate the quality of the micropipette's opening under the microscope. The micropipette should not have fissures; the edges should be relatively smooth. Micropipettes with too big an opening can draw an excess of collection drop solution in them, which can in turn later inhibit the original amplification. A too small opening can complicate the transfer of the dissected chromosomal fragment into the micropipette.

5. Place the micropipette in a pre-cleaned stainless steel instrument tray with a lid. The tray should have a cushion roll made of aluminum foil and placed in the middle of the tray; it keeps the micropipettes in inclined slightly upward position and prevents the breaking of the pipette tip. The tray should always be closed with a lid to avoid contamination.
6. Micropipettes can be prepared ahead of time. The following siliconization and sterilization procedures are time-consuming, so it is recommended to prepare ~200 pipettes at a time.

Micropipette Siliconization and Sterilization

All solutions should be prepared in 1.5 ml Safe-Lock Eppendorf BIOPUR® tubes using sterile filter tips:

- 1 % dichlorodimethylsilane in carbon tetrachloride. The solution should be prepared strictly under a fume hood.
- 1 mM NaEDTA. The solution should be prepared in a sterile PCR hood. For every tube with 1 % dichlorodimethylsilane, prepare two tubes with 1 mM NaEDTA.

Siliconization of Micropipettes Should Take Place Under the Fume Hood Equipped with Faucet Aspirator Vacuum Pump (Steps 1–6)

1. Connect the micropipette to the working vacuum pump. The vacuum pump will aspirate tiny volumes of siliconizing solutions through the micropipette thus distributing them evenly on the micropipette's walls.
2. Dip the tip of the micropipette in the 1 % dichlorodimethylsilane twice.
3. Dip the tip of the micropipette in the first tube with 1 % NaEDTA.
4. Dip the tip of the micropipette in the second tube with 1 % NaEDTA.
5. Wait until the micropipette is almost dry (the working vacuum pump facilitates drying). If big liquid drops are present inside the micropipette near its tip, it can destroy the micropipette's tip during the sterilization.
6. Disconnect the micropipette from the faucet aspirator vacuum pump, and place it in a different stainless steel instrument tray with lid.
7. Sterilization of micropipettes should take place immediately after all of them were siliconized. Place the instrument tray with the micropipettes in a heating and drying lab oven. Incubate as follows:
   (a) 60 °C—1 h
   (b) 100 °C—30 min
   (c) 60 °C—1 h

3.2 Preparation of FISH Probe

The FISH probe which is used to identify target chromosome/chromosomal regions should be prepared or purchased ahead, tested on control slides, and visualized on the microdissection microscope. The fluorochrome should be chosen based on the available filters; the use of directly labeled probes is recommended. If the probe requires prehybridization with Cot-1 DNA, it should be precipitated together with Cot-1 DNA. For the preparation of homemade probes, see the chapter by Thomas Liehr "Homemade Locus-Specific FISH Probes: Bacterial Artificial Chromosomes".

3.3 Pre-cleaning of Coverslips

Use of an inverted microscope makes the use of slides for microdissection impossible—the only exception is microdissection of the giant lampbrush chromosomes [3]. Fixed suspensions of mitotic chromosomes should be "dropped" on coverslips (see Sect. 3.4). Use of special cover glass forceps enables easy handling of coverslips.

Coverslips should be soaked in 10 % SDS solution in a sterile 50 ml tube for several days prior to use.

3.4 Preparation of Chromosome Spreads: FISH and Post-hybridization Wash (Day 1)

Preparation of chromosomes spreads should take place in the microdissection room. Gloves should be worn at all times. The following prearrangements should be done:

1. Sterile water can be prepared ahead by autoclaving the Milli-Q® water in clean glass bottles with lid. After that, sterile water can be stored in the microdissection room for 2–3 months.
2. All Coplin jars should be washed and autoclaved prior to use.
3. The coverslips should be dehydrated and washed in sterile 50 ml tubes (two coverslips can be put in the tube "back to back"), see Sect. 3.4.1. and 3.4.2.
4. All the solutions, including ethanol dilutions and post-hybridization wash solutions, should be prepared with sterile autoclaved water.
5. The use of sterile filter tips is recommended.

3.4.1 Preparation of Chromosome Spreads

1. Take a coverslip out of 10 % SDS solution, and wash it by rubbing both of its surfaces with your fingers.
2. Rinse the coverslip with tap water by holding it under the running faucet.
3. Rinse the coverslip with sterile autoclaved water, and place it in a Coplin jar with sterile water. Prepare a required amount of coverslips. Close the Coplin jar with a lid or Parafilm M®, and let it cool down for about 20–30 min at 4 °C.
4. Prepare the chromosome spreads on the pre-cleaned coverslips; the exact method may depend on the conditions in the lab (temperature, humidity) [11]. The quality of chromosome spreads is one of the key factors for the success of microdissection.

Ideally, there should be enough metaphases with good spreading (*see* **Note 2**).

5. Place the coverslip on a paper tissue, and carefully break a tiny piece from one of the corners of the coverslip. This will help to always correctly identify the side of the coverslip with the chromosomes on it. It is not possible to subscribe the coverslip—diamond pen will damage the glass, and the inscriptions done with the laboratory marker will be gone after the slide pretreatment. It is not recommended to work with more than two different cases at the same time (the coverslips with another suspension can be marked by three broken corners).
6. Dehydrate the preparations in the ethanol series 70–95–100 % 5 min each, and air-dry them.
7. Age the preparations by incubating the coverslips in a heating and drying lab oven for 1.5–2 h. Meanwhile, aliquot autoclaved Milli-Q® water for the post-hybridization wash. For every two coverslips, prepare two 50 ml sterile tubes—one with 40 ml of water and one with 49 ml of water. Bring these tubes to the FISH room for the preparation of post-hybridization wash solutions.

3.4.2 FISH on Chromosome Spreads

FISH procedure should be done in the FISH room. Gloves should be worn at all times.

Pretreatment

1. The use of pepsin pretreatment is not recommended for FISH-microdissection; however, it can be done in some exceptional cases. For pepsin pretreatment, see chapter by Thomas Liehr et al. "The Standard FISH Procedure".
2. Soak the coverslip in a Coplin jar in phosphate buffer pH 6.88 for 2–3 min.
3. Blot the excess phosphate buffer on a paper tissue, and apply 100 μl of Postfix solution on the coverslip. Cover the coverslip with a slide, and turn them over so the slide is on the bottom "covered" with the coverslip. Incubate at RT for 10 min. This step should be performed under the fume hood.
4. Carefully remove the slide from the coverslip, and soak the coverslip in phosphate buffer pH 6.88 for 5 min.
5. Briefly dip the coverslip in a Coplin jar with autoclaved Milli-Q® water.
6. Dehydrate the preparation in 70–95–100 % ethanol series, 5 min each.
7. Air-dry the coverslip.

Denaturation and FISH

1. Apply 20–25 μl of the probe mix on the coverslip, cover it with a slide, turn the slide with the coverslip over, and seal the edges

of the coverslip with rubber cement (Fixogum). Avoid air bubbles being trapped between the slide and the coverslip.

2. Place the slide on the heating plate pre-heated to 73–75 °C for 5–6 min. (The chromosome preparation is on the coverslip soaked in a hybridization buffer with the excess of the probe.)
3. Take the slide from the heating plate and place it in a humid chamber; incubate at 37 °C for 6 h. Incubation time might vary depending on the probes being used.

Post-hybridization wash

Post-hybridization wash should be done in 50 ml sterile tubes with lid (see Sect. 3.4.1, step 7). Two coverslips can be washed together in one tube (placed "back-to-back").

1. Prepare 0.4 × SSC solution by adding 1 ml of 20 × SSC solution to 49 ml of water, and place the tube with solution in a water bath at 62–65 °C.
2. Prepare 4 × SSC/TWEEN solution by adding 10 ml of 20 × SSC solution and 25 μl of TWEEN® 20–40 ml of water.
3. Take the slide with the coverslip out of the humid chamber, and remove the rubber cement.
4. Carefully remove the slide from the coverslip. Alternatively, the slide with the coverslip can be soaked for some minutes in 2 × SSC solution (prepared with autoclaved Milli-Q® water).
5. Put the coverslip in 0.4 × SSC solution at 62–65 °C for 3–4 min.
6. Transfer the coverslip in 4 × SSC/TWEEN solution and wash at RT with agitation.
7. Transfer the coverslip in a Coplin jar with phosphate buffer pH 6.88, and let it soak for 10 min.
8. Rinse the coverslip in a Coplin jar with autoclaved Milli-Q® water.
9. Air-dry the coverslip; then put it in cell culture rectangular dishes with lid. Coverslip should be kept at +4 °C prior to microdissection (*see* **Note 1**).

3.5 Preparation of Collection Drop Solution, PCR Mixes, and Microdissection (Day 2)

Preparation of collection drop solution and PCR mixes should take place in a sterile hood for PCR in the pre-PCR room; clean lab coats and gloves should be worn at all times (*see* **Notes 3–5**). There should be a set of "DNA-free" pipettes in the hood dedicated only to the preparation of PCR stocks and collection drop solution; sterile filter tips should be used. All the solutions must be prepared in Safe-Lock BIOPUR® Eppendorf tubes.

3.5.1 Preparation of Collection Drop Solution and PCR Mixes

1. Prepare the saline solution by mixing:

100 mM NaCl	20 μl
100 mM tris-HCl pH 7.5	20 μl
1 % triton X-100	20 μl
1 % SDS	20 μl
10 mM NaEDTA	20 μl
Water, PCR grade	14 μl

   Saline solution can be prepared ahead and stored at −20 °C for several months.
2. Aliquot proteinase K and glycerol in small volumes. Proteinase K should be stored at +4 °C, and glycerol should be stored at −20 °C (*see* **Note 6**).
3. Prepare collection drop solution by mixing:

Saline solution	5.7 μl
Proteinase K	1.0 μl
Glycerol	3.3 μl

   Collection drop solution should be prepared fresh and used the same day. First, proteinase K should be added to the saline solution; only after that glycerol is added. Mix the collection drop solution gently by shaking the tube and spin the tube briefly. The tube with the collection drop can be stored at room temperature in the microdissection room during dissection (*see* **Note 7**).
4. Prepare 2.5 mM dNTPs mix from 10 mM individual dNTPs:

dATP	100 μl
dTTP	100 μl
dGTP	100 μl
dCTP	100 μl

5. Prepare label mix from 10 mM individual dNTPs:

dATP	100 μl
dGTP	100 μl
dCTP	100 μl
dTTP	50 μl
Water, PCR grade	150 μl

6. Aliquot in small volumes dNTPs mix, label mix, water for PCR, 5 × Platinum® Tfi Reaction Buffer and 50 mM magnesium chloride (supplied with Platinum® *Tfi* Exo(-) DNA Polymerase), GeneAmp® 10 × PCR Buffer II and 25 mM magnesium chloride (supplied with AmpliTaq® DNA Polymerase), and Sequenase Dilution Buffer and Sequenase Reaction Buffer (supplied with Sequenase Version 2.0 DNA Polymerase). Aliquots can be stored at −20 °C for several months (*see* **Note 8**).
7. Dilute DOP primer in PCR grade water to the concentration of 40 μM; aliquot in small volumes and store at −20 °C for several months (*see* **Note 8**).
8. Prepare Master Mix solution for the low-temperature cycles of DOP-PCR (LTC-Master Mix). For one sample, mix:

Water, PCR grade	3.37 μl
DOP primer, 40 μM	0.63 μl
dNTPs mix, 2.5 mM	0.40 μl

Bigger quantities of LTC-Master Mix can be prepared, and 4.4 μl of it can be aliquoted in 0.5 Safe-Lock BIOPUR® Eppendorf tubes. Aliquots can be stored at −20 °C for 2–3 months.

3.5.2 FISH-Microdissection

All the steps should be performed in the microdissection room. Gloves must be worn for steps 1–3 and 26–29.

1. Take a siliconized and sterilized micropipette and check its tip under the microscope. As the micropipette is connected to a vacuum aspiration pump during siliconization, it can happen that the tip of the micropipette is blocked by dust particles, etc. Only micropipettes that show no such contamination can be used.
2. Immerse the tip of the micropipette in the collection drop solution, and allow the collection drop solution to fill the tip of the micropipette by capillary forces. The time when the tube with the collection drop solution is open should be minimized (*see* **Note 7**).
3. Place the micropipette in a stainless steel instrument tray with some autoclaved Milli-Q® water on the bottom. The tray should have a special holder for micropipettes that can be carved with scalpel from used filter tip boxes or lids (Fig. 1e). The micropipettes should remain in the tray between dissections.
4. Turn on the microscope, the epifluorescence block, and the computer with the image analysis software. The image analysis software should be set for "manual filter change."
5. Load a microneedle onto a holder of the micromanipulator so that the tip of the needle is close to the center of the objective.
6. Using 10× objective and light microscopy, find the needle in the view field, and center it. Use additional 2.5× lens to increase the magnification and reconfirm the needle position. Never switch to the 40× objective if there is no coverslip between the needle and the objective.
7. Remove the 2.5× lens and put the coverslip on the rotatable specimen stage, and find a suitable metaphase spread using 10× objective and phase contrast (the preparation is not counterstained). Switch to 40× objective. Capture the phase-contrast image with the image analysis software—this image substitutes the DAPI image of chromosomes.
8. Turn off the halogen light, remove the manual shutter blocking the fluorescence illumination, install the appropriate filter, and capture the image with FISH signals.

9. As soon as an image with FISH signals is acquired, block further illumination of the chromosomes with the shutter to prevent signal bleaching.
10. Analyze the combined phase-contrast (representing the position of chromosomes) and FISH image. Identify the target chromosome (Fig. 1a–d). Because no antifade solution is used, signals might appear significantly weaker as compared to normal FISH slides that are mounted in antifade medium.
11. Turn on the halogen light and visualize the metaphase using phase contrast. Insert the additional 2.5× lens.
12. Rotate the specimen holder/rotatable microscope table in order to position the target chromosome at a right angle to the needle. The needle dissects only from right to left if the motor-drive micromanipulator is attached to the microscope from the right side.
13. The three-axis motor-drive micromanipulator usually has a manual coarse manipulator which allows moving the pipette holder up and down. Using this manual manipulator, bring the microneedle closer to the preparation leaving 2 mm between the tip of the needle and the surface of the coverslip.
14. Use the motor-drive micromanipulator (joystick) to bring the microneedle down until it is just above the chromosomes. Place

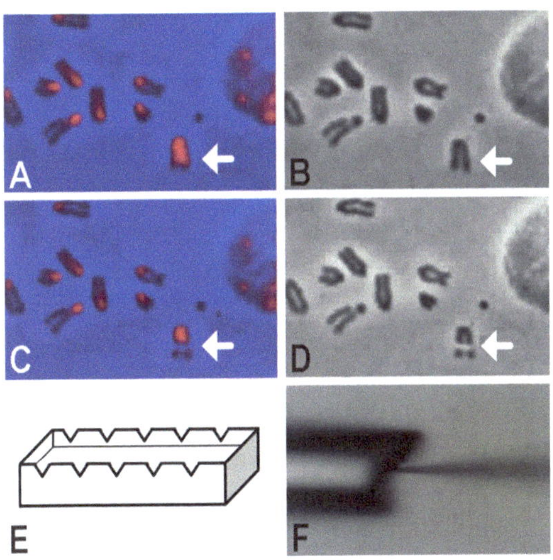

Fig. 1 (**a–d and f**) represent the snapshots from a previously published video of FISH-microdissection [8]. (**a**) and (**c**) show combined phase-contrast (in blue) and FISH signal images identifying the target chromosome before and after dissection. (**b**) and (**d**) demonstrate the corresponding phase-contrast image in black and white. Image (**f**) illustrates the transfer of the dissected material into a micropipette. (**e**) A self-made micropipette holder for an instrument tray (see Sect. 3.5.2, step 3). Little triangles are excised with scalpel from a lid of a filter tip box. In this case, five micropipettes can be placed on such holder

the needle right to the target chromosome/chromosomal region, and bring it down until you see the tip of the needle sharp and sliding left.

15. Dissect the target chromosome/chromosomal fragment by moving the needle from right to left; collect the dissected fragment by carefully trying to make the dissected fragment "roll over" the needle tip.
16. Use the motor-drive micromanipulator (joystick) to bring the microneedle up leaving 2 mm between the tip of the microneedle and the surface of the coverslip.
17. Use the manual manipulator to bring the microneedle up to the top.
18. Change to the 10× objective, and remove the coverslip with the preparation from the microscope table.
19. Install the micropipette in the pipette holder onto a three-axis manual coarse manipulator attached to the microscope from the left side. The tip of the micropipette should be close to the center of the objective. Use a coarse manipulator to find the micropipette in the view field and center it. By moving the 10× objective up and down, visualize the "upper" and "lower" walls of the micropipette, and try to focus in between of these two positions.
20. Use the manual manipulator to bring the microneedle down leaving 2 mm between the tip of the microneedle and the tip of the micropipette.
21. Use the motor-drive micromanipulator (joystick) to bring the microneedle down until you see the tip of the needle "sharp" (indicating that the tip of needle and the micropipette are located at the same level).
22. Use the motor-drive micromanipulator (joystick) to bring the microneedle left to dip it in the collection drop solution in the micropipette (Fig. 1e). Small chromosomal fragments dissolve relatively fast. If a whole chromosome was dissected, one can sometimes see it on the tip of the needle and then dissolving in the collection drop solution (visual control of the transfer).
23. Use the motor-drive micromanipulator (joystick) to bring the microneedle right out of the micropipette, center it, and lift it up 2 mm away from the tip of the micropipette. Use the manual manipulator to bring the microneedle up to the top.
24. Remove the micropipette from the pipette holder, and place it back in the instrument tray.
25. Collect 10–20 more chromosomes/chromosomal fragments, and transfer them in the same micropipette (use the same microneedle): repeat steps 7–24.

26. Put the micropipette in the instrumental tray (with the self-made holder for the micropipettes and some autoclaved Milli-Q® water) that is placed in a water bath at +60 °C. Two Coplin jars can be used as a pedestal for the instrumental tray, so the tray does not touch the water in the water bath. Incubate the micropipettes for 2 h at +60 °C.
27. Insert the transfer tube (the side with the self-made cotton wool filter) into the micropipette until it reaches the narrower part of the micropipette and is fixed there tightly. Insert a sterile 10 μl filter tip in another opening of the transfer tube.
28. Dip the tip of the micropipette in 4.4 μl aliquot of LTC-Master Mix in a 0.5 ml Eppendorf tube (see Sect. 3.5.1). By blowing in the 10 μl filter tip, transfer the collection drop solution with the dissected chromosomes in the Master Mix (the transfer should be visually controlled). As you keep blowing, you will see a little air bubble going into the Master Mix from the micropipette indicating that the transfer is finished. Immediately take the tip of the micropipette out of the Master Mix, and close the Eppendorf tube. Avoid breaking the tip of the needle. Use of two filters (the self-made cotton wool filter in the transfer tube and a filter in the sterile filter tip) prevents contamination of the microdissected material.

    If mouth pipetting is not allowed in your lab, other pipetting aid devices suitable for Pasteur pipettes can be used. However, make sure there is a filter system that can prevent contamination of your sample.
29. Spin the Eppendorf tube to consolidate the LTC-Master Mix with dissected chromosomes. At this point, the tubes can be frozen and kept at −20 °C for up to 2 weeks prior to DOP-PCR. Alternatively, you can continue with DOP-PCR on fresh samples.

3.6 DOP-PCR

All the solutions should be prepared in the sterile PCR hood with the set of DNA-free pipettes and sterile filter tips. Use Safe-Lock Eppendorf BIOPUR® tubes (*see* **Notes 4** and **5**). All the PCR mix calculations are provided per sample.

1. Prepare Sequenase mix by mixing 2.8 μl Sequenase Dilution Buffer and 0.4 μl of Sequenase Version 2.0 DNA Polymerase.
2. Prepare Master Mix for high-temperature cycles of DOP-PCR (HTC-Master Mix) by mixing:

Water, PCR grade	26.53 μl
5× Platinum® Tfi Reaction Buffer	10.00 μl
$MgCl_2$, 50 mM	2.50 μl
dNTPs mix	4.40 μl
DOP primer, 40 μM	1.37 μl
Invitrogen Platinum® *Tfi* Exo(-) DNA Polymerase	0.20 μl

3. Return all the enzymes and PCR stocks back to the freezer leaving only the prepared mixes and an aliquot of Sequenase Reaction Buffer in the sterile hood. Spin the tube with the LTC-Master Mix and microdissected material, and add 0.6 μl of Sequenase Reaction Buffer to it (place a little drop on the tube's wall, and do not touch the mix with the filter tip). Consolidate the mix by spinning the tube. Discard the leftover Sequenase Reaction Buffer aliquot.
4. DOP-PCR with Sequenase Version 2.0 DNA Polymerase requires adding the enzyme every cycle. In case the thermal cycler can be cleaned and placed in the sterile hood, it can remain there during all low-temperature cycles of DOP-PCR. Alternatively, the thermal cycler can be placed near the sterile PCR hood, and the tubes can be taken out of the thermal cycler and placed in the sterile hood every time the enzyme should be added.
5. Start the LTC program of DOP-PCR (use heated lid):

   (1) **92 °C 5 min**

   This step is important for inactivation of proteinase K in the mix.

   (2) **25 °C 2 min 20 s**

   During step 2, the thermal cycler can be paused; the tube can be taken out of the thermal cycler and briefly centrifuged if necessary. 0.24–0.30 μl of Sequenase mix is added directly in the LTC-Master Mix. Place the tube back in the thermal cycler.

   (3) **34 °C 2 min**

   (4) **90 °C 1 min**

   (5) **Repeat steps 2–4 for 7 more cycles.**

   (6) **30 °C 2 min 20 s**

   During step 6, the thermal cycler can be paused; the tube can be taken out of the thermal cycler and briefly centrifuged if necessary. 45 μl of HTC-mix are then added to the tube. Mix the tube content by inverting the tube a couple of times, centrifuge it briefly, and place it back in the thermal cycler. If the thermal cycler was in the sterile hood during low-temperature cycles of DOP-PCR, move it out of the sterile hood.

   (7) **94 °C 2 min**

   This step is required to activate Invitrogen Platinum® *Tfi* Exo(-) DNA Polymerase.

   (8) **56 °C 2 min**

   (9) **70 °C 2 min**

   (10) **0.1°C s^{-1} to 74 °C**

(11) **92 °C** **1 min**

(12) **56 °C** **2 min**

(13) **70 °C** **2 min**

(14) **0.1°C s^{-1} to 74 °C**

Repeat steps 11–14 for 30 more cycles.

(15) **72 °C** **10 min**

(16) **4 °C** **hold**

6. Add 1 μl of 100 mM NaEDTA to the amplified sample and store it at −20 °C.
7. Three μl of the amplified sample should be run on 2 % agarose gel. The product of DOP-PCR amplification looks like a smear with a size of fragments varying between 100 bp and 800 bp with most of the fragments being 300–400 bp long. However, the quantity of the amplified product is very low, and often the smear is barely visible on the gel (Fig. 2). Always amplify a negative control when performing DOP-PCR (Master Mix without any chromosomal material and Master Mix with the collection drop solution incubated at +60 °C without any collected chromosomal fragments).

3.7 Re-amplification

Original DOP-PCR can be followed by several rounds or re-amplification. 1 μl of DOP-PCR product can be used as a template for the first re-amplification, after which 1 μl of the first re-amplification product can be used as a template for the second re-amplification, and so on. We recommend using the product of the third and the fourth re-amplification as a template for labeling reaction for probe preparation. However, for some applications

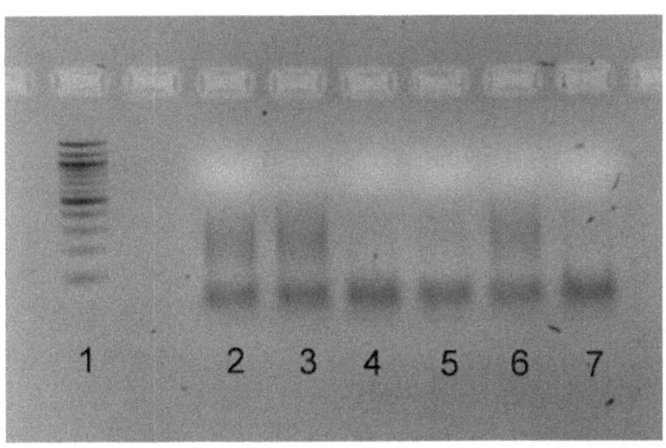

Fig. 2 Agarose gel electrophoresis of DOP-PCR products after microdissection and FISH-microdissection. *Lane 1*: Quick-Load 100 bp DNA Ladder (New England BioLabs Inc.). *Lane 2*: DOP-PCR product based on 20 copies of half of p-arm of chromosome 1, microdissection of trypsin-Giemsa-stained chromosomes. *Lane 3*: DOP-PCR product based on 10 copies of half of q-arm of chromosome X, FISH-MD. *Lane 4*: DOP-PCR product based on collection drop solution with no collected chromosomal fragments (collection drop negative control). *Lane 5*: DOP-PCR product based on 4 copies of half of q-arm of chromosome 7, FISH-MD. *Lane 6*: DOP-PCR product based on 10 copies of half of p-arm of chromosome 2, FISH-MD. *Lane 6*: DOP-PCR Master Mix negative control

(e.g., sequencing), use of original DOP-PCR product or the product of the first re-amplification is preferable. Both Invitrogen Platinum® *Tfi* Exo(-) DNA Polymerase and Applied Biosystems™ AmpliTaq™ DNA Polymerase can be used for re-amplification.

1. Re-amplification with Invitrogen Platinum® *Tfi* Exo(-) DNA Polymerase

Water, PCR grade	31.05 μl
5× Platinum® Tfi Reaction Buffer	10.00 μl
$MgCl_2$, 50 mM	2.50 μl
dNTPs mix	4.00 μl
DOP primer, 40 μM	1.25 μl
Invitrogen Platinum® *Tfi* Exo(-) DNA Polymerase	0.20 μl

2. Re-amplification with Applied Biosystems™ AmpliTaq™ DNA Polymerase

Water, PCR grade	33.50 μl
GeneAmp® 10 × PCR Buffer II	5.00 μl
$MgCl_2$, 25 mM	5.00 μl
dNTPs mix	4.00 μl
DOP primer, 40 μM	1.25 μl
Applied Biosystems™ AmpliTaq™ DNA Polymerase	0.25 μl

3. Add 1 μl of DNA template to the PCR mix, and place the tube in the thermal cycler (use heated lid):
   1. **94 °C 3 min**
   2. **91 °C 1 min**
   3. **56 °C 1 min**
   4. **70 °C 2 min**
   5. **0.1 °C s^{-1} to 74 °C**
   6. **Repeat steps 2–5 for 30 more cycles.**
   7. **72 °C 5 min**
   8. **4 °C hold**
4. Store the product of re-amplification at −20 °C.

3.8 Label-PCR

The specificity and the quality of the microdissected probe are evaluated by reverse-FISH (FISH of the generated probe on control metaphases). The product of original DOP-PCR or re-amplification can be labeled by PCR with DOP primer and a modified nucleotide (biotinylated/digoxigenylated or linked to a fluorochrome).

1. Label-PCR with Applied Biosystems™ AmpliTaq™ DNA Polymerase.

Prepare Label-PCR buffer by adding 0.5 μl of NP-40 to 1 ml of GeneAmp® 10 × PCR Buffer II.
Prepare PCR mix as follows:

Water, PCR grade	10.00 μl
Label-PCR buffer	2.00 μl
$MgCl_2$, 25 mM	2.00 μl
Label mix	2.00 μl
DOP primer, 40 μM	1.00 μl
Modified nucleotide (concentration according to manufacturer's instruction)	1.00 μl
Applied Biosystems™ AmpliTaq™ DNA Polymerase	0.12 μl

2. Add 2 μl of template DNA and place the tube in the thermal cycler. Use the same program as for re-amplification.

3.9 Probe Precipitation and Reverse-FISH

See chapter by Thomas Liehr "Homemade Locus-Specific FISH Probes: Bacterial Artificial Chromosomes" and the chapter by Thomas Liehr et al. "The Standard FISH Procedure".

4 Notes

1. If hybridization of metaphase spreads was done overnight, the person who performed the post-hybridization wash cannot perform chromosome microdissection the same day. Microdissection should either be postponed until the next day or another person who did not work in FISH room can do microdissection. Carryover contamination can happen even due to pens, mobile phones, jewelry, etc. being brought to the microdissection room from the post-PCR area/FISH room. Separation of microdissection and FISH equipment is extremely important to avoid carryover contamination. The microdissection room should have its own required equipment and supplies like a water bath, pipettes, tube racks, clean lab coats, gloves, etc.
2. Only freshly prepared metaphase spreads are suitable for successful microdissection. Dissection should be performed on 2 subsequent days after metaphase spreads were prepared.
3. Keep all collection drop and PCR reagents separated from DNA templates.
4. Always use sterile filter tips that are longer than the tubes so that the pipette itself does not touch the inside of the tubes.
5. Carefully open and close the Eppendorf tubes; never touch the tube's lid from inside. Always spin the tubes briefly before opening them.

6. Proteinase K activity may significantly reduce over time resulting in failed amplification of the microdissected fragments. Monitoring of the expiration date is required.
7. As only few chromosomal fragments are used as a template for the original amplification, there is a serious danger of sample contamination that can occur at any step before the material is amplified with DOP-PCR. Preparation of collection drop in a contamination-free environment is pivotal, as any cell/genetic material that accidently gets in the tube with the collection drop solution can be subsequently amplified.
8. All the solutions should be aliquoted in small volumes, so even if one tube gets contaminated, other tubes will remain intact.

References

1. Backx L, Van Esch H, Melotte C et al (2007) Array painting using microdissected chromosomes to map chromosomal breakpoints. Cytogenet Genome Res 116:158–166
2. Weise A, Timmermann B, Grabherr M et al (2010) High-throughput sequencing of microdissected chromosomal regions. J Hum Genet 18:457–462
3. Zlotina A, Kulikova T, Kosyakova N et al (2016) Microdissection of lampbrush chromosomes as an approach for generation of locus-specific FISH-probes and samples for high-throughput sequencing. BMC Genomics 17:126
4. Weimer J, Kiechle M, Arnold N (2000) FISH-microdissection (FISH-MD) analysis of complex chromosome rearrangements. Cytogenet Cell Genet 88:114–118
5. Weimer J, Koehler MR, Wiedemann U et al (2001) Highly comprehensive karyotype analysis by a combination of spectral karyotyping (SKY), microdissection, and reverse painting (SKY-MD). Chromosome Res 9:395–402
6. Weimer J, Cohen M, Wiedemann U et al (2006) Proof of partial imbalances 6q and 11q due to maternal complex balanced translocation analyzed by microdissection of multicolor labeled chromosomes (FISH-MD) in a patient with Dandy-Walker variant. Cytogenet Genome Res 114:235–239
7. Jancuskova T, Plachy R, Stika J et al (2013) A method to identify new molecular markers for assessing minimal residual disease in acute leukemia patients. Leuk Res 37:1363–1373
8. Kosyakova N, Hamid AB, Chaveerach A et al (2013) Generation of multicolor banding probes for chromosomes of different species. Mol Cytogenet 6:6
9. Telenius H, Carter NP, Bebb CE et al (1992) Degenerate oligonucleotide-primed PCR: general amplification of target DNA by a single degenerate primer. Genomics 13:718–725
10. Rabbitts P, Impey H, Heppell-Parton A et al (1995) Chromosome specific paints from a high resolution flow karyotype of the mouse. Nat Genet 9:369–375
11. Claussen U, Michel S, Mühlig P et al (2002) Demystifying chromosome preparation and the implications for the concept of chromosome condensation during mitosis. Cytogenet Genome Res 98:136–146

Homemade Locus-Specific FISH Probes: Bacterial Artificial Chromosomes

Thomas Liehr

Abstract

Besides the well-known applications of bacterial artificial chromosomes (BACs) in classical molecular genetics, these probes are also well suited for molecular cytogenetic studies. BACs are nowadays the most often applied locus-specific probes in FISH. Various applications are possible like gene mapping, FISH banding, determination of chromosomal breakpoints, characterization of derivative chromosomes, studies on the interphase architecture, or the karyotypic evolution. Here the basic principle how BACs can be hybridized in situ on chromosome preparations is outlined. Moreover, an overview is given on possible questions to be studied using BACs as FISH probes.

Keywords Bacterial artificial chromosomes (BACs), Molecular genetics, Applications of BAC–FISH, Preparation of BACs for FISH, Labeling of BACs for FISH

1 Introduction

The hybridization of nucleic acid sequences immobilized on a nitrocellulose membrane is a method that is well known to those working in the field of molecular genetics. Southern blotting is designed to locate a particular sequence of DNA within a complex mixture, e.g., to locate a particular gene within an entire genome [1]. In contrast, the hybridization and visualization of nucleic acid sequences directly within a tissue or on chromosomes is a technique that is known and applied in specialized areas, mainly or even exclusively in molecular cytogenetic working, i.e., (human) genetic or pathologic laboratories. This is somewhat surprising, as FISH provides various unique options, like direct localization of DNA sequences within each genome (chapter by Amanda Larracuente "FISH in Drosophila"; chapter by Thomas Liehr "FISH on Insect Cells Transfected with Heterologous DNA"; chapter by Ekaterina Badaeva et al. "In Situ Hybridization to Plant Chromosomes"; chapter by Cassia Yano et al. "Fish-FISH: Molecular Cytogenetics in Fish Species"; chapter by Anna Zlotina and Alla Krasikova "FISH

Thomas Liehr (ed.), *Fluorescence In Situ Hybridization (FISH)*, Springer Protocols Handbooks,
DOI 10.1007/978-3-662-52959-1_8,

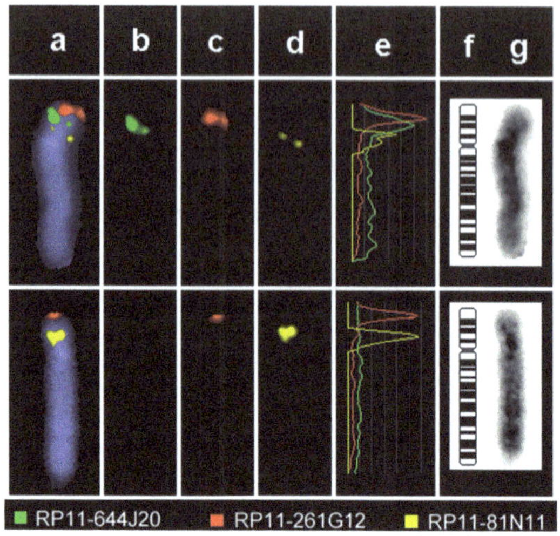

Fig. 1 Clinical case of an interstitial deletion in the short arm of chromosome 4, as confirmed by the application of region-specific BAC clones. The normal chromosome 4 (*top*) shows three expected signals in the expected order. The chromosome with the deletion (*bottom*) lacks the green BAC signal, indicating a deletion of the corresponding region. In detail: (**a**) merged fluorescence image, (**b**) BAC RP11-644 J20 in FITC, (**c**) BAC RP11-261G12 in Spectrum Orange, (**d**) BAC RP11-81 N11 in Cy5, (**e**) fluorescence profiles along the chromosome, (**f**) ideogram of chromosome 4, and (**g**) inverted DAPI image

in Lampbrush-Chromosomes"); also two or more DNA sequences can be visualized and distinguished at the same time (chapter by Thomas Liehr "Classification of FISH probes"; chapter by Thomas Liehr et al. "cenM-FISH Approaches"; chapter by Anja Weise and Thomas Liehr "Subtelomeric and/or subcentromeric probe sets"; chapter by Thomas Liehr et al. "Bar-Coding Is Back"; chapter by Sandra Louzada et al. "Fluorescence In Situ Hybridization onto DNA Fibers Generated Using Molecular Combing"; chapter by Anja Weise and Thomas Liehr "Parental Origin Determination FISH: pod-FISH"; chapter by Thomas Liehr and Sven Hauke "Interphase FISH in Diagnostics") or even RNA may be traced (chapter by Bin Ma and Naoko Tanese "RNA Imaging in Living Cells"; chapter by Bin Ma and Naoko Tanese "RNA-Directed FISH and Immunostaining"). Besides commercially available probes (chapter by Thomas Liehr "Commercial FISH Probes") and probe sets (chapter by Thomas Liehr and Nadezda Kosyakova "Multiplex FISH and Spectral Karyotyping"; chapter by Thomas Liehr et al. "FISH Banding Techniques"), locus-specific probes as BAC clones are nowadays also at hand via different commercial suppliers (chapter by Thomas Liehr and Anja Weise "Back ground"). BAC probes have already been used successfully in FISH for the following applications: gene mapping (e.g., [2]), the creation of FISH banding probe sets (e.g., [3]), the determination of chromosomal breakpoints (e.g., [4]), the characterization of derivative chromosomes (e.g., [3], Fig. 1), determining the three-dimensional structure of the interphase nucleus (e.g., [5]), and interspecies comparative studies, called Zoo-FISH, which are performed in order to find out more about (karyotypic) evolution (e.g. [2]). BACs are also applied to visualize translocations (i.e., one of the signals is located on a nonhomologous chromosome) or duplications of the target region (i.e., a double or enlarged BAC signal is obtained [6, 7]). BAC clones are ideal tools for breakpoint mapping. Because of their known sequences, it is possible to localize and characterize the breakpoint region very precisely. In the case of a BAC clone that spans the breakpoint region, you will see signal splitting (chapter by Thomas Liehr et al. "Two- to Three-Color FISH"). Otherwise, BACs can be defined to be breakpoint flanking (e.g., [8]). With the previously emerging, unfortunately already outdated use of genome-wide BAC array platforms for screening submicroscopic genome alterations, BAC–FISH is often used to confirm the array results; still also, e.g., SNP-aCGH results may be checked by BAC–FISH (e.g., [9–11]).

The most important advantages of using BAC probes in FISH are their defined sequences and their size, which leads in general to very bright, intense, and easy-to-evaluate FISH results.

The basic method applied when using BACs in any kind of FISH study is outlined in the following. Moreover, a shortened and more efficient FISH protocol using microwave treatment can be found in this book as well (chapter by Anja Weise and Thomas Liehr "Microwave Treatment for Better FISH Results in a Shorter Time").

2 Materials

Apart from the standard cell biological and molecular cytogenetic equipment, including standard solutions (such as ethanol, methanol, formamide, formaldehyde, etc.), the following more specialized items are needed. The equipment needed for FISH itself is listed in chapter by Thomas Liehr et al. "The Standard FISH Procedure".

2.1 Chemicals and Other Materials

- Glycerol (Cat. No. G5516, Sigma, St. Louis, MO, USA)
- Nick translation kit (Cat. No. 1745808, Roche, Basel, Switzerland)
- Peptone (Cat. No. 1.07214, Merck, Darmstadt, Germany)
- tRNA (Cat. No. 0109541, Roche)
- Yeast extract (Cat. No. 1.03753, Merck)

2.2 Solutions to Be Prepared

- Use and set up the antibiotics as required by the special BAC clone resistance; corresponding information can be obtained from the individual BAC clone suppliers.
- 0.5 M EDTA (pH 8).
- LB (Luria–Bertani) medium: To 800 ml aqua dest, add 10 g peptone, 5 g yeast extract, and 10 g NaCl. Adjust the pH to 7.5 with NaOH. Adjust the volume to 1 l with aqua dest. Sterilize by autoclaving. Unopened bottles can be stored at room temperature (RT).
- 3 M sodium acetate: 24.6 g sodium acetate is added to 100 ml aqua des; the pH is adjusted to 5.2 with acetic acid.

3 Methods

3.1 Selection and Ordering of BACs

The BAC clone is a modified F-plasmid containing a human DNA sequence of ~30,000–500,000 bp and a resistance gene. BACs that are anchored in the human DNA sequence can be selected from different genome browsers (chapter by Thomas Liehr and Anja Weise "Background"). Aside from fully sequenced clones, there are also BAC end-sequenced (BES) clones that are assigned by their end sequences only. When searching for BAC clones with no cross hybridization caused by sequence homology, we recommend that you should either BLAST the BAC sequence (e.g., in the NCBI genome browser) or use eFISH (chapter by Thomas Liehr and Anja Weise "Background"). Once selected, there are several sources of BAC clones (chapter by Thomas Liehr and Anja Weise "Background").

3.2 Cultivation of E. coli with BACs

Normally, BAC clones are shipped as *E. coli* clones cultivated on LB agar stabs. The latter can be stored at 4 °C for several weeks, avoiding evaporation. There are several methods and kits available for BAC/plasmid isolation. We prefer the QIA prep® Miniprep kit (Cat. No. 27106, Qiagen, Venlo, The Netherlands), because it purifies large but also low-copy plasmids, like BACs. The protocol is modified from the supplier's instructions. The following cultivation and DNA isolation is described for a so-called Miniprep that results in ~20 μg plasmid DNA from 1 to 5 ml overnight cultures of *E. coli*:

1. Prepare 5 ml LB medium with the appropriate antibiotic in an adequate sterile jar (*see* **Note 1**).
2. Use a sterile pipette tip to pick *E. coli* from the LB agar stab and transfer the tip into the LB medium.
3. Cover the sterile jar, but make sure that air can be exchanged; cultivate on a shaker with 200 rpm at 37 °C for ~16 h (*see* **Note 2**).

3.3 Plasmid DNA Extraction from E. coli

Before starting the plasmid preparation, mix 0.5 ml of the *E. coli* culture with 0.5 ml glycerol and store at −80 °C. The BAC clone is stable for several years and can be used for repeated cultivation and plasmid preparation. For plasmid DNA extraction, follow the detailed instruction manual.

3.4 Labeling of BAC DNA by Nick Translation

There are several options for labeling BAC DNA directly with fluorochromes or indirectly by biotin or digoxigenin. If you want to do a lot of FISH experiments with one BAC clone, we recommend plasmid amplification followed by label PCR with degenerated oligonucleotide primers (DOP) (chapter by Nadezda Kosyakova et al. "FISH-Microdissection"). Another possibility is labeling by a light- or heat-activated chemical reaction using Photoprobe® Biotin (Cat. No. SP-1000, Vector, Burlingame, CA, USA; [11]). The most common method of labeling DNA is Nick translation, which is described in the following protocol for a kit from Roche:

1. 0.5–1 μg plasmid DNA in 16 μl aqua dest.
2. Add 4 μl of the nick translation kit of your choice (direct or indirect labeling) and mix well.
3. Incubate the mixture for 90 min at 15 °C in a thermocycler.
4. Stop the reaction by adding 1 μl 0.5 M EDTA (pH 8) and incubate for 10 min at 65 °C.
5. Precipitate the labeled DNA by adding 10 μl tRNA, 20 μl aqua dest, 5 μl 3 M sodium acetate, and 110 μl ethanol. Mix well and incubate at −20 °C for 1 h or at −80 °C for 20 min and centrifuge at 15,000 rpm for 15 min. Discard the supernatant and dry the DNA pellet by vacuum or in a heating oven.
6. Resuspend the pellet carefully in 20 μl of hybridization buffer (chapter by Thomas Liehr et al. "The Standard FISH Procedure").

3.5 Slide Pretreatment and FISH

As described in chapter by Thomas Liehr et al. "The Standard FISH Procedure".

4 Notes

1. In some cases, for example, when contamination of the stab with another clone cannot be excluded, a subcultivation of the *E. coli* BAC clone should be performed (i.e., plate bacteria from the stab on an agar plate to get single colonies after cultivation overnight at 37 °C). The initial culture for plasmid preparation should be done by picking a single clone from this plate.

2. Respect the rules in different countries when working with genetically modified organisms; for example, in Germany, you must work in S1 laboratories that are authorized to perform cultivation, isolation, and storage.

References

1. Southern EM (1975) Detection of specific sequences among DNA fragments separated by gel electrophoresis. J Mol Biol 98:503–517
2. Weise A, Starke H, Mrasek K et al (2005) New insights into the evolution of chromosome 1. Cytogenet Genome Res 108:217–222
3. Liehr T, Weise A, Heller A et al (2002) Multicolor chromosome banding (MCB) with YAC/BAC-based probes and region-specific microdissection DNA libraries. Cytogenet Genome Res 97:43–50
4. Gisselsson D, Palsson E, Hoglund M et al (2002) Differentially amplified chromosome 12 sequences in low- and high-grade osteosarcoma. Genes Chromosomes Cancer 33:133–140
5. Othman MAK, Lier A, Junker S et al (2012) Does positioning of chromosomes 8 and 21 in interphase drive t(8;21) in acute myelogenous leukemia? BioDiscovery 4:4
6. Weise A, Gross M, Mrasek K et al (2008) Parental-origin-determination FISH (pod-FISH) distinguishes homologous human chromosomes on a single cell level. Int J Mol Med 21:189–200
7. Rodríguez L, Bhatt SS, García-Castro M et al (2014) A unique case of a discontinuous duplication 3q26.1-3q28 resulting from a segregation error of a maternal complex chromosomal rearrangement involving an insertion and an inversion. Gene 535:165–169
8. Weise A, Kosyakova N, Voigt M et al (2015) Comprehensive analyses of white-handed gibbon chromosomes enables access to 92 evolutionary conserved breakpoints compared to the human genome. Cytogenet Genome Res 145:42–49
9. Othman MA, Melo JB, Carreira IM et al (2015) High rates of submicroscopic aberrations in karyotypically normal acute lymphoblastic leukemia. Mol Cytogenet 8:45
10. Varvagiannis K, Papoulidis I, Koromila T et al (2014) De novo 393 kb microdeletion of 7p11.2 characterized by aCGH in a boy with psychomotor retardation and dysmorphic features. Meta Gene 2:274–282
11. Vlaikou AM, Manolakos E, Noutsopoulos D et al (2014) An interstitial 4q31.21q31.22 microdeletion associated with developmental delay: case report and literature review. Cytogenet Genome Res 142:227–238

Part II

FISH Procedure

The Standard FISH Procedure

Thomas Liehr, Katharina Kreskowski, Monika Ziegler, Katja Piaszinski, and Katharina Rittscher

Abstract

Molecular cytogenetics originally comprised of two basic approaches: fluorescence in situ hybridization (FISH) and primed in situ hybridization (PRINS). Nowadays FISH is the one routine approach still used in research and routine molecular cytogenetics field. Here the basic protocol in how to do FISH using commercial and/or homemade DNA probes is described. For a protocol for FISH on tissue sections, please refer to chapter by Thomas Liehr "Characterization of Archived Formalin-Fixed/Paraffin-Embedded or Cryofixed Tissue, Including Nucleus Extraction." Besides direct-labeled probes, also indirect-labeled probes can be used, and both variants are outlined here.

Keywords Molecular cytogenetics, Fluorescence in situ hybridization (FISH), Core protocol, Biotin, Digoxigenin, Directly labeled, Indirectly labeled, Hapten, Commercial probes, Homemade probes

1 Introduction

The history of human cytogenetics is marked mainly by different technical developments and can be divided into three major time periods: the prebanding era (1879–1970), the pure chromosomal banding era (1970–1986), and the molecular cytogenetic era (since 1986) (chapter by Thomas Liehr and Anja Weise "Background"). Prebanding and pure banding era ended (at least for human chromosomes; chapter by Cassia Yano et al. "Fish-FISH: Molecular Cytogenetics in Fish Species"; chapter by Ana Paula Alves-Silva et al. "General Protocol of FISH for Insects"; chapter by Ekaterina Badaeva et al. "In Situ Hybridization to Plant Chromosomes") in 1986 with the first molecular cytogenetic experiment on human chromosomes [1], which was also the starting point for the youngest discipline in human genetics: molecular cytogenetics. The major techniques used in molecular cytogenetics are fluorescence in situ hybridization (FISH; chapter by Thomas Liehr and Anja Weise "Background"; Fig. 1) and primed in situ hybridization (PRINS; chapter by Thomas Liehr and Anja Weise "Background";

Thomas Liehr (ed.), *Fluorescence In Situ Hybridization (FISH)*, Springer Protocols Handbooks,
DOI 10.1007/978-3-662-52959-1_9,

Fig. 2). In situ hybridization is an approach that allows nucleic acid sequences to be examined inside the cells or on chromosomes and was described first in 1969 as a radioactive variant [2]. Nonradioactive probe labeling, such as biotin detected by avidin coupled to a fluorochrome, was invented in 1981 [3]. In 1989, the primed in situ hybridization (PRINS) labeling technique was introduced as an alternative to conventional FISH for in situ chromosomal detection as the second basic approach of molecular cytogenetic field [4]. PRINS is based on the principles of the polymerase chain reaction (PCR); it uses oligonucleotide primers and a *Taq* DNA polymerase for the in situ detection of target DNA sequences [5]. While FISH has been continuously developed and improved and is now the most widely used technique for in situ localization of nucleic acids, as illustrated by the great variety of applications in research and diagnosis [5], PRINS has almost no practical application any more nowadays—details on PRINS protocol thus can only be found in the previous edition of this book [6].

FISH, like other DNA-based approaches, takes advantage of the ability of nucleic acids to de- and renature, and the most relevant feature of nucleic acids is that, in single-stranded DNA, homologous sequences find each other and build a double helix again. In a regular FISH experiment, the formation of DNA–DNA hybrids is normally intended. In other words, the target DNA is fixed on a slide, the probe DNA is labeled, and both of these DNAs are unified in a hybridization mixture for reaction. There are, however, exceptions where PNA or RNA is used as the probe and/or target (chapter by Thomas Liehr "Classi fication of FISH Probes"; chapter by Nikolay Rubtsov and Natalya Zhdanova "The Replicative Detargeting FISH (ReD-FISH) Tech nique in Studies of Telomere Replication"; chapter by Gordana Joksic et al "Telomere Length Measurement by FISH"; chapter by Bin Ma and Naoko Tanese "RNA Imaging in Living Cells"; chapter by Bin Ma and Naoko Tanese "RNA-Directed FISH and Immunostaining").

The principle of DNA-DNA FISH is as follows (see also chapter by Thomas Liehr and Anja Weise "Background"; Fig. 1):

- Fix the target DNA onto a slide surface. The target DNA can be cells, nuclei, metaphase chromosomes, or pure DNA.
- Label the probe DNA. Labeling can be directly or indirectly. Direct labeling means that the fluorochromes are to be detected in the microscope (chapter by Ivan Iourov "Microscopy and Imaging Systems"; chapter by Michael Sommerauer et al. "Opti cal Filters and Light Sources for FISH") are directly bound to the probe DNA. An indirect label refers to the incorporation of a hapten that is not visible under a fluorescence microscope into the probe DNA. However, the hapten can be detected immunohistochemically by a fluorophore-tagged antibody against the

hapten: biotin and digoxigenin are the most frequently used haptens for FISH.

- Denature the target and probe DNA—most often formamide is applied here to reduce the melting point of DNA (for some variants, see chapter by Anja Weise and Thomas Liehr "Micro wave Treatment for Better FISH Results in a Shorter Time"; chapter by Gábor Méhes et al "One Day Quick-FISH"; chapter by Emanuela Volpi "Formamide-Free Fluorescence In Situ Hybridization (FISH)").
- In most cases, an excess of unlabeled repetitive DNA is added to the labeled probe DNA and a prehybridization is allowed in order to block repetitive elements (for alternatives, see chapter by Vladimir Trifonov et al. "FISH with and Without COT1 DNA").
- Renature the target and probe DNA together.
- Perform post-hybridization washes.
- When applying indirectly labeled probes, fluorophore-tagged antibodies should now be used for detection.
- Perform detection washes.
- Add the counterstain, antifade, and coverslip to finish the procedure.
- Evaluate at the fluorescence microscope (for alternatives, see chapter by Hannes Schmidt, Thilo Eickhorst "Gold-FISH: In Situ Hybridization of Microbial Cells for Combined Fluores cence and Scanning Electron Microscopy").

For a protocol for FISH on tissue sections, please refer to chapter by Thomas Liehr "Characterization of Archived Forma lin-Fixed/Paraffin-Embedded or Cryofixed Tissue, Including Nucleus Extraction".

2 Materials

2.1 For FISH Probe Labeling

- Biotin nick translation kit (Cat. No.: 11745824910, Roche Diagnostics, Basel, Switzerland).
- Cot1-DNA human (Cat. No.: 1373242, Roche Diagnostics, Basel, Switzerland; store at −20 °C)
- Double-distilled water = Aqua ad iniectabilia (Cat. No.: 235 1544, Braun; aliquot and store at −20 °C).
- EDTA 0.5 M (e.g. Merck; store at −20 °C).
- Ethanol 100 % [e.g. Merck; store at room temperature (RT)].
- Hybridization buffer: Dissolve 2 g dextran sulfate in 10 ml 50 % deionized formamide/2 × SSC/50 mM phosphate buffer for 3 h

at 70 °C. pH adjusted to 7 with phosphate buffer; hydrochloric acid destabilizes buffer solution. Aliquot and store at −20 °C.

- Sodium acetate solution (3 M, pH 5.2; e.g. Merck; store at −20 °C).

2.2 For Pretreatment

- Ethanol 100 %, 90 % and 70 %, 100 ml, each (e.g. Roth K928.4; store at RT).
- PBS 1× (phosphate buffered saline—Cat. No.: L1825, Biochrom; store at RT).
- Pepsin stock solution (Cat. No.: P-7012, Sigma, Germany).
- Pepsin solution: Add 1 volume of 0.2 M HCl to 19 volumes distilled water and heat at 37 °C, then add 0.5 volumes of pepsin stock solution 2 % (w/v) just before digestion step, and leave HCl-pepsin solution at 37 °C; make fresh as required.
- Postfix solution (10 ml, 1 % paraformaldehyde—toxic!—needs to be discarded after use as hazardous waste): mix 5 ml of 2 % paraformaldehyde (e.g. Merck, Germany) with 4.5 ml of 1×PBS and 0.5 ml 1 M $MgCl_2$ (make fresh as required).

2.3 For FISH Procedure Itself

- Antifade Vectashield (Cat. No.: H1000, Vector Laboratories/Biozol; store at +4 °C).
- Biotinylated antiavidin (Cat. No.: BA0300, CAMON Vector Laboratories; store at +4 °C).
- Bovine serum albumin (BSA) (Cat. No.: 10 735 078 001, Roche Diagnostics; store at −20 °C).
- DAPI (4,6-diamidino-2-phenylindol.2HCl) stock solution (Cat. No.: 124653, Merck; store at −20 °C).
- DAPI solution: Dissolve 1.5 μl of 1 M DAPI stock solution in 1 ml Vectashield antifade (store at +4 °C; can be used at least for 3 months).
- Denaturation buffer: 70 % (v/v) formamide (teratogen!), 20 % (v/v) filtered double distilled water, 10 % (v/v) 20×SSC; make fresh as required.
- Ethanol 100 %, 90 % and 70 %, 100 ml, each (e.g. Merck; store at RT).
- Fluorescein avidin DCS (Cat. No.: A2011, Vector Laboratories/Biozol; store at +4 °C).
- Fluorochrome-labeled nucleotides at 1 mM, such as FITC-12-dUTP (Cat. No.: 11 373 242 910, Roche Diagnostics; store at −20 °C) and TRITC-6-dUTP (Cat. No.: 11534 378 910, Roche Diagnostics; store at −20 °C).
- Formamide (Cat. No.: 1 09684 2500, Merck, Germany; aliquot and store at −20 °C; please remember to discard the formamide solution as hazardous waste).

- Formamide solution (please remember to discard the formamide solution as hazardous waste): 2 × SSC/50 % formamide, pH 7.0.
- PBS 1×(phosphate buffered saline—Cat. No.: L1825, Biochrom; store at RT).
- Phosphate buffer: prepare 0.5 M Na_2HPO4 and 0.5 M NaH_2PO4, mix these two solutions (1:1) to get pH 7.0, and then aliquot and store at −20 °C.
- Rubber cement: Fixogum™ (Marabu, Tamm, Germany; store at RT).
- SSC 20×= saline sodium citrate (Cat. No.: 15557-036; Gibco BRL; store at RT); set up 1 × and 2 × SSC before use.
- Solution I (blocking solution): add 2 ml of fluorescein avidin/ 4×SSC/0.05 % Tween/5 % bovine serum albumin (BSA) to 0.1 g of Marvel; (pH 7–7.5); make fresh as required.
- Solution II (antibody solution): Biotinylated antiavidin/4×SSC/ 0.05 % Tween/5 % BSA (1:20:100); make fresh as required.
- Tween 20 = polyoxyethylene-sorbitan monolaurate (Cat. No.: 10670-1000, Sigma, Germany, store at RT).
- Washing buffer (diluted from stock 20×SSC): 4×SSC, 0.05 % Tween 20; make fresh as required.

3 Methods

3.1 FISH Probes

3.1.1 Commercially Available Probes

Commercially available FISH probes (chapter by Thomas Liehr "Commercial FISH Probes") are offered as fluorescence- or hapten-labeled probe DNA. Examples of companies that provide such probes are (in alphabetical order) Applied Spectral Imaging (ASI), Cytocell, Kreatech/Leica, MetaSystems, Vysis/Abbott, and ZytoVision, although there are others, too. No addresses are provided here, as providers of labeled FISH probes easily be identified using the Internet.

1. Treat probes prior to use in FISH according to the manufacturer's instructions (*see* **Note 1**); often this means dilute a certain amount of labeled probe solution to a hybridization solution provided by the manufacturer (*see* **Note 2**).
2. Denature the corresponding solution in a 0.5 or 1.5 ml reaction cup at 75 °C for 5 min (*see* **Note 3**). If necessary for the probe, a prehybridization of 15–30 min at 37 °C can be added (*see* **Note 3**).
3. Store probe on ice until applied to the denatured slide (*see* Sect. 3.3 and **Note 4**).

3.1.2 Homemade and/or Self-Labeled Probes

When homemade or other unlabeled probes are to be used for FISH, the most feasible way to label these probes is to use nick translation [7]; for application of PCR-based approaches (chapter by Thomas Liehr "Commercial FISH Probes"; chapter by Feng-tang Yang et al "Generation of Paint Probes from Flow-Sorted and Microdissected Chromosomes"; chapter by Nadezda Kosyakova et al. "FISH-Microdissection"):

1. The probe DNA is labeled (for example) with a hapten by nick translation using the corresponding kit (kits for direct and indirect labeling available). Here we use the biotin nick translation kit from Roche (11745824910) as an example: dilute 1 μg of the probe DNA in 16 μl of double distilled water and add 4 μl of the nick translation solution.
2. Mix carefully using the tip of a 20 μl pipette, and incubate the 0.5 ml microtube at 15 °C for 90 min.
3. Add 0.5 μl of 0.5 M EDTA and incubate at 60–65 °C for 10 min to stop the reaction.
4. For each slide to be hybridized, precipitate 200 ng of the biotin-labeled probe together with 1–4 μg human Cot1 DNA with 2.5 vol ethanol (100 %) and 0.1 vol sodium acetate (3 M, pH 5.2). Precipitation can be done for either 20 min at −80 °C or 12–20 h at −20 °C.
5. Pellet the DNA by centrifugation at 14,000–15,000 rpm for 20 min at 4 °C, discard the supernatant, and dry the DNA pellet at RT or using a speed vac.
6. Dissolve the pellet from step 5 in 20 μl of hybridization buffer, vortex, and spin down.
7. Denature the probe solution at 75 °C for 5 min, and do a prehybridization step at 37 °C for 30 min (*see* **Note 3**).
8. Store probe on ice until applied to the denatured slide (*see* Sect 3.3 and **Note 4**).

3.2 Slide Pretreatment

In a conventional FISH approach, pretreatment of the slides with pepsin followed by postfixation with formalin buffer is required to reduce the background. Pretreatment with RNase A is also suggested in some protocols [8]. However, according to our experience, this step does not lead to any significant effects and can be skipped (*see* **Notes 5** and **6**):

1. Dehydrate slides with metaphase spreads and/or interphase nuclei on them in an ethanol series (70 %, 90 %, 100 %, 3 min each) and air-dry.
2. Put slides for 5–10 min in pepsin solution at 37–38 °C in a coplin jar; it is also possible to use 300 μl of pepsin solution, only, and to incubate for 5–10 min covered with a 24 × 60 coverslip on a 37 °C heating plate.

3. Remove the coverslip by shaking it off with momentum.
4. Incubate slides in 100 μl 1×PBS (RT) for 5 min using a 24 × 60 coverslip to spread the solution on whole slide surface (*see* **Note 7**).
5. Repeat Step 3.
6. Postfix nuclei on the slide surfaces by replacing with postfix solution for 10 min (RT, under the hood) by covering the slide with a 24 × 60 coverslip—alternatively also use of coplin jar and 100 μl of postfix solution is possible—however, in terms of avoiding toxic waste we prefer the variant with the coverslip (*see* **Note 7**).
7. In case of using a coverslip for step 5, repeat step 3 again here, but collect coverslip together with paraformaldehyde solution as hazardous waste. In case of applying a coplin jar and postfix solution in the range of milliliters, the solution may be used for up to 2 days by storing at 4 °C in between. Afterward it also needs to be discarded as hazardous waste.
8. Dehydrate slides in an ethanol series (70 %, 90 %, 100 %, 3 min each) and air-dry.

3.3 Fluorescence In Situ Hybridization

1. Add 100 μl of denaturation buffer to each slide and cover with 24 × 60 mm coverslip.
2. Incubate slides on a warming plate for 2–4 min at 75 °C (*see* **Note 8**).
3. Remove the coverslip immediately and place slides in a coplin jar filled with 70 % ethanol (−20 °C; 3 min) to conserve target DNA as single strands.
4. Dehydrate slide in ethanol (90 %, 100 %, RT, 3 min each) and air-dry.
5. Add 20 μl of probe solution onto each denatured slide, put a 24 × 50 mm coverslip on the drops, and seal with rubber cement. It is also possible to hybridize different probes on the same slide using smaller coverslips. The amount of probe/probe solution must be reduced according to the coverslip size.
6. Incubate slides for 8–16 h (up to 3 days; *see* **Note 9**) at 37 °C in a humid chamber.
7. Take the slides out of the 37 °C chamber and remove the rubber cement and coverslips with forceps (optional: letting them swim off in washing buffer [RT, 100 ml coplin jar)]. Go on with step 8 (smaller probes, i.e., below 5 Mb in size) or 9 (larger and commercially available probes).
8. Postwash the slides for 2 to 3 × 5 min in formamide solution (42 °C) followed by 3 × 5 min in 2 × SSC (42 °C) in a 100 ml coplin jar, with gentle agitation (*see* **Note 10**)—please

remember to discard the formamide solution as hazardous waste. Go to step 10.

9. Postwash the slides 5 min in 1 × SSC solution (62–64 °C) with gentle agitation (*see* **Notes 10** and **11**). Go to step 10.
10. Put the slides in washing buffer (100 ml, RT) for 5 min on a shaker. When using directly labeled probes exclusively, go straight to step 15. In the case of indirectly labeled probes, perform steps 11–14, which explain how to detect biotinylated probe DNA.
11. Add 100 μl of solution I to each slide, cover with a 24 × 50 mm coverslip, and incubate at 37 °C for 15 min in a humid chamber.
12. Remove the coverslip and wash for 1 min in washing buffer (RT, with gentle agitation).
13. Add 100 μl of solution II to each slide, cover with a 24 × 50 mm coverslip, and incubate at 37 °C for 35 min in a humid chamber. Steps 10–14 may be repeated for signal amplification—remember that also background is becoming stronger.
14. Remove coverslip and wash 5 min in washing buffer (RT, with gentle agitation; optional to repeat wash step once).
15. Wash slides briefly in 1 × PBS (RT).
16. Dehydrate slides in an ethanol series (70 %, 90 %, 100 %, 3 min each) and air-dry.
17. Counterstain the slides with 20 μl of DAPI solution (antifade already included), cover with a coverslip 24 × 60, and evaluate the results under a fluorescence microscope.

4 Notes

1. Many commercially available probes are nowadays supplied in a ready-to-use format. Diluting or combining these probes with other probes might lead to the loss of a certain regulatory status (e.g., the use for in vitro diagnostic applications) of such probes. This, as well as using other protocols than recommended by the manufacturer subsequently, requires an in-house validation before use.
2. For most commercially available labeled probes, hybridization solutions of any supplier will be suited for FISH. Thus, also probes derived from different manufacturers may be mixed and used together in hybridization. However, for specific applications and especially diagnostics, this has to be tested before application of a new probe combination in "a real case." In the author's lab, the here described homemade hybridization

buffer works for the probes offered by the mentioned companies in this chapter and also for and in combination with homemade probes.

3. Denaturation of a probe can be done in waterbath or, if available more easily in a thermocycler. The latter is more convenient, as there it is possible also to add a prehybridization step at 37 °C if needed without further hand-on-time.
4. Denatured (and prehybridized) probes should be applied to the denatured slide within 15–30 min; FISH went also still successful if applied after 60 min. Alternatively DNA probes and slides may be denatured simultaneously. However, not always denaturation time of probe and sample are alike, and additionally common denaturation excludes prehybridization.
5. Some FISH protocols do not do any pretreatment. According to our experience, this may be indicated in case of chromosome preparations derived from amnion or chorion cells. However, for blood-, bone marrow-, and fibroblast-derived samples, pretreatment as described here definitely leads to better results, which are more easily to evaluate, have brighter signals, and have less background.
6. The pepsin treatment time must be adapted in each lab. The success of the pretreatment must be controlled by microscopic inspection. A balance between tissue preservation and tissue/chromosome digestion must be found. If the tissue/chromosomes are preserved too well, the DNA probes may not be able to pass through, so that no result is obtained from FISH. In the case of too much tissue/chromosome digestion, FISH signals may still be obtained, but it may not be possible to correlate them to a specific tissue/chromosomal region. Complete loss of the tissue/chromosomes during the FISH procedure might occur. It is recommended that beginners should start with target samples that are not limited in availability.
7. Washing or incubating of slides here and at other points of the described protocol can also be done in coplin jars; however, as this needs much more material, we prefer the here described way.
8. Denaturation times of only 2–4 min are suggested for the maintenance of available metaphase chromosomes. When working with tissues without metaphase spreads, this aspect is of no significance.
9. 1–3 days of FISH hybridization is recommended. Stopping the incubation after 48 h may result in weaker signals, while stopping after 96 h may lead to some cross hybridization problems.
10. During FISH washing steps, it is important to stop the slide surfaces from drying out; otherwise, background problems may arise.

11. Manufacturer-specific FISH protocols of commercially available probes mainly differ in the recommended stringency wash conditions (temperature and/or salt concentration of the wash buffers). While in principal most probes of one manufacturer work with protocols of other manufacturers as well, care should be taken not to combine temperatures recommended by one manufacturer with buffers of another manufacturer. Additionally, probes requiring high-stringency washing conditions should not be used with protocols optimized for probes (such as of ZytoVision) which, due to their stringent and specific composition, require only low-stringency washing conditions. In contrast to this, these low-stringency washing probes can easily be combined with high-stringency washing protocols without negatively affecting the hybridization results.

References

1. Pinkel D, Straume T, Gray JW (1986) Cytogenetic analysis using quantitative, high sensitivity, fluorescence hybridization. Proc Natl Acad Sci USA 83:2934–2938
2. Gall G, Pardue ML (1969) Formation and detection of RNA-DNA hybrid molecules in cytogenetical preparations. Proc Natl Acad Sci USA 63:378–381
3. Langer PR, Waldrop AA, Ward DC (1981) Enzymatic synthesis of biotin-labeled polynucleotides: novel nucleic acid affinity probes. Proc Natl Acad Sci USA 78:6633–6637
4. Koch JE, Kolvraa S, Petersen KB et al (1989) Oligonucleotide-priming methods for the chromosome-specific labelling of alpha satelliteDNA in situ. Chromosoma 98:259–265
5. Liehr T (2015) Basics and literature on multicolor fluorescence in situ hybridization application. http://fish-tl.com/mfish.html. Accessed 06/01/2016
6. Liehr T, Pellestor F (2009) Molecular cytogenetics: the standard FISH and PRINS procedure. In: Liehr T (ed) Fluorescence in situ hybridization (FISH)—application guide, 1st edn. Springer, Berlin
7. Rigby PW, Dieckmann M, Rhodes C, Berg P (1977) Labeling deoxyribonucleic acid to high specific activity in vitro by nick translation with DNA polymerase I. J Mol Biol 113:237–251
8. Liehr T, Thoma K, Kammler K et al (1995) Direct preparation of uncultured EDTA-treated or heparinized blood for interphase FISH analysis. Appl Cytogenet 21:185–188

Microwave Treatment for Better FISH Results in a Shorter Time

Anja Weise and Thomas Liehr

Abstract

Molecular cytogenetic approaches applying smaller, locus-specific probes like cDNA, plasmids, cosmids, fosmids, P1 clones, bacterial artificial chromosomes (BACs), or yeast artificial chromosomes (YACs) sometimes may be hampered by inefficient hybridization. Also, especially in diagnostics, FISH results may be required within a few hours. Here a FISH protocol using microwave treatment is presented, leading to better hybridization efficiency in case of smaller probes and evaluable results within a few hours.

Keywords Microwave pulses, Microwave treatment, Hybridization efficacy, Locus-specific probes, Bacterial artificial chromosomes (BACs), cDNA, Plasmids, Cosmids, Fosmids, P1 clones

1 Introduction

Molecular cytogenetics can be applied in diagnostics and research, as FISH is a straightforward method for the direct localization of DNA sequences within a given genome ([1]; chapter by Thomas Liehr and Anja Weise "Background"). Probes generated by molecular genetics approaches and applied in FISH may be oligonucleotides/PNAs (chapter by Gordana Joksic et al. "Telomere Length Measurement by FISH"; chapter by Nikolay Rubtsov and Natalya Zhdanova "The Replicative Detargeting FISH (ReD-FISH) Tech nique in Studies of Telomere Replication"), cDNA [2], plasmids [3], cosmids ([3]; chapter by Anja Weise and Thomas Liehr "Sub telomeric and/or Subcentromeric Probe Sets"; chapter by Cassia Yano et al. "Fish-FISH: Molecular Cytogenetics in Fish Species"; chapter by Ana Paula Alves-Silva et al. "General Protocol of FISH for Insects"), P1 clones ([4]; chapter by Amanda Larracuente "FISH in Drosophila"; chapter by Ekaterina Badaeva et al. "In situ Hybridization to Plant Chromosomes"; chapter by Benedetta Bottari et al. "FISHing for Food Microorganisms"; chapter by Alexander Swidsinski and Vera Loening-Baucke "Evaluation of Polymicrobial Involvement Using Fluorescence In Situ

Thomas Liehr (ed.), *Fluorescence In Situ Hybridization (FISH)*, Springer Protocols Handbooks,
DOI 10.1007/978-3-662-52959-1_10,

Hybridization (FISH) in Clinical Practice"), fosmids ([5]; chapter by Amanda Larracuente "FISH in Drosophila"), yeast artificial chromosomes (YACs) [6, 7], and bacterial artificial chromosomes (BACs) ([6–8]; chapter by Thomas Liehr "Homemade Locus-Specific FISH Probes: Bacterial Artificial Chromosomes"; chapter by Thomas Liehr "Classification of FISH Probes"; chapter by Anja Weise and Thomas Liehr "Subtelomeric and/or Subcentromeric Probe Sets"; chapter by Thomas Liehr et al. "Bar-Coding Is Back"; chapter by Sandra Louzada et al. "Fluorescence In Situ Hybrid ization onto DNA Fibers Generated Using Molecular Combing"; chapter by Anja Weise and Thomas Liehr "Parental Origin Deter mination FISH: pod-FISH"; chapter by Cassia Yano et al. "Fish-FISH: Molecular Cytogenetics in Fish Species"; chapter by Anna Zlotina and Alla Krasikova "FISH in Lampbrush-Chromosomes"; chapter by Amanda Larracuente "FISH in Drosophila"; chapter by Ekaterina Badaeva et al. "In Situ Hybridization to Plant Chromosomes"). The latter are mainly applied in diagnostics as BAC probes have defined sequences, and due to their size (~30 to ~500 kb), they lead in general to very bright, intense, and easy-to-evaluate FISH results. Still in some cases, the BAC probe needed to solve a diagnostic problem does not provide evaluable signals, due to peculiarities of the sample and/or due to low hybridization yield. In such cases and also in case a quick result is required, the following protocol may be helpful [9].

In general, when applying BACs for FISH in most cases, one can expect two small signals on each chromosome, i.e., one signal on each chromatid. These may, if the BAC is >100 kb in size, merge into one large signal. BACs can also be evaluated, if strong and specific enough, in interphase FISH. A microwave treatment established by us [9] is presented here which leads to higher hybridization efficiency especially during first 2 h of hybridization time and can be used to shorten this time and/or to induce stronger signals on the studied sample. Furthermore our microwave enhanced FISH was also able to improve the quality of FISH results on difficult samples like tumor sections or isolated tumor nuclei ([10], chapter by Thomas Liehr "Characterization of Archived Formalin-Fixed/Paraffin-Embedded or Cryofixed Tissue, Includ ing Nucleus Extraction") (Fig. 1).

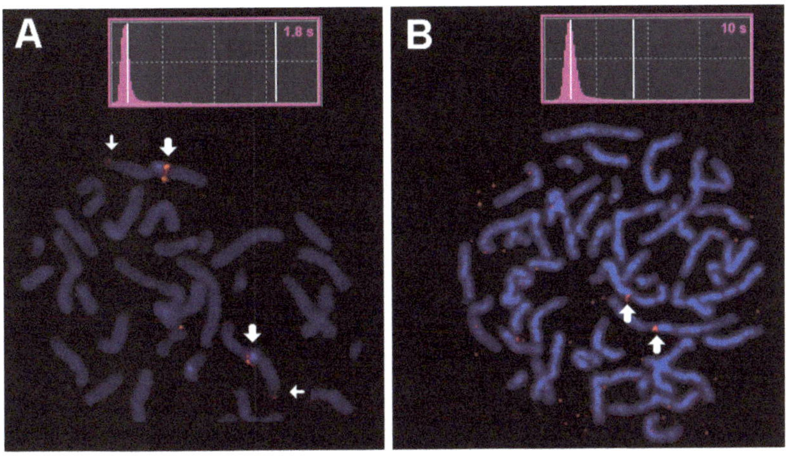

Fig. 1 FISH efficiency after 20 min of hybridization time with (**a**) and without (**b**) microwave treatment. A locus-specific BAC probe RP11-35B4 for 1q21 was applied (*thick arrows*). Due to segmental duplications, this clone gives additional weaker signals in 1p36.1 (*thin arrows*), which are only delectable in the microwave-treated sample (**a**). Please note also the 5 times reduced autoimaging time for the FISH picture, which leads to a clear and specific FISH signal and less background in the microwave-treated sample (**a**)

2 Materials

The standard cell biological and molecular cytogenetic equipment, including standard solutions (such as ethanol, methanol, formamide, formaldehyde, etc.), are needed as outlined in chapter by Thomas Liehr et al. "The Standard FISH Procedure."

3 Methods

3.1 Slide Pretreatment and Denaturation

As described in chapter by Thomas Liehr et al. "The Standard FISH Procedure."

3.2 Fluorescence In Situ Hybridization (FISH)

1. Place the pretreated and denatured slides in a Gene Frame® (Cat. No. AB-0577, Abgene, Epsom, UK) with a maximum volume of 65 μl on the slide.
2. For each slide dissolve the DNA-probe with appropriate amount of COT1 DNA in 30 μl hybridization buffer. Denature this probe solution at 75 °C for 5 min, cool down to 4 °C for 2 min, and prehybridize at 37 °C for 30 min in a thermocycler or corresponding water bath.
3. Add the denatured probe solution to the Gene Frame® area and seal with the polyester cover provided.
4. Put the prepared slide in a coplin jar within a water bath placed in a microwave oven (e.g., M 752, Miele, Gütersloh,

Germany). Place the microwave temperature sensor in the water bath to control the hybridization temperature. Apply 4–5 microwave beams of 600 W within 30 min; the hybridization temperature during this time must be between 25 and 37 °C. The microwave irradiation will lead to an enhanced water temperature, so it might be necessary to cool the water bath with ice (*see* **Note 1**).

5. Remove the Gene Frame® from the slide and continue with washing, and if necessary detection and sealing according to the FISH protocols as described in chapter by Thomas Liehr et al. "The Standard FISH Procedure."

4 Notes

1. It is crucial that the water bath in the microwave oven should not be overheated. An alternative to ice cubes and measuring the temperature is to adjust a defined volume of water with a certain starting temperature such that the water bath ends up at 37 °C. As an example, for 750 ml of water at 17 °C, 4 min at 360 W are required to end up with 37 °C. This kind of calibration curve must be adapted for each particular microwave oven.

References

1. Ferguson-Smith MA (2015) History and evolution of cytogenetics. Mol Cytogenet 8:19
2. von Deimling F, Scharf JM, Liehr T et al (1999) Human and mouse RAD17 genes: identification, localization, genomic structure and histological expression pattern in normal testis and seminoma. Hum Genet 105:17–27
3. Tsuchiya D, Matsumoto A, Covert SF et al (2002) Physical mapping of plasmid and cosmid clones in filamentous fungi by fiber-FISH. Fungal Genet Biol 37:22–28
4. Mark H, Wyandt H, Huang X et al (2005) Delineation of a supernumerary marker chromosome utilizing a multimodal approach of G-banding, fluorescent in situ hybridization, confirmatory P1 artificial chromosome fluorescent in situ hybridization, and high-resolution comparative genomic hybridization. Clin Genet 68:146–151
5. Birren BW, Tachi-iri Y, Kim UJ et al (1996) A human chromosome 22 fosmid resource: mapping and analysis of 96 clones. Genomics 34:97–106
6. Liehr T, Weise A, Heller A et al (2002) Multicolor chromosome banding (MCB) with YAC/BAC-based probes and region-specific microdissection DNA libraries. Cytogenet Genome Res 97:43–50
7. Liehr T (2006) Application of YACs in fluorescence in situ hybridization (FISH). In: Alesandair M (ed) Methods in molecular biology YAC protocols, 2nd edn. Humana, Totowa, NJ, pp 175–186. ISBN 1-58829-612-1
8. Mrasek K, Wilhelm K, Quintana LG et al (2015) BAC-probes applied for characterization of fragile sites (FS). In: Narayanan K (ed) Bacterial artificial chromosomes, Methods in molecular biology. Springer, New York, pp 289–298
9. Weise A, Liehr T, Claussen U et al (2005) Increased efficiency of fluorescence in situ hybridization (FISH) using the microwave. J Histochem Cytochem 53:1301–1303
10. Woelfel C, Liehr T, Weise A et al (2011) Molecular cytogenetic characterization of epithelioid hemangioendothelioma. Cancer Genet 204:671–676

FISH with and Without COT1 DNA

Vladimir A. Trifonov, Nadezhda V. Vorobieva, Natalia A. Serdyukova, and Willem Rens

Abstract

Complex FISH probes comprising large spans of genomic DNA always contain a high amount of dispersed repetitive sequences hampering the visualization of specific signals. To overcome this problem, different approaches have been elaborated that depend on experiment type and probe quality. A classical way to suppress repetitive sequences is to use unlabelled competitor DNA (sheared total genomic DNA or repeated sequences enriched DNA fractions). Here we present two protocols—the first one describes a rapid COT DNA isolation and peculiarities of its use in different FISH experiments, and the second is elaborated for COT-free FISH with complex probes and is based on a special software tool for image enhancement.

Keywords Repetitive DNA, DNA reassociation kinetics, Genome composition, COT1-DNA

1 Introduction

Fluorescence in situ hybridization (FISH) is a versatile tool for localizing DNA sequences on fixed chromosomes (chapter by Anja Weise and Thomas Liehr "Pre- and Postnatal Diagnostics and Research on Peripheral Blood, Bone Marrow, Chorion, Amniocytes, and Fibroblasts") or interphase nuclei (chapter by Eyad Alhourani et al. "Tumorcytogenetic Diagnostics and Research on Blood and Bone Marrow Smears or Effusions"; chapter by Thomas Liehr and Sven Hauke "Interphase FISH in Diagnostics"; chapter by Ivan Iourov et al. "Interphase FISH for Detection of Chromosomal Mosaicism"; chapter by Thomas Liehr and Nadezda Kosyakova "Three-Dimensional Interphase Analysis Enabled by Suspension FISH"; chapter by Galina Hovhannisyan and Rouben Aroutiounian "Comet-FISH"; chapter by Galina Hovhannisyan et al. "Micronucleus FISH"). The first in situ hybridization experiments used radioactively labeled fractions of repetitive DNA (ribosomal clusters, alphoid sequences) as probes ([1] chapter by Thomas Liehr and Anja Weise "Background"). Later,

Thomas Liehr (ed.), *Fluorescence In Situ Hybridization (FISH)*, Springer Protocols Handbooks, DOI 10.1007/978-3-662-52959-1_11,

nonradioactive versions of in situ hybridization were developed, and it became possible to detect unique DNA sequences [2]. Eukaryotic genomes contain high amounts of repetitive sequences that can be concentrated in specific chromosomal regions (heterochromatic blocks) or dispersed throughout the genomes. To locate a unique sequence, a probe needs to be large enough to give a detectable signal, and thus it inevitably contains small or large portions of dispersed repetitive DNA sequences. During the hybridization process, those sequences anneal to complementary genomic DNAs distributed throughout the genome. This results in a background that may have almost the same intensity as the target genomic locus. There are, however, a few ways to solve this background problem.

Sealey et al. [3] first suggested adding unlabeled, sheared total genomic DNA as a competitor to the hybridization reaction before the formation of duplexes with target genomic DNA. Landegent et al. [4] used a fraction of repetitive DNA sequences (COT1) to suppress the nonspecific hybridization of probes derived from BAC clones. Lichter et al. [5] used sheared total genomic DNA to block background signal in FISH experiments: the painting probes were preannealed with the sheared DNA prior to hybridization. Since then, most authors have used commercially available COT1 DNA or sheared genomic DNA in FISH experiments. However, sheared genomic DNA is less efficient than COT1 DNA at improving the signal-to-background ratio, since unique sequences of the probe are also suppressed during the prehybridization process, albeit at a lower extent than the repetitive sequences. COT is a product of time (seconds) and the DNA concentration (moles of nucleotides per liter). It expresses the reciprocity between the concentration and duration of second-order reactions and is approximately the optical density at 260 nm × hours 2^{-1}. COT1 is postulated in the classical work of Britten et al. [6] as the DNA fraction that reanneals in 1.2× SSC at 60 °C with starting DNA concentration of 83 μg ml^{-1} for 1 h. Thus, it is a characteristic of the DNA fraction in reassociation kinetics.

Another approach is to use probes without dispersed repetitive sequences. Rogan et al. [7] suggested designing probes for relatively large chromosomal regions by choosing PCR segments that lack repetitive DNA. However, this can only be done if the region of interest is sequenced and repetitive elements can be recognized. Alternatively, one can remove repeats from the complex probes by using special methods based on affinity capture [8, 9] or PCR-mediated suppression [10].

If the relative amount of repetitive sequences within the probe is limited, preannealing of the probe itself may be sufficient to obtain decent signal-to-background ratios [11]. This is often the case when complex chromosome-specific probes are derived from hundreds of flow-sorted copies by DOP-PCR ([12]; chapter by Fengtang Yang et al. “Animal Probes and ZOO-FISH”). Examples include paints of species such as human, horse, camel, and cat. The relative amount of repetitive sequences within a probe may be reduced by using a different primer: special primers were designed for mouse (*Mus musculus*) and Guinea pig (*Cavia porcellus*) in order to decrease the amount of pericentromeric heterochromatic sequences in the paints [13, 14].

Background may be less of an issue when FISH is performed using complex whole-chromosomal probes (paints) between diverged species (cross-species painting). According to personal communications with Dr Yang (The Sanger Institute, UK) and Dr Graphodatsky (The Institute of Molecular and Cellular Biology, Russia), flow-sorting-derived probes give very low background signal when hybridized on chromosomes of different species, due to rapid sequence evolution and degeneration of repetitive DNA.

Another method that was described recently involves removing the background in silico by employing special software [15]. The software is based on the following logic. The background is caused by the binding of the repetitive component of the probe to regions that are different from the target. This repetitive component should also be present in a different probe. If these two probes are labeled with different colors, regions corresponding to this repetitive component exhibit both colors, and so these signals can be removed by the software based on this dual color feature. This image enhancement procedure produces high-contrast chromosome paint images and is well suited for images where brightness vs. contrast enhancement is subjective. It is very efficient at removing nonspecific hybridization signals from the chromosome paint image. The procedure is very simple to use, it removes background in a controlled and defined manner, and it can be used when tissue for making COT1 is not available for the species of interest or when COT1 insufficiently blocks nonspecific (background) hybridization.

Here we provide protocols for two methods that can be used in FISH experiments that require suppression of repetitive DNA (if one is interested in the localization of specific repetitive sequences, then obviously these protocols are not used). The first method relies on COT fraction isolation (see Sect. 1.1) and requires a sufficient amount of the genomic DNA of the target species. The second uses special software and requires at least two differently labeled probes (see Sect. 1.2).

1.1 COT Isolation

The choice of repetitive DNA fraction (COT1, COT2, COT3, or higher) depends on the type of dispersed repetitive sequences that constitute the main fraction of the genome. Although COT1 DNA isolated from the same or closely related species is normally used in most FISH protocols, we recommend isolating the COT2–COT10 fraction, which was found to be more efficient in some cases. Note that COT1 DNA of certain species, such as human, mouse, and bovine are commercially available.

Here we describe a method that can be used to get the maximum amount of competitor DNA from different animal tissues quickly and efficiently. The protocol is simpler than high molecular weight DNA isolation, as it is not essential to maintain DNA integrity. The starting amount of DNA needs to be quite high when performing many FISH experiments (*i.e.*, ten FISH experiments may consume up to ~0.1 mg of competitor DNA, so it is recommended to start the isolation with at least 1 g of tissue), unless the method described by Dugan et al. [15] is used, which needs only 1 μg of COT per PCR reaction.

We successfully isolated and applied in FISH experiments different COT DNA fractions (usually COT2–COT30) from over 20 species of eutherians and two species of monotremes and several species of birds, squamates, fishes, including acipenserids, showing that the protocol is generally species independent. Tissues of heterogametic sex are preferable for COT isolation to block Y- or W-specific repetitive sequences. Here we describe some protocol modifications for animals (such as sturgeons and mollusks) and tissues, characterized by a high content of mucopolysaccharides.

The amount of COT1 DNA added to hybridization mixture depends on the probe type and varies between different laboratories, as the optimal amount is measured empirically. Still most protocols use about 1–20 μg of COT1 DNA per slide.

In cross-species experiments that require suppression of repetitive DNA, it is recommended that either the COT DNA fraction of target species (i.e., species whose metaphases are used in the hybridization experiment) or the COT fraction of the probe species (i.e., species whose DNA was used to generate the probe). Theoretically, the target species COT is better, since some repetitive sequences can be highly represented in the target species and underrepresented in the probe species, in which case probe-specific COT will not block them.

The quality of COT fraction is tested by the following FISH experiments. A pure COT fraction does not produce any additional signals, background, or autofluorescence. It blocks signals of repetitive sequences in both heterochromatic (C-positive) and euchromatic (C-negative) chromosomal regions. Some C-positive blocks composed of highly repetitive sequences may be difficult to suppress, even when using larger amounts of COT fraction, especially if the COT is isolated from a different species.

1.2 Removal of Background Using Image Enhancement Tool

In order to use this technique, a standard dual-color FISH experiment is conducted where both complex probes contain a repetitive DNA fraction. As after any dual-color FISH, we will get images with three main colors. For example, if we label chromosome of interest with Cy3 and any second chromosome with FITC, red chromosome regions will be enriched in sequences that hybridized only to the Cy3-chromosome paint and thus are specific for the chromosome of interest. Green chromosome regions are enriched in sequences that hybridized only to the FITC chromosome paint and thus are specific for the other chromosome. Yellow chromosome regions contain sequences that were hybridized by both paints and thus are nonspecific sequences shared by both chromosomes (of course in some cases we will see different grades of yellow varying from greenish to reddish respective to repetitive sequences representation). By removing all pixels in green or yellow program, it reveals in red the regions specific for the chromosome of interest. By removing just the pixels in yellow, it reveals in red those regions specific for the chromosome of interest and in green those regions specific for the FITC-labeled chromosome (Fig. 1).

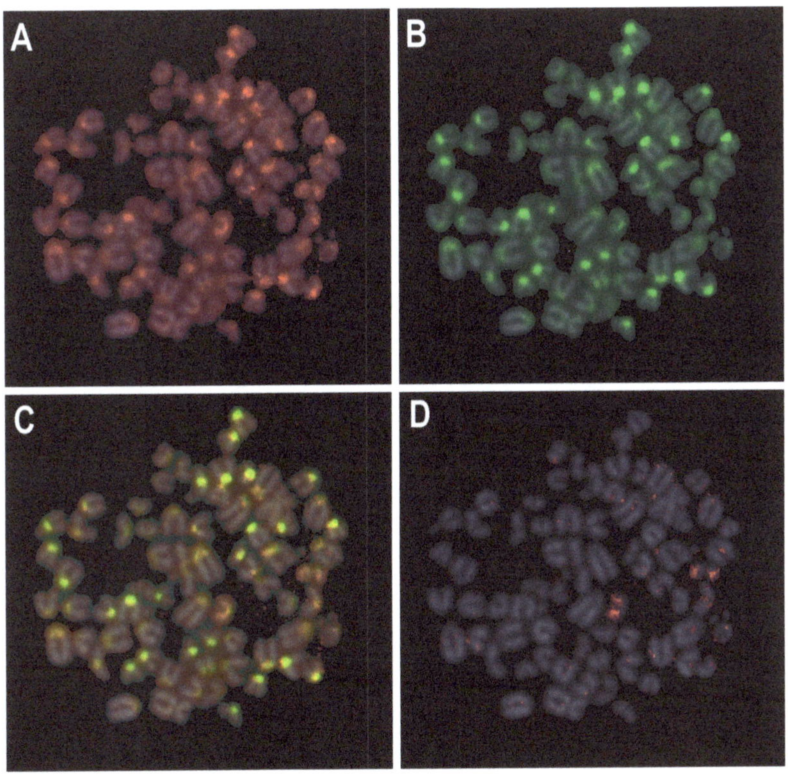

Fig. 1 An example of the use of the image enhancement tool is shown here. Chromosome paints from flow-sorted roe deer chromosomes hybridize strongly to centromeric heterochromatic DNA, resulting in images with bright centromeres and weak (just above background) chromosome-specific signals. Upper left is an image of the hybridization of an autosome-specific chromosome paint (red) to a roe deer metaphase. The chromosome pair is difficult to recognize due to the bright heterochromatic regions. Upper right is an image of hybridization to the same metaphase of another chromosome paint (green), which is merged with the upper left image, resulting in the image shown at the lower left. In this image, none of the yellow regions are chromosome-specific, as they are hybridized by both autosome paints. The lower right is the image after removing green and yellow pixels; the autosome pair corresponding to the Cy3-labeled paint is easy to recognize. Note that the centromeres of this pair are not labeled, as their DNA sequence is not specific for this pair

2 Materials

Apart from the standard cell biological and molecular cytogenetic equipment, including standard solutions (e.g., ethanol, isopropanol, phenol, chloroform, etc.), the following more specialized reagents are needed. The equipment needed for FISH is listed in chapter by Thomas Liehr et al. "The Standard FISH Procedure".

2.1 Chemicals

- S1 nuclease from *Aspergillus oryzae* (Sigma-Aldrich, St. Louis, MO, USA)
- Hexadecyltrimethylammonium bromide or CTAB (Sigma-Aldrich, St. Louis, MO, USA)

2.2 Solutions to be Prepared

- 10 × Buffer for S1 nuclease (0.33 M NaAc, 0.5 M NaCl, 10 mM $ZnSO_4$, pH 5.0)
- 20 × SSC buffer (3 M NaCl, 0.3 M sodium citrate, pH 7.0)
- TE buffer (10 mM Tris-Cl, pH 7.5, 1 mM EDTA)
- 10 % SDS in water
- Buffer A (0.35 M sucrose, 0.05 M Tris–HCl, pH 7.5, 0.066 M EDTA, 0.003 M $CaCl_2$, 0.025 M KCl)
- Buffer B (0.05 M Tris-Cl, pH 7.5, 0.066 M EDTA, 0.1 M NaCl)
- Buffer C (for tissues with a high content of mucopolysaccharides) (0.1 M Tris-Cl, pH 7.5, 0.1 M EDTA, 1.4 M NaCl)

3 Methods

3.1 COT DNA

3.1.1 DNA Isolation from Tissues with Relatively Low Concentration of Mucopolysaccharides

1. Take 10 g of fresh tissue (the liver, spleen, kidney, or placenta are particularly good to use), cut into small pieces, and homogenize in 60 ml of ice-cold buffer A (*see* **Note 1**).
2. Slowly add Triton X100, stirring with a glass rod, to get a final concentration of 1 %. (We recommend adding 10 % solution of Triton X100 in buffer A.)
3. Centrifuge at 800× *g* (at 4 °C) for 5 min and discard the supernatant.
4. To wash, add 50 ml of ice-cold buffer A to the pellet and resuspend by gentle vortexing.
5. Centrifuge at 800× *g* (at 4 °C) for 5 min and discard the supernatant.
6. Attention! The wash steps can be repeated 2–3 times, but make sure that the pellet does not get viscous.

7. Resuspend the pellet in 5 ml of buffer A and add 20 ml of buffer B (at room temperature = RT) and RNase A (final concentration 10 mg ml^{-1}).
8. Add equal amount (25 ml) of 0.4 % SDS and 0.1 mg ml^{-1} of proteinase K in buffer B and mix.
9. Incubate at 60 °C, vortexing every 10–20 min (the difference from the high molecular weight DNA isolation is that the DNA does not need to be treated gently; it can be vortexed and even homogenized if dense pieces are observed during lyses). Proceed to 3.1.4.

3.1.2 DNA Isolation from Mucopolysaccharide-Rich Tissues

1. Take 10 g of fresh or frozen tissue; grind tissues using liquid nitrogen, mortar and pestle. Homogenize in 60 ml of ice-cold buffer C (*see* **Note 1**).
2. Slowly add CTAB, stirring with a glass rod, to get a final concentration of 2 %. (We recommend adding 10 % solution of CTAB in buffer C.)
3. Add proteinase K to a final concentration 0.25 mg ml^{-1} and incubate at 60 °C for 2 h, vortexing every 10–20 min.
4. Add an equal volume of phenol: chloroform (1:1) and vortex. Centrifuge for 5 min at max speed at RT.
5. Remove the aqueous fraction to a new tube. Repeat step 4 and proceed to step 6.
6. Remove the aqueous fraction to a new tube. Add equal volume of chloroform and vortex. Centrifuge for 5 min at max speed at RT.
7. Remove the aqueous fraction to a new tube.
8. Proceed to ultrasound DNA fragmentation and ethanol precipitation as described above. Proceed to 3.1.4.

3.1.3 DNA Isolation from High Molecular Weight DNA

1. Sonicate the high molecular weight DNA in TE buffer until the fragment size is ~500 bp. Precipitate the sonicated DNA in ethanol and dissolve in TE buffer at a higher concentration (0.1–0.5 µg μl^{-1}). Determine the final DNA concentration, 260 nm/280 nm and 260 nm/230 nm absorbance ratios using NanoDrop (low DNA purity may result in inhibition of S1 nuclease). Proceed to 3.1.4.

3.1.4 Further DNA Workup

1. Sonicate the DNA using an ultrasonic homogenator (Sonoplus HD 2070, Bandelin, Berlin, Germany) until the fragment size is approx. 500 bp (controlled electrophoretically).
2. Add NaCl to a final concentration of 0.4 M (5 ml of 4 M NaCl) and two volumes of ethanol (100 ml).

Table 1
DNA concentrations and reannealing times required for the isolation of different COT fractions

COTX	DNA concentration (μg μl^{-1})	Time (min)
COT1	0.083	60
COT1	0.1	50
COT1	0.5	10
COT2	0.1	100
COT2	0.5	20
COT5	0.5	50
COT10	0.5	100
COT20	1	100

3. Wash the pellet twice with 70 % ethanol at RT and dissolve in 10 ml of TE buffer.
4. Determine the DNA concentration (usually the amounts mentioned above result in a final DNA concentration of 0.1–0.5 mg ml^{-1}).
5. Calculate the incubation time, which will depend on DNA concentration and the desired fraction. Britten et al. [6] postulated the COT1 as the DNA fraction that reanneals in 1.2× SSC at 60 °C with a starting DNA concentration of 83 μg ml^{-1} for 1 h. Thus, the incubation time (t, min) is calculated according to the formula:

$$t = \text{COTX} \times 4.98/C_0$$

where C_0 is the initial DNA concentration in μg μl^{-1}, t is the incubation time in minutes, X corresponds to the COT fraction (COT1 = 1, COT2 = 2, etc.), 4.98 = 60 min × 0.083.
6. Denature DNA at 95 °C for 5–10 min.
7. Add 1/10 volume of 12 × SSC.
8. Incubate for t min at 60 °C. For COT1 isolation with starting DNA concentration of 83 μg ml^{-1}, incubate the mixture for 1 h; alternatively use Table 1.
9. Place the tube with DNA on ice.
10. Add 10 × S1-nuclease buffer and S1-nuclease (final concentration: 100 units per 1 mg of DNA).
11. Incubate for 1 h at 42 °C.
12. Precipitate DNA by adding 0.8 volume of isopropanol. Centrifuge DNA at 10,000× g.

13. Wash twice in 70 % ethanol.
14. Air-dry and dissolve the pellet in TE buffer (1 ml).
15. Determine DNA concentration. (Usually the final amount of COT1 DNA is 10–15 % from original DNA concentration; it depends on repetitive DNA content of the genome) (*see* **Note 2**).

3.1.5 Amount of COT DNA in FISH Probe

The normally used amount of COT1 DNA is 2–10 μg per slide. It should be higher for probes of chromosomes with large heterochromatic blocks and for probes derived from microdissection or flow-sorting with low copy number.

The usual way of combining the probe and competitor DNA is mutual precipitation followed by dilution in hybridization buffer. For example, you may start with the combination of 50 ng of probe and 10 μg of COT DNA in 12 μl of hybridization buffer and then increase the amount of COT DNA if background blocking is insufficient (*see* **Note 3**).

Alternatively, the competitor DNA can be aliquoted in appropriate amounts, lyophilized, and stored at −20 °C. In this case, the probe and hybridization buffer just need to be added into the tube and mixed properly.

3.2 FISH Without COT: Using Special Software

3.2.1 Probe Labeling and FISH

1. Label the probes (two or more) with different fluorochromes (or haptens) either by nick translation (BACs, YACs) or PCR reaction (whole-chromosome probes (paints) obtained by DOP-PCR). For example, the chromosome-specific paint of interest can be labeled with Cy3, while a second chromosome paint that produces a high background image can be labeled with FITC. Fifty nanograms of each probe are dissolved in the hybridization mixture.
2. Perform FISH according to standard protocol for hybridization and signal detection ([16, 17], chapter by Ivan Iourov "Micros copy and Imaging Systems"; chapter by Michael Sommerauer et al. "Optical Filters and Light Sources for FISH"; chapter by Thomas Liehr et al. "The Standard FISH Procedure").

3.2.2 Image Registering and Background/Combined Color Removal

The procedure is explained using Leica instrumentation (Leica Microsystems, Wetzlar, Germany), as the software module used is incorporated into the Leica image processing software (Leica CW4000 Karyo):

1. Capture the images using the LeicaQFISH software and a cooled CCD camera mounted on a fluorescence microscope equipped with an automated filter wheel with DAPI-, FITC-, and Cy3-specific filter sets and a 63X objective.
2. Capture FITC, Cy3, and DAPI signals separately as 8-bit black-and-white images, and then normalize and merge them to a 24-bit color image (*see* **Note 4**).
3. Signal-to-background enhancement is performed by a software module incorporated in the Leica CW4000 Karyo software. The software module compares the separate red and green image and creates a histogram of red vs. green ratios. The relative number of exclusively green pixels is at the origin and the number of exclusively red pixels is at the end; intermediate values are in-between. Two sliding bars provide gates that can be set to control the green vs. red ratio of the final image. By moving the left bar to the left, only pixels that are "pure" green will be shown. By moving this bar to the right, pixels with a relatively low red intensity will be shown in addition to the "pure" green pixels. Similarly, by moving the right bar to the right, only pixels that are "pure" red will be shown. By moving this bar to the left, pixels with a relatively low green intensity will be shown in addition to the "pure" red pixels. This flexibility is needed to account for a possible difference in intensity between the Cy3 and FITC images. The positions of the bars are quantified to make the enhancement tool reproducible.

4 Notes

1. All procedures should be done on ice or in the cold room.
2. If nuclease S1 did not work properly, the amount of isolated COT DNA is almost equal to the initial amount of DNA; please check the buffer pH and enzyme activity. You can precipitate DNA again and repeat steps starting from Sect. 3.1.4, point 1. Sometimes, if the isolated DNA is not pure enough, traces of chemicals may inhibit the nuclease activity.
3. If bright dots are formed on the preparation after adding COT fraction, check the size of the COT DNA fraction; the presence of long DNA may produce this kind of background. Alternatively, probe that is not dissolved properly produces bright dots.
4. If one of the probes is too weak and the resulting combination of colors is similar to the original, increase the amount of weak probe and reduce the amount of bright probe (ideally, probes should be of equal brightness).

Acknowledgments

This work was supported by grant of RSF (№14-14-00275) to VAT and by Welcome Trust grant to WR.

References

1. Pardue ML, Gall JG (1975) Nucleic acid hybridization to the DNA of cytological preparations. Methods Cell Biol 10:1–16
2. Hopman AHM, Raap AK, Landegent JE et al (1988) Nonradioactive in situ hybridization. Mol Neuroanat 10:43–64
3. Sealey PG, Whittaker PA, Southern EM (1985) Removal of repeated sequences from hybridisation probes. Nucleic Acids Res 13:1905–1922
4. Landegent JE, Jansen in de Wal N, Dirks RW et al (1987) Use of whole cosmid cloned genomic sequences for chromosomal localization by non-radioactive in situ hybridization. Hum Genet 77:366–370
5. Lichter P, Cremer T, Borden J et al (1988) Delineation of individual human chromosomes in metaphase and interphase cells by in situ suppression hybridization using recombinant DNA libraries. Hum Genet 80:224–234
6. Britten RJ, Graham DE, Neufeld BR (1974) Analysis of repeating DNA sequences by reassociation. In: Methods in Enzymology. Academic Press, New York, pp 363–406
7. Rogan PK, Cazcarro PM, Knoll JH (2001) Sequence-based design of single-copy genomic DNA probes for fluorescence in situ hybridization. Genome Res 11:1086–1094
8. Craig JM, Kraus J, Cremer T (1997) Removal of repetitive sequences from FISH probes using PCR-assisted affinity chromatography. Hum Genet 100:472–476
9. Bolzer A, Craig JM, Cremer T et al (1999) A complete set of repeat-depleted, PCR-amplifiable, human chromosome-specific painting probes. Cytogenet Cell Genet 84:233–240
10. Dugan LC, Pattee MS, Williams J et al (2005) Polymerase chain reaction-based suppression of repetitive sequences in whole chromosome painting probes for FISH. Chromosome Res 13:27–32
11. Wienberg J, Adamski E, Yang F et al (1997) Chromosome painting without competitor DNA. Trends Genet Technical Tips Online. http://www.elsevier.nl/locate/tto
12. Telenius H, Carter NP, Bebb CE et al (1992) Degenerate oligonucleotide-primed PCR: general amplification of target DNA by a single degenerate primer. Genomics 13:718–725
13. Rabbitts P, Impey H, Heppell-Parton A et al (1995) Chromosome specific paints from a high resolution flow karyotype of the mouse. Nat Genet 9:369–375
14. Romanenko SA, Perelman PL, Trifonov VA et al (2015) A first generation comparative chromosome map between guinea pig (*Cavia porcellus*) and humans. PLoS ONE 10:e0127937
15. Rens W, Moderegger K, Skelton H et al (2006) A procedure for image enhancement in chromosome painting. Chromosome Res 14:497–503
16. Rens W, Grützner F, O'Brien PC et al (2004) Resolution and evolution of the duck-billed platypus karyotype with an X1Y1X2Y2X3Y3X4-Y4X5Y5 male sex chromosome constitution. Proc Natl Acad Sci U S A 101:16257–16261
17. Yang F, O'Brien P, Milne B et al (1999) A complete comparative chromosome map for the dog, red fox, and human and its integration with canine genetic maps. Genomics 62:189–202

Formamide-Free Fluorescence In Situ Hybridization (FISH)

Emanuela V. Volpi

Abstract

Formamide is an ionising solvent which is widely used in molecular biology research for its thermodynamic effects on the DNA double-helix stability. In fluorescence in situ hybridization (FISH), the addition of formamide to aqueous buffers solutions of DNA enables key procedural steps—such as the prehybridization denaturation, the reannealing step and the post-hybridization stringency washes—to be carried out at lower, less harsh temperatures without compromising the overall efficiency and specificity of the hybridization. However, formamide is toxic and a potential teratogen, and its use in research laboratories demands implementation of specific safety measures and the introduction of precautionary steps which can potentially complicate the logistical flow. This chapter outlines two simple, effective alternatives to the standard FISH protocol in which the use of formamide in sizable volumes is essentially eliminated and replaced by the utilisation of 'safer' chemicals. These developments might be of particular interest to research scientists considering the implementation in their laboratories of safety solutions and long-term sustainability strategies aimed at simplifying procedures, reducing exposure risks for the workers and minimising production of toxic waste.

Keywords Fluorescence in situ hybridization, FISH, DNA, Formamide, Formamide-free, Alkaline, denaturation

1 Introduction

The thermodynamic effects of formamide on the DNA double-helical stability have been extensively studied. An indicator of DNA duplex stability is its 'melting point' (Tm) or the temperature at which half of the DNA is present in a single-stranded or denatured form. Normally DNA 'melts' (denatures) at +90 °C to +100 °C in 0.1–0.2 M Na^+. Addition of formamide to aqueous buffer solutions of DNAs lowers their stability linearly by 2.4–2.9 °C mol^{-1} of formamide depending on the (G + C) composition, the helix conformation and the state of hydration [1]. Accordingly, formamide is widely used in molecular biology research, particularly in experiments concerned with nucleic acid studies. Indeed, high concentrations of this organic solvent, normally of molecular biology 'ultrapure' grade (99.5 %), are routinely used in DNA fluorescence

Thomas Liehr (ed.), *Fluorescence In Situ Hybridization (FISH)*, Springer Protocols Handbooks,
DOI 10.1007/978-3-662-52959-1_12,

in situ hybridization (FISH) experiments to allow for lower, less harsh denaturation temperatures prior to hybridization and higher stringency conditions at lower temperatures during the hybridization and post-hybridization washes.

Efficient denaturation is key in FISH to ensure complete dissassociation of the DNA semi-helices and nucleotide sequences full 'unmasking' to facilitate specific and efficient annealing of the DNA probes to their DNA targets by way of base complementarity. Wang et al. [2] have quite recently published a systematic study on DNA denaturation and renaturation characteristics of double-stranded DNA, including a comprehensive comparison of physical and chemical denaturation methods used for DNA hybridization-based applications. Traditionally, in FISH analysis, a short incubation in a 70 % formamide/2 × SSC solution at 70–75 °C—as described in the standard FISH procedure protocol (chapter by Thomas Liehr et al. "The Standard FISH Procedure")—is the denaturation method of choice.

However, formamide—an amide derived from formic acid with an ammonia-like odour—is toxic (by inhalation or skin contact) and a potential teratogen (it may cause birth defects via a toxic effect on embryo or foetus). As such, handling of formamide in the research laboratory requires precautionary safety measures beyond the customary use of personal protective equipment, to include the use of a chemical hood—specifically required when handling formamide solutions at high temperature in connection to toxic fumes inhalation risk—and dedicated hazardous waste disposal routes.

There is an increasing drive within modern research towards the implementation of 'safer' and more sustainable or 'environmentally friendly' solutions in the laboratory. This is partly the context within which variations on the FISH protocol aimed at minimising or totally eliminating the use of formamide have been explored by different laboratories over the last few years. The content of this chapter is specifically based on a formamide-free DNA denaturation protocol for simultaneous detection of FISH signals and bromodeoxyuridine incorporation previously devised and published by Moralli and Monaco [3, 4] and other formamide-free FISH adaptations relying on alkaline denaturation, used for research and students' projects purposes in my laboratory [5, 6].

2 Materials

Apart from standard FISH equipment and solutions as described in chapter by Thomas Liehr et al. "The Standard FISH Procedure", the specialised items required are as follows:

2.1 For Protocol A

- Tris–HCl pH 8.0
- KCl

- Glycerol
- 0.1 × SSC
- Absolute ethanol
- Thermocycler plate at 95 °C

2.2 For Protocol B

- NaOH 0.07 N (obtained from serial dilution of 10 N stock solution)
- 0.1 × SSC
- 2 × SSC
- Tween 20
- Absolute ethanol

3 Methods

3.1 Protocol A

(adapted from [3] and [4])

1. Prepare the following denaturation buffer: 10 mM Tris–HCl pH 8.0, 50 mM KCl, 5 % glycerol.
2. Prepare a Coplin jar with 0.1 × SSC for first post-denaturation wash at room temperature (RT).
3. Prepare ethanol dilutions at 70 % and 90 % in Coplin jars, for post-denaturation dehydration series.
4. Set thermocycler at 95 °C.
5. Once the thermocycler has reached the set temperature, position the slides, cell side up, on the thermocycler plate.
6. Add 150–200 μl of denaturation mix to each slide, and cover with a medium-/large-sized (22 × 40 or 22 × 50 mm) coverslip.
7. Close the thermocycler lid and incubate for 5–10 min at 95 °C (*see* **Note 1**).
8. Remove the slides form the thermocycler and immerse them in the Coplin jar with 0.1 × SSC at RT.
9. Dehydrate the slides in ethanol series (70 %, 90 % and 100 % for 2 min each, at RT).
10. Let the slides air-dry before proceeding with setting up the hybridization (probes will have to be denatured separately—*see* **Note 2**).

3.2 Protocol B

(adapted from [5] and [6])

All steps are to be carried out in Coplin jars:

1. Prepare a Coplin jar with 0.1 × SSC and a Coplin jar with 2 × SSC and put them at 4 °C in preparation of post-denaturation washes.

2. Place slides in prewarmed 2 × SSC in water-bath at 70 °C for 30 min (*see* **Note 3**).
3. Take the Coplin jar out of the water-bath and let the temperature in the Coplin jar cool down to 37 °C. Monitor drop in temperature with a thermometer. This step should take approximately 20 min.
4. Transfer slides to 0.1 × SSC at RT for 1 min.
5. Proceed with the denaturation step by immersing the slides in 0.07 N NaOH at RT for 1 min (*see* **Note 4**).
6. Quickly transfer the slides to 0.1 × SSC at 4 °C for 1 min.
7. Transfer the slides to 2 × SSC at 4 °C for 1 min.
8. Dehydrate the slides in an ethanol series (70 %, 90 % and 100 % ethanol for 1 min each, at RT).
9. Let the slides air-dry before proceeding with setting up the hybridization (probes will have to be denatured separately—*see* **Note 5**).

4 Notes

1. Given the high temperature required, it is advisable for the optimal length of the denaturation incubation to be established empirically in order to reach a satisfactory balance between adequate DNA 'melting' and preservation of chromosome morphology. A number of factors and variables can potentially contribute to differences in this respect, from specimen dehydration and 'ageing' to G + C content of target sequence, slight pH variations and differential DNA packaging within chromosomes.
2. In this protocol, post-hybridization washes are carried out in 0.1 × SSC buffer at 65 °C (3 washes in Coplin jars × 5 min each). The combination of low salt and high temperature ensures high stringency even in absence of formamide. The use of three prewarmed Coplin jars, to be used sequentially, prevents possible drops in temperature during the procedure. As before, it is advisable to optimise empirically the stringency required for best—as in most specific and accurate—hybridization results.
3. It is advisable to place the Coplin jar with 2 × SSC in the water-bath before the temperature starts to rise significantly in order to avoid sudden changes in temperature which might cause the glass to crack.
4. Exact timing of this stage is critical.

5. In this protocol, post-hybridization washes are carried out in 0.4 × SSC buffer at 72 °C for 2 min, followed by a wash in 2 × SSC, 0.5 % Tween20 at 42 °C for 30 s. Same considerations as in **Note 2** on the need for empirical optimisation of stringency conditions apply here.

References

1. Blake RD, Delcourt SC (1996) Thermodynamic effects of formamide on DNA stability. Nucleic Acids Res 24:2095–2103
2. Wang X, Lim HJ, Son A (2014) Characterization of denaturation and renaturation of DNA for DNA hybridization. Environ Health Toxicol 29:e2014007
3. Moralli D, Monaco ZL (2009) Simultaneous detection of FISH signals and Bromo-Deoxyuridine incorporation in fixed tissue culture cells. PLoS One 4(2):e4483
4. Moralli D, Monaco ZL (2010) Simultaneous visualisation of FISH signals and Bromo-deoxyuridine incorporation by Formamide-free DNA denaturation. In: Bridger JM, Volpi EV (eds) Fluorescence in situ hybridization (FISH) protocols and applications. Springer, Berlin
5. Yusuf M, Bauer DLV, Lipinski DM et al (2011) Combining M-FISH and Quantum Dot technology for fast chromosomal assignment of transgenic insertions. BMC Biotechnol 11:121
6. Moralli D, Yusuf M, Mandegar MA et al (2011) An improved technique for chromosomal analysis of human ES and iPS cells. Stem Cell Rev 7 (2):471–477

One-Day Quick FISH

Gábor Méhes, Tamás Csonka, and Katalin Hegyi

Abstract

The classical FISH approach requires overnight incubation for proper hybridization result. Tissue morphological features are varying due to aggressive pretreatment and high temperatures applied. To increase the speed of the molecular cytogenetic finding and eliminate some of the technical limitations, an alternative one-day FISH method was recently introduced. This procedure allows the completion and evaluation of a FISH reaction within one day by the reduction of the hybridization time to 60–120 min. Moreover, the low denaturation temperature significantly contributes to better tissue and cell morphology of the patient samples. The utility of the instant quality FISH (IQFISH) was carefully evaluated for the determination of *HER2* copy number status in breast carcinoma and *ALK* translocation status in lung adenocarcinoma samples in a clinical setting. In summary, the one-day IQFISH diagnostic kits point with fast and stable hybridization reaction in the diagnostic practice without any major loss compared to the conventional (overnight) FISH procedure.

Keywords IQFISH, Same-day FISH, One-day FISH, *HER2* amplification, *ALK* translocation, Breast cancer, Lung adenocarcinoma

1 Introduction

A number of structural or numerical chromosomal aberrations gained special attention in the clinical oncological practice as predictive factors contributing to the decision of targeted therapy applications. Cytogenetic testing is increasingly based on the fluorescence in situ hybridization (FISH) analysis of surgically removed samples or biopsies of cancer tissue specimen following the histopathological analysis and diagnosis (chapter by Eyad Alhourani et al. "Tumorcytogenetic Diagnostics and Research on Blood and Bone Marrow Smears or Effusions"). In frequent cancer types, e.g., breast, gastric, or lung adenocarcinoma, the relevant genetic aberration can be defined in the same setting right after histology. FISH is a broadly used method for determining *HER2* gene amplification in breast and gastric cancer using formalin-fixed, paraffin-embedded (FFPE) material ([1, 2]; chapter by Thomas Liehr and Sven Hauke "Interphase FISH in Diagnostics"; chapter by Thomas Liehr "Commercial FISH Probes"). Similarly, the translocation of

Thomas Liehr (ed.), *Fluorescence In Situ Hybridization (FISH)*, Springer Protocols Handbooks, DOI 10.1007/978-3-662-52959-1_13,

the *ALK* gene (*ALK/EML4*) is now widely tested by FISH using tumor tissue sections [3, 4]. The increased clinical need for predictive cytogenetic parameters significantly induced the workload in the cytogenetic or pathology laboratories routinely applying the FISH technique. Urgent FISH results may be more frequently required for fast diagnosis or therapeutic decisions. In addition, the general laboratory setup may also significantly benefit from the acceleration of the conventional FISH procedure. Considering the needs, the one-day FISH method using the instant quality FISH DNA probes (IQFISH pharmDx™, DAKO, Glostrup, Denmark) was developed to allow a same-day application by the significant shortening of the specific DNA-probe binding times.

The preparation of clinical samples is difficult to standardize due to many reasons (tissue type and composition, ischemia time, differences in the fixation and tissue processing, etc.). Therefore, tissue integrity is variably influenced by aggressive pretreatment and hybridization reagents, DNA annealing, and hybridization times during the FISH procedure [5]. The IQFISH method relies on the modification of the hybridization chemistry resulting in two qualitative achievements: (1) reduced hybridization time (60–120 min) and (2) reduced denaturation temperature (only 66 °C) significantly improved the sample handling together with the better preservation of the tissue morphology [6]. Since its first appearance, the one-day FISH approach was profoundly tested in a clinical setting compared in parallel with the conventional overnight FISH approaches [7, 8]. Its performance was found satisfactory for the routine clinical application with the clear benefit of time saving [6]. We found an almost perfect accordance regarding *HER2* (17q32) copy number and *ALK* (2p23) translocation testing between the one-day and the conventional FISH testing indicating the absence of bias between the two methods.

In summary, the one-day IQFISH approach is well applicable with the *HER2* and *ALK* dual color DNA-probe kits (DAKO) for the everyday-determination molecular cytogenetic markers offering some relevant advances further to the short incubation time, including enhanced tissue integrity and preserved fluorescent signal intensity compared to the conventional FISH kits. All these features make fluorescence detection more competitive in the area of light microscopy-based in situ hybridization approaches.

2 Materials

2.1 Equipment

- Glass slides (Superfrost + or poly-L-lysine coated)
- Coverslips (18 × 18 mm or larger)
- Staining cuvettes

- Hybridization chamber (StatSpin ThermoBrite, Abbott Molecular)
- Water bath (37 °C and 95 °C)
- Fluorescence microscope equipped with DAPI, FITC, Spectrum Orange (SpO), and Texas Red (TR) filters
- Fluorescence image capturing workstation (microscope camera and software)

2.2 Reagents for Tissue Treatment and FISH

- Fluorescence mounting medium with DAPI (DAKO, Glostrup, Denmark)
- Fixogum rubber cement (Marabu, Tamm, Germany)
- Ethanol series (100 %, 85 %, and 70 %)
- Histology FISH Accessory Kit (K5799, DAKO, Glostrup, Denmark)
- Pretreatment solution (20x), containing MES (2-[*N*-morpholino]ethanesulfonic acid) buffer
- Pepsin solution, pH 2.0
- Pepsin diluents (10x concentrated), pH 2.0
- Stringent wash buffer (20x), containing concentrated SSC/Tween-20
- Wash buffer (20x concentrated), containing Tris/HCl buffer
- FISH probes/kits (IQFISH, pharmDx™)

3 Methods

3.1 Sections

1. Cut 5 μm thick sections from FFPE material and pull them on the pretreated glass slide (*see* **Note 1**).
2. Incubate paraffin sections at 65 °C to mature (optimally 1 h).
3. Remove paraffin from sections in 2 changes of xylene (*see* **Note 2**) in a Coplin jar for 5 min each at room temperature (RT).
4. Rehydrate sections in descending series of ethanol (100, 85, and 70 % ethanol in distilled water, 2 min each) at RT.
5. Wash sections in 1× wash buffer for 2 min at RT.

3.2 Slide Pretreatment

1. Preheat the staining cuvette with the pretreatment solution MES (2-[*N*-morpholino]ethanesulfonic acid) buffer to 95 °C in the water bath and put slides into the jar for 10 min.
2. Cool down the jar with the slides to RT by removing it from the water bath for 15 min.
3. Rinse slides in two changes of wash buffer at RT for 3 min each.

4. Apply proteolytic digestion using diluted pepsin solution in a cuvette for 15 min at 37 °C (provided by the Histology FISH Accessory kit).
5. Wash in 1× wash buffer 2 times for 3 min at RT.
6. Dehydrate through ascending series of ethanol.

3.3 FISH Procedure

1. Apply IQFISH pharmDx™ (DAKO) ready-to-use probe mix to the slides following vortexing.
2. Coverslip and seal with rubber cement.
3. Co-denature probe and genomic DNA at 66 °C for 10 min in hybridization chamber.
4. Perform hybridization 60–120 min at 45 °C for any of the DNA-probe pairs in the same hybridization chamber.
5. Remove rubber cement (using a forceps) and coverslip following wash in 1× stringent wash solution for 5 min at RT.
6. Perform stringent wash at 63 °C for 10 min.
7. Wash in two changes of 1× wash buffer for 3 min each at RT.
8. Dehydrate slides in ascending ethanol series (70 %, 85 %, and 100 %) for 2 min each at RT.
9. Air-dry (in the dark).
10. Cover with fluorescent mounting medium and coverslip (allow 30 min to diffuse).
11. Store at 4 °C (in the fridge) until evaluation.

3.4 Evaluation of FISH Results

FISH signals can be counted using a fluorescence microscope (e.g., Zeiss Axio Imager Z2) equipped with DAPI, FITC, SpO, and TR filters considering only tumor cells (*see* **Note 3**). Evaluation should follow the actual guidelines defining the number and quality of the cells to be evaluated (*see* **Note 4**). Back up fluorescence images using an appropriate image capturing device (e.g., ISIS fluorescence imaging system, MetaSystems, Altlussheim, Germany) (*see* **Note 5**).

4 Notes

1. The experience described here is based on the routine use of the HER2 IQFISH pharmDx (K5731) probe and the ALK IQFISH pharmDx (G111600) break-apart probe (DAKO).
2. Xylene can be generally replaced by the nontoxic NeoClear solution (Merck, Darmstadt) which was also tested and found appropriate for deparaffinization of FFPE sections.
3. Microscopic evaluation for *HER2* amplification or *ALK* translocation status was carried out on the basis of current

guidelines, including the overview of the whole hybridization area to determine relevant tumor-containing areas and to consider heterogeneity using a 20x objective followed by the detailed evaluation of FISH signals in tumor cell nuclei using a 63x immersion oil objective. Recommendations are issued and continuously reviewed by the College of American Pathologists (CAP), Association for Molecular Pathology (AMP), and cancer organizations.

4. Technical issues repeatedly observed in routine FISH diagnostics are frequently associated with fixation and tissue handling problems. High fluorescence background interfering with the FISH signals appears in a small fraction of routine samples. However, reaction interpretation was virtually always possible in case of the one-day approach in the evaluated series of cases [6]. Ambiguous values could be cleared after exhaustive counting of additional tumor cell nuclei and recalculation of the red/green ratios as recommended [9, 10]. The IQFISH approach significantly reduced autofluorescence, and cell and nuclear contours were more defined. These observations led us to conclude that the modified pretreatment and hybridization conditions favor tissue morphology and prevent the generation of tissue-related fluorescence.
5. The stability of the FISH signals during longer-term storage was also evaluated. IQFISH performed clearly better in the signal intensity and reproducibility after 2 months storage than the conventional FISH approach that may gain special significance, e.g., reassessment or quality assurance purposes [6]. It is of special interest that FISH signal reading proved to be easier during reevaluation when IQFISH was used.

References

1. Penault-Llorca F, Bilous M, Dowsett M et al (2009) Emerging technologies for assessing HER2 amplification. Am J Clin Pathol 132:539–548
2. Saez A, Andreu FJ, Segui MA et al (2006) HER-2 gene amplification by chromogenic in situ hybridisation (CISH) compared with fluorescence in situ hybridisation (FISH) in breast cancer—a study of two hundred cases. Breast 15:519–527
3. Yi ES, Chung JH, Kulig K, Kerr KM (2012) Detection of anaplastic lymphoma kinase (ALK) gene rearrangement in non-small cell lung cancer and related issues in ALK inhibitor therapy: a literature review. Mol Diagn Ther 16:143–150
4. V Lafert M, Warth A, Penzel R et al (2013) Anaplastic lymphoma kinase (ALK) gene rearrangement in non-small cell lung cancer (NSCLC): results of a multi-centre ALK-testing. Lung Cancer 81:200–206
5. Walker RA, Bartlett JM, Dowsett M et al (2008) HER2 testing in the UK: further update to recommendations. J Clin Pathol 61:818–824
6. Hegyi K, Lønborg C, Mónus A, Méhes G (2013) One-day FISH approach for the high-speed determination of HER2 gene copy status in breast carcinoma. Appl Immunohistochem Mol Morphol 21:567–571
7. Tafe LJ, Steinmetz HB, Allen SF et al (2015) Rapid fluorescence in situ hybridisation (FISH) for HER2 (ERBB2) assessment in breast and gastro-oesophageal cancer. J Clin Pathol 68:306–308

8. Pala EE, Bayol U, Ozguzer A et al (2013) HER2 status in gastric cancer: a comparison of two novel in situ hybridization methods (IQ FISH and dual color SISH) and two immunohistochemistry methods (A0485 and HercepTest™). Pathol Res Pract 209: 548–554
9. Yang YL, Fan Y, Lang RG et al (2012) Genetic heterogeneity of HER2 in breast cancer: impact on HER2 testing and its clinicopathologic significance. Breast Cancer Res Treat 134:1095–1102
10. Wolff AC, Hammond EH, Hicks DG et al (2014) Recommendations for human epidermal growth factor receptor 2 testing in breast cancer: American Society of Clinical Oncology/College of American Pathologists clinical practice guideline update. Arch Pathol Lab Med 138:241–256

Telomere Length Measurement by FISH

Gordana Joksic, Ivana Joksic, Jelena Filipović, and Thomas Liehr

Abstract

Telomere length influences numerous cellular processes such as senescence, carcinogenesis, and aging. Quantitative FISH (Q-FISH) is a comprehensive method that allows measuring of individual chromosome telomere length in single cell with the resolution of 200 base pairs. The method is based on the use of a peptide nucleic acid (PNA) telomere oligonucleotide probe and appropriate digital image software for capture and quantification of fluorescence signals. The length of telomere is directly related to its integrated fluorescence intensity, as PNA probes are assumed to hybridize quantitatively to telomeric repeats. For the accuracy of Q-FISH measurement, it is important to use adequate internal controls such as fluorescence beads of defined size to avoid imprecisions due to lamp intensity variations. Fluorescence intensity of beads is then used to correct fluorescence intensities of telomere signals. Telomeric/centromeric ratio or calibration relative to cultured cell of known telomere length allows fluorescence intensity values to be converted into units of DNA length (RTLU). However, this step is not essential as fluorescence measurements in arbitrary units (TFU) will yield accurate results given that internal control (i.e., fluorescence beads) is used properly.

Keywords Telomere length, Age-related changes, Tumor, Cell line, Quantitative FISH (Q-FISH), Telomere shortening, Peptide nucleic acid (PNA) probe

1 Introduction

Adequate telomere length maintenance is essential for cell survival, and length is one of the most important indicators of telomere function [1]. Telomere length is maintained by dynamic processes of shortening and lengthening. Elongation of telomeres is accomplished by action of telomerase (reverse transcriptase that adds base pairs to chromosome ends) and rarely by homologous recombination. Loss in telomeric length and function can be due to regular telomere shortening (in cells lacking telomerase), loss of shelter in protective function, or altered DNA repair [2]. Although telomere length is variable among different species and is even heterogeneous in different cell types in one individual, a certain critical number of telomeric $(TTAGGG)_n$ repeats is needed to "cap" chromosomes and thus avoid activation of DNA damage response.

Thomas Liehr (ed.), *Fluorescence In Situ Hybridization (FISH)*, Springer Protocols Handbooks,
DOI 10.1007/978-3-662-52959-1_14,

Uncapped telomeres are recognized as double-strand breaks (DSBs) that in turn activate two main DNA damage response mechanisms: homologous recombination (HR) and nonhomologous end joining (NHEJ) [2]. Telomere length regulation is considered to be one of the molecular mechanisms capable of counting cell divisions and inducing cell cycle arrest. Critically shortened chromosome ends can lead to telomere-initiated senescence or apoptosis ([3]; chapter by Nikolay Rubtsov and Natalya Zhdanova "The Replicative Detargeting FISH (ReD-FISH) Technique in Studies of Telomere Replication"). Telomere shortening, by limiting cellular proliferation, act as tumor suppressor mechanism. Indeed, loss of proper telomere length maintenance leads to genome instability and is recognized as one of the crucial steps in cancerogenesis. One of the hallmarks of tumor cells is upregulation of telomerase activity or homologous telomere recombination that leads to telomere elongation [3]. Telomere attrition has been implicated in ageing and many age-related diseases. Numerous studies have demonstrated telomere shortening with age. In addition, several premature aging syndromes such as *dyskeratosis congenita* or inherited bone marrow failure syndromes are caused by mutations in telomerase or proteins that maintain telomere stability [4]. Measurement of telomere length was previously most often done by Southern blot method, but this approach gives only a crude estimate of telomere length in all cells of the sample. Development of quantitative FISH (Q-FISH) has made possible to measure individual chromosome telomere length in single cells with the resolution of 200 base pairs [5]. Also, intra-chromosome distribution of telomere length (p-arms vs. q-arms) and shortest/longest telomeres in given cell can be determined. In addition, Q-FISH may be used to estimate telomere length in species containing interstitial telomeric sites in their genomes as well as in species that have ultra-long and heterogeneous telomeres such as mice [6]. Q-FISH is based on the use of peptide nucleic acid (PNA) telomere oligonucleotide and appropriate digital image capture software for quantification of fluorescence signals. PNAs are resistant on degradation by nucleases and peptidases and show highly specific hybridization with DNA [6]. For the accuracy of Q-FISH measurement, it is important to use adequate fluorescence intensity calibration in all experiments [output of the system may vary over time; e.g., aging of the lamp, alignment of the optics (chapter by Ivan Iourov "Microscopy and Imaging Systems"; chapter by Michael Sommerauer et al. "Optical Filters and Light Sources for FISH")]. To avoid erroneousness due to lamp power variations, fluorescence beads of defined size must be used at the same time when sample images are picked up [7, 8]. Fluorescence intensity values of beads are then used to correct fluorescence intensities of telomere samples. Results of telomere measurement are then expressed in arbitrary units. The two most commonly used approaches to quantify telomere fluorescence intensity as length of TTAGGG repeats have included (1) telomeric/centromeric ratio or (2) calibration relative to cultured cell of known telomere length. Analysis of 15 to 20 metaphases per sample is recommended to obtain reliable results. One of the drawbacks of this technique is inability to use it in nondividing cells.

The principle of telomere length measurement is outlined here in short and detailed below:

- Fix the target metaphase chromosomes onto a slide surface.
- Denature the target and probe DNA together.
- Renature probe and target DNA together.
- Perform post-hybridization washes.
- Add counterstain, antifade, and coverslip to finish the procedure.
- Analyze using image capture software and telomere measurement tools (ISIS, MetaSystems, Altlussheim, Germany).
- The length of telomere is directly related to its integrated fluorescence intensity, as PNA probes are assumed to hybridize quantitatively to telomeric repeats.

2 Material

Apart from standard cell biological and molecular cytogenetic equipment, including standard solutions (e.g., ethanol, methanol, formamide, formaldehyde, xylene, etc.), only on more specialized items as mentioned below is required. The equipment needed for FISH is listed in chapter by Thomas Liehr et al. “The Standard FISH Procedure”.

- Fluorescent beads (FluoSpheres Polystyrene 1.0 µm, orange fluorescent (540/560) cat Nu F13082).

3 Methods

3.1 FISH Probes

1. PNA telomeric Cy3-labeled probe (commercially available PNA FISH Kit/Cy3 K5326 from DAKO Glostrup Denmark).
2. Centromeric probe for chromosome 2 (available on request from DACO Glostrup Denmark).

3.2 Procedure

3.2.1 Slide Preparation

1. Age slides 1.5 h at 60 °C or overnight at 55 °C.
2. Pretreat the slide (procedure as for conventional FISH).

3.2.2 Fluorescence In Situ Hybridization (PNA FISH)

1. Put 20 μl slide^{-1} of hybridization mixture [ready-to-use PNA telomeric Cy3-labeled probe (commercially available PNA FISH Kit/Cy3 (K5326 from DAKO, Glostrup Denmark) supplemented by centromeric probe for chromosome 2 (valuable on request from DACO) in final concentration 2 ng ml^{-1} and then put on the coverslip (22 × 50 mm))].
2. Heat slides on 70 °C for 2 min only.
3. Leave in dark, damp container for 2 h.

3.2.3 Post-hybridization Washes in the Dark

1. Put slides two times 15 min in 70 % formamide solution (in dark coupling jars and no shaking).
2. Put slides three times 5 min in 1 × PBS (in dark jars and on the shaker).
3. Dehydrate slides in ethanol series (70 %, 90 %, 100 %) 5 min each.
4. Add 15 μl of VECTASHIELD/DAPI and cover with 22 × 60 mm coverslip and fix it on nail varnish.
5. Slides are now ready to be observed under the microscope.

3.3 Evaluation

For quantification of telomere length by Q-FISH, fluorescent reference beads of defined size (i.e., 1 μm) are imaged before target samples using CCD camera linked to the ISIS software (MetaSystems, Altlussheim, Germany) as internal control of fluorescent intensity. Afterwards metaphases are captured with an individual excitation filter sets for DAPI and Cy3 and automatically saved with each image individual color channels. Background correction is performed by subtracting the mean background level of the Cy3 image from each pixel within the measuring area. Chromosomes of merged images are separated and transferred to the karyotype window. Fluorescence intensity values of beads are then used to correct fluorescence intensities of telomere samples. This allows fluorescence intensity values to be converted into units of DNA length. It can be done with Q-FISH software, automatically, or Peter Landsrops' software (TFL-TELO, download from the internet) and analyze manually (*see* **Note 1**). Alternatively the centromere probe 2 can be used as reference (*see* **Note 2**; Fig. 1).

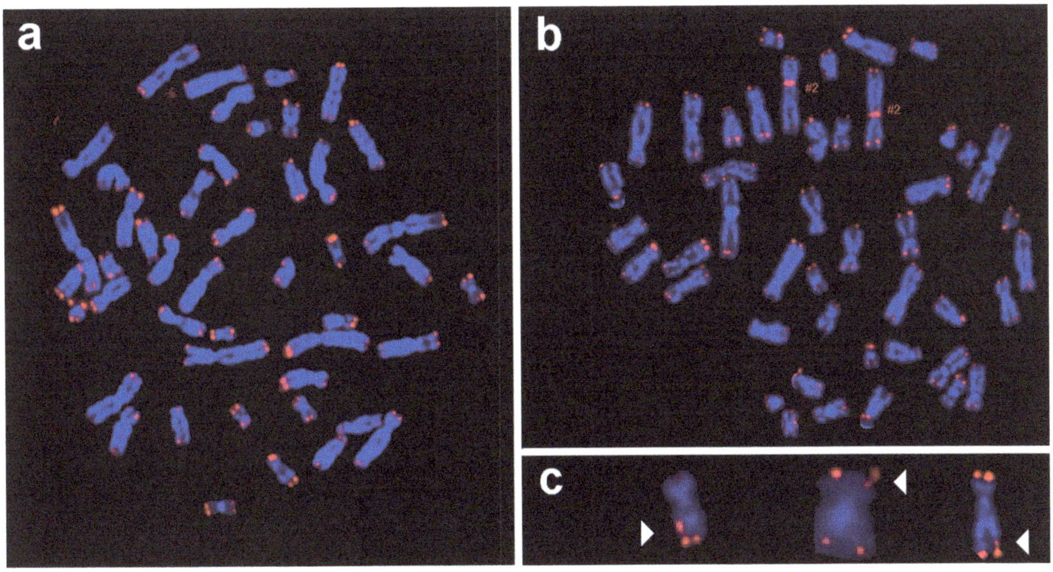

Fig. 1 (**a**) Metaphase spread hybridized with telomeric PNA probe suitable for analysis using (TFL-TELO) software. (**b**) Metaphase spread hybridized with telomeric PNA probe and reference centromeric PNA probe suitable for analysis using Q-FISH software. (**c**) Fragile telomeres and interstitial telomeric signals indicated by arrows (altered from [9])

4 Notes

1. If TFL-TELO software is used, results are expressed in telomere fluorescence units (TFU), with each unit corresponding to 1 kb of telomere repeats.
2. It likely has best resolution when used with centromere normalization. The use of a very long telomere length "high" calibration standard (if telomere length distribution is wide) for telomere intensity control can interfere with correlations and makes it difficult to obtain precise results. Centromere/telomere normalization method helps avoid these obstacles.

References

1. Hemann MT, Strong MA, Hao LY et al (2001) The shortest telomere, not average telomere length, is critical for cell viability and chromosome stability. Cell 107:67–77
2. Longhese M (2008) DNA damage response at functional and dysfunctional telomeres. Genes Dev 22:125–140
3. Collado M, Blasco M, Serrano M (2007) Cellular senescence in cancer and aging. Cell 130:223–233
4. Wong J, Collins K (2003) Telomere maintenance and disease. Lancet 362:983–988
5. Slijepcevic P (2001) Telomere length measurement by Q-FISH. Methods Cell Sci 23:17–22
6. Poon S, Lansdorp PM (2001) Quantitative fluorescence in-situ hybridization. Curr Protoc Cell Biol 18.4.1–18.4.21
7. Perner S, Bru S, Derlein C (2003) Quantifying telomere lengths of human individual chromosome arms by centromere-calibrated

fluorescence in situ hybridization and digital imaging. Am J Pathol 163:1751–1756

8. Aida J, Izumiyama-Shimomura N, Nakamura K et al (2014) Determination of telomere length by the quantitative fluorescence in situ hybridization (Q-FISH) method. Am J Anal Chem 5:775–783

9. Joksic I, Vujic D, Guc-Scekic M et al (2012) Dysfunctional telomeres in primary cells from Fanconi anemia FANCD2 patients. Genome Integr 3:6

RNA Imaging in Living Cells

Bin Ma and Naoko Tanese

Abstract

Insight into the dynamics of RNA biosynthesis, processing, and cellular activity is highly valuable because it will deepen our understanding of cell physiology and help explain how mRNA-misregulation contributes to the development of many diseases. To date, the study of mRNA inside cells is a challenge; many analytical approaches focus only on quantifying expression levels of transcripts and are not capable of reporting their intracellular locations, which have emerged as a critical determinant of RNA function. Similarly, techniques capable of probing RNA localization merely offer snapshots in fixed cells. Herein, we describe a protocol based on the bacteriophage MS2 genetic system for use in detecting mRNA in living cells.

Keywords mRNA, RNA imaging, MS2-GFP system, Live cell imaging

1 Introduction

RNA molecules are responsible for a wide range of functions in living cells, as carriers of genetic information, catalysts of biochemical reactions, adapter molecules in protein synthesis, and structural molecules in ribosomes. These functions are controlled by the expression level and stability, both temporally and spatially, of specific RNAs in a cell. Therefore, the study of complete spatial-temporal profiles of RNA synthesis (chapter by Tiphaine Aguirre-Lavin and Nathalie Beaujean "Three-Dimensional Immunofluorescence In Situ Hybridization in Preimplantation Mouse"; chapter by Bin Ma and Naoko Tanese "RNA-Directed FISH and Immunostaining"), processing, and transport is critical to our understanding of cell function and behavior under physiological and pathological conditions. In addition, RNA-based research has also found applications in modern drug development, molecular diagnostics, and forthcoming RNA-based therapeutic strategies [1].

A variety of approaches have been developed to identify length, sequence, and structure of RNA molecules and to measure gene expression levels within one cell population or between different cell populations. Several of the more widely adopted methods

Thomas Liehr (ed.), *Fluorescence In Situ Hybridization (FISH)*, Springer Protocols Handbooks,
DOI 10.1007/978-3-662-52959-1_15,

include the polymerase chain reaction (PCR), DNA microarrays, RNA pulldown assays, Northern hybridization (or Northern blotting), and fluorescence in situ hybridization (FISH). Although each of the aforementioned RNA detection methods can provide information concerning relative changes in gene expression for a population of cells, they generally do not provide a measure of dynamic RNA expression at the single-cell level. Therefore, imaging of RNA in living cells is crucial for obtaining detailed spatial and temporal data about RNA dynamics, including expression, localization, storage, and degradation of RNA molecules [2]. Live cell imaging not only eliminates the need to handle RNA but also provides an opportunity to analyze gene expression at the single-cell level without arduous fixation, permeabilization, and washing steps. However, successful execution of live cell imaging is a challenge, since RNA imaging probes must exhibit a high degree of specificity, sensitivity, and signal-to-background ratio, especially for low abundance transcripts.

Several classes of molecular probes have been developed for RNA detection in living cells, including fluorescently labeled antisense oligonucleotide probes [3], linear fluorescence resonance energy transfer (FRET) probes [4], molecular beacons [5], and fluorescent RNA-binding protein-based labeling techniques [2, 6–8]. The fluorescent RNA-binding protein-based systems are based on naturally occurring high-affinity interactions between specific RNA structural motifs and their binding proteins. Both the RNA and protein components in these systems are typically engineered to optimize for tighter binding. These systems include the MS2 system, λN22 system, BglG system, PP7, and eIF4A system, among others [2, 7, 8].

In this protocol, we describe the MS2-GFP system, a powerful and broadly used fluorescent RNA-binding protein-based labeling technique for RNA imaging in living cells developed by Robert Singer and colleagues [9, Fig. 1]. The MS2 RNA-binding protein forms the viral capsid of the bacteriophage MS2. During the late phase of phage replication, an MS2 dimer binds a 19-nucleotide stem-loop structure on the phage genomic RNA, which initiates phage capsid assembly and encapsidation of the genome [10]. This system has two features that enable the imaging of mRNA with a high signal-to-noise ratio in the cytoplasm. First, the high affinity of MS2 coat protein for MS2 stem loops results in specific recognition of reporter RNA. Second, the nuclear localization sequence (NLS) in MS2-GFP allows newly synthesized RNA to be bound by the reporter cotranscriptionally, leaving the unbound MS2-GFP in the nucleus [11]. We have used Lipofectamine for transfection of DNA constructs in cell culture. Other transfection methods can also be utilized for the delivery of constructs to the cells, as long as high transfection efficiency and cell survival can be achieved.

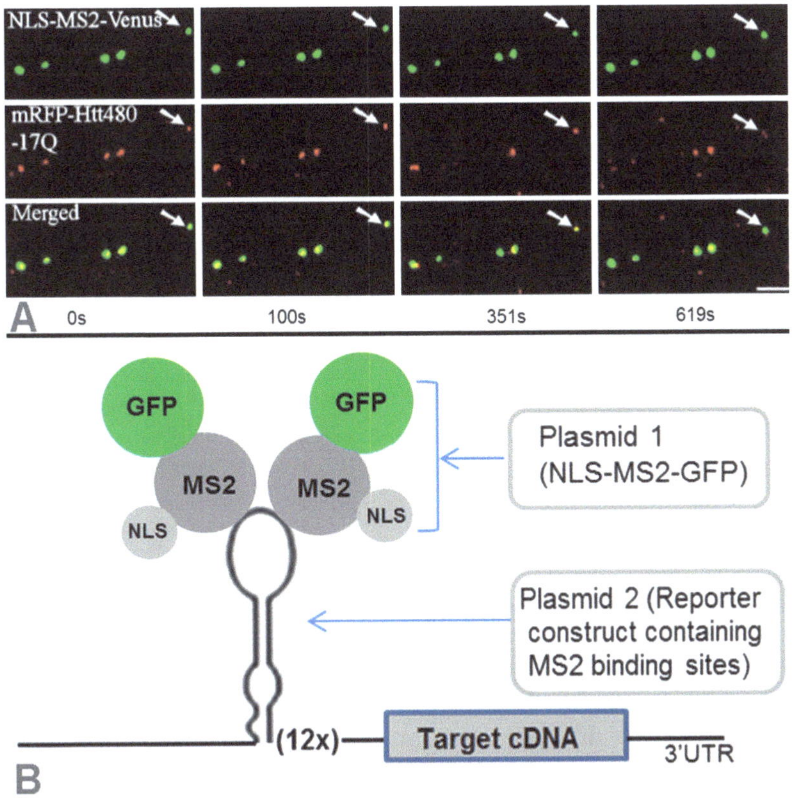

Fig. 1 (**a**) Co-trafficking of β-actin mRNA with huntingtin (Htt) protein in rat cortical neurons [7]. β-actin mRNA (detected by the MS2 system) is shown in *green* and transiently expressed mRFP-Htt480-17Q protein in *red*. Four cropped images from a time-lapse series captured over 652.5 s are shown. The left part of each image is the proximal part of the dendrite. The *arrows* indicate one retrogradely moving RNA granule with Htt in the dendrite. The distance that the granule traveled was 2.78 μm. Scale bar, 5.0 μm. (**b**) Principles of the MS2-GFP system [9]. The system is based on high-affinity interactions between MS2 dimer (from the MS2 bacteriophage) and a 19-nucleotide stem-loop structure derived from phage genomic RNA. Two plasmids (Plasmid 1 and Plasmid 2) are transfected into the cells for mRNA imaging in living cells. *NLS* nuclear localization signal; *3′UTR* 3′ untranslated region

2 Materials

- Plasmid 1: NLS-MS2-GFP (or other fluorescent proteins, such as Red fluorescent protein (RFP), Venus). Plasmids containing multiple MBS (MS2-binding sequence) cassettes and fluorescent protein-fused MS2 coat proteins are available from Addgene (www.addgene.org).
- Plasmid 2: MS2 reporter construct (also available from Addgene) with 12 × MS2 binding sequences [generated by first subcloning target cDNA into a CMV-based mammalian expression plasmid, followed by insertion of the 12 × MS2-binding sequence upstream of the target cDNA (*see* **Note 1**)].

- Opti-MEM Reduced Serum Medium (Cat. No.: 3198507, Thermo Fisher Scientific).
- Lipofectamine 2000 (Cat. No.:11668–027, Thermo Fisher Scientific).
- Cell culture media (prepared according to the need of cells examined) with/without serum and antibiotics.
- Falcon snap-cap round-bottom tubes (5 ml, Cat. No.: 352063, BD Bioscience).
- Nunc™ Lab-Tek™ II Chambered Coverglass (Cat. No.: 155379, Nalge Nunc International).
- Coverslips Thermanox for 24-well multi-dish 13-mm diameter (Cat. No.: 10252961, Fisher Scientific).
- 24-well flat-bottom Corning® Costar® cell culture plates (Cat. No.: CLS3524, Sigma).

3 Methods

3.1 Transfection

1. Grow cells on 13 mm diameter round glass coverslips in 24-well culture plates to 70–90 % confluence at the time of transfection.
2. Remove cell culture media, replace it with media without serum and antibiotics, and return the cells to the incubator.
3. Prepare two tubes for the cells in each well to be transfected. In Tube A, combine 0.5 μg Plasmid 1 and 0.5 μg Plasmid 2 in 50 μl Opti-MEM (*see* **Note 2**).
4. In Tube B, dilute 2 μl Lipofectamine 2000 reagents in 50 μl Opti-MEM and incubate for 5 min at room temperature (*see* **Note 3**).
5. Gently combine contents of tubes A and B and incubate for 20 min at room temperature.
6. Add the Lipofectamine/DNA mixture to each well and gently swirl to mix.
7. Incubate at 37 °C for 1 to 2 h and then replace the cell culture media with fresh media containing serum and antibiotics.
8. Check for transfection efficiency by viewing the cells by fluorescence microscopy. Allow the cells to grow for 1 to 2 days before examination.

3.2 Imaging

1. Set up the microscope. A confocal microscope should be adequately aligned and calibrated.
2. Pick up the coverslips with cells from culture plates and put them upside down in the chamered coverglass prefilled with medium. Image the cells by using an inverted confocal microscope equipped with a 37 °C microscope incubator (Pecon, Germany). 40× or 60× objective lens is used for imaging.

Laser exciting at 488 nm is used for the detection of GFP (or Venus) inside the cells (*see* **Note 4**).

4 Notes

1. An MS2-GFP detection system should be designed with several considerations. (a) The MS2-GFP construct must include appropriate 3′ untranslated region (UTR), which plays an essential role in mRNA localization by mediating the interaction of the mRNA with the cytoskeleton or RNA-binding proteins; (b) the MS2 repeats must be inserted in a carefully selected location—verification of the proper trafficking of mRNA by means of FISH is highly recommended in order to avoid potential problems; (c) the reporter plasmid must be sequenced, since the number of stem loops may decrease during plasmid preparation in *E. coli*; (d) the use of 24 × MBS (instead of 12 × MBS) for imaging of mRNA is not recommended because too many RNA-bound MS2-GFP can alter normal mRNA trafficking.
2. (a) Normally no more than 1 μg of plasmid DNA in total for each well is used for transfections; (b) achieving an optimal ratio of MS2-GFP to RNA might require titration of different amounts of each expression construct; (c) the localization of free MS2-GFP in the nucleus may partially obscure visualization of nuclear RNAs, although punctate intranuclear MS2-GFP signal can be eliminated by carefully titrating MS2-GFP levels.
3. (a) DNA-Lipofectamine 2000 complexes must be made in serum-free medium such as Opti-MEM Reduced Serum Medium and can be added directly to cells in culture medium in the absence of serum/antibiotics; (b) the amount of Lipofectamine 2000 Reagent required for successful transfection varies depending on the cell type. Transfection conditions should be optimized starting with the four concentrations of Lipofectamine 2000 Reagent recommended by the manufacturer; (c) Lipofectamine 3000 Reagent can be used to improve transfection efficiency and cell viability in a broad range of cell types.
4. The 488-nm laser (or other lasers) should be used at low power (e.g., as low as 5 %) to avoid photobleaching of GFP during the imaging process.

References

1. Penchovsky R (2013) Present and future RNA-based approaches to medical genomics. Int J Genomic Med 1:2
2. Urbanek MO, Galka-Marciniak P, Olejniczak M et al (2014) RNA imaging in living cells—methods and applications. RNA Biol 11:1083–1095
3. Dirks RW, Tanke HJ (2006) Advances in fluorescent tracking of nucleic acids in living cells. Biotechniques 40:489–496

4. Endoh T, Mie M, Kobatake E (2008) Direct detection of RNA transcription by FRET imaging using fluorescent protein probe. J Biotechnol 133:413–417
5. Monroy-Contreras R, Vaca L (2011) Molecular beacons: powerful tools for imaging RNA in living cells. J Nucleic Acids 2011:741723
6. Bann DV, Parent LJ (2012) Application of live-cell RNA imaging techniques to the study of retroviral RNA trafficking. Viruses 4:963–979
7. Ma B, Savas JN, Yu MS, Culver BP et al (2011) Huntingtin facilitates β-actin mRNA transport in dendrites of cortical neurons. Sci Rep 1:140
8. Ma B, Baj G, Tongiorgi E, Chao MV et al (2010) Localization of BDNF mRNA with the Huntington's disease protein in rat brain. Mol Neurodegener 5:22
9. Chao JA, Yoon YJ, Singer RH (2012) Imaging translation in single cells using fluorescent microscopy. Cold Spring Harb Perspect Biol 4. doi:10.1101/cshperspect.a012310
10. Ni CZ, Syed R, Kodandapani R, Wickersham J et al (1995) Crystal structure of the MS2 coat protein dimer: implications for RNA binding and virus assembly. Structure 3:255–263
11. Shav-Tal Y, Darzacq X, Shenoy SM et al (2004) Dynamics of single mRNPs in nuclei of living cells. Science 304:1797–1800

The Replicative Detargeting FISH (ReD-FISH) Technique in Studies of Telomere Replication

Nikolay Rubtsov and Natalya Zhdanova

Abstract

Based on the chromosome orientation-FISH (CO-FISH) procedure, the replicative detargeting FISH (ReD-FISH) was developed as a unique tool to study the replicative patterns of telomeres located on individual chromosomal arms. This method is also suited for examination of telomeres of species belonging to different classes of animals, for which well-proliferated cell cultures can be established and maintained. ReD-FISH is based on pulsed inclusion of labeled brominated nucleotides in replicating DNA, destruction of regions with brominated analogs, and standard FISH with single-strand-specific probe/probes.

Keywords Molecular cytogenetics, Telomere, DNA replication, Replicative detargeting FISH (ReD-FISH), Chromosome orientation-FISH (CO-FISH), PNA and LNA probes, BrdU/BrdC

1 Introduction

Replicative detargeting FISH (ReD-FISH) is based on the chromosome orientation-FISH (CO-FISH) procedure that differs from standard FISH by its ability to label only one of the two homologous DNA strands on a chromosome. To achieve this, the second DNA strand was fully replicated in the presence of 5-bromodeoxyuridine (BrdU) and destroyed by UV-light treatment followed by EXOIII enzyme incubation (EXOIII binds at the UV-induced gaps and degrades the strand). CO-FISH experiments require the development of special single-stranded probes [1], and in standard CO-FISH experiments, only head-to-tail-oriented repeated sequences are detected, although CO-FISH may also be applied to identify long unique DNA sequences. Furthermore, CO-FISH has already been used to localize repeats showing identical orientation. In case of irregular repeat arrangement, like in Alu repeats, CO-FISH cannot be applied, while it allows answering the question of repeat orientation in repeat clusters. Thus, telomeric repeats being arranged in clusters were intensively studied by CO-FISH. To distinguish this CO-FISH application from others

Thomas Liehr (ed.), *Fluorescence In Situ Hybridization (FISH)*, Springer Protocols Handbooks,
DOI 10.1007/978-3-662-52959-1_16,

(chapter by Gordana Joksic et al. "Telomere Length Measurement by FISH"), it was denominated replicative retargeting FISH (ReD-FISH). Telomeric DNA generally consists of repeats in head-to-tail orientation and are either G- or C-rich [2].

For ReD-FISH experiments, highly specific peptide-nucleic acid (PNA) and locked nucleic acid (LNA) telomeric probes are used instead of DNA probes as nucleic acid analogs (chapter by Thomas Liehr "Classification of FISH Probes"). In PNA, the sugar-phosphate backbone is replaced by a synthetic peptide formed from N-(2-amino-ethyl)-glycine monomers [3]. PNA oligonucleotides form hybrid duplexes with single-stranded DNA which show higher thermal stabilities than the duplex of normal DNA as PNA does not contain negatively charged phosphate groups [4, 5]. In LNA, the ribose ring is "locked" by a methylene bridge connecting the 2′-O-atom with the 4′-C-atom. Like PNA, LNA can harbor the identical four bases being present in DNA and RNA and thus are able to hybridize as PNA/DNA or LNA/DNA, however, with increased thermal stability. These properties make PNA and LNA to excellent suited probes for the detection of single-stranded sequences, combining their high binding affinities with short hybridization times [6].

ReD-FISH experiments may be performed using one of the two brominated analogs, BrdU or 5-bromo-2-deoxycytidine (BrdC) ([7]; chapter by Anja Weise and Thomas Liehr "Pre- and Postnatal Diagnostics and Research on Peripheral Blood, Bone Marrow, Chorion, Amniocytes, and Fibroblasts"). The scheme on how to perform ReD-FISH is shown in Fig. 1. In case of complete telomere replication during BrdU treatment, ReD-FISH signal should occur only on one of two sister chromatids. If telomere is replicated in another window of time, ReD-FISH signals will be visible on both sister chromatids.

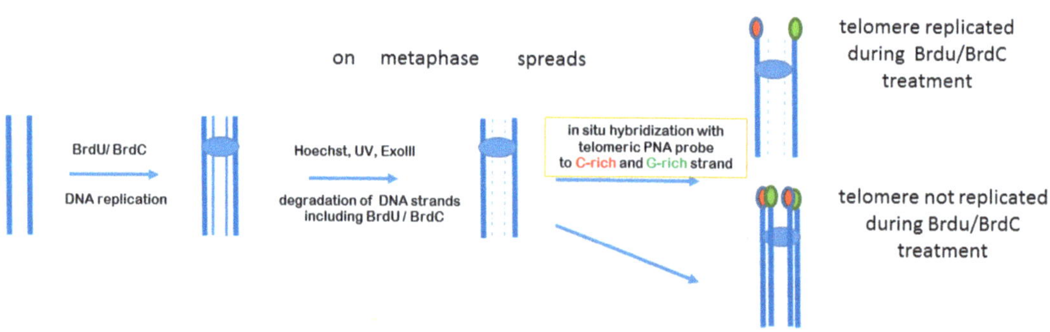

Fig. 1 Scheme of ReD-FISH protocol allowing to study telomeric replication. ReD-FISH procedure consists of three stages: (1) pulsed inclusion of brominated analogs (BrdU (5-Bromo-deoxyuridine) or BrdC (5-Bromo-deoxycytidine)) in replicating telomeres; (2) destruction of included the brominated analogs (DNA strands by UV treatment and enlargement of single-strand breaks by treatment with ExoIII (exonuclease III)); (3) application of FISH using single-stranded G-/C-rich probes for differential detection of telomeres not including BrdU or BrdC

1.1 Applications of the CO-FISH and ReD-FISH Techniques

CO-FISH or ReD-FISH are applied in specialized studies where desired information cannot be obtained by other methods. These techniques are widely applied in studies devoted to investigation of telomere organization and replication. Telomeres are nucleoprotein structures on the chromosome ends which preserve chromosomal ends from shortening and fusion, playing a role in genome stability. The G-rich lagging-strand of telomeric DNA contains the repeats TTAGGG and is finalized by a single-stranded 3′ overhang of 30–600 nucleotides in size [8, 9]. The latter introduces double-strand telomere region capping and *t*-loop formation [10, 11]. Together with the specialized telomere protein complex shelterin, *t*-loop preserves chromosomes termini from undesirable DNA recombination, degradation, and impairment by the cellular repair machinery. With rare exceptions, telomeres of mammalian chromosomes consist of a few kb to a few dozen kb of telomeric repeats. Human telomeres normally have a length of 5–15 kb [12]. For maintaining of telomere length, the dividing cells use two replicative mechanisms. To replicate telomere region, the conventional replicative machinery based on Okazaki fragments is used, while to resolve end replication problems and extend the overhang, the enzyme-RNA-protein complex called telomerase is used [13–15].

As abovementioned ReD-FISH is suited for studying simultaneously the replication patterns of telomeres on individual chromosomal arms. Its application in mammalian led to major advances in the knowledge of telomere replication. Earlier, only data on yeast telomeres replication was available, which is different to mammals, takes place in late S-phase, and consists of synchronous replication of the two telomeres of individual yeast chromosomes [16, 17]. The authors describing the ReD-FISH procedure initially [7] tested it on Indian muntjac (*Muntiacus muntjak*) fibroblasts cell lines, which were established from male and female individuals. *M. muntjak* cell lines were chosen as the karyotype of this species comprises six (male) and seven (female) well-identifiable chromosomes only [18]. ReD-FISH applied in Indian muntjac [7] and human [19] revealed some common features of the telomere replication program in mammals: (1) mammalian telomeres replicate throughout the entire S-phase of cell cycle, (2) each individual telomere has a preferential time window for replication, (3) replication timing of homologous telomeres in different cells of individuals and between individuals are more coordinated than the replication timing of telomeres located on p and q arms of the same chromosome, (4) at least in humans, late telomere replication is associated with satellite-like repeats and some other sequences of subtelomeric regions, and (5) late replicating telomeres often localized on the nuclear periphery, while the early replicating ones are preferentially located in the center of the nucleus; thus, subtelomeric regions appear to control the replication pattern of telomeres and position of chromosome arms in the nucleus [19].

By now, replication timing of individual telomeres can exclusively be studied by the ReD-FISH procedure. Besides CO-FISH can be applied to determine the level of telomeric sister chromatid exchanges (T-SCE), both in the first and in the second mitosis after treatment of cells with BrdU and BrdC [20, 21].

ReD-FISH is not informative for studies on exact origins of telomere replication, or how replication fork moves through the telomere. Therefore, another approach is called "single-molecule analysis of replicated DNA" (SMARD)-FISH [22]. Using SMARD-FISH, it has been shown that, as a rule, human telomeres are replicated from origins located in subtelomeric regions and replication forks usually move through the telomere to the chromosome end. Only in a few telomeres replication starts within clusters of telomere repeats. Overall, human telomeres demonstrate a chromosomal arm-specific replication, rather than a universal program [23]. It should be noted that telomeric DNA is an inconvenient substrate for replication due to loop structures; in front of the loops, the replicative fork pauses or stalls [15].

Here the protocol on how to perform ReD-FISH is outlined.

2 Materials

Apart from standard cell biological and molecular cytogenetic equipment, including standard solutions (e.g., ethanol, methanol, formamide, formaldehyde, xylene), no more specialized items are required. Solutions and equipment needed for ReD-FISH itself are listed below and in chapter by Thomas Liehr et al. "The Standard FISH Procedure".

- Phosphate-buffered saline tablets (PBS) (Cat. No.: P-4417, dissolve 1 tablet in 200 ml of deionized water to obtain 0.01 M phosphate buffer, 0.0027 M potassium chloride, and 0.137 M sodium chloride, pH 7.4, store at 0–5 °C, Sigma-Aldrich, Germany).
- Thymidine powder (Cat. No.: T1895, store at −20 °C, Sigma-Aldrich, Germany).

2.1 For Cell Culture Synchronization

- Eagle's minimum essential medium (EMEM) (Cat. No.:564-19C-10 L, US, Sigma-Aldrich, Germany).
- Fetal bovine serum (FBS) (Cat. No.: SH30071.03HI, heat inactivated, store at −20 °C, HyClone, USA).
- Aphidicolin (Cat. No.: A4487, store at −20 °C, soluble in ethanol up to 1 mg/ml, stable at least a week at 4 °C, Sigma-Aldrich, Germany).

2.2 For Pulsed Inclusion of Brominated Analogs in Newly Synthesized DNA, Metaphase Collection

- 5-Bromo-2′-deoxyuridine (5BrdU) (Cat. No.: B5002, store at −20 °C, soluble in EMEM, store at −20 °C, Sigma-Aldrich, Germany).
- Colcemid (Cat. No.: 477-30-5, Solution 10 μg/ml, Roche, Sigma-Aldrich, Germany).
- Dulbecco's phosphate-buffered saline (DPBS): 0.02 % KCl, 0.02 % KH_2PO4, 0.144 % $Na_2HPO_4 \times 2H_2O$, 0.8 % NaCl, (Cat. No.: D8537, liquid, Sigma-Aldrich, Germany).
- Trypsin-EDTA solution (Cat. No.: T4049, 0.25 %, sterile-filtered, BioReagent, 2.5 g porcine trypsin, and 0.2 g EDTA × 4Na per liter of Hanks' balanced salt solution with Phenol red, Sigma-Aldrich, Germany).

2.3 For the Slide Pretreatment for ReD-FISH Procedure

- Ribonuclease A from bovine pancreas (RNase A) (Cat. No.: R4642, in a solution of 50 % glycerol containing 10 mM Tris–HCl (pH 8.0), DNase free, Sigma-Aldrich, Germany).
- Pepsin (Cat. No.: P7000, powder, ≥250 units/mg, dissolve in 10 mM HCl, store at 2–8 °C, Sigma-Aldrich, Germany).
- Hoechst 33258 (Cat. No.: 94403, Solution, Store in dark at 0–5 °C, Sigma-Aldrich, Germany).
- Exonuclease III (Cat. No.: M0206S, From E. coli K-12, New England Biolabs, UK).

2.4 For FISH Procedure

- PNA telomere probe fluorescence for in situ hybridization (Cat. No.: 4337107, 1 vial 100 μl for 10 reactions, Applied Biosystems, probe on request). Other manufacturers of PNA probe: PNA Bio, Cambridge Research Biochemicals, and Panagene.

3 Methods

3.1 Cell Culture Synchronization

1. Cultivate the cell lines under optimal conditions. It is advisable to use EMEM supplemented by 10 % FBS as growth medium and incubate cells at 37 °C, 5 % CO_2, and 95 % humidified air (heating cabinet).
2. Seed cells in 25 cm^3 flask containing 5 ml growth medium; after 1 day of cultivation, the cells should form a monolayer. The number of flasks depends on the duration of S-phase of selected cell type. Usually eight to ten flasks are sufficient for ReD-FISH experiments (*see* **Note 1**). Below is given the scheme of synchronization by a double block of thymidine/aphidicolin, which we and others successfully used for mouse primary embryonic fibroblasts and other cells [23] (*see* **Note 2**).

3. Add thymidine at a final concentration of 2 mM to each flask and cultivate for 16 h in the heating cabinet.
4. Wash the cells three times with 5 ml PBS prewarmed at 37 °C.
5. Add 5 ml growth medium prewarmed at 37 °C and incubate for 8 h in the heating cabinet.
6. Add 1 μg/ml aphidicolin and incubate for 16 h in the heating cabinet.
7. Repeat step 4.
8. Add 5 ml growth medium; this is the start of the experiment.

3.2 Pulsed Inclusion of Brominated Analogs in Newly Synthesized DNA, Metaphase Collection

1. Treat cell cultures with 10 μM BrdU (or mixture of BrdU and 3.3 μM BrdC) for one hour. Add corresponding amounts of brominated analogues into the flasks after cell synchronization every hour on the hour up to the end of S-phase. If the duration of S-phase is 9 h, add bromated analogues through the first up to the ninth hour.
2. Wash the cells two times with 5 ml PBS prewarmed at 37 °C.
3. Add 5 ml growth medium containing 100 μM thymidine.
4. Repeat step 2 and add 5 ml of growth medium per flask.
5. If the length of G_2-phase in designated culture is 4 h, the cells in different flasks are cultivated 3 h after addition of growth media.
6. Add 20 ng/ml colcemid for one hour in each flask to accumulate metaphases.
7. Rinse with prewarmed DPBS.
8. Collect the cells by treatment with solution 0.25 % trypsin and 0.01 M EDTA at 37 °C and pellet cells by centrifugation at 1,200 rpm for 7 min at room temperature (RT).

3.3 Slide Preparation

1. Add freshly prepared prewarmed at 37 °C hypotonic solution 0.075 M (0.56 %) KCl in H_2O for 20 min (*see* **Note 3**), resuspend gently, and prefix cells by adding carefully 0.5 ml of fixative (= methanol – acetic acid 3:1; always make up fresh), and then mix gently.
2. Pellet by centrifugation at 1,200 rpm for 7 min at RT.
3. Remove supernatant and add fixative; store on ice for 20 min.
4. Repeat steps 2 and 3 three times.
5. Store suspension in fixative over night at −20 °C.
6. Change fixative by centrifugation and drop the cell suspension on cooled microscopic glass slides and air-dry.

3.4 Slide Pretreatment for ReD-FISH Procedure

1. Rinse slides in 1 × PBS.
2. Incubate in moist camera with 100 μg/ml RNase A for 1 h at 37 °C.
3. Wash in 1 × PBS three times for 5 min, each, at RT.
4. Post-fix with 3.7 % formaldehyde for 2 min at RT.
5. Repeat step 3.
6. Treat with 200 μl pepsin (40 units/ml in 10 mM HCl) under coverslip for 10 min at 37 °C.
7. Repeat step 3.
8. Dehydrate in an ethanol series (75 %, 85 %, 100 %, v/v in H_2O) for 5 min, each.
9. Stain with 0.5 μg/ml Hoechst 33258 in 2 × SSC for 15 min at RT in the dark.
10. Rinse in water.
11. Irradiate by 312 nm UV light for 15 min at RT at distance of 2 cm under previously heated-up lamp Vilber Lourmat – 115. M France. Do not allow slides to dry out during the procedure (*see* **Note 4**).
12. Incubate slides for 5 min in 3 U/μl exonuclease III in the recommended buffer (New England Biolabs) at RT. The enzyme extends UV-induced preferentially nicks in the BrdU-/BrdC-substituted DNA strands.
13. Rinse the slides in water and air-dry.

3.5 FISH Procedure

1. Denature slides in 70 % formamide, 30 % 2 × SSC prewarmed to 72 °C for 2 min.
2. Dehydrate in an ice-cold ethanol series (75 %, 85 %, and 100 %) for 2 min in each and air-dry.
3. Prepare 50 μl hybridization mixture per slide to be hybridized (0.2 μg/ml conjugated with fluorochrome PNA probe, 70 % (v/v) formamide, 12 mM Tris–HCl pH 7.2, 5 mM KCl, 1 mM $MgCl_2$).
4. Denature for 5 min at 72 °C.
5. Place probe-mix on ice for 5 min.
6. Add 50 μl of mix per denaturated slide.
7. Add 24 × 60 mm coverslip, place in a dark moist chamber for 3–6 h at 37 °C for hybridization (*see* **Note 5**).
8. Wash in 50 % formamide in 2 × SSC 3 times for 5 min at 37 °C.
9. Wash in 2 × SSC 3 times for 5 min at 37 °C.
10. Counterstain chromosomes by antifade with DAPI (Vectashield, Vector).

3.6 Analyses

1. For analysis, select only cells which passed through S-phase once. They differ from those who passed two times by the absence of sister chromatid exchanges. For visualization and processing of images, it is sufficient to have, e.g., an Axioplan 2 imaging microscope (Zeiss) equipped with a charge-coupled device (CCD) camera (CoolCube 1; Meta Systems), a Chroma filter set, and Isis 3 software (MetaSystems, Altlussheim, Germany).

4 Notes

1. If the duration of S-phase and G_2-phase of your cells is not known, before initiation of your experiment, these parameters must be determined. Therefore, FACS analyses may be used. For determination of S-phase length, collect cells each hour after release from aphidicolin block; when the amount of DNA reaches a plateau, the duration of S-phase is done; G_2-phase can be determined as the time between S-phase and mitosis.
2. Although Zou et al. [7] showed that similar replication timing of individual telomeres characterizes both synchronized and unsynchronized cells, synchronized cells are more easy to evaluate after ReD-FISH. The cells cultured in the presence of such inhibitors as hydroxyurea, aphidicolin, mimosine, or an excess of thymidine are arrested at the entrance to the S-phase, and upon release from the block, they synchronously pass through S-, G_2-, and M-phase. The inhibitory mechanism of hydroxyurea involves the enzyme ribonucleotide reductase, and thus, the synthesis of deoxyribonucleotides from ribonucleotides is inhibited. Aphidicolin is an inhibitor of DNA polymerase alpha. The plant amino acid mimosine inhibiting the initiation of DNA replication can also stop cells in S stage. Normal, nontumor cells can also be synchronized in G_0/G_1 by removal of growth factors, i.e., by "serum starvation" or by contact inhibition. Whichever inhibitor you choose, it should provide a high level of synchronization, cell survival, and stable results. These results are often achieved by using of twofold inhibition of DNA replication. For example, mouse primary embryonic fibroblasts can be successfully synchronized by a double block thymidine/aphidicolin [19].
3. For a better separation of sister chromatids, 1.6 % sodium citrate or a mixture of 1:1 of 0.075 M KCl and 1.6 % sodium citrate may be applied instead of 0.075 M KCl for hypotonic shock.
4. To induce single-strand breaks in BrdU-/BrdC-substituted DNA strands, it is also possible to irradiate at 365 nm

(Stratalinker 1800 UV Irradiator, Stratagene) under coverslip for 25–30 min.

5. Another recommended variant of hybridization procedure is to shortly denature the air-dried slides at 80 °C for 4 min in the presence of the strand-specific PNA telomeric probe conjugated with fluorochrome (Applied Biosystems) dissolved as 50 nM in 70 % formamide, 25 mM Tris (pH 7.4), and incubated in moist chamber for 2 h at RT [23].

Acknowledgments

This work was supported by the budget project VI.53.1.4. of the Federal Research Center Institute of Cytology and Genetics of SB RAS.

References

1. Goodwin E, Meyne J (1993) Strand-specific FISH reveals orientation of chromosome 18 alphoid DNA. Cytogenet Cell Genet 63:126–127
2. Bailey SM, Goodwin EH, Cornforth MN (2004) Strand-specific fluorescence in situ hybridization: the CO-FISH family. Cytogenet Genome Res 107:14–17
3. Nielsen PE, Egholm M, Berg RH et al (1991) Sequence-selective recognition of DNA by strand displacement with a thymine-substituted polyamide. Science 254:1497–1500
4. Tomac S, Sarkar M, Ratilainen T et al (1996) Ionic effects on the stability and conformation of peptide nucleic acid (PNA) complexes. J Am Chem Soc 118:5544–5552
5. Weiler J, Gausepohl H, Hauser N et al (1997) Hybridisation based DNA screening on peptide nucleic acid (PNA) oligomer arrays. Nucleic Acids Res 25:2792–2799
6. Silahtaroglu AN, Tommerup N, Vissing H (2003) FISHing with locked nucleic acids (LNA): evaluation of different LNA/DNA mixmers. Mol Cell Probes 17:165–169
7. Zou Y, Gryaznov SM, Shay JW et al (2004) Asynchronous replication timing of telomeres at opposite arms of mammalian chromosomes. Proc Natl Acad Sci U S A 101:12928–12933
8. Chai W, Shay JW, Wright WE (2005) Human telomeres maintain their overhang length at senescence. Mol Cell Biol 25:2158–2168
9. Makarov VL, Hirose Y, Langmore JP (1997) Long G tails at both ends of human chromosomes suggest a C strand degradation mechanism for telomere shortening. Cell 88:657–666
10. de Lange T (2005) Shelterin: the protein complex that shapes and safeguards human telomeres. Genes Dev 19:2100–2110
11. Palm W, de Lange T (2008) How shelterin protects mammalian telomeres. Annu Rev Genet 42:301–334
12. Riethman H (2008) Human telomere structure and biology. Annu Rev Genomics Hum Genet 9:1–19
13. Watson JD (1972) Origin of concatemeric T7 DNA. Nat New Biol 239:197–201
14. Olovnikov AM (1973) A theory of marginotomy: the incomplete copying of template margin in enzymic synthesis of polynucleotides and biological significance of the phenomenon. J Theor Biol 41:181–190
15. Gilson E, Géli V (2007) How telomeres are replicated. Nat Rev Mol Cell Biol 10:825–838
16. Wellinger RJ, Wolf AJ, Zakian VA (1993) Origin activation and formation of single-strand TG1-3 tails occur sequentially in late S phase on a yeast linear plasmid. Mol Cell Biol 13:4057–4065
17. Raghuraman MK, Winzeler EA, Collingwood D et al (2001) Replication dynamics of the yeast genome. Science 294:115–121
18. Lee C, Sasi R, Lin CC (1993) Interstitial localization of telomeric DNA sequences in the Indian muntjac chromosomes: further evidence for tandem chromosome fusions in the karyotypic evolution of the Asian muntjacs. Cytogenet Cell Genet 63:156–159

19. Arnoult N, Schluth-Bolard C, Letessier A et al (2010) Replication timing of human telomeres is chromosome arm-specific, influenced by subtelomeric structures and connected to nuclear localization. PLoS Genet 6, e1000920
20. Londoño-Vallejo JA, Der-Sarkissian H, Cazes L et al (2004) Alternative lengthening of telomeres is characterized by high rates of telomeric exchange. Cancer Res 64:2324–2327
21. Zhdanova NS, Draskovic I, Minina JM et al (2014) Recombinogenic telomeres in diploid *Sorex granarius* (Soricidae, Eulipotyphla) fibroblast cells. Mol Cell Biol 34:2786–2799
22. Sfeir A, Kosiyatrakul ST, Hockemeyer D et al (2009) Mammalian telomeres resemble fragile sites and require TRF1 for efficient replication. Cell 138:90–103
23. Drosopoulos WC, Kosiyatrakul ST, Yan Z et al (2012) Human telomeres replicate using chromosome-specific, rather than universal, replication programs. J Cell Biol 197:253–266

Part III

Material Suited for FISH Applications in Humans

Pre- and Postnatal Diagnostics and Research on Peripheral Blood, Bone Marrow, Chorion, Amniocytes, and Fibroblasts

Anja Weise and Thomas Liehr

Abstract

Peripheral blood, amniocytic fluid, chorion, and fibroblasts are the most frequently used tissues for chromosome studies. All four of them are relatively easy to obtain and can simply be brought into short-term culture, and metaphase spreads can be prepared within rather short time. Such metaphase spreads, as well as the huge amounts of previously superfluous interphase nuclei in cytogenetic preparations, are a material very well suited for FISH analyses. All kinds of FISH probes can be used to analyze metaphase spreads, while the interphase nuclei actually can be analyzed for routine purposes only by satellite and locus-specific probes. As quality and quantity of metaphases are important in research and routine diagnostics, the preparation procedure of the chromosomes is outlined in more detail. Moreover, a standard two-color FISH approach including the possibility to enhance the DAPI-banding pattern by BrdU during cultivation is described here.

Keywords Peripheral blood, Bone marrow, Amniocytic fluid cells, Chorionic villi tissue, Fibroblast cells, Chromosome preparation, Air-drying method, Cytogenetics, Metaphase, Interphase, Bromo-deoxyuridine (BrdU)

1 Introduction

In cytogenetic studies, the most frequently applied human tissues are peripheral blood, amniocytic fluid, chorion tissue, and skin fibroblasts. Peripheral blood lymphocytes lead this group, as they are easy to obtain as well as cultivated in a short-term culture, and metaphase spreads can be prepared in high quality within a short time [1]. Amniocytes and chorion cells are important in prenatal diagnostics [2], and 1–15 days of cell culture is needed to obtain metaphase cells in this case [3]. Prenatal and postnatal skin fibroblasts as well as those from the aborted tissue can be studied, too; here, the cultivation time is again in the range of 15 days. Metaphase spreads, as well as the huge amounts of previously superfluous interphase nuclei in cytogenetic preparations, are materials that are very well suited for fluorescence in situ hybridization (FISH)

Thomas Liehr (ed.), *Fluorescence In Situ Hybridization (FISH)*, Springer Protocols Handbooks,
DOI 10.1007/978-3-662-52959-1_17,

analyses. Each type of FISH probe can be used to analyze metaphase spreads, while the interphase nuclei can only be analyzed successfully by satellite and locus-specific probes and not (at least not in routine approaches) by whole or partial chromosome painting probes (Part IV). However, the quality of the metaphases is important in routine diagnostics, as well as in research approaches (chapter by Thomas Liehr and Anja Weise "Background"). Thus, the preparation procedure for the chromosomes themselves is discussed in more detail in the following passage, using the example of preparing peripheral blood lymphocytes. The currently used method of cell culture and chromosome preparation is a combination of many single-step procedures that were introduced into the current protocol by different researchers. An aliquot of heparinized peripheral blood is added to the cell culture medium, mixed with 10–20 % fetal calf serum, penicillin/streptomycin to avoid contamination from prokaryotic cell growth, and phytohemagglutinin. The latter is a mitogen that stimulates the in vitro growth of T lymphocytes [4, 5]. After 72 h of incubation at 37 °C/5 % CO_2, mimicking the conditions in human veins, the cells are harvested. Colcemid (diacetylmethylcolchicine), which acts as a mitotic spindle inhibitor, is added, and a cell cycle block is introduced between the metaphase and the anaphase [6, 7]. The "air-drying method" of chromosome preparation from Moorhead and Hsu [8] includes a hypotonic treatment with 0.075 M KCl [9, 10], a fixation step and several washing steps using Carnoy's fixative (methanol/glacial acetic acid 3:1), and, finally, the dropping of the suspension onto the slide surface.

Even though human chromosomes have been prepared like this for ~70 years, the structure and process of chromosome spreading were not completely understood for a long time. Recent studies revealed that the spreading is not based on a "bursting" process for the metaphase cell, as was suggested for years, but that fixed lymphocytes at the metaphase stage spread after being attached to the slide surface [11]. This surprisingly slow process is humidity dependent [12] and is driven by the evaporation of Carnoy's fixative; first methanol evaporates, followed by acetic acid. As acetic acid is hydrophilic, water is acquired from the atmosphere, and the chromosomes elongate due to a stretching or swelling process [11, 13]. A standard preparation procedure for human lymphocytes, chorion cells, amniocytes, and fibroblasts is outlined here. The flow chart in Fig. 1 shows which tissue is available at which time during the human life cycle and in which section the corresponding preparation protocol is discussed.

The here described methods are aimed at cultivating living cells from tissue specimens and stopping the cell cycle at the metaphase stage at the time of preparation, which results in metaphase spreads. As previously mentioned, interphase nuclei are also prepared as a by-product using this approach. Both metaphases and interphase

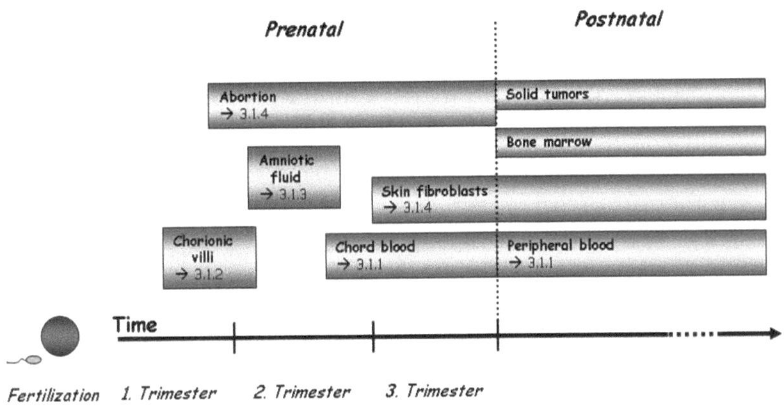

Fig. 1 Prenatal and postnatal specimens suited to chromosome preparations. The corresponding sections are mentioned in the figure

are easily accessible to applications of the FISH technique and analysis (for uncultivated amnion cells, see also chapter by Anja Weise et al. "FISH in Uncultivated Amniocytes"). Nevertheless, the quantity and quality of metaphase spreads can vary from patient to patient. In general, the longest and best-spread chromosomes are achieved from blood and fibroblasts, followed by amniocytes and long-term chorion cultures. The poorest quality is observed in short-term chorion or bone marrow preparations. One reason for this is that only spontaneous mitotic cells of the chorion and bone marrow are prepared with a resolution of less than 300 bands per haploid karyotype normally. However, this is sufficient to detect aneuploidies, sex chromosomes, and larger chromosomal rearrangements or to do FISH analysis.

When in situ culture of the chorion is performed, metaphase spreads appear in the mitotic active growth zone in the periphery of the cell clone. It can be assumed that all of the cells of a clone originate from a single cell, with a few exceptions that are described in more detail in [14]. Alternatively, there are also protocols that can be applied for amniocyte/chorion/fibroblast flask cultures in order to obtain cell pellets like those obtained after blood preparation [3]. The advantages of in situ cultures are their rapid proliferation and the more reliable analysis of mosaicism [14].

2 Materials

Apart from a standard cell biological and molecular cytogenetic equipment, including standard solutions (e.g., ethanol, methanol, formamide, formaldehyde, etc.), the following more specialized items are needed. The equipment needed for FISH is listed in chapter by Thomas Liehr et al. "The Standard FISH Procedure".

2.1 Chemicals, Other Materials, and Solutions to Be Prepared

- AmnioMAX® (Basal Medium Cat. No. 17 001–140; Supplement Cat. No. 12 556–031, Gibco, Grand Island, NY, USA)
- BrdU = bromodeoxyuridine (Cat. No. B5002–100 MG, Sigma, St. Louis, MO, USA)
- Carnoy's fixative, methanol/glacial acetic acid 3:1, freshly prepared, at 4 °C
- Chang C® medium (Cat. No. T 101–019, Laboserv, Giessen, Germany)
- Collagenase (Cat. No. C2674, Sigma)
- Colcemid (Cat. No. L 6221, Biochrom, Berlin, Germany)
- Fetal bovine serum (Cat. No. S 0113, Biochrom)
- Glacial acetic acid (Cat. No. 1.00063.2500, Merck, Darmstadt, Germany)
- Fixative for short-term chorion culture, methanol/glacial acetic acid 6:1, freshly prepared, at 4 °C
- Hyaluronidase (Cat. No. H 3506, Sigma)
- Hypotonic solution for amniocytes: 15 ml 0.33 % $MgCl_2$, 10 ml 1 % sodium citrate, and 1 ml hyaluronidase, freshly prepared
- Hypotonic solution for blood preparation: 0.075 M KCl, freshly prepared
- Hypotonic solution for chorion preparation: 1 % sodium citrate, freshly prepared
- KCl (Cat. No. 1.04936.1000, Merck)
- L-Glutamine (Cat. No. K 0282, Biochrom)
- Methanol (Cat. No. 1.060092500, Merck)
- $MgCl_2$ (Cat: No. 1.05833.0250, Merck)
- PBS (1 ×) (Cat. No. L 1825, Biochrom)
- Penicillin/streptomycin (Cat. No. A 2212, Seromed, Berlin, Germany)
- Phytohemagglutinin (Cat. No. M 5030, Biochrom)
- QuadriPERM® plate (960 77 308, Greiner, Frickenhausen, Germany)
- RPMI 1640 medium with glutamine (Cat. No. 72400–021, Gibco BRL)
- Sodium citrate (Cat. No. 1.06448.500, Merck)
- Trypsin (Cat. No. L 2143, Biochrom)
- 0.1 % trypsin-collagenase mix: 100 mg trypsin in 50 ml 1 × PBS mixed with 100 mg collagenase in 50 ml PBS, sterile filtered and can be aliquoted and stored at −20 °C

3 Methods

3.1 Preparation Metaphase Spreads Suited for FISH

In this section, the preparation of metaphase spreads derived from peripheral blood T lymphocytes, chorion cells, amniocytes, and fibroblasts is described (*see* **Note 1**). For more specific preparation protocols for bone marrow and solid tumors, the reader should refer to more specialized handbooks [3].

3.1.1 Peripheral Blood Lymphocytes: BrdU Treated

1. Add 1 ml of heparinized blood (*see* **Note 2**) to 9 ml of cell culture medium (e.g., RPMI 1640 medium/20 % fetal calf serum (300 μg ml^{-1})/ L-glutamine (1 U ml^{-1})/ penicillin (1 μg ml^{-1})/ streptomycin (1 μg ml^{-1})/0.1 ml phytohemagglutinin) (*see* **Note 3**), mix the suspension carefully, and incubate for 72 h at 37 °C/5 % CO_2. Steps 1–3 must be performed under sterile conditions (*see* **Note 4**).
2. 16 h before harvesting (*see* **Note 5**), add 180 μg of BrdU (*see* **Note 6**) to 10 ml of the cell culture.
3. 30 min before harvesting the cells, add 1 μg of colcemid, mix gently, and incubate at 37 °C/5 % CO_2.
4. Transfer the fluid into a 15 ml tube; sterile conditions are no longer necessary.
5. Centrifuge the solution at room temperature (RT) for 8 min at 1,000 rpm, and discard the supernatant by sucking it off carefully with a glass pipette (1 ml of supernatant is left in the tube to avoid loss of material).
6. For hypotonic treatment, the pellet is resuspended in 1 ml 0.075 M KCl (37 °C) and incubated at 37 °C for 20 min.
7. Slowly add 0.6 ml of Carnoy's fixative (4 °C) and mix the solution carefully.
8. Repeat step 5.
9. Resuspend the pellet in 10 ml of fixative (4 °C) and incubate at 4 °C for 20 min.
10. Repeat step 5.
11. Resuspend the pellet in 5 ml of fixative (4 °C) and repeat step 5.
12. Repeat step 11 twice.
13. Depending on the density of the suspension, the pellet is finally resuspended in 0.3–1 ml of fixative (remove as much of the suspension as necessary after step 12).
14. Drop 1–2 drops of the suspension onto a clean and humid slide using a glass pipette and let the slide dry at RT.
15. After incubation overnight at RT, the slides can be subjected to the pretreatment (see chapter by Thomas Liehr et al. "The Standard FISH Procedure"), stored dust-free at RT for several weeks, or frozen at −20 °C for several months.

3.1.2 Chorion Cells

1. Transfer (under sterile conditions) the specimen from the transport tube to a 60-mm Petri dish containing 5 ml of RPMI 1640 (*see* **Note 4**).
2. Wash the chorionic villi with fresh medium to remove blood cells.
3. Using an inverted microscope, carefully dissect and remove any remaining clots or decidua from the chorionic villi.
4. Allocate two equal portions of the tissue and apply one as described in Sect. "Chorion Cells: Short-Term Culture" and the other as described in Sect. "Chorion Cells: Long-Term Culture" (*see* **Notes** 7 and **8**).

Chorion Cells: Short-Term Culture

1. The cleaned specimen is cultured in 5 ml of AmnioMAX® medium for 24 h at 37 °C/5 % CO_2 in a culture flask (*see* **Note 3**). Steps 1–3 must be performed under sterile conditions (*see* **Note 4**).
2. 90 min before harvesting the cells, add 1 μg of colcemid to the culture flask, mix gently, and incubate at 37 °C/5 % CO_2.
3. Discard the medium by sucking it off carefully with a glass pipette, and add 1 % hypotonic solution for chorion preparation (=1 % sodium citrate) for 10 min.
4. Repeat step 3.
5. Add cold fixative (methanol/glacial acetic acid 6:1) to the hypotonic solution for 30 s.
6. Discard the fixative and repeat step 5.
7. Discard the fixative and add Carnoy's fixative (methanol/glacial acetic acid 3:1) for 30 s.
8. Discard Carnoy's fixative, repeat step 7, and maintain for 2 h at −20 °C.
9. Discard Carnoy's fixative and hydrate in an ethanol series starting with 100 %, 75 %, and 50 % for 2 min each.
10. Dry the cells and add some drops of 60 % glacial acid. After approximately 5 min, monitor the tissue under an inverse microscope; when single cells start to detach, take up the cell suspension using a glass pipette and move it onto cleaned, wet slides.
11. After incubation overnight at 70 °C or for 1 h at 90 °C, the slides can be subjected to the pretreatment (see chapter by Thomas Liehr et al. "The Standard FISH Procedure"), stored dust-free at RT for several weeks, or frozen at −20 °C for several months.

Chorion Cells: Long-Term Culture

1. The cleaned specimen is cut in a sterile Petri dish with sterile scissors (*see* **Note 4**). All of the steps described here must be performed under sterile conditions.
2. Transfer the material into a tube with 5 ml 0.1 % trypsin-collagenase mix and incubate for 45 min at 37 °C.
3. Centrifuge at 900 rpm for 8 min and discard the supernatant with a sterile pipette.
4. Resuspend in about 2 ml AmnioMAX® medium (*see* **Note 3**), and transfer to one or two culture flasks.
5. Incubate at 37 °C/5 % CO_2 for 10–14 days and check the cell proliferation. Exchange the medium after about 5 days.
6. Discard the medium with a pipette, add 2 ml of prewarmed trypsin solution, carefully shake the culture flask, and check for cell disaggregation from the bottom of the flask under an inverse microscope.
7. To perform in situ culture, now add cleaned and sterile glass slides to a quadriPERM® plate with about 5 ml AmnioMAX®, and carefully drop the trypsin-treated cells onto the glass slides.
8. Incubate at 37 °C/5 % CO_2 for 5–7 days and check the cell proliferation. The culture flask from step 7 can be refilled with medium and used for another preparation starting from step 5.
9. In situ chromosome preparation is described under Sect. 3.1.3, step 6.

3.1.3 In Situ Culture of Amniocytes

1. Centrifuge 15 ml amniotic fluid at 900 rpm for 8 min. Discard the supernatant and leave about 4 ml to resuspend the pellet. Sterile conditions must be maintained for steps 1–6 (*see* **Note 4**).
2. To perform in situ culture, put four cleaned and sterile glass slides into a quadriPERM® plate (*see* **Note 9**) with 5 ml Amnio-MAX® medium in each chamber (*see* **Note 3**).
3. Add the sediment from step 1 to the four slides and keep 1 ml for a backup flask culture with 4 ml Chang C® medium. Incubate both cultures at 37 °C/5 % CO_2 for 4–5 days; check the cell proliferation under an inverse microscope.
4. When the cells start to proliferate as clones, exchange the medium every 2 days and check the proliferation every day.
5. Start the preparation of the in situ slides when three large clones are visible per slide.
6. Once a slide has been selected for preparation, transfer the chamber medium and the slide with sterile tweezers to another quadriPERM® plate. Incubate the other slides at 37 °C/5 % CO_2 for later preparation.

7. 90 min before harvesting the cells, 0.25 μg of colcemid is added to the in situ culture and incubated at 37 °C/5 % CO_2.
8. Set up the preparation solutions, prewarm the hypotonic solution (37 °C), and cool Carnoy's fixative (−20 °C).
9. Discard the medium from the chamber with a pipette.
10. Carefully add 3 ml prewarmed hypotonic solution; incubate for 10–14 min at RT.
11. Carefully add 1.5 ml Carnoy's fixative (−20 °C) to the chamber with the hypotonic solution and incubate for 10–12 min at RT.
12. Discard fluid from the chamber, add 3 ml Carnoy's fixative (−20 °C), and incubate for 5–7 min at RT.
13. Repeat step 12 but incubate for 10–15 min at RT.
14. Pick up the slide with tweezers and carefully rinse the slide with Carnoy's fixative.
15. Dry the backside of the slide with tissue and air-dry the front by leaning the slide at a 45° angle on a wet tissue.
16. After incubation overnight at 70 °C or for 1 h at 90 °C, the slides can be subjected to pretreatment (see chapter by Thomas Liehr et al. "The Standard FISH Procedure"), stored dust-free at RT for several weeks, or frozen at −20 °C for several months.

3.1.4 In Situ Culture of Fibroblasts by Mechanical and Trypsin-Collagenase Treatment

1. Bloody material (abortions, skin, etc.) (*see* **Note 10**) should be washed with medium (*see* **Note 3**).
2. Cut the specimen in a sterile Petri dish with sterile scissors (*see* **Note 4**).
3. Allocate two portions of the tissue and use one as described in Sect. "Mechanical Treatment" and the other as described in Sect. "Trypsin-Collagenase Treatment".

Mechanical Treatment

1. Transfer one third of the material onto the bottom of a culture flask with sterile tweezers, and wait until the material slowly starts to dry out.
2. Add enough Chang C® medium to cover the bottom of the flask (*see* **Note 3**).
3. Incubate at 37 °C/5 % CO_2 for 14–18 days and check the cell proliferation. Exchange the medium after 5 days.
4. To perform in situ culture, now go to Sect. 3.1.2.2, step 6.

Trypsin-Collagenase Treatment

1. Transfer two thirds of the material into a tube with 2 ml 0.1 % trypsin-collagenase mix using sterile tweezers. Incubate at 37 °C/5 % CO_2 for a minimum of 45 min.
2. Centrifuge at 900 rpm for 8 min.
3. Discard the supernatant with a sterile pipette and resuspend with 2 ml AmnioMAX® medium (*see* **Note 3**).

4. Transfer the cells into two culture flasks and incubate at 37 °C/5 % CO_2 for 14–18 days and check the cell proliferation. Exchange the medium after 5 days.
5. To perform in situ culture, go to Sect. 3.1.2., Subsect. "Chorion Cells: Long-Term Culture", step 6.

3.2 Slide Pretreatment and Fluorescence In Situ Hybridization (FISH)

As described in chapter by Thomas Liehr et al. "The Standard FISH Procedure".

4 Notes

1. These procedures are very specialized and require some experience, especially when adopting laboratory-specific conditions, like optimizing the time taken for hypotonic treatment, the fixation steps, and the sensitive final drying of the metaphase spreads, which is dependent on the surrounding humidity and temperature, including the weather conditions or air condition. When working with different specimens at the same time, always make sure that they do not get mixed up. This is especially true when they are used for diagnostic purposes.
2. It is a well-known fact that EDTA- or sodium acetate-treated blood samples cannot be successfully cultured for chromosome preparation. In case EDTA blood is sent for cytogenetic analysis, urgently rewash the cells with medium and set up cell culture as described.
   For chromosome preparation from prenatal blood samples, always clarify the potential maternal cell contamination in the case of female or mixed female/male karyotype using methods like the Kleihauer-Betke test [15] or microsatellite [16] testing.
3. Before using them, check all culture media by eye for possible contamination (color changes, cloudiness).
4. Sterile cell culture conditions must be maintained when handling living cells.
5. In urgent cases, the blood culture can be stopped after 48 h. This normally results in less metaphase spread.
6. BrdU is a thymidine analog which is readily incorporated into chromosomes. Cultures containing BrdU should be protected from light, as this will result in chromosome breakage. Moreover, the yield of metaphase may be reduced compared to cultures without BrdU, as this chemical is cytotoxic.
7. If no result is obtained from the short-term culture and the long-term culture was found to be normal female, keep a

possible maternal cell contamination in mind. In this case, test the maternal DNA (from EDTA blood) and chorion culture DNA by, for example, microsatellite analysis [16].

8. Also, be aware of confined placental mosaicism, which is discussed in more detail in Gardner and Sutherland [14].
9. To minimize the chance of microbiological contamination for in situ cultures, it is generally advisable to work with a minimum of two different culture media (one for the in situ culture and one for the backup flask culture) and with two different media flasks for the four quadriPERM® chambers.
10. Fibroblasts can be cultivated from different sources, like abortion material or skin biopsies, and are sometimes also cultivated in amniotic fluid specimens.

References

1. Verma RS, Babu A (1994) Human chromosomes: principles and techniques, 2nd edn. McGraw-Hill, New York
2. Eisenberg B, Wapner RJ (2002) Clinical procedures in prenatal diagnosis. Best Pract Res Clin Obstet Gynaecol 16:611–627
3. Wegner RD (1999) Diagnostic cytogenetics. Springer, Berlin
4. Nowell PC (1960) Phytohemagglutinin: an initiator of mitosis in cultures of normal human leukocytes. Cancer Res 20:462–466
5. Moorhead PS, Nowell PC, Mellman WJ (1960) Chromosome preparations of leukocytes cultured from human peripheral blood. Exp Cell Res 20:613–616
6. Ford CE, Hamerton JL (1956) The chromosomes in man. Nature 178:1020–1030
7. Tjio JH, Levan A (1956) The chromosome number of man. Hereditas 42:6–9
8. Moorhed PS, Hsu TC (1956) Cytologic studies of HeLa, a strain of human cervical carcinoma. III. Durations and characteristics of the mitotic phases. J Natl Cancer Inst 16:1047–1066
9. Hsu TC (1952) Mammalian chromosomes in vitro: 1. Karyotype of man. J Hered 43:167–172
10. Hughes A (1952) Some effects of abnormal toxicity on dividing cells in chick tissue cultures. Q J Microsc Sci 93:207
11. Hliscs R, Mühlig P, Claussen U (1997) The spreading of metaphases is a slow process which leads to a stretching of chromosomes. Cytogenet Cell Genet 76:167–171
12. Spurbeck JL, Zinsmeister AR, Meyer KJ et al (1996) Dynamics of chromosome spreading. Am J Med Genet 61:387–393
13. Claussen U, Michel S, Mühlig P et al (2002) Demystifying chromosome preparation and the implications for the concept of chromosome condensation during mitosis. Cytogenet Genome Res 98:136–146
14. Gardner RJM, Sutherland GR (2004) Chromosome abnormalities and genetic counseling, 3rd edn. Oxford University Press, New York
15. Kohne E (2007) 50 years Kleihauer–Betke test. Klin Padiatr 219:252–253
16. Jarrett KL, Michaelis RC, Phelan MC et al (2001) Microsatellite analysis reveals a high incidence of maternal cell contamination in 46,XX products of conception consisting of villi or a combination of villi and membranous material. Am J Obstet Gynecol 185:198–203

Application of FISH to Previously GTG-Banded and/or Embedded Cytogenetic Slides

Thomas Liehr and Monika Ziegler

Abstract

GTG-banded slides are often considered to be worthless for further molecular cytogenetic analyses. In this chapter we describe a simple protocol on how to reactivate such GTG-banded slides for further use in molecular cytogenetics. The slides can be previously Eukitt® or "Canada Balsam" embedded or previously not covered with a coverslip at all. With this possibility at hand, archival material of rare cytogenetic aberrations as well as more profound studies on actual cases with limited access to material can be performed.

Keywords Archival cytogenetic slides, Reevaluation, GTG banded, Eukitt®, Canada Balsam, Xylene, Interphase, Metaphase

1 Introduction

Countless laboratories around the world produce and store cytogenetic slides (for cytogenetic preparations, see chapter by Anja Weise and Thomas Liehr "Pre- and Postnatal Diagnostics and Research on Peripheral Blood, Bone Marrow, Chorion, Amniocytes, and Fibroblasts"). Such slides are usually archived and then later discarded without being used again. However, these stored archival cytogenetic slides could be made available for interphase and/or metaphase FISH studies by a simple recovery technique; previously the usefulness of such slides for PCR application was also shown [1]. Unique clinical and tumorgenetic cases (chapter by Eyad Alhourani et al. "Tumorcytogenetic Diagnostics and Research on Blood and Bone Marrow Smears or Effusions") are here of particular interest, since they could be (re-)evaluated for their chromosomal anomalies. New questions could be raised with present knowledge as well as the probe sets that are currently available. Retrospective studies on diagnosis, prognosis, and therapeutic response are also possible [2]. Also in our laboratory, GTG-

Thomas Liehr (ed.), *Fluorescence In Situ Hybridization (FISH)*, Springer Protocols Handbooks,
DOI 10.1007/978-3-662-52959-1_18, © Springer-Verlag Berlin Heidelberg 2017

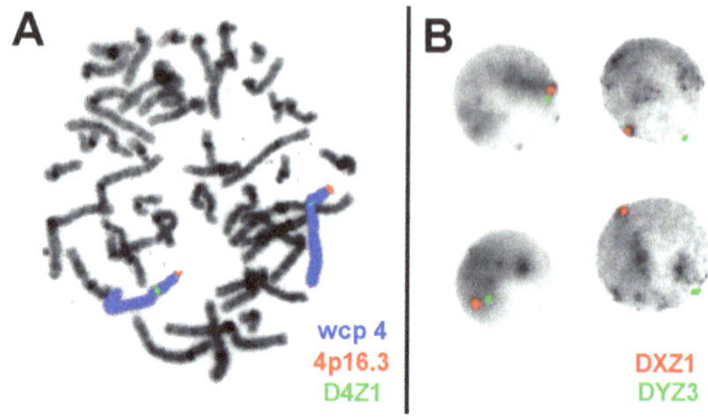

Fig. 1 FISH results obtained from previously GTG-banded slides. (**a**) Three-color FISH results using a whole-chromosome paint (wcp) for chromosome 4, a centromeric probe for chromosome 4 (D4Z1), and a probe for the subtelomeric region in 4p16.3 revealed a normal result in this case with suspicion of Wolf-Hirschhorn syndrome and limited amount of lymphocyte suspension. (**b**) Interphase FISH applied in this prenatal case revealed a normal male signal pattern using centromeric probes for X chromosome (DXZ1) and Y chromosome (DYZ3)

banded slides (embedded or not) are routinely used for subsequent FISH in case of limited available material (Fig. 1).

Here a rather uncomplicated technique for reactivating slides made permanent by embedding in Eukitt® or "Canada Balsam" [3] and also for using GTG-banded slides for FISH is reported. The presented technique consists of (1) a coverslip removal procedure, (2) a pretreatment step for excluding any vestiges of protein, and (3) a FISH protocol. The technique can be applied to slides obtained from human lymphocytes, bone marrow, or human solid tumors or cell lines.

2 Materials

Apart from standard cell biological and molecular cytogenetic equipment, including standard solutions (e.g., ethanol, methanol, formamide, formaldehyde, xylene, etc.), no more specialized items are required. Solutions and equipment needed for FISH itself are listed in the chapter by Thomas Liehr et al. ("The Standard FISH Procedure").

3 Methods

The following protocol includes environmental toxins (like xylene or formamide). Please ensure that these substances are collected and treated as hazardous waste after use. Perform all steps where xylene and formamide or formaldehyde are included under a hood.

3.1 Removal of the Coverslips from Embedded Slides

1. Put the slides embedded in Eukitt® or "Canada Balsam" in xylene (room temperature = RT).
2. Leave the slides in xylene until the coverslip floats off spontaneously, i.e., after 2–5 days. During this time, change the xylene everyday if the embedding medium becomes coated in a rather thick layer.
3. Check for coverslip removal 1 to 2 times a day; excessive xylene exposure may destroy metaphases and nuclei. Control and eventually support the detachment of the coverslip from the slide very carefully using forceps (*see* **Note 1**).
4. Air-dry the slides at RT immediately after coverslip removal.
5. Check the slide quality by microscopic inspection under phase contrast.
6. When the slides are providing metaphases and/or nuclei not (too) damaged, dry the slides at RT overnight.

3.2 Giemsa Destaining

1. Incubate previously embedded (from Sect. 3.1) as well as previously not embedded GTG-banded slides for 10–15 min in 100 ml 2-propanol (e.g., Roth 6752.1).
2. Wash 5 min in 1 × PBS.
3. Incubate 5–10 min in methanol/glacial acetic acid (3:1) and rehydrate in an ethanol series (100 %, 95 %, 70 %) and air-dry.
4. After at least 12 h at RT, dehydrate in an ethanol series (70 %, 95 %, 100 %) and control slide quality under phase contrast (*see* **Note 2**).

3.3 Slide Pretreatment and FISH

As described in chapter by Thomas Liehr et al. ("The Standard FISH Procedure").

Variations from standard protocol:

- Avoid pepsin pretreatment and start with PBS immersion of the slides.
- Denature the slides for 5–15 min at 75 °C (*see* **Note 3**).
- Increase the probe concentration by 1.5–5 × compared to fresh slides.

3.4 Evaluation

For metaphase FISH evaluation, weaker signals than normally accustomed to may be available; this might be compensated by longer exposure times of the image acquisition program. In interphase FISH, a semistatistical evaluation of the FISH results has to be performed (chapter by Thomas Liehr and Sven Hauke "Inter phase FISH in Diagnostics"; chapter by Ivan Iourov et al. "Inter phase FISH for Detection of Chromosomal Mosaicism"). At least 50, and optimally 100–300, nuclei are evaluated for the number of specific FISH signals per cell. If an automated evaluation system is available, this number can easily be expanded up to 1,000 or more cells.

The protocol presented here was successfully used in examinations of archival embedded cytogenetic slides prepared, e.g., from breast carcinomas (cancerous effusions), rectal tumors [3], or bone marrow samples of human leukemia. We mainly used interphase FISH, although in principle metaphase FISH could also be used, as shown previously [3]. Also an example on how metaphase FISH results can look like on previously GTG-banded but not embedded slides is presented in Fig. 1.

4 Notes

1. The percentage of slides suitable for FISH after the removal of the coverslip is dependent on many factors, such as the type of embedding medium used, the amount of embedding medium applied onto the individual slide, the age of the slide, and storage conditions. In our hands, 50–80 % of slides embedded in Eukitt® were recoverable, as were 70–100 % of those embedded in "Canada Balsam."
2. This step provides stability to chromosome and nucleus shape.
3. Due to the fact that DNA in archival tissues has undergone some fixation steps and has been stored for up to several years, a prolonged denaturation time appears useful. Moreover, in other FISH protocols with denaturation times of only 2–5 min, the maintenance of the available metaphase chromosomes is the main aspect; however, this is of no significance if only interphase FISH is foreseen.

References

1. Sukpanichnant S, Vnencak-Jones CL, McCurley TL (1993) Detection of clonal immunoglobulin heavy chain gene rearrangements by polymerase chain reaction in scrapings from archival hematoxylin and eosin-stained histologic sections: implications for molecular genetic studies of focal pathologic lesions. Diagn Mol Pathol 2:168–176
2. Schad CR, Dewald GW (1995) Building a new clinical test for FISH. Appl Cytogenet 21:1–14
3. Gebhart E, Hofbeck P, Hofmann Y et al (1996) Recovering of archival tumor cytogenetic slides for two-colour-interphase FISH. Appl Cytogenet 22:146–148

FISH in Uncultivated Amniocytes

Anja Weise, Monika Ziegler, and Thomas Liehr

Abstract

In prenatal diagnostics, interphase FISH (iFISH) may be applied for the detection of certain numeric chromosomal aneuploidies. iFISH in uncultivated amniocytes can be performed from 12 weeks of gestation to the third trimester. Alternatively, the described aneuploidy screening set here is also suitable on short-term or uncultivated chorionic villus sampling (CVS) without metaphases as a fast screening test for the most common aneuploidies. Also this test helps to distinguish between culture-induced or real fetal triploidy or tetraploidy (mosaicism), which is found from time to time in the parallel cell culture; in this setting native cells are evaluated and they reflect the real fetal status. Commercially available chromosome-specific DNA probes for chromosomes 13, 18, 21, X, and Y are routinely applied for this. Thus, trisomies of chromosomes 13, 18, and 21, gonosomal aberrations, as well as tri- or tetraploidy can be identified. False-positive results are possible, but rare.

Keywords Prenatal diagnostics, Interphase FISH (iFISH), Second trimester, Trisomy 13, Trisomy 18, Trisomy 21, Turner syndrome, Klinefelter syndrome, Triple-X-syndrome, Triploidy, Tetraploidy, Heteromorphism

1 Introduction

Obtaining a fetal karyotype from amniotic fluid takes nowadays ~12.5 days ([1], chapter by Anja Weise and Thomas Liehr "Pre- and Postnatal Diagnostics and Research on Peripheral Blood, Bone Marrow, Chorion, Amniocytes, and Fibroblasts"), while it was 3–4 weeks in the 1980s [2]. However, 2 weeks of waiting cause psychological distress for pregnant women [2], and this was one of the main reasons for the introduction of timesaving molecular (cytogenetic) methods for prenatal diagnosis. The most common chromosome disorders of the second trimester, such as trisomy 13, trisomy 18, and trisomy 21, gonosomal abnormalities (like monosomy X, trisomy X, trisomy XXY, trisomy XYY, etc.), and triploidy and tetraploidy, are tested.

Quantitative fluorescence polymerase chain reaction (QF-PCR) or FISH can be used and lead to informative results in less than 24 h [2]. Most often applied in FISH is the only FDA-

Thomas Liehr (ed.), *Fluorescence In Situ Hybridization (FISH)*, Springer Protocols Handbooks,
DOI 10.1007/978-3-662-52959-1_19, © Springer-Verlag Berlin Heidelberg 2017

approved FISH test for rapid aneuploidy screening in uncultivated amniotic cells, the AneuVysion Kit (commercially available at Abbott/Vysis, Downers Grove, IL; chapter by Thomas Liehr and Sven Hauke "Interphase FISH in Diagnostics"). The latter consists of three alpha-satellite DNA probes for chromosomes X, Y, and 18 (cep X, cep Y, and cep 18) and two locus-specific probes for 13q14 (LSI 13) and 21q22.13 ~ 22.2 (LSI 21). According to manufacturer's instructions, the three centromeric and the two locus-specific probes are applied in two different hybridizations. However, a "normal" result of the rapid aneuploidy screening cannot exclude the following possible conditions in a fetus:

- Balanced structural rearrangements
- Unbalanced structural rearrangements
- Numerical aberrations beside 13, 18, 21, X, or Y
- Mosaicism of any numerical aberration (also maybe below cut-off level)
- Discrimination between free trisomy vs. translocation trisomy and partial or full aneuploidy
- Microdeletions/microduplications
- Uniparental disomy
- Mutations detectable only by molecular genetics

Thus, the rapid aneuploidy screening must be accompanied by banding cytogenetic analyses in Germany. Only if sonographic signs and result of the rapid aneuploidy screening fit together an abortion can be done if requested by the parents before a fetal karyotype is available. This carefulness is also supported by the fact that among the ~2,200 cases studied by this approach in our laboratory, there were five false-positive or false-negative cases after rapid aneuploidy screening test due to centromeric heteromorphisms, the presence of a small supernumerary marker chromosome, or a dicentric marker chromosome not detectable in the interphase [3]. Similar limitations were also reported by others [4].

As it is necessary to introduce laboratory-specific cutoff levels for each iFISH test, this is especially important in the case of this sensitive prenatal screening test (for our cutoff levels, see [3]). Interestingly, in principle this test is also suited to distinguish free trisomy 21 from translocation trisomy t(21;21) [5]. The rapid aneuploidy screening test is routinely applied in amniocytes [3, 6] and may also be used in samples of chorionic villi [7].

As previously suggested, all observed pitfalls in prenatal aneuploidy screening by FISH could easily be obviated by the exclusive use of locus-specific probes [8]. The advantage of such probes, such as probes LSI 13 and LSI 21, has been proven, e.g., by the fact that the LSI 13 probe is suitable for the detection of free trisomy 13 and small duplications of the tested region [5, 8]. Interestingly, a similar probe set is applied in the QF-PCR test [2, 9]; however, the latter is less or not suited at all to detect reliably monosomy X or tri-/tetraploids. Besides, prenatal aneuploidy screening by FISH helps to distinguish between culture-induced and real fetal triploidy or tetraploidy (mosaicism), which is found from time to time in the parallel cell culture; in this setting native cells are evaluated and reflect the real fetal status (own unpublished data). In urgent cases the whole FISH procedure may be finished within one day (own unpublished data).

As stated by us before: "In summary, the rapid prenatal aneuploidy test is, if applied with the necessary caution and with careful explanation of its possibilities and limitations, a powerful tool for the clinician in the care of pregnant women" [3].

2 Materials

Apart from a standard cell biological and molecular cytogenetic equipment, including standard solutions (e.g., ethanol, methanol, formamide, formaldehyde, etc.), the following more specialized items are needed. The equipment and consumables needed for FISH are listed in chapter by Thomas Liehr et al. ("The Standard FISH Procedure").

2.1 Chemicals

- AneuVysionTM EC DNA Kit (300 μl 13/21; 300 μl 18/X/Y) including NP 40, Abbott Molecular, Order No.: 5J 3710 Abbott; aliquots of Vysis LSI (13 Spectrum Green/21 Spectrum Orange) and Vysis CEP (18 Spectrum Aqua/X Spectrum Green/Y Spectrum Orange); use 10 μl per slide, subsequently refreeze, and, when required, thaw it.

3 Methods

3.1 Preparation of Amnion Cells

1. 2–3 ml of amniotic fluid is centrifuged (1,000–1,500 rpm, 5 min).
2. The pellet is suspended in 3 ml trypsin/EDTA and incubated for 15 min at 37 °C.
3. Stop trypsinization by addition of 5 ml 1 × PBS and 1 ml fetal calf serum at 37 °C.
4. After centrifugation, the pellet is resuspended in 5 ml 0.075 M KCl and incubated at 37 °C for 20 min.
5. Add slowly 2 ml of Carnoy fixative (methanol/acetic acid 3:1), centrifuge (1000–1500 rpm, 5 min), and reduce supernatant to ~400 μl.
6. Resuspend and add ~2.6 ml of Carnoy fixative and incubate at −20 °C for 5 min.

7. Centrifuge (1,000–1,500 rpm, 5 min), discard the supernatant, and dilute the cells in 100 μl of the remaining supernatant.
8. Place the fluid on two spots on a dry slide and air-dry to increase the amount of cells on the spots to be hybridized with the two different probe mixes and later evaluated.

3.2 Slide Pretreatment and Fluorescence In Situ Hybridization (FISH)

As described in chapter by Thomas Liehr et al. ("The Standard FISH Procedure").

3.3 Evaluation

1. 50 interphase nuclei per case and probe combination are evaluated under the fluorescence microscope.
2. Thus a semistatistical evaluation is done, and results are compared to the cutoff levels ([3], chapter by Ivan Iourov et al. "Interphase FISH for Detection of Chromosomal Mosaicism").

References

1. Held KR (2003) QS Zytogenetik Bericht 2002/2003. Med Genet 15:420–421
2. Hulten MA, Dhanjal S, Pertl B (2003) Rapid and simple prenatal diagnosis of common chromosome disorders: advantages and disadvantages of the molecular methods FISH and QF-PCR. Reproduction 126:279–297
3. Liehr T, Ziegler M (2005) Rapid prenatal diagnostics in the interphase nucleus: procedure and cut-off rates. J Histochem Cytochem 53:289–291
4. Braha E, Martiniuc V, Panzaru M et al (2013) Prenatal diagnosis of gonosomal anomalies: limitations of the FISH method and genetic counseling difficulties in 15 cases. Rev Med Chir Soc Med Nat Iasi 117:450–456
5. Liehr T, Starke H, Beensen V et al (1999) Translocation trisomy dup(21q) and free trisomy 21 can be distinguished by interphase-FISH. Int J Mol Med 3:11–14
6. Eiben B, Trawicki W, Hammans W et al (1999) Rapid prenatal diagnosis of aneuploidies in uncultured amniocytes by fluorescence in situ hybridization. Evaluation of 3,000 cases. Fetal Diagn Ther 14:193–197
7. Goumy C, Bonnet-Dupeyron MN, Cherasse Y et al (2004) Chorionic villus sampling (CVS) and fluorescence in situ hybridization (FISH) for a rapid first-trimester prenatal diagnosis. Prenat Diagn 24:249–256
8. Liehr T, Schreyer I, Neumann A et al (2002) Two more possible pitfalls of rapid prenatal diagnostics using interphase nuclei. Prenat Diagn 22:497–499
9. Cirigliano V, Voglino G, Cañadas MP et al (2004) Rapid prenatal diagnosis of common chromosome aneuploidies by QF-PCR. Assessment on 18,000 consecutive clinical samples. Mol Hum Reprod 10:839–846

Tumorcytogenetic Diagnostics and Research on Blood and Bone Marrow Smears or Effusions

Eyad Alhourani, Moneeb A.K. Othman, Shaymaa S. Hussein Azawi, and Thomas Liehr

Abstract

Blood smears, bone marrow smears, and effusions can be very helpful for quick molecular cytogenetic diagnostics in leukemia, lymphoma, or even solid tumors. Such samples can be studied by suited interphase FISH probes or probe sets for diagnostics, prognostics, therapeutic decisions, follow-up and control of effects of medication, or success of bone marrow transplantation. Here two protocols are provided on how to prepare smears or cell suspension in methanol/acetic acid (3:1) for subsequent FISH analyses.

Keywords Tumorcytogenetics, Blood smears, Bone marrow smears, Effusions, Interphase FISH, Leukemia, Lymphoma, Solid tumors

1 Introduction

Molecular cytogenetic tumor diagnostics and research are based either on primary samples derived from solid tumors (surgery or effusion) or from leukemia (blood or bone marrow); besides there are lymphomas where samples may be derived from surgery of the lymphoma tissue, from effusions, or from bone marrow aspiration [1]. While from solid tumors (surgery or effusions), metaphases can hardly be obtained, the mitotic index in leukemia and lymphoma samples derived from the blood or bone marrow is normally much better.

Cytogenetic diagnostics of leukemia are currently routinely performed by conventional banding analysis as well as by FISH (chapter by Anja Weise and Thomas Liehr "Pre- and Postnatal Diagnostics and Research on Peripheral Blood, Bone Marrow, Chorion, Amniocytes, and Fibroblasts"). A lack of metaphase spreads in cytogenetic preparations can in many cases be compensated for by using interphase cytogenetics applying specific probes, such as commercially available ones or homemade ones (chapter by

Thomas Liehr (ed.), *Fluorescence In Situ Hybridization (FISH)*, Springer Protocols Handbooks,
DOI 10.1007/978-3-662-52959-1_20,

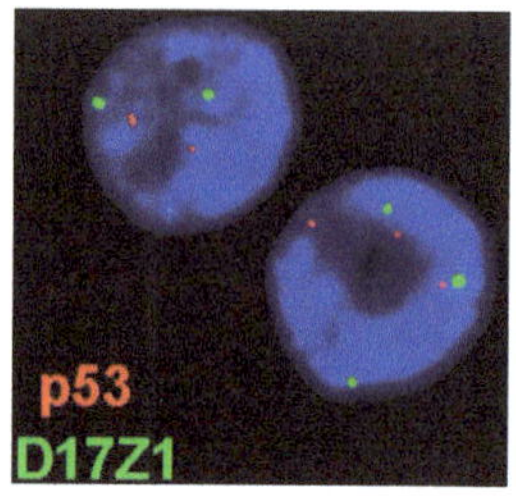

Fig. 1 Bone marrow smear of a chronic lymphocytic leukemia case with an acquired mosaic trisomy 17 in the malignant cells as detected by a probe for *TP53* (p53) and centromere of chromosome 17 (D17Z1). *Left depicted cell* shows two signals, each, being most likely a normal lymphocyte, while the *right one*, most likely due to a trisomy 17, has three specific signals, each

Anja Weise et al. "FISH in Uncultivated Amniocytes"; chapter by Thomas Liehr "Commercial FISH Probes"; chapter by Thomas Liehr "Homemade Locus-Specific FISH Probes: Bacterial Artificial Chromosomes"; chapter by Thomas Liehr and Sven Hauke "Inter phase FISH in Diagnostics"; chapter by Ivan Iourov et al. "Inter phase FISH for Detection of Chromosomal Mosaicism"). Special approaches, i.e., the monitoring of bone marrow transplants, can be performed on interphase nuclei without any need for time-consuming cytogenetic preparation [2]. Moreover, for blood or bone marrow samples of leukemia, it is worth remembering that the percentage of subpopulations (even after a 24-h short-term culture) does not necessarily represent the in vivo situation. Thus, interphase cell analysis of uncultured bone marrow aspirate yields usually more reliable information for the clinician upon repeated analysis of bone marrow aspirates during the course of a leukemic process [3] (Fig. 1) or when monitoring reverse sex bone marrow transplantation; however there are some exceptions form that rule, like, e.g., in chronic lymphocytic leukemia, where addition of specific substances can promote specifically the growth of malignant cells [4]. Interphase nuclei from peripheral blood or bone marrow can be prepared directly [5]—or more rapidly—by generating a blood or bone marrow smear on a slide [6]. The latter technique has the further advantages that (1) the cellular structure can be maintained and (2) that tumor cells can be distinguished from non-tumorous ones [7].

For solid tumors either sections from the malignant tissue itself (chapter by Thomas Liehr "Characterization of Archived Formalin-Fixed/Paraffin-Embedded or Cryofixed Tissue, Including Nucleus Extraction") or cells derived from effusions are studied. Here the corresponding protocols are presented for interphase FISH on blood smears, bone marrow smears, and effusions.

2 Materials

Apart from the standard cell biological and molecular cytogenetic equipment, including standard solutions (e.g., ethanol, methanol, formamide, formaldehyde, etc.), no more specialized items are required. The equipment needed for FISH itself is listed in chapter by Thomas Liehr et al. ("The Standard FISH Procedure").

The following protocol comprises environmental toxins (like sodium thiocyanate, formamide, and formaldehyde). Please ensure that these substances are collected after use and treated as hazardous waste.

2.1 Special Solution to Be Prepared

- 1 M sodium thiocyanate: dissolve 0.162 g sodium thiocyanate (Cat. No.: 71938, Sigma–Aldrich, St. Louis, MO, USA) in 2 ml filtered double-distilled water; make fresh as required.

3 Methods

3.1 Smear Preparation

1. In case of effusion, pellet the cells to 0.5 to 2 ml acc. to cell concentration in the sample. Blood or bone marrow sample (*see* **Note 1**) can be immediately processed in step 2.
2. Drop 100–200 μl of the sample fluid onto one end of a clean and dry slide.
3. Spread the fluid over the whole slide surface using the small edge of a 24 × 60 mm coverslip. The edge of the coverslip is dipped into the fluid and moved slowly—once only—over the slide and without touching the slide surface, since this could disrupt the cells (*see* **Note 2**).
4. Let the fluid dry out at room temperature (RT) for approx. 12 h, before performing the slide pretreatment (*see* **Note 3**).

3.2 Pretreatment of the Smears

Three different variants for this step are presented and must be chosen according to the studied question.

Variant 1

1. Do not do any pretreatment if the cellular structure needs to be maintained and tumor cells have to be distinguished from non-tumorous ones!
2. Perform a 10 min incubation of the slides in 2 × SSC at RT, prior to FISH.

Variant 2

As described in chapter by Thomas Liehr et al. ("The Standard FISH Procedure") (*see* **Note 4**).

Variant 3

1. Incubate the slide in a Coplin jar with fixative (= methanol/glacial acetic acid 3:1) for 10 min at RT and let the slide dry by air after this time.
2. Add 200 μl of 1 M sodium thiocyanate solution to the slide (*see* **Note 5**), cover with a 24 × 60 mm coverslip, and incubate at 37 °C for 40 min in a humid chamber.
3. Wash the slide in 1 × PBS in a 100 ml Coplin jar (RT) for 5 min with agitation, dehydrate in an ethanol series (70 %, 95 %, 100 %), and air-dry.

3.3 Direct Blood, Bone Marrow, or Effusion Preparation

Instead of preparing smears from the blood, bone marrow, or effusions, this material may be worked up as described below; this leads to interphase cell suspension in methanol/glacial acetic acid (3:1) which can be stored for years at −20 °C [8]:

1. Add 1 ml of blood, bone marrow, or concentrated effusion to 9 ml of cell culture medium (e.g., RPMI 1640 medium) (*see* **Note 6**), and mix the solution carefully (*see* **Note 7**).
2. Centrifuge the solution in a 15 ml tube at RT for 8 min at 1,000 rpm, and discard the supernatant by sucking it off carefully with a glass pipette (1 ml of supernatant is left in the tube to avoid loss of material).
3. For hypotonic treatment, the pellet is resuspended in 10 ml 0.4 % KCl (37 °C) and incubated at 37 °C for 20 min.
4. Slowly add 0.6 ml of fixative (= methanol/glacial acetic acid 3:1) (4 °C) and mix the solution carefully.
5. Repeat step 2.
6. Resuspend the pellet in 10 ml of fixative (4 °C) and incubate at 4 °C for 20 min.
7. Repeat step 2.
8. Resuspend the pellet in 5 ml of fixative (4 °C) and repeat step 2.
9. Repeat step 8 twice.
10. The pellet is finally resuspended in 0.3–1 ml (depending on the density of the suspension) of fixative (suck off as much of the suspension as necessary after step 9).
11. Place 1–2 drops of the suspension onto a clean and humid slide using a glass pipette and let the slide dry at RT.
12. After incubation overnight at RT, the slides can be subjected to the pretreatment as described in chapter by Thomas Liehr et al. ("The Standard FISH Procedure"), stored dust-free at RT for several weeks, or frozen at −20 °C for several months.

3.4 Fluorescence In Situ Hybridization (FISH)

As described in chapter by Thomas Liehr et al. ("The Standard FISH Procedure").

4 Notes

1. EDTA-, sodium acetate-treated, or heparinized blood or bone marrow sample may be used. Sodium citrate-treated blood or bone marrow should be used to prepare a "smear slide" if possible, as it spreads the best. However, blood or bone marrow treated with other anticoagulants can also be used.
2. Cell density is a critical parameter in the evaluation of cell smears. Too many cells hamper evaluation, while too few cells make it difficult to analyze a sufficient number of nuclei. The best cell density is normally found in the region of the smear which looks like a banner.
3. Slides with smears can be used 4 weeks after preparation if stored at 4 °C.
4. In case of blood or bone marrow during this procedure, the slides lose the red erythrocytes and they become more and more transparent. At the end, the nuclei of the blasts and lymphocytes are the only remaining cellular components on the slide surface.
5. Sodium thiocyanate solution may be used in case of samples, which do not give evaluable FISH results when using pretreatment variants 1 or 2. This step can help to permeabilize the cells/nuclei and thus enable probe DNA to reach its goal.
6. Medium also may be replaced by isotonic NaCl.
7. In the case of very high density, the cells may clump after the addition of fixative.

Acknowledgments

Supported in parts by KAAD (Catholic Academic Exchange Service), DAAD (German Academic Exchange Service), and the Graduate Academy of Jena.

References

1. Liehr T, Othman MA, Rittscher K et al (2015) The current state of molecular cytogenetics in cancer diagnosis. Expert Rev Mol Diagn 15:517–526
2. Cotelingam JD (2003) Bone marrow biopsy: interpretive guidelines for the surgical pathologist. Adv Anat Pathol 10:8–26
3. Bentz M, Cabot G, Moos M et al (1994) Detection of chimeric BCR-ABL genes on bone marrow samples and blood smears in chronic myeloid and acute lymphoblastic leukemia by in situ hybridization. Blood 83:1922–1988
4. Delgado J, Aventin A, Briones J et al (2010) The use of tetradecanoylphorbol acetate-stimulated

peripheral blood cells enhances the prognostic value of interphase fluorescence in situ hybridization in patients with chronic lymphocytic leukemia. Genes Chromosomes Cancer 49:327–332

5. Liehr T, Thoma K, Kammler K et al (1995) Direct preparation of uncultured EDTA-treated or heparinized blood for interphase FISH analysis. Appl Cytogenet 21:185–188
6. Gebhart E, Liehr T, Harrer P et al (1995) Determination by interphase FISH of the clonality of aberrant karyotypes in human hematopoietic neoplasias. Leukemia Lymphoma 17:295–302
7. Haferlach T, Loffler H, Nickenig C et al (1998) Cell lineage specific involvement in acute promyelocytic leukaemia (APL) using a combination of May-Grunwald-Giemsa staining and fluorescence in situ hybridization techniques for the detection of the translocation t(15;17)(q22;q12). Br J Haematol 103:93–99
8. Liehr T, Grehl H, Rautenstrauss B (1997) Molecular diagnosis of PMP22-associated neuropathies using fluorescence in situ hybridization (FISH) on archival peripheral nerve tissue preparations. Acta Neuropathol 94:266–271

Characterization of Mosaicism in Different Easy-to-Acquire Body Tissues Such As Buccal Smears, Skin Abrasions, Hair Root Cells, or Urine

Thomas Liehr and Nadezda Kosyakova

Abstract

Mosaic karyotypes are present in at least 0.3 to 1 % of clinical cases analyzed by banding cytogenetics. It is well known that the pattern of mosaicism can be extremely variant in different tissue types of the same patient. However, normally in maximum two different tissues of a child or an adult are cytogenetically studied, i.e., peripheral blood and skin fibroblasts. Here preparation protocols for four easily acquirable further tissues are presented. The resulting preparations of interphase nuclei can be used in routine interphase FISH applications. Here it is summarized how to treat buccal smears, skin abrasions, hair root cells, or urine before the FISH procedure.

Keywords Buccal smear, Skin abrasion, Hair root cells, Urine, Interphase FISH, Mosaicism

1 Introduction

Patients with inborn somatic chromosomal mosaicism are detectable in about 0.3–1 % of the cases analyzed by banding cytogenetics ([1], chapter by Ivan Iourov et al. "Interphase FISH for Detection of Chromosomal Mosaicism"). In such cases, mosaic karyotypes in connection with gonosomes are the most frequently observed [2]. However, autosomes can also be involved, as e.g. also reported for cases with small supernumerary marker chromosomes [3]. Such types of mosaicism can easily be detected by interphase FISH using centromere-specific or locus-specific probes of the corresponding chromosomes. It is a well-known fact that the pattern of mosaicism can be extremely variant in different kinds of tissues from the same patient [4, 5]. Thus, the analysis of different tissue types by molecular cytogenetic methods can be of great interest with respect to the expression of clinical signs and/or prognosis. This has been demonstrated, for example, in a patient with microduplication on chromosome 17p11.2, which appeared

Thomas Liehr (ed.), *Fluorescence In Situ Hybridization (FISH)*, Springer Protocols Handbooks,
DOI 10.1007/978-3-662-52959-1_21,

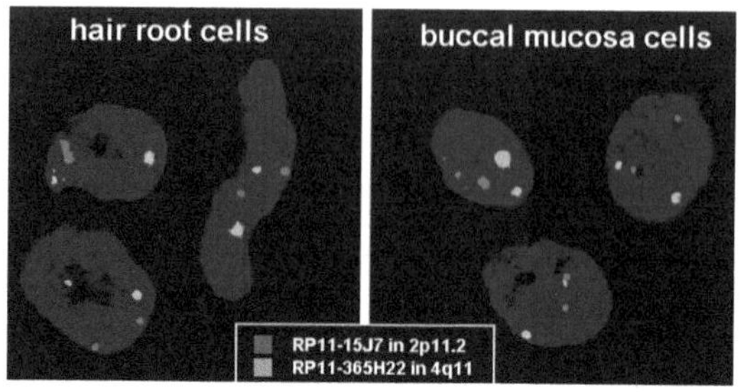

Fig. 1 A chromosome 2-specific probe and a chromosome 4-specific probe were hybridized on nuclei derived from hair root and buccal mucosa cells from the same healthy person. The clones applied were located in variant regions of the human genome. Interestingly, different signal intensities were observed in different tissues and even within the same cells. Note that different cell types that differ in terms of the size and shape of the nuclei were present in hair root cells

in different patterns of mosaicism in peripheral blood [49 %], in buccal mucosa [51 %], in nerve tissue [74 %], and in hair root cells [66 %] [6]. Characterization of mosaicism is the main reason for studying different body tissues by cytogenetics and/or molecular cytogenetics. Apart from mosaicism of rearranged or marker chromosomes and aneuploidies [5, 6], another reason for such studies is to find possible differences in copy number variants in different tissue types [7] (Fig. 1). By applying BAC probes to variant regions of the human genome, we have found hints that these variant regions may show somatic inter-tissue differences ([7], chapter by Anja Weise and Thomas Liehr "Parental Origin Determination FISH: Pod-FISH").

The following protocols describe how to pretreat buccal smears [8], skin abrasions (yet unpublished protocol), hair root cells [9], or urine [8] before FISH. It may be useful in analyses of mosaic cases, as well as in cases where other types of cells are difficult to obtain.

2 Materials

Apart from standard cell biological and molecular cytogenetic equipment, including standard solutions (e.g., ethanol, methanol, formamide, formaldehyde, etc.), the following more specialized items are needed. The equipment needed for FISH itself is listed in chapter by Thomas Liehr et al. ("The Standard FISH Procedure").

2.1 Chemicals and Solutions to Be Prepared

- Trypsin/EDTA (Cat. No.: L 2143, Biochrom, Berlin, Germany).

- Fixative: mix methanol and acetic acid in a 3:1 ratio; make fresh every working day and store at 4 °C.
- Hypotonic solution: 0.075 M KCl.
- Tris-HCl (pH 7.4): 1.21 g Tris-aminomethane, add 1 l aqua dist.; adjust to pH 7.4 with HCl.
- Otto's solution: 100 ml Tris-HCl (pH 7.4), 220 mg KCl, 100 mg $MgCl_2$, and 46 g dithioerythritol.

3 Methods

3.1 Cell Preparation

3.1.1 Buccal Cell Preparation

The direct use of a buccal cell smear is presented, as well as cell preparation. In our hands, the latter led to more evaluable FISH results:

1. Collect cells from buccal mucosa and salvia using a wooden tongue depressor; the first smear should be discarded as it contains mainly dead cells, bacteria, and fungi. Go on with step 2 for direct use of buccal cell smear acc. to [6] or step 7 for buccal cell preparation.
2. Make a second smear and spread it on a slide; let it air-dry for at least 5 min.
3. At this stage the slide can also be stored (or shipped to another lab) for ~2–3 days at room temperature (RT) (*see* **Note 1**).
4. Put the slide in a 100 ml Coplin jar with Otto's solution (37 °C) for 30 min (*see* **Note 2**).
5. Exchange Otto's solution for fixative (RT) and incubate for 5 min.
6. Wash two more times with fixative (RT, 5 min) and then air-dry. The slide is ready for FISH.
7. Make a second smear and wash the wooden tongue depressor or cotton pad (which can also be used) on a suitable amount of cell culture medium (any kind of sterile cell culture medium is acceptable, e.g., RPMI-1640). At this stage the cells can be stored (or shipped to another lab) for ~2–3 days at RT, or better at 4 °C (*see* **Note 1**).
8. Pellet the cells by centrifugation at 1,500 rpm, 5 min.
9. Remove the supernatant and resuspend the cells in the rest of the fluid.
10. Add 3 ml trypsin/EDTA (37 °C) and incubate for 15 min (37 °C).
11. Add 5 ml 1 × PBS and centrifuge at 1,500 rpm, 5 min.

12. Repeat step 9, add 5 ml hypotonic solution (37 °C), and incubate for 20 min (37 °C). Slowly add 2 ml fixative (4 °C) by slightly shaking the tube.
13. Repeat steps 8 and 9.
14. Add 3 ml fixative (4 °C) and incubate for 5 min at −20 °C.
15. Repeat step 8 and remove the supernatant, leaving about 100 μl of fluid in the tube. Resuspend the pellet.
16. Put the fluid on a dry and clean slide by dipping all drops at the same spot. Let the slide air-dry; now the slide is ready for FISH.

3.2 Skin Abrasions and Cell Preparation

1. A scheme on how to obtain skin abrasion without hurting the proband is depicted in Fig. 2; the first step is to apply ~200 of μl fluid (e.g., water) on the skin of the arm or other body parts.
2. Abrade superficial skin cells by scratching them off together with the fluid by the edge of a slide. The fluid must be collected by the slide and then the slide is brought into a horizontal position.
3. Take a second clean slide and make a cell smear as shown in Fig. 2f and let it air-dry for at least 5 min; now the slide is ready for FISH (*see* **Note 1**).

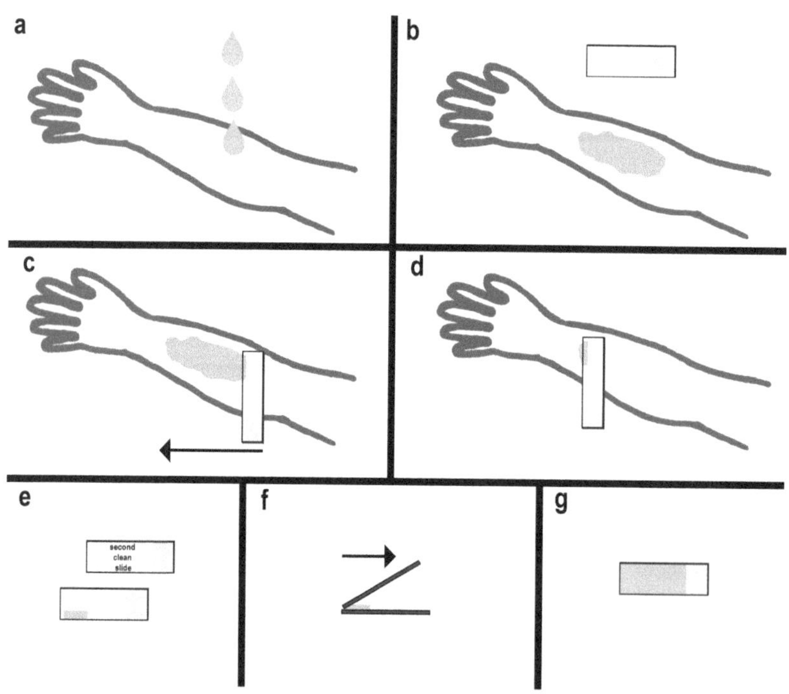

Fig. 2 Scheme on how to obtain skin abrasion without hurting the proband. (**a**) Drop some fluid (e.g., water) on the skin. (**b–d**) Abrade skin cells by the edge of a slide. (**e–g**) Take a second clean slide and make a cell smear

3.3 Hair Root Cell Preparation According To Lampel et al. [9]

1. Freshly extracted hair can be used at once for hair root cell preparation, or it can be transferred immediately into any kind of sterile cell culture medium (e.g., RPMI-1640) and stored and/or shipped for up to 48 h at RT.
2. Dip the tips of 3–6 hair roots into 60 ml of 50 % acetic acid in a 1.5 ml tube; cut the end of the hair so that it is sufficiently short that the tube can be closed, and then incubate for 10 min at RT.
3. Centrifuge for 5 min at 4,000 rpm to detach the hair root cells from the hair.
4. Remove the hair from the tube, add 30 ml of methanol to the suspension, and incubate for 30 min at RT.
5. Put the fluid on a dry and clean slide by dipping all drops at the same spot and the slide is ready for FISH (*see* **Note 1**).

3.4 Urine Cell Preparation According to Rauch et al. [8]

1. Transfer 10–40 ml urine [fresh or up to 2 h old (*see* **Note 3**)] into a 50 ml tube and centrifuge at 1000 rpm.
2. Remove the supernatant, leaving 200–300 μl, and resuspend the pellet.
3. Put the fluid on a dry and clean slide and let it air-dry for at least 15 min.
4. Put the slide in a 100 ml Coplin jar with Otto's solution (37 °C) for 30 min (*see* **Note 2**).
5. Exchange Otto's solution by fixative (RT) and incubate for 5 min.
6. To finalize the preparation of slide, wash twice more with fixative (RT, 5 min) and then air-dry (*see* **Note 1**).

3.5 Slide Pretreatment and Fluorescence In Situ Hybridization (FISH)

As described in chapter by Thomas Liehr et al. ("The Standard FISH Procedure").

3.6 Evaluation

As is standard in interphase FISH, a semistatistical evaluation of the FISH results has to be performed. At least 50, and optimally 100–300, nuclei are evaluated for the number of specific FISH signals per cell. If an automated evaluation system is available, this number can easily be expanded up to 1,000 or more cells, if applicable (chapter by Thomas Liehr and Sven Hauke "Interphase FISH in Diagnostics"; chapter by Anja Weise et al. "FISH in Uncultivated Amniocytes"; chapter by Ivan Iourov et al. "Inter phase FISH for Detection of Chromosomal Mosaicism").

4 Notes

1. It is possible to send (using Express Mail only) slides or medium with cells to other labs. The longest transportation/storage time that gave successful results for slides and cells in medium was 4 days; the longer the storage, the more critical the temperature during this time. The optimal temperature range is +4 to +8 °C. Avoid freezing and too much heat; the former damages cells, while the latter leads to the growth of contaminating microorganisms.
2. Otto's solution can also be replaced by regular hypotonic solution. However, we found the swelling of the nuclei to be more efficient when using Otto's solution.
3. It is not possible to send native urine to other places, as cells in the fluid degrade within 10 and 120 min at most.

References

1. Liehr T, Klein E, Mrasek K et al (2013) Clinical impact of somatic mosaicism in cases with small supernumerary marker chromosomes. Cytogenet Genome Res 139:158–163
2. Liehr T, Mrasek K, Hinreiner S et al (2007) Small supernumerary marker chromosomes (sSMC) in patients with a 45,X/46,X,+mar karyotype - 17 new cases and a review of the literature. Sex Dev 1(6):353–362
3. Liehr T (2016) sSMC homepage. http://ssmc-tl.com/sSMC.html
4. Pfeiffer RA, Schulze T (1994) Mosaicism in three cases 47,XY(or XX),+i(18)(p10) detected by interphase FISH of buccal mucosa. Ann Genet 37:210–214
5. Fickelscher I, Starke H, Schulze E et al (2007) A further case with a small supernumerary marker chromosome (sSMC) derived from chromosome 1–evidence for high variability in mosaicism in different tissues of sSMC carriers. Prenat Diagn 27:783–785
6. Liehr T, Rautenstrauss B, Grehl H et al (1996) Mosaicism for the Charcot-Marie-Tooth disease type 1A duplication suggests somatic reversion. Hum Genet 98:22–28
7. Mkrtchyan H, Gross M, Hinreiner S et al (2010) The human genome puzzle – the role of copy number variation in somatic mosaicism. Curr Genomics 11:426–431
8. Rauch A (2001) Epithelial cells from buccal smears and urine. In: FISH technology (Springer Lab Manual). Springer, Berlin, pp 97–108, ISBN: 3-540-67276-1
9. Lampel S, Steilen H, Zang KD et al (1993) Sex chromatin in hair roots-25 years later: fluorescence in situ hybridization of hair root cells for detection of numerical chromosome aberrations. Cytogenet Cell Genet 63:244–246

Characterization of Archived Formalin-Fixed/Paraffin-Embedded or Cryofixed Tissue, Including Nucleus Extraction

Thomas Liehr

Abstract

Interphase cytogenetics using archival tissue samples is a straightforward approach to obtain "cytogenetic information" from nuclei of solid tissue samples. It is the major tool to investigate specific numerical chromosomal aberrations, chromosomal translocations, amplification of oncogenes, or deletion of tumor suppressor genes in archival tissue samples on a single-cell level, as no metaphase spreads can be obtained, obviously. Here a collection of protocols for archival sectioned and mounted material as well as extracted nuclei (from formalin-fixed/paraffin-embedded or cryofixed tissues) is presented.

Keywords Formalin-fixed/paraffin-embedded tissue, Cryofixed tissues, Interphase cytogenetics, Solid tumor, Archived tissue, Buffered formalin, Tissue integrity

1 Introduction

Interphase cytogenetics on formalin-fixed/paraffin-embedded tissue is a well-established technique that makes it possible to obtain "cytogenetic information" from interphase nuclei, especially those of solid tumors [1–3], although also during the postmortem analysis of aborted fetuses [4]. It is the major tool used to investigate specific numerical chromosomal aberrations, chromosomal translocations, amplification of oncogenes, or deletion of tumor suppressor genes in archival samples on a single-cell level. In contrast to metaphase FISH, only locus-specific and satellite probes are appropriate for routine interphase FISH (for probes, chapter by Thomas Liehr "Commercial FISH Probes"; chapter by Thomas Liehr "Homemade Locus-Specific FISH Probes: Bacterial Artificial Chromosomes").

Interphase FISH studies on formalin-fixed/paraffin-embedded tissue can be done either directly on sectioned and mounted material [5] or on extracted interphase nuclei, a technique first described by Hedley et al. [6]. Both methods are comparable and reliable for the detection of chromosomal changes in archived tissue; however,

Thomas Liehr (ed.), *Fluorescence In Situ Hybridization (FISH)*, Springer Protocols Handbooks,
DOI 10.1007/978-3-662-52959-1_22,

each of them has advantages and disadvantages [7]. The first approach is recommended when the tissue architecture must be preserved, e.g., in the case of small and/or invasive tumors, while the second technique can be applied successfully when more or less homogeneous (tumor)-tissue samples are being studied [8]. During the evaluation of tissue sections, the problems of (1) nuclei that overlap due to the presence of several cell layers on top of each other cannot be evaluated and (2) cut nuclei that lead to artificial signal loss in interphase cytogenetic studies often arise [9, 10]. Such problems are not present in nuclear extraction techniques [4, 7, 8, 11–14].

Formalin-fixed/paraffin-embedded tissue is readily available, as this kind of tissue fixation is the most common standard technique in clinical practice. However, there are some disadvantages of this material, and an increasing number of laboratories now collect archival formalin-fixed/paraffin-embedded and cryofixed tissue samples. Formalin-fixed/paraffin-embedded tissue is, for example, not suitable for all kinds of immunohistochemical approaches, as specific antigens can be destroyed during the fixation procedure. Moreover, if formalin fixation is performed in unbuffered formalin and/or the incubation period is too long, the tissue becomes unsuitable for any kind of FISH study, because the DNA is degraded and washed out of the cells [15]. These problems can be solved using cryofixed tissues. In the following, different ways of handling formalin-fixed/paraffin-embedded [4, 11–14] and cryofixed material for FISH [12–14, 16] are described (see Fig. 1). To enhance FISH signals, a microwave treatment as described in chapter by Anja Weise and Thomas Liehr ("Microwave Treatment for Better FISH Results in a Shorter Time") is recommended.

2 Materials

Apart from standard cell biological and molecular cytogenetic equipment, including standard solutions (e.g., ethanol, methanol, formamide, formaldehyde, xylene, etc.), no more specialized items are required. Solutions and equipment needed for FISH itself are listed in chapter by Thomas Liehr et al. ("The Standard FISH Procedure").

2.1 Additional Material and Solutions to Be Prepared

- Nytal 55 (55 μm nylon mesh) (Cat. No.: 3A07-0049-102-00, SEFAR-AG, Heiden, Switzerland).
- Polylysine-coated slides (Cat. No.: G312P-W, ProSciTec, Thuringow, Australia).
- Proteinase K solution (PK solution): 5 mg proteinase K (Cat. No.: 03115887001, Roche, Basel, Switzerland); 50 μl 1 M Tris–HCl (pH 7.5), 20 μl 0.5 M EDTA (pH 7.0), and 2 μl 5 M NaCl make up to 1 ml with filtered double-distilled water; make fresh as required.

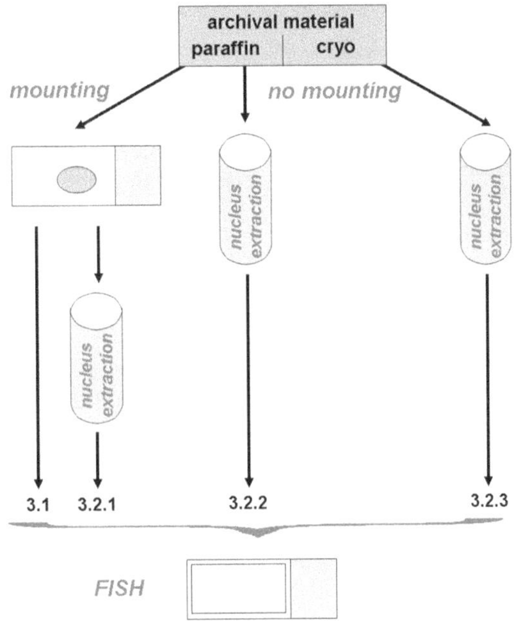

Fig. 1 Outline of the approaches described in this chapter which are used to obtain slides suitable for FISH. Archival material, either paraffin embedded or cryofixed (cryo), can be mounted on a slide (or not). How to get material suited for FISH directly on the mounted tissue or after nucleus extraction is described in detail in the text in Sects. 3.1 and 3.2.1–3.2.3

3 Methods

The following protocol, outlined in Fig. 1, comprises environmental toxins (like xylene and sodium thiocyanate = NaSCN). Please ensure that these substances are collected after use and treated as hazardous waste.

3.1 Slide Preparation for FISH Directly on Mounted Formalin-Fixed/Paraffin-Embedded Tissue Sections

1. Mount 3–6 μm sections of tissue on the polylysine-coated slides (*see* **Note 1**).
2. Dry slides at 58–60 °C overnight.
3. The mounted section is dewaxed in 100 ml xylene in a Coplin jar (2 × 5 min) and rehydrated in an ethanol series (100 %, 90 %, 50 %, 30 %, 3 min each).
4. Bake slides/sections for 1 h at 80 °C to preserve tissue structure.
5. Incubate slides for 1–5 min in 1 M NaSCN at 80 °C in a Coplin jar (*see* **Note 2**).
6. Rinse in 1 × PBS in a Coplin jar for 2 min.
7. Continue with the protocol under Sect. 3.3.

3.2 Nuclear Extraction

3.2.1 Nuclear Extraction from Mounted Formalin-Fixed/Paraffin-Embedded Tissues

1. A 5–50 μm section of paraffin-embedded tissue is mounted on a glass slide. Uncoated or coated slides (e.g., those with amino-triethoxysilane or polylysine) can be used.
2. The mounted section is dewaxed in 100 ml xylene in a Coplin jar (2 × 5 min) and rehydrated in an ethanol series (100 %, 90 %, 70 %, 50 %, 3 min each) and 0.9 % NaCl solution (2 × 2 min).
3. If necessary, undesirable parts of the tissue can be removed at this point by scratching it from the slide by scalpel. This may be helpful when analyzing tumor-infiltrating normal tissue.
4. After this, the tissue is covered with proteinase K solution and incubated at 37 °C for approx. 1 h in a moist chamber. The use of a coverslip should be avoided.
5. Collect fluid with disaggregated tissue and continue with the protocol under Sect. 3.2.4.

3.2.2 Nuclear Extraction from Unmounted Formalin-Fixed/Paraffin-Embedded Tissues

1. According to the diameter of the studied tissue piece, collect 2 to 20 10–20 μm sections of paraffin-embedded tissue produced by a microtome in a glass tube.
2. Dewax the tissue by adding 10 ml of xylene (100 %) for 10 min at room temperature (RT).
3. Sediment the tissue by centrifugation (1,000 rpm, 3 min) and discard the supernatant (1–2 ml of supernatant is left in the tube to avoid the loss of small tissue pieces).
4. Repeat steps 2 and 3.
5. Remove the xylene rests by adding 10 ml of ethanol (100 %), and incubate for 10 min at RT.
6. Repeat steps 3 and 5.
7. Rehydrate the tissue by adding 10 ml of 90 % ethanol (5 min at RT), 10 ml ethanol 70 % (5 min at RT), 10 ml ethanol 50 % (5 min at RT), and 10 ml 0.9 % NaCl (5 min at RT). Remove the corresponding supernatants by repeating step 3.
8. Wash the rehydrated tissue in 10 ml 0.9 % NaCl (2 min at RT) and repeat step 3.
9. Put the tissue, together with approx. 1 ml of the NaCl solution from the glass tube, into a 1.5 ml microtube. This can easily be done using a 1 ml Eppendorf pipette with a cut blue tip, thus enhancing the diameter of the tip.
10. Remove the NaCl solution from the microtube using a 200 μl Eppendorf pipette, add (depending on the amount of tissue) 0.2–1 ml of PK solution, and vortex the microtube.

11. Incubate the microtube at 37 °C for 30 min. During this time, vortex the microtube every five minutes to promote tissue disaggregation.
12. Collect fluid with disaggregated tissue and continue with the protocol under Sect. 3.2.4.

3.2.3 Nuclear Extraction from Cryofixed Tissue

1. Transfer cryofixed tissue from −80 °C to a freezer at −20 °C for 1 h.
2. Transfer the tissue to a glass dish on ice and cut into small pieces (not larger than 1 μm^3) using a scalpel and forceps precooled to +4 °C.
3. Add 1 ml of formalin buffer at RT to the cold tissue pieces and transfer them together with the buffer into a 1.5 ml microtube. The tissue should thaw on the addition of formalin buffer.
4. Incubate the tissue in the formalin buffer for 1–3 h at RT with or without agitation.
5. Pellet the tissue pieces by centrifugation (3,800*g*, 30 s, RT). Repeat this step if necessary.
6. Remove the fluid using a 200 μl Eppendorf pipette, add 1 ml of sterile 0.9 % NaCl (w/v), and vortex the microtube.
7. Repeat steps 5–7.
8. Remove the fluid using a 200 μl Eppendorf pipette, add (depending on the amount of tissue) 0.2–1 ml of PK solution, and vortex the microtube.
9. Incubate the microtube at 37 °C for 30 min. During this time, vortex the microtube every five minutes to promote tissue disaggregation.
10. Collect fluid with disaggregated tissue and continue with the protocol under Sect. 3.2.4.

3.2.4 Purification of the Released Nuclei

1. Transfer the fluid with disaggregated tissue onto a 55 μm nylon mesh. Fluid and nuclei will pass through the mesh by the force of gravity, and they are collected in a 15 ml plastic tube. Nuclei remaining in the mesh are washed out with 4 ml 1 × PBS, passed through the mesh, and collected in the 15 ml plastic tube.
2. Pellet the extracted nuclei by centrifugation (850 × *g*, 8 min); remove the supernatant with the exception of about 300 μl.
3. Resuspend the pellet in 4 ml 1 × PBS and repeat step 2.
4. Resuspend the remaining 300 μl of 1 × PBS.
5. Place one drop of the suspension on a clean and dry slide; allow to dry out on a 40 °C warming plate and afterward at RT overnight (*see* **Note 3**).

6. Store the rest of the suspension at 4 °C overnight.
7. Fix slide in 100 ml formalin buffer in a Coplin jar for 10 min (RT).
8. Replace formalin buffer in the Coplin jar with 1 × PBS. After 5 min of incubation at RT, 1 × PBS is replaced with distilled water.
9. Remove the water after 1 min, perform an ethanol series (70 %, 90 %, 100 %, 3 min each) to dehydrate the slides, and air-dry.
10. The success of the nuclear extraction can be evaluated by phase-contrast light microscopy.
11. Based on the results, the density of the fluid mentioned in step 6 can be adapted by removing or adding PBS. Further slides can be produced by following steps 5 and 7–10.

3.3 Slide Pretreatment

Start with step 2 of the protocol described in chapter by Thomas Liehr et al. ("The Standard FISH Procedure") (*see* **Note 4**).

3.4 Fluorescence In Situ Hybridization (FISH)

As described in chapter by Thomas Liehr et al. ("The Standard FISH Procedure").

Variations

- Denature the slides for 5–15 min at 75 °C (*see* **Note 5**).
- Increase the probe concentration by 2–5 times compared to fresh slides (*see* **Note 6**).

3.5 Evaluation

As is standard in interphase FISH, a semistatistical evaluation of the FISH results must be performed. At least 50, and better 100–300, nuclei are evaluated for the number of specific FISH signals per cell. If an automated evaluation system is available, this number can easily be expanded up to 1,000 or more cells (chapter by Thomas Liehr and Sven Hauke "Interphase FISH in Diagnostics"; chapter by Anja Weise et al. "FISH in Uncultivated Amniocytes"; chapter by Ivan Iourov et al. "Interphase FISH for Detection of Chromo somal Mosaicism").

The evaluation of neoplastic tissue specimens requires involvement of an anatomical or surgical pathologist. Depending on the neoplasia, marking relevant areas on an adjacent slide stained with hematoxylin and eosin might be sufficient. In more complex specimens, such as specimens with only focal nests of tumor cells or in which it is necessary to distinguish, e.g., carcinoma in situ from invasive cancer, this might not be adequate and a pathologist should be directly involved in the evaluation procedure.

FISH on tissue sections will frequently lead to nuclei showing pseudo-aberrant signal patterns, caused by truncating nuclei when cutting specimen sections. These cutting artifacts are dependent on the tissue/cell types, the thickness of the sections, and the probe which is used. When implementing a FISH test for diagnostic purposes, cutoff criteria have to be established distinguishing non-aberrant from aberrant situations. For further information, please refer to [17–19].

4 Notes

1. The thinner the sections are cut, the higher is the number of nuclei which can be clearly discriminated during evaluation and, on the other hand, the more the cutting artifacts are observed which have to be taken into consideration when establishing cutoff criteria.
2. The time needed for NaSCN treatment has to be adapted for each tissue type and each sample separately. The success of the pretreatment must be monitored by microscopic inspection. A balance between tissue preservation and tissue digestion must be found. If the tissue is preserved too well, the DNA probes may not be able to pass through and so no FISH result will be obtained. In the case of too much tissue digestion, one may still get FISH signals, but it may prove impossible to correlate them to a specific tissue region. Complete loss of the tissue during the FISH procedure may also occur. It is recommended that beginners start with tissue samples that are not limited in availability.
3. An evaluation of the quantity and quality of the extracted nuclei is not possible directly after step 6 mentioned in Sect. 3.2 due to the crystallization of PBS salts on the slide surface.
4. The pepsin treatment time must also be adapted; similar things can happen here to those mentioned in **Note 1**.
5. Due to the fact that DNA in archival tissues has undergone some fixation steps and has been stored for up to several years, a prolonged denaturation time appears useful. Moreover, in other FISH protocols with denaturation times of just 2–5 min, the maintenance of available metaphase chromosomes is the main aspect, but this is of no significance in the present protocol.
6. The most commercially available probes for use on FFPE tissue sections are nowadays supplied in a ready-to-use format. In this case, proceed with step 5 (application of the probe) of the FISH protocol as described in chapter by Thomas Liehr et al. ("The Standard FISH Procedure").

References

1. Wolfe KQ, Herrington CS (1997) Interphase cytogenetics and pathology: a tool for diagnosis and research. J Pathol 181:359–361
2. Tibiletti MG, Bernasconi B, Dionigi A et al (1999) The applications of FISH in tumor pathology. Adv Clin Path 3:111–118

3. Fuller CE, Perry A (2002) Fluorescence in situ hybridization (FISH) in diagnostic and investigative neuropathology. Brain Pathol 12:67–86
4. Fickelscher I, Starke H, Schulze E et al (2007) A further case with a small supernumerary marker chromosome (sSMC) derived from chromosome 1–evidence for high variability in mosaicism in different tissues of sSMC carriers. Prenat Diagn 27:783–785
5. Neubauer S, Liehr T, Tulusan HA et al (1994) Interphase cytogenetics by FISH on archival paraffin material and cultured cells of human ovarian tumors. Int J Oncol 4:317–321
6. Hedley DW, Friedlander ML, Taylor IW et al (1983) Method for analysis of cellular DNA content of paraffin-embedded pathological material using flow cytometry. J Histochem Cytochem 31:1333–1335
7. Qian JQ, Bostwick DG, Takahashi S et al (1996) Comparison of fluorescence in situ hybridization analysis of isolated nuclei and routine histological sections from paraffin-embedded prostatic adenocarcinoma specimens. Am J Pathol 149:1193–1199
8. Köpf I, Hanson C, Delle U et al (1996) A rapid and simplified technique for analysis of archival formalin-fixed, paraffin-embedded tissue by fluorescence in situ hybridization (FISH). Anticancer Res 16:2533–2536
9. Dhingra K, Sahin A, Supak J et al (1992) Chromosome in situ hybridization on formalin-fixed mammary tissue using non-isotopic, non-fluorescent probes: technical considerations and biological implications. Breast Cancer Res Treat 23:201–210
10. Liehr T, Stübinger A, Thoma K et al (1994) Comparative interphase cytogenetics using FISH on human ovarian carcinomas. Anticancer Res 14:183–188
11. Liehr T, Grehl H, Rautenstrauss B (1995) FISH analysis of interphase nuclei extracted from paraffin-embedded tissue. Trends Genet 11:377–378
12. Liehr T, Grehl H, Rautenstrauss B (1997) Molecular diagnosis of PMP22-associated neuropathies using fluorescence in situ hybridization (FISH) on archival peripheral nerve tissue preparations. Acta Neuropathol 94:266–271
13. Liehr T, Claussen U, Gebhart E (1999) Nucleus extraction from single mounted tissue sections. Genet Anal 15:65–69
14. Winkler S, Hempel M, Brückner S et al (2015) Mouse white adipose tissue-derived mesenchymal stem cells gain pericentral and periportal hepatocyte features after differentiation in vitro, which are preserved in vivo after hepatic transplantation. Acta Physiol (Oxf) 215:89–104
15. Long AA, Komminoth P, Lee E et al (1993) Comparison of indirect and direct in-situ polymerase chain reaction in cell preparations and tissue sections. Detection of viral DNA, gene rearrangements and chromosomal translocations. Histochemistry 99:151–162
16. Liehr T, Grehl H, Rautenstrauss B (1996) A rapid method for FISH analysis on interphase nuclei extracted from cryofixed tissue. Trends Genet 12:505–506
17. Wayne PA (2013) Fluorescence in situ hybridization methods for clinical laboratories—Approved guidelines, 2 edn. Clinical Laboratory Standards Institute MM07-A2, ISBN 1-56238-885-1
18. Wiktor AE, Van Dyke DL, Stupca PJ et al (2006) Preclinical validation of fluorescence in situ hybridization assays for clinical practice. Genet Med 8:16–23
19. Wolff DJ, Bagg A, Cooley LD et al (2007) Guidance for fluorescence in situ hybridization testing in hematologic disorders. J Mol Diagn 9:134–143

FISH on Sperm, Spermatocytes and Oocytes

Maria Oliver-Bonet

Abstract

It is well known that chromosome in situ hybridization allows the unequivocal identification of targeted human somatic chromosomes. Different fluorescence in situ hybridization (FISH) techniques have been developed throughout the years, and following the mitotic studies, meiotic analyses have been performed using these different techniques. The application of FISH protocols on meiotic cells requires adaptation of standard protocols to the particularities of these cells. Specific sample fixation is usually required, and in some cases samples need to go through particular pretreatments to guarantee a successful FISH. The application of FISH to meiotic cytogenetic research has proved very useful and has provided us with significant information about many of the processes that take place during human gametogenesis. The protocols described in this chapter illustrate the processing of different meiotic samples as well as the FISH procedures.

Keywords FISH, M-FISH, Meiosis, Sperm, Oocyte, Spermatocyte, Gamete

1 Introduction

Methods for visualizing several target chromosomes simultaneously by fluorescence in situ hybridization (FISH) have been developed and widely applied to the analysis of somatic and meiotic processes (chapter by Thomas Liehr and Anja Weise "Back ground"). Meiotic analysis, for instance, has greatly benefited from the application of FISH techniques to meiotic cells. Indeed, FISH permits rapid identification of chromosomes throughout meiosis, allowing the characterization of meiotic anomalies in both human males and females.

Information on the frequency of meiotic aneuploidy in humans, for instance, has been obtained mainly, thanks to the application of FISH on sperm and on oocytes or polar bodies [1]. FISH technique on sperm head has been applied in several studies: it has been used to asses aneuploidy in controls and infertile men, [2] to assess the effects of lifestyle on the production of chromosome abnormalities, [3, 4] or to determine the segregation pattern of chromosomal rearrangements [5–7]. The application of FISH

Thomas Liehr (ed.), *Fluorescence In Situ Hybridization (FISH)*, Springer Protocols Handbooks,
DOI 10.1007/978-3-662-52959-1_23,

on the first polar body (1 PB) and meiosis 2 (MII) oocytes is indicated in the analysis of transmitted chromosomal alteration of maternal origin [8, 9]. In addition, FISH on 1 PB and 2 PB has also been used to detect the origin of female meiotic alterations (MI or MII) [10].

FISH has also been a very useful technique for the characterization of meiotic processes: it has allowed researchers to study synapsis initiation and homology search [11] and the behavior of specific chromosomes during meiotic prophase I [12]. FISH on meiotic chromosomes spread by classical techniques was first performed by Goldman and Hulten [13]. Since then, whole-chromosome painting (WCP), centromere (CEN), locus-specific (LS) probes, and M-FISH techniques have been used to identify particular bivalents at different stages of meiosis and to study diverse aspects of the meiotic process and chromosomal behavior during meiosis in controls and in infertile men and women [14–17]).

Finally, FISH technique and multi-FISH technique have been used on synaptonemal complex (SC) spreads of control and infertile males and female, providing insights about the SC-associated chromatin organization and about meiotic recombination and synapsis [18–22].

In order to successfully use FISH in the analysis of meiotic cells, protocols required special adaptations to overcome difficulties inherent in working with these cells. For instance, in order for DNA probes to have enough space to access targeted DNA in the sperm, the sperm heads must be decondensed, breaking the large numbers of disulfide cross-links between protamine molecules. In other meiotic samples, the conditions under which cells are fixed are critical in order to obtain good chromosome extensions. The protocols described in this chapter illustrate the processing of different meiotic cells in preparation for FISH, as well as FISH procedures for successful hybridization of these cells.

The different procedures presented in this chapter have been outlined in Fig. 1.

2 Materials

Apart from standard cell biological and molecular cytogenetic equipment, including standard solutions (e.g., ethanol, methanol, formamide, formaldehyde, xylene, etc.), more specialized items are listed below. The equipment needed for FISH is listed in chapter by Thomas Liehr et al. ("The Standard FISH Procedure").

- MII oocytes and PB FISH
- Silane-prepared slides (Sigma, cat # S4651)

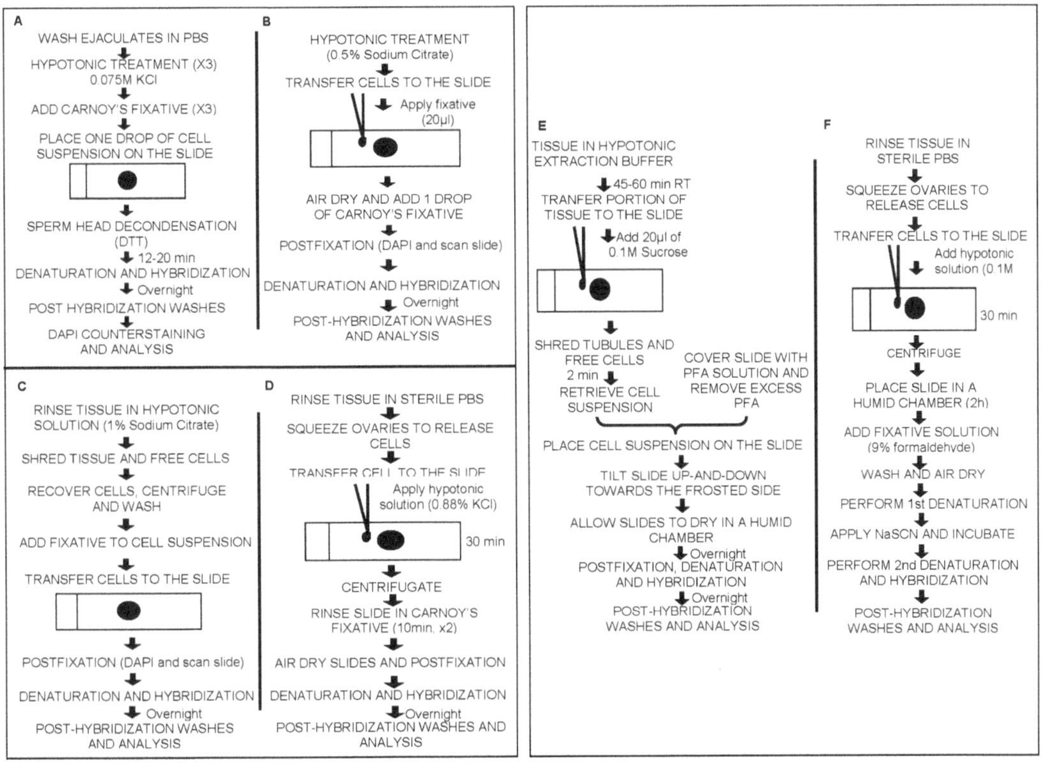

Fig. 1 Outline of the procedures described in this chapter. (**a**) Sperm FISH, (**b**) MII and PB FISH, (**c**) FISH on male meiotic chromosomes, (**d**) FISH on female meiotic chromosomes, (**e**) FISH on male SC, (**f**) FISH on female SC

- Stereomicroscope
- Meiotic chromosomes and SC (both male and female)
- Fine scissors and fine forceps (×2)
- Stereomicroscope
- Watch glasses
- Meiotic chromosomes and SC (female)
- Cytocentrifuge

2.1 Chemicals

- 1,4-Dithiothreitol (DTT) (Roche, cat # 708984)
- Boric acid (Sigma, cat # B7660)
- Paraformaldehyde, prilled 95 % (Sigma, cat #441244)
- Photo-Flo (Kodak Photo-Flo 200 solution, cat #146 4510)
- Triton-X (Sigma, cat # T8787)
- Trizma base (Sigma, cat # T1503)

2.2 Solutions to Be Prepared

2.2.1 Stock Solutions

- 20 × SSC (1 l): 175.32 g NaCl, 88.22 g sodium citrate, up to 1 l ddH_2O. Store at room temperature (RT).
- PBS (1 l) pH = 7.0: 8.0 g NaCl, 0.2 g HCl, 1.15 g Na_2HPO_4, 0.2 g KH_2PO_4. Up to 1 l ddH_2O. Store at RT.
- 0.075 M KCl (1 l): 5.6 g KCl. Up to 1 l ddH_2O. Store at RT.

2.2.2 Washing Solutions (Make Fresh as Required and Discard After Use)

- 2 × SSC/0.1 % Tween 20 (PBD solution), pH = 7.0, 500 ml: 50 ml 20 × SSC, 450 ml ddH_2O, 500 μl Tween 20
- 4 × SSC/0.05 % Tween 20, pH = 7.0, 500 ml: 100 ml 20 × SSC, 450 ml ddH_2O, 250 μl Tween 20
- 2 × SSC, pH = 7.0, 500 ml: 50 ml 20 × SSC, 450 ml ddH_2O
- 0.4 × SSC, pH = 7.0, 500 ml: 10 ml 20 × SSC, 490 ml ddH_2O
- 0.04 × SSC pH = 7.0, 500 ml: 1 ml 20 × SSC, 499 ml ddH_2O
- 0.04 % Photo-Flo: dilute 20 μl of Photo-Flo in 50 ml of ddH_2O
- 1 % Photo-Flo: dilute 500 μl of Photo-Flo in 50 ml of ddH_2O

2.2.3 Decondensation

- 1 M DTT (use it as stock solution to prepare 5 mM DTT): dissolve 0.308 g DTT in 2 ml ddH_2O. Store 250 μl aliquots in microcentrifuge tubes at −20 °C.
- 5 M DTT in 0.1 M Tris/1 % Triton-X, pH = 7.4, 50 ml (make fresh as required): 250 μl 1 M DTT, 5 ml 1 M Tris, 500 μl Triton-X. Up to 50 ml of ddH_2O. Adjust pH with concentrated HCl.
- 1 M Tris (use it as stock solution to prepare 0.1 M Tris when needed) pH = 8, 1 l: 121.1 g Tris base, 800 ml ddH_2O, 42 ml (approximately) concentrated HCl. Dissolve Tris in water, add concentrated HCl, and bring pH to 8.0. Bring the total volume to 1 l with ddH_2O. Aliquot, autoclave, and store at RT.

2.2.4 Hypoextraction Buffer (Synaptonemal Complex Protocol)

Prepare the following stock solutions. Aliquot them in 1 ml cup, and store them at RT, unless otherwise specified differently:

- 600 mM Tris, pH 8.2: 2.52 g in 25 ml ddH_2O.
- 500 mM sucrose: 1.71 g in 10 ml ddH_2O (keep in fridge).
- 170 mM citric acid (trisodium citrate dihydrate): 0.5 g in 10 ml ddH_2O.
- 250 mM EDTA: 2.3 g in 25 ml ddH_2O. Adjust pH to 8.0 to facilitate dissolving.
- 500 mM DTT. 0.077 g in 1 ml ddH_2O (keep in freezer).
- 50 mM PMSF: 0.0087 g in 1 ml isopropanol (keep in freezer).
- To prepare 10 ml of hypoextraction buffer (to use within 2 h of DTT addition):
  - 500 μl of 600 mM Tris (final concentration 30 mM).
  - 1 ml of 500 mM sucrose (final concentration 50 mM).

- 1 ml of 170 mM citric acid (final concentration 17 mM).
- 100 μl of 500 mM EDTA (final concentration 5 mM).
- 50 μl of 500 mM DTT (final concentration 2.5 mM).
- 100 μl of 50 mM PMSF (final concentration 0.5 mM).
- Add 7.25 ml of ddH_2O.
- Adjust pH to 8.2–8.4 before use.

- To prepare 500 μl of 0.1 M sucrose, mix 100 μl of 500 mM sucrose with 400 μl of ddH_2O.

2.2.5 Fixative Solutions

- 1 % paraformaldehyde (PFA), pH = 9.2–9.4, 25 ml: resuspend 25 g of paraformaldehyde in 22.4 ml ddH_2O. Add 1 drop of 1 N NaOH, and incubate for 20–60 min at 37 °C (up to 60 °C, do not exceed this temperature) to dissolve paraformaldehyde. Cool at RT. Adjust pH to 9.2 using 1 M sodium borate buffer (to prepare it, dissolve 3.09 g of boric acid in 50 ml ddH_2O water by heating, and then bring RT solution to pH 8.0 with 10 N NaOH). Add 50 μl of Triton-X and mix well.
- 9 % formaldehyde in ddH_2O pH = 10.0 (10 ml): 2.5 ml 35–40 % formaldehyde, 7.5 ml ddH_2O.

2.2.6 Postfixative Solutions

- PBS/50 mM $MgCl_2$ (50 ml): 2.5 ml 1 M $MgCl_2$, 47.5 ml PBS
- 1 % formaldehyde in PBS/50 mM $MgCl_2$ (50 ml): 2.5 ml 1 M $MgCl_2$, 1.5 ml 35–40 % formaldehyde, 46 ml PBS

2.2.7 Denaturation Buffer

- 70 % formamide/2 × SSC (50 ml) pH = 7.0, 50 ml: 35 ml formamide, 5 ml 20 × SSC, 10 ml ddH_2O

3 Methods

3.1 FISH on Gametes

3.1.1 Sperm FISH (Adapted from [23])

Slide Preparation

1. If frozen sperm is to be used, thaw one 0.5 ml sperm cryotube (containing 0.25 ml semen).
2. Suspend 0.5 ml of semen in 3.5 ml of 1 × PBS (10 ml centrifuge tube).
3. Centrifuge at 600 × *g* for 5 min. Discard supernatant.
4. Resuspend the pellet, and add about 5 ml of 0.075 M KCl (make sure that the first twenty drops are slowly added).
5. Centrifuge at 600 × *g* for 5 min. Discard supernatant and repeat wash twice.
6. Resuspend pellet and add about 5 ml of Carnoy's fixative. Initially (first twenty drops or so), add the fixative one drop at

a time, allowing it to mix well with cell suspension before adding the next drop.

7. Centrifuge at 600 × *g* for 5 min. Discard supernatant and repeat step 6 twice.
8. Bring pellet to the desired concentration.
9. Using a glass Pasteur pipette, place one drop of the sperm suspension on a clean slide. Allow to dry at RT.
10. Place slides in 2 × SSC for 3 min at RT.
11. Dehydrate slides in an ethanol series, 70 %, 85 %, and 100 %, 2 min each. Allow slides to dry at RT.
12. Using a diamond marking pencil, circle on the underside of the slides the area containing the sperm cells.

Sperm Head Decondensation

1. Place slides in a Coplin jar containing 5 mM DTT in 0.1 M Tris/1 % Triton-X for at least 12 min at 37 °C. Decondensation time varies among samples.
2. Rinse slides in 2 × SSC for 3 min at RT, and dehydrate it in an ethanol series.

In order to check decondensation, slides can be observed under phase-contrast microscope (40 × magnification).

Denaturation and Hybridization (*See* **Note 1**)

1. Place decondensed slides in a Coplin jar containing 50 ml of denaturation buffer at 72 °C for 3 min.
2. Remove slides (using forceps), dehydrate them in a cold ethanol series (2 min each), and let them dry at RT.
3. Meanwhile, cut 20 × 20 mm^2 coverslips to obtain 15 × 15 mm^2 coverslips.
4. Prepare DNA probe mix according to manufacturer's instructions; denature approximately 5 μl of probe mix per slide at 75 °C for 5 min; transfer it to the slide area over the sperm; and place the 15 × 15 mm^2 coverslips over it. Apply rubber cement to seal the coverslip onto the glass.
5. Incubate slides in a dark, humid chamber at 37 °C for 3–12 h. For small DNA probes, such as subtelomeric or locus-specific probes, longer hybridization times might be required.

Post-hybridization and Counterstaining

1. Gently remove rubber cement and coverslip from slides.
2. Wash slides in a 0.4 × SSC solution at 75 °C for 2 min, followed by a wash in 2 × SSC/0.1% Tween 20 (NP-40 can also be used) at RT for 1 min (agitate slide during this last wash).

3. Air-dry slides and apply DAPI and antifade.
4. Slides can be stored at −20 °C for a long time.

Evaluation

1. The number of spermatozoa to be scored per individual and per set of probes will depend on the margin of error that can be acceptable in the design of a given experiment. For instance, in case of aneuploidy screening, the frequency of aneuploidy for a given chromosome is usually so low that the size of the sample will affect the validity of the results. In these cases, and whenever the number of spermatozoa present in the sample does not represent a limitation, it would be convenient to count 1×10^4 sperm per slide.
2. Scoring criteria: in order to minimize interobserver variability and eliminate subjective factors as much as possible, strict scoring criteria have to be followed [24]:
   - Only individual (nonoverlapped), well-delineated, and intact spermatozoa will be evaluated.
   - A sperm head will be scored as having two or more signals of the same color only when the signals are of similar size, shape, and intensity and are separated by at least one fluorescence domain.

3.1.2 MII Oocytes and PB FISH (Adapted from [25])

Slide Preparation (*See* **Note 2**)

1. Place the MII oocyte or the PB in a hypotonic solution (sodium citrate 0.5 %) for 4 min.
2. Using a diamond marking pencil, circle on the underside of a clean slide (coated with silane) a tiny round area. Transfer the PB or the MII oocyte (plus a small amount of hypotonic solution) to the marked area.
3. Wait until the drop of hypotonic solution containing the PB starts to evaporate, and add 20 μl of fixative (methanol or ethanol: acetic acid 1:1) (*see* **Note 3**). This step is critical: the fixative should be added not directly over the hypotonic solution, but half cm away from it. It should be dropped gradually, in a continuous way, so a fixative flow is created toward the hypotonic drop. The idea is to displace the hypotonic drop without disturbing the cell position. If 20 μl of fixative is not enough, add a second drop, or even a third, if needed.
4. Let the fixative evaporate, and observe slides under phase-contrast microscope (40 ×). In case cell membrane or cytoplasm is observed, proceed with a pepsin treatment (50 μg ml^{-1} in 10 mM HCl) for 0.5–3 min at 37 °C, and rinse slide in ddH_2O at RT.
5. Air-dry slides and apply a drop of Carnoy's fixative. Slides can be used immediately or be frozen until use.

Denaturation and Hybridization

1. If slides have been frozen, let them defrost for 5 min at RT, and dehydrate them in an ethanol series (70 %, 85 %, and 100 %, 2 min each)
2. Transfer slides into a Coplin jar containing PBS/50 mM $MgCl_2$, and rinse the slides for 4 min at RT.
3. Postfix cells using 1 % formaldehyde in PBS/50 mM $MgCl_2$ for 8 min at RT.
4. Wash slides in 1 × PBS for 4 min at RT.
5. Dehydrate slides in an ethanol series and air-dry.
6. Prior FISH, it is convenient to dye chromosomes with DAPI (33 ng ml^{-1}) and to scan slides to capture chromosome extensions using a fluorescence microscope and capture system. This will provide us with an image of chromosome morphology before denaturation and hybridization that will be very useful when analyzing the FISH results.
7. Remove DAPI by rinsing the slides in two successive washes of 2 × SSC/0.1% Tween 20 at RT.
8. Dehydrate slides in an ethanol series (70 %, 85 %, and 100 %, 2 min each).
9. Air-dry slides.
10. Place slides on to a hot plate warmed to 45 °C for 4 min.
11. Drop 0.5 μl of probe mix in denaturation buffer on the small area containing the cell. Place a 6 × 6 mm coverslip and seal with rubber cement.
12. Codenaturate slide and probe for 3–5 min at 73 °C.
13. Incubate slides in a dark, humid chamber at 37 °C overnight.

Post-hybridization and Counterstaining

1. Gently remove rubber cement and coverslip from slides.
2. Wash slides in a 0.4 × SSC solution at 72 °C for 2 min, followed by a wash in 2 × SSC/0.1 % Tween 20 (NP-40 can also be used) at RT for 1 min (agitate slide during this last wash).
3. Dehydrate slides in an ethanol series.
4. Air-dry slides and apply DAPI and antifade. Capture using the appropriate filter set and capture software.
5. If a second FISH round has to be applied, proceed from step 8 of the previous section.

Slide Preparation (Obtaining Meiotic Chromosomes) (Adapted from [26])

1. Rinse the testicular tissue in a watch glass containing a small volume of hypotonic solution (1 % sodium citrate).

2. Using scissors or fine forceps, torn the tissue to very small pieces to extract the maximum amount of meiotic cells from the tubules.
3. Transfer the cell suspension obtained into a 10 ml centrifuge tube.
4. Add 5 ml of 1 % sodium citrate into the tube. With the help of a plastic Pasteur pipette, draw the suspension up and down to flush out the remainder of the cells from tubules.
5. Let it stand for 3–4 min or until all the tubular remnants have settled.
6. Transfer the supernatant (containing the meiotic cells) to a clean 10 ml centrifuge tube.
7. Spin for 5 min at 300 × *g*. Discard supernatant, and gently resuspend pellet in 3–5 ml of 1 % sodium citrate.
8. Repeat step number 7.
9. Resuspend pellet and add about 5 ml of Carnoy's fixative. Initially (first 20 drops or so), add the fixative one drop at a time, allowing it to mix well with cell suspension before adding the next drop.
10. Centrifuge at 300 × *g* for 5 min. Discard supernatant and repeat step 9 twice.
11. Bring pellet to the desired concentration, and, using a glass Pasteur pipette, place one drop of the final cell suspension on a clean slide. Allow to dry at RT. Slides can be used immediately or be frozen until use.

3.2 FISH on Meiotic Chromosomes (MI and MII)

3.2.1 Male Studies

Post-fixation, Denaturation, and Hybridization (Adapted from [27])

1. If slides have been frozen, let them defrost for 5 min at RT, and dehydrate them in an ethanol series (70 %, 85 %, and 100 %, 2 min each).
2. Transfer slides into a Coplin jar containing PBS/50 mM $MgCl_2$, and rinse the slides for 5 min at RT.
3. Postfix cells using 1 % formaldehyde in PBS/50 mM $MgCl_2$ for 8 min at RT.
4. Wash slides in 1 × PBS for 4 min at RT.
5. Dehydrate slides in an ethanol series and air-dry.
6. Prior FISH, it is convenient to dye chromosomes with DAPI (33 ng/ml) and to scan slides to capture chromosome extensions using a fluorescence microscope and capture system. This will provide us with an image of chromosome morphology before denaturation and hybridization that will be very useful when analyzing the FISH results.

7. Remove DAPI by rinsing the slides in two successive washes of 2 × SSC/0.1% Tween 20 at RT.
8. Dehydrate slides in an ethanol series (70 %, 85 %, and 100 %, 2 min each).
9. Apply 5–10 μl of probe mix (the amount will depend on the spreading area) containing denaturation buffer. Apply a cover-slip and seal with rubber cement.
10. Codenaturate at 70 °C, for 2 min. Incubate slides in a dark, humid chamber at 37 °C overnight.

Post-hybridization and Counterstaining

Post-hybridization washes and counterstaining follow the standard FISH protocol (see chapter by Thomas Liehr et al. "The Stan dard FISH Procedure".

3.2.2 Female Studies

Slide Preparation (Obtaining Meiotic Prophase I Chromosomes) (Adapted from [12])

1. Once the fetal ovaries have been collected, place them in a watch glass containing sterile PBS.
2. Rinse ovaries and remove any extra somatic tissue (*see* **Note 2**).
3. Transfer ovaries to a second watch glass containing sterile PBS.
4. Using fine needles, gently squeeze the ovaries so the prophase cells are released in to the PBS (*see* **Note 2**).
5. Collect the PBS and place it in a 10 ml centrifuge tube.
6. Add new sterile PBS and repeat the process (do it until no more cells are being freed from the ovaries).
7. Centrifuge at 600 × *g* for 7 min.
8. Discard supernatant and resuspend pellet in sterile PBS.
9. Prewarm hypotonic solution (0.88 % KCl) at 37 °C.
10. Assemble slides to the cytocentrifuge complex
11. Add eight drops of hypotonic solution and one drop of cell suspension in PBS per slide
12. Let it stand for 30 min.
13. Cytocentrifuge the slide at 115 × *g* for 15 min.
14. Let rest for 10 min.
15. Remove slides from cytocentrifuge complex, and let slides stand (horizontally) for 10 min.
16. Rinse slides twice in Carnoy's fixative, 10 min each.
17. Air-dry slides. Slides can be used immediately or be frozen until use.

Post-fixation, Denaturation, and Hybridization

Post-fixation protocol follows the procedure described in Sect. 3.2.1, steps 1–6 of the post-fixation protocol.

1. Place slides on a hot plate warmed to 69 °C (±1 °C).
2. Add 100 ml of denaturing solution (70 % formamide/2 × SSC), and cover with a 24 × 60 mm coverslip. Let it stand for 3 min.
3. Remove slides, pass them through a cold ethanol series (1 min each), and allow to air-dry.
4. Meanwhile, denature 10 μl of probe according to manufacturer's instructions (or, in case of self-made probes, according to supplier's instructions).
5. Add denatured probe to the desired area on the slide, apply coverslip. and seal with rubber cement.
6. Incubate slides in a humid chamber at 37 °C for, at least, 24 h.

Post-hybridization and Counterstaining

Post-hybridization washes and counterstaining follow the standard FISH protocol (see chapter by Thomas Liehr et al. "The Standard FISH Procedure").

3.3 FISH on Synaptonemal Complexes

3.3.1 Male

Slide Preparation (Obtaining Synaptonemal Complexes (SCs)) (Adapted from [28])

1. Rinse the testicular tissue in a watch glass containing a small volume of hypoextraction buffer.
2. Transfer the tissue to a second watch glass of hypoextraction buffer. Use forceps (or needles) to gently separate the tubules, until all tubules are spread and exposed to the hypotonic solution (*see* **Note 2**).
3. Incubate the tissue for 60 min in the hypoextraction buffer at RT.
4. When incubation time is finished, place 20 μl of sucrose 0.1 M on a clean slide.
5. Transfer a small amount of the tissue to the sucrose on the slide. Shred the tubules using forceps and/or needles until the sucrose turns cloudy (this means meiotic cells have been released in to the sucrose from the tubules) (*see* **Note 2**).
6. Using a pipette, retrieve the sucrose solution containing the meiotic cells. Tilting the slide toward you will help to recover the maximum volume possible.
7. Spread PFA solution over the surface of a clean slide.
8. Drop cells onto the right end of the PFA slide. Tilt the slide up and down and gradually move the cells toward the frosted side of the slide. This movement helps to spread the cells evenly over the slide. To control the position of the cells, an air bubble can be formed using a pipette.

9. Place slides in a humid chamber.
10. Add 20 μl of sucrose 0.1 M to the tissue sample on the slide, and repeat the shredding and transfer processes (the process can be repeated on the same tissue sample as long as cells are being released into the sucrose solution. Add 20 μl of sucrose 0.1 M as needed in order to prevent the tissue from drying out).
11. Replace new tissue sample, and repeat process from step 4.
12. Allow slides to dry overnight in the humid chamber at RT. If slides dry too quickly, cell extension will be poor, and the analysis will be difficult.
13. Wash slides in 0.04 % Photo-Flo for 4 min at RT. Do not wash more than four slides per Coplin jar to allow a good circulation of the Photo-Flo and a better removal of the PFA.
14. Air-dry slides.
15. Use phase-contrast microscopy to visualize meiotic cells.
16. Process the slides immediately for the best results for immunostaining.

Post-fixation, Denaturation, and Hybridization (Adapted from [19])

1. If necessary, remove DAPI from SC spreads using 4 × SSC/0.05 % Tween-20 solution (soak slides in three changes of this solution for 5 min each, at RT).
2. Transfer slides into a Coplin jar containing PBS for 5 min at RT.
3. Replace PBS with PBS/50 mM $MgCl_2$, and rinse the slides for 5 min at RT.
4. Postfix cells using 1 % formaldehyde in PBS/50 mM $MgCl_2$ for 10 min at RT.
5. Wash slides in 1 × PBS for 5 min at RT.
6. Dehydrate slides in an ethanol series (70 %, 85 %, and 100 %, 2 min each) and air-dry.
7. Place slides on a hot plate warmed to 70 °C (±1 °C).
8. Add 100 ml of denaturing solution (70 % formamide/2 × SSC), and cover with a 24 × 60 mm coverslip. Let it stand for two min.
9. Remove slides, pass them through a cold ethanol series (70 %, 85 %, and 100 %, 1 min each), and allow air-drying.
10. Meanwhile, denature the appropriate amount of probe according to manufacturer's instructions (or, in case of self-made probes, according to supplier's instructions).
11. Add probe mix to SC slides, apply coverslips, and seal using rubber cement.
12. Incubate slides in a humid chamber at 37 °C overnight.

Post-hybridization Washes and Counterstaining

1. Wash slides in 0.4 × SSC at 70 °C for 2 min, followed by 2 min in 4 × SSC/0.05 % Tween 20 at RT.
2. Air-dry slides and apply DAPI and antifade.
3. Slides can be stored at −20 °C for a long time.

3.3.2 Female

Slide preparation (Obtaining Synaptonemal complexes (SCs)) (Adapted from [29])

1. Follow steps 1–8 in Sect. 3.2.2, slide preparation
2. Assemble slides to the cytocentrifuge complex.
3. Add 0.5 ml of hypotonic solution (0.1 M sucrose) plus one drop of cell suspension in PBS.
4. Centrifuge at 115 × *g* for 10 min.
5. Disassemble slides from cytocentrifuge complex, and let them rest (horizontal position) in a humid chamber for 2 h.
6. Add 0.6 ml of fixative solution (9 % formaldehyde in ddH_2O, pH = 10) onto the slides, and let them stand in a humid chamber for 10 min.
7. Remove slides from humid chamber, and let them almost dry at RT.
8. Place slides in a Coplin jar containing a 1 % Photo-Flo solution (in ddH_2O) and wash for 1 min.
9. Repeat previous wash 3 ×.
10. Use phase-contrast microscopy to visualize meiotic cells.
11. Process the slides immediately for the best results for immunostaining.

Post-fixation, Denaturation, and Hybridization (Adapted from [12])

1. If necessary, remove DAPI from SC spreads using 4 × SSC/ 0.05 % Tween-20 solution (soak slides in three changes of this solution for 5 min each, at RT).
2. Rinse slides in ddH_2O for 2 min at RT.
3. Air-dry slides.
4. Perform a first denaturation of the slide: place slides on a hot plate warmed to 70 °C (±1 °C), add 100 μl of denaturing solution (70 % formamide/2 × SSC), and cover with a 24 × 60 mm coverslip. Let it stand for 5 min.
5. Remove coverslip and rinse slide in ddH_2O for 1 min.
6. Add 100 μl of 1 M NaSCN (sodium thiocyanate) to each slide. Incubate for 3 h in a humid chamber at 65 °C.
7. Air-dry slides.

8. Perform a second denaturation: again place slides on a hot plate warmed to 70 °C (±1 °C), add 100 μl of denaturing solution (70 % formamide/2 × SSC), and cover with a 24 × 60 mm coverslip. Let it stand for five min. In case small DNA probes (such as LSI probes) are used, denaturation temperature can be increased up to 75 °C.
9. Meanwhile, denature probe according to manufacturer's instructions.
10. Remove slides from hot plate, dehydrate them using a cold ethanol series, and allow air-drying.
11. Add probe mix to SC slides, apply coverslip, and seal using rubber cement.
12. Incubate slides in a humid chamber at 37 °C for at least 48 h.

Post-hybridization Washes and Counterstaining

1. Wash slides in 0.04 × SSC at 45 °C for 5 min.
2. Air-dry slides and apply DAPI and antifade.
3. Slides can be stored at −20 °C for a long time.

4 Notes

1. There are other different protocols that can be applied to perform FISH on sperm heads. One in particular uses microwave to achieve decondensation and codenaturation. The procedure has been described in detail in [30]. The microwave protocol is considerably shorter, and the volume of decondensation reagents is also significantly lower than the time and volume of the protocol presented in this chapter. Nevertheless, the power and time suggested by the authors has to be adjusted according to voltage, intensity, and resistance of the microwave that is being used. In our lab in Barcelona, we have set the decondensation power and time to 75 W for 10 s, respectively (versus the 550 W for 15 s suggested in the abovementioned work), and have also lowered DTT concentration to a final 1 mM.
2. The whole process has to be controlled under stereomicroscope.
3. Fixation is a crucial step in order to obtain good chromosome extensions. In our experience [31], better extensions are obtained when using cold (−20 °C) methanol:acetic acid (1:1). Before performing MII or PB fixation, it is recommended to test the fixative on a clean, empty slide. If the fixative does not spread satisfactorily, prepare new, fresh fixative.

Acknowledgments

The author wish to thank Evelyn Ko for her inestimably help and comments on the procedures described in this manuscript.

References

1. Pacchierotti F, Adler I-D, Eichenlaub-Ritter U et al (2007) Gender effects on the incidence of aneuploidy in mammalian germ cells. Environ Res 104:46–69
2. Martin RH, Greene C, Rademaker A et al (2000) Chromosome analysis of spermatozoa extracted from testes of men with non-obstructive azoospermia. Hum Reprod 15:1121–1124
3. Smith JL, Garry VF, Rademaker AW et al (2004) Human sperm aneuploidy after exposure to pesticides. Mol Reprod Dev 67:353–359
4. Jurewicz J, Radwan M, Sobala W et al (2014) Lifestyle factors and sperm aneuploidy. Reprod Biol 14:190–199
5. Benet J, Oliver-Bonet M, Cifuentes P et al (2005) Segregation of chromosomes in sperm of reciprocal translocation carriers: a review. Cytogenet Genome Res 111:281–290
6. Moradkhani K, Puechberty J, Bhatt S et al (2006) Rare Robertsonian translocations and meiotic behaviour: sperm FISH analysis of t (13;15) and t(14;15) translocations: a case report. Hum Reprod 21:3193–3198
7. Kirkpatrick G, Ren H, Liehr T et al (2015) Meiotic and sperm aneuploidy studies in three carriers of Robertsonian translocations and small supernumerary marker chromosomes. Fertil Steril 103:11621169.e7
8. Munné S, Escudero T, Sandalinas M et al (2000) Gamete segregation in female carriers of Robertsonian translocations. Cytogenet Cell Genet 90:303–308
9. Pujol A, Boiso I, Benet J et al (2003) Analysis of nine chromosome probes in first polar bodies and metaphase II oocytes for the detection of aneuploidies. Eur J Hum Genet 11:325–336
10. Kuliev A, Verlinsky Y (2004) Meiotic and mitotic nondisjunction: lessons from preimplantation genetic diagnosis. Hum Reprod Update 10:401–407
11. Scherthan H, Weich S, Schwegler H et al (1996) Centromere and telomere movements during early meiotic prophase of mouse and man are associated with the onset of chromosome pairing. J Cell Biol 134:1109–1125
12. Roig I, Robles P, Garcia R et al (2005) Chromosome 18 pairing behavior in human trisomic oocytes. Presence of an extra chromosome extends bouquet stage Reproduction 129:565–575
13. Goldman AS, Hultén MA (1992) Chromosome in situ suppression hybridisation in human male meiosis. J Med Genet 29:98–102
14. Armstrong SJ, Goldman AS, Speed RM et al (2000) Meiotic studies of a human male carrier of the common translocation, t(11;22), suggests postzygotic selection rather than preferential 3:1 MI segregation as the cause of liveborn offspring with an unbalanced translocation. Am J Hum Genet 67:601–609
15. Oliver-Bonet M, Benet J, Sun F et al (2005) Meiotic studies in two human reciprocal translocations and their association with spermatogenic failure. Hum Reprod 20:683–688
16. Robles P, Roig I, Garcia R et al (2007) Pairing and synapsis in oocytes from female fetuses with euploid and aneuploid chromosome complements. Reproduction 133:899–907
17. Uroz L, Liehr T, Mrasek K et al (2009) Centromere-specific multicolour fluorescence in situ hybridization on human spermatocyte I and II metaphases. Hum Reprod 24:2029–2033
18. Barlow AL, Hultén MA (1998) Combined immunocytogenetic and molecular cytogenetic analysis of meiosis I oocytes from normal human females. Zygote 6:27–38
19. Oliver-Bonet M, Liehr T, Nietzel A et al (2003) Karyotyping of human synaptonemal complexes by cenM-FISH. Eur J Hum Genet 11:879–883
20. Sun F, Mikhaail-Philips M, Oliver-Bonet M et al (2008) The relationship between meiotic recombination in human spermatocytes and aneuploidy in sperm. Hum Reprod 23:1691–1697
21. Codina-Pascual M, Campillo M, Kraus J et al (2006) Crossover frequency and synaptonemal complex length: their variability and effects on human male meiosis. Mol Hum Reprod 12:123–133
22. Lynn A, Koehler KE, Judis L et al (2002) Covariation of synaptonemal complex length

and mammalian meiotic exchange rates. Science 296:2222–2225
23. Vidal F, Moragas M, Català V et al (1993) Sephadex filtration and human serum albumin gradients do not select spermatozoa by sex chromosome: a fluorescent in-situ hybridization study. Hum Reprod 8:1740–1743
24. Blanco J, Egozcue J, Vidal F (1996) Incidence of chromosome 21 disomy in human spermatozoa as determined by fluorescent in-situ hybridization. Hum Reprod 11:722–726
25. Durban M, Benet J, Sarquella J et al (1998) Chromosome studies in first polar bodies from hamster and human oocytes. Hum Reprod 13:583–587
26. Evans EP, Breckon G, Ford CE (1964) An air-drying method for meiotic preparations from mammalian testes. Cytogenetics 3:289–294
27. Goldman AS, Hultén MA (1993) Analysis of chiasma frequency and first meiotic segregation in a human male reciprocal translocation heterozygote, t(1;11)(p36.3;q13.1), using fluorescence in situ hybridisation. Cytogenet Cell Genet 63:16–23
28. Peters AH, Plug AW, van Vugt MJ et al (1997) A drying-down technique for the spreading of mammalian meiocytes from the male and female germline. Chromosome Res 5:66–68
29. Martínez-Flores I, Cabero LL, Egozcue J et al (2003) Synaptic process in the rat (Rattus norvegicus): influence of methodology on results. Microsc Res Tech 60:450–457
30. Ko E, Rademaker A, Martin R (2001) Microwave decondensation and codenaturation: a new methodology to maximize FISH data from donors with very low concentrations of sperm. Cytogenet Cell Genet 95:143–145
31. Durban M, Benet J, Boada M et al (2001) PGD in female carriers of balanced Robertsonian and reciprocal translocations by first polar body analysis. Hum Reprod Update 7:591–602

Part IV

Multicolor-FISH-Probe Sets (mFISH) and Immunostaining

Two- to Three-Color FISH

Thomas Liehr, Sven Hauke, and Britta Meyer

Abstract

Two-color FISH is normally done to apply simultaneously a specific and a control probe. Three-color FISH can be done also as a combination of control and specific probes; however, also three specific target regions may be addressed at the same time. Here the application of commercially available two- and three-color FISH probes is described, and applications in diagnostics and research are discussed.

Keywords Two-color FISH, Three-color FISH, Whole-chromosome painting probes, Locus-specific probes, Centromeric satellite probes, Locus-specific probes, Translocation, Deletion, Amplification

1 Introduction

While sophisticated multicolor-FISH approaches (see e.g., chapter by Thomas Liehr and Nadezda Kosyakova "Multiplex FISH and Spectral Karyotyping"; chapter by Thomas Liehr et al. "FISH Banding Techniques"; chapter by Thomas Liehr et al. "cenM-FISH Approaches"; chapter by Thomas Liehr et al. "Heterochro matin Directed M-FISH (HCM-FISH)"; chapter by Thomas Liehr et al. "Bar-Coding Is Back"; chapter by Anja Weise and Thomas Liehr "Parental Origin Determination FISH: Pod-FISH") are highly informative in cases with complex rearranged chromosomes [1] or in detection of cryptic aberrations [1], two- to three-color FISH is routinely applied in the majority of molecular cytogenetic tests (chapter by Thomas Liehr "Commercial FISH Probes"; chapter by Thomas Liehr "Classification of FISH Probes"; chapter by Thomas Liehr and Sven Hauke "Interphase FISH in Diagnostics"). One-color FISH experiments are normally the exception, as the use of a control probe is highly recommended, and thus the two-color FISH is that what is most often done. Examples for one-color FISH are replicative detargeting-FISH (ReD-FISH) (chapter by Nikolay Rubtsov and Natalya Zhdanova "The Replicative Detargeting FISH (ReD-FISH) Technique in Studies of Telomere Replica tion"), chromosome orientation-FISH (CO-FISH) ([2], chapter

Thomas Liehr (ed.), *Fluorescence In Situ Hybridization (FISH)*, Springer Protocols Handbooks, DOI 10.1007/978-3-662-52959-1_24, © Springer-Verlag Berlin Heidelberg 2017

by Nikolay Rubtsov and Natalya Zhdanova "The Replicative Detargeting FISH (ReD-FISH) Technique in Studies of Telomere Replication"), chromosome orientation and direction-FISH (COD-FISH) [2], nuclease in situ digestion-FISH (NU-FISH) [3], and DNA breakage detection-FISH (DBD-FISH) [4]. Different probe types as outlined in the chapter by Thomas Liehr ("Clas sification of FISH Probes") may be used in one- to three-color FISH approaches, and they may be based on DNA, PNA, or LNA (chapter by Thomas Liehr "Classification of FISH Probes").

For two- to three-color FISH here, we discuss (1) the use of whole-chromosome painting (WCP) probes for detection of radiosensitivity by means of three-color FISH [5] and (2) commercially available probes.

1. Early investigations in radiation cytogenetics (chapter by Galina Hovhannisyan et al. "Micronucleus FISH"; chapter by Galina Hovhannisyan et al. "Micronucleus FISH") applied classical cytogenetic analysis of Giemsa-stained or GTG-banded chromosomes and revealed a broad spectrum of induced chromosomal aberrations. The use of three-color FISH and simultaneous detection of chromosomes 1, 2, and 4 by WCPs allow reliable detection of translocations, insertions, and complex rearrangements additionally to the well-known radiation-induced acentric fragments, dicentric, and ring chromosomes. The mentioned three selected chromosomes represent about 22 % of the DNA content of the human genome. The principal idea of this three-color FISH method is to have a simple and rapid assay for the detection of increased radiosensitivity in cancer patients at hand [5] (Fig. 1a).
2. Other applications of two- to three-color FISH are locus-specific and centromeric satellite probes or a combination thereof which is widely used in cancer diagnostics and research [6–10]. The number of targets addressed with one FISH probe is mainly restricted by the number of different fluorochromes that can be reliably distinguished using standard filter sets (chapter by Michael Sommerauer et al. "Optical Filters and Light Sources for FISH"). The most commonly used multi-band filter sets are the green/orange dual-color filter set and the DAPI/green/orange triple-color filter set suitable to detect two-color FISH probes labeled in green and orange together with the DAPI counterstain, respectively. To detect numerical aberrations, i.e., copy number changes, for example, gene amplification or deletion, a combination of a locus-specific and the corresponding centromeric satellite probe is used. Here the ratio of locus specific to centromeric signals provides information on whether amplification or aneuploidy has occurred. A combination of two locus-specific probes

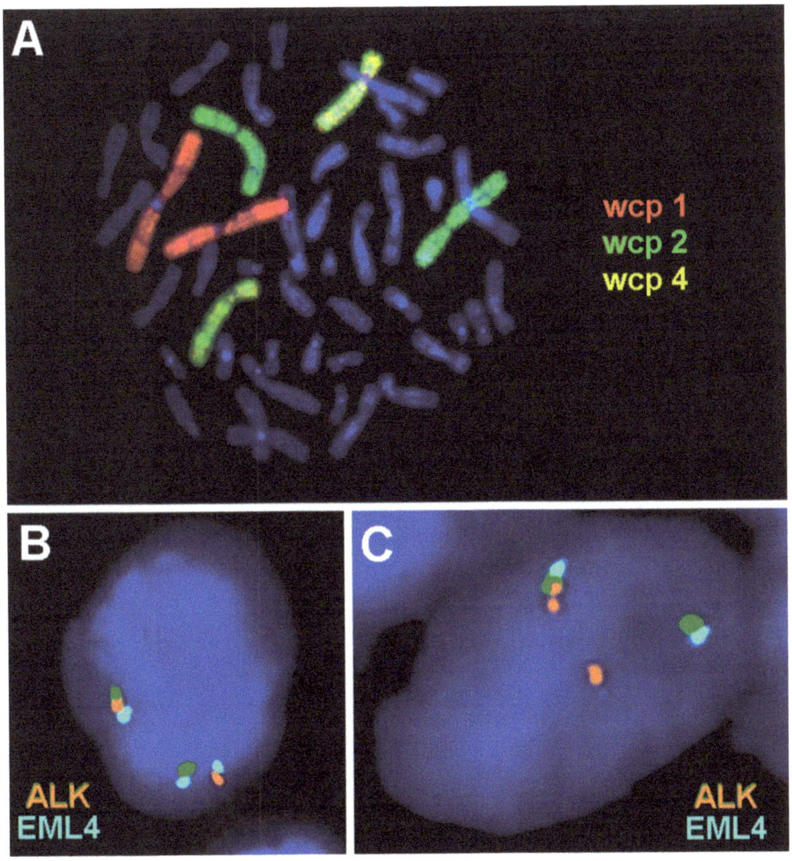

Fig. 1 (**a**) Three-color FISH using whole-chromosome painting (WCP) probes for chromosomes 1, 2, and 4 labeled with FITC (*green*) and SpectrumOrange (SO, *red*) only. The *yellow* color is derived from mix of FITC and SO. Thus, overall only three filters (including DAPI channel) or one triple filter is necessary for evaluation. A normal metaphase is depicted. (**b**) Zyto*Light* SPEC ALK/EML4 TriCheck™ probe hybridized to interphase nucleus showing an *ALK/EML4* fusion due to an inversion inv(2)(p21p23) as indicated by separated *green* and *orange ALK* signals each co-localizing with an aqua *EML4* signal. (**c**) Zyto*Light* SPEC ALK/EML4 TriCheck™ probe hybridized to interphase nucleus showing an *ALK* translocation without involvement of *EML4* as indicated by separated *green* and *orange ALK* signals not co-localizing with aqua *EML4* signals

either as break apart or as single- or dual-fusion probe can be used to detect structural aberrations, for example, chromosomal translocations (Fig. 1 in chapter by Thomas Liehr and Sven Hauke "Interphase FISH in Diagnostics").

Likewise, three-color probes can either be designed to detect numerical aberrations (e.g., enumeration of chromosomes 13/18/21) or to detect structural aberrations (e.g., *ALK/EML4* inversions in lung cancer). Compared to the mere break-apart probe design, the latter three-color probe also targets the fusion partner of interest usually labeled in aqua which can be detected by either an additional aqua single band-pass filter set or an aqua/green/orange triple-color filter set. This probe concept allows for the simultaneous determination whether a break has occurred in a specific gene and, additionally, whether a specific fusion partner is involved. A corresponding example of such a commercially available probe is the Zyto*Light* SPEC ALK/EML4 TriCheck™ probe by ZytoVision. By using the three-color probe, *ALK/EML4* fusions due to an inversion inv(2)(p21p23) (Fig. 1b) can reliably be discriminated from *ALK* translocations not affecting the *EML4* gene (Fig. 1c).

In the following we describe the application of two- and three-color FISH probes provided by ZytoVision (Bremerhaven, Germany).

2 Materials

Apart from standard cell biological and molecular cytogenetic equipment, including standard solutions (e.g., ethanol, methanol, formamide, formaldehyde, xylene, etc.), no more specialized items are required. The equipment needed for FISH is listed in chapter by Thomas Liehr et al. ("The Standard FISH Procedure").

3 DNA Probes

Probes from ZytoVision were used—recommended protocols are principally all the same for one- to three-color FISH probes and ready to use.

4 Methods

4.1 Slide Pretreatment

In principal, the protocol as described in chapter by Thomas Liehr ("Characterization of Archived Formalin-Fixed/Paraffin-Embedded or Cryofixed Tissue, Including Nucleus Extraction") was followed with the exception of using a citric acid-based heat pretreatment instead of the hazardous NaSCN.

4.2 Fluorescence In Situ Hybridization (FISH)

After application of ready-to-use probes, cover slips were applied and sealed with rubber cement. Probe and target DNA were co-denatured, followed by hybridization overnight at 37 °C. Stringency wash was done at 37 °C using SSC-based buffers according to ZytoVision protocols. Slides were then dehydrated and counterstained using DAPI/antifade (*see* **Note 1**).

4.3 Evaluation

The main advantage using two-color probes is that, when using suitable dual-color filter sets, signals of both colors and their location toward each other can be observed in each nucleus or on each metaphase spread without the need to change filter sets and/or to take and merge pictures. Orange (or red) and green signals located in close proximity will appear as yellow signals. To confirm that a yellowish signal is indeed composed of a green and an orange signal, single-color filter sets specific for each fluorochrome can be used (see chapter by Michael Sommerauer et al. "Optical Filters and Light Sources for FISH").

Gain or loss of gene regions is mainly defined by an increased or decreased number of signals targeting this region in comparison to signals derived by control regions located on the same chromosome. Translocations or fusions are detected by altered distances of signals flanking or encompassing possible breakpoints (see chapter by Thomas Liehr and Sven Hauke "Interphase FISH in Diagnostics", Fig 1). On tissue sections even a certain proportion of normal/nonneoplastic cells and nuclei will show signal patterns other than expected due to cutting artifacts affecting the nuclei (chapter by Thomas Liehr "Characterization of Archived Formalin-Fixed/Paraffin-Embedded or Cryofixed Tissue, Including Nucleus Extraction", Sect. 3.5), while there is no such problem when working with metaphase spreads, blood or bone marrow smears.

Additionally, signal patterns other than the patterns which can be expected based on the type of probe might be observable due to complex genomic rearrangements. In case of break-apart probes, unbalanced translocations might lead to loss of signals, while when using dual-color dual-fusion probes, extra signals but no fusions might indicate translocations involving other fusion partners than expected. Thus, it is important to know about the target addressed by the FISH probe and potential mechanisms of rearrangements that can affect the target before evaluating FISH slides (chapter by Thomas Liehr "Commercial FISH Probes").

5 Notes

1. Due to their unique composition, ZytoVision FISH probes can be used with stringency washing temperatures as low as 37 °C. Alternatively, probes of ZytoVision can (and probes of other manufacturers have to) be used with high-stringency protocols such as that described in chapter by Thomas Liehr et al. ("The Standard FISH Procedure", Sect. 3.3).

References

1. Houge G, Liehr T, Schoumans J et al (2003) Ten years follow up of a boy with a complex chromosomal rearrangement: going from >5 to 15-breakpoint CCR. Am J Med Genet A 118A:235–240
2. Meyne J, Bailey SM, Goodwin EH (2002) Strand-specific fluorescence in situ hybridization: CO-FISH and COD-FISH. In: Liehr T, Rautenstrauss B (eds) FISH-technology, Springer-lab manual. Springer, Berlin, pp 262–271. ISBN 3-540-67276-1
3. Gosalvez J, de la Torre J, Lopez-Fernandez C et al (2002) The combined use of nuclease in situ digestion and FISH (NU-FISH). In: Liehr

T, Rautenstrauss B (eds) FISH-technology, Springer-lab manual. Springer, Berlin, pp 272–281. ISBN 3-540-67276-1

4. Fernandez JL, Goyanes V, Gosalvez J (2002) DNA breakage detection-FISH (DBD-FISH). In: Liehr T, Rautenstrauss B (eds) FISH-technology, Springer-labmanual. Springer, Berlin, pp 282–290. ISBN 3-540-67276-1
5. Neubauer S (2002) Detection of radiosensitivity by means of three-color FISH. In: Liehr T, Rautenstrauss B (eds) FISH-technology, Springer-labmanual. Springer, Berlin, pp 291–301. ISBN 3-540-67276-1
6. Othman MA, Grygalewicz B, Pienkowska-Grela B et al (2015) Novel cryptic rearrangements in adult B-cell precursor acute lymphoblastic leukemia involving the MLL gene. J Histochem Cytochem 63:384–390
7. Othman MA, Melo JB, Carreira IM et al (2015) High rates of submicroscopic aberrations in karyotypically normal acute lymphoblastic leukemia. Mol Cytogenet 8:45
8. Alhourani E, Rincic M, Othman MA et al (2014) Comprehensive chronic lymphocytic leukemia diagnostics by combined multiplex ligation dependent probe amplification (MLPA) and interphase fluorescence in situ hybridization (iFISH). Mol Cytogenet 7:79
9. Fountzilas G, Christodoulou C, Bobos M et al (2012) Topoisomerase II alpha gene amplification is a favorable prognostic factor in patients with HER2-positive metastatic breast cancer treated with trastuzumab. J Transl Med 10:212
10. Popper HH, Gruber-Moesenbacher U, Müllauer L et al (2013) Recommendations of the Austrian Working Group on Lung Pathology and Oncology for predictive molecular and immunohistochemical testing in non-small cell lung cancer. memo 6:83–91

Multiplex FISH and Spectral Karyotyping

Thomas Liehr and Nadezda Kosyakova

Abstract

Multicolor fluorescence in situ hybridization (mFISH) assays are indispensable for a precise description of complex chromosomal rearrangements and marker chromosomes. Routine application of such techniques on human chromosomes started in 1996 with the simultaneous use of all 24 human whole-chromosome painting (WCP) probes in multiplex-FISH (M-FISH) and spectral karyotyping (SKY). Here we present a review on the available mFISH approaches using WCP probes and describe a basic protocol for M-FISH and SKY.

Keywords Multicolor fluorescence in situ hybridization (mFISH), Whole-chromosome paints, Multiplex-FISH (M-FISH), Spectral karyotyping (SKY), Combinatorial labeling, Ratio labeling, Human probe sets, Murine probe sets

1 Introduction

Chromosome banding (e.g., by GTG banding, G-bands by trypsin using Giemsa) is still the gold standard for all routine techniques in human cytogenetics (chapter by Anja Weise and Thomas Liehr "Pre- and Postnatal Diagnostics and Research on Peripheral Blood, Bone Marrow, Chorion, Amniocytes, and Fibroblasts"; chapter by Thomas Liehr and Monika Ziegler "Application of FISH to Previously GTG-Banded and/or Embedded Cytogenetic Slides"). However, the chromosome morphology and the black-and-white banding pattern are the only two parameters that are evaluated in this approach [1]. Thus, the origin of additional material in a structurally altered chromosome often remains vague. In order to overcome such limitations, FISH approaches were introduced into cytogenetics in the 1980s, and then the new field of molecular cytogenetics was born [2]. One of the main sources of progress of the field in recent years has been the introduction and further development of multicolor FISH (mFISH) approaches into molecular cytogenetics. mFISH is defined as the simultaneous use of at least three different ligands or fluorochromes for the specific

Thomas Liehr (ed.), *Fluorescence In Situ Hybridization (FISH)*, Springer Protocols Handbooks,
DOI 10.1007/978-3-662-52959-1_25, © Springer-Verlag Berlin Heidelberg 2017

labeling of DNA, excluding the counterstain [3]. According to this definition, the first successful mFISH experiments were done in 1989 by Nederlof and coworkers [4] by visualizing three differently labeled nucleic acid sequences, simultaneously, in blue [amino methyl coumarin acetic acid (AMCA)], red [tetramethylrhodamine isothiocyanate (TRITC)], and green [fluorescein isothiocyanate (FITC)]. The first mFISH probe sets were put together 7 years later in 1996 [5–7].

The staining of each of the 24 human chromosomes in different colors at the same time using whole-chromosome painting (WCP) probes has been described on several occasions. Different names have been introduced for more or less the same probe sets: multiplex-FISH (M-FISH) [6], spectral karyotyping (SKY) [7], multicolor FISH [8, 9], combined binary ratio labeling-FISH (COBRA-FISH) [10], 24-color FISH [11], as well as recently again (!) in 2015 (!) as karyotype identification via spectral separation (KISS) [12]. Moreover, the basic mFISH probe set using WCP probes has been modified through either inducing molecular changes in the probes themselves or by the addition of supplementary probes. The so-called IRS-PCR multiplex-FISH (IPM-FISH) method uses whole-chromosome painting probes which are modified by an interspersed polymerase chain reaction (IRS), which leads to a 24-color FISH painting plus an R-band-like pattern [13]. For special questions, other probes have also been added to the basic 24-color FISH probe set, like single-copy probes (e.g., a probe for human papillomavirus [14]) or subtelomeric probes [15], chromosome region-specific probes (e.g., a probe for the short arms of all acrocentric chromosomes [16]), or chromosome arm-specific probes for all human chromosomes [17].

The applications of mFISH using WCP probes cover the whole spectrum of human cytogenetics and are summarized in [18]. Four to seven different fluorescence dyes can be used to label the WCP probes for M-FISH, SKY, COBRA-FISH, and so on. For most of the aforementioned approaches, the principle of combinatorial labeling is applied. However, the required 24 (or more)-color combinations can also be achieved using ratio labeling. In the latter case, used, for example, by Tanke et al. [10], only four fluorochromes are necessary to achieve 96 possible color combinations or pseudocolors. However, for the combinatorial approach, very exact adjustments of differently labeled probes are necessary. Thus, despite the undeniable advantage of having more pseudocolors available and fewer necessary color channels in ratio labeling, combinatorial labeling is still the labeling approach most frequently used in mFISH. CCD camera-based image acquisition and computer-based image analysis are normally required for mFISH (chapter by Ivan Iourov "Microscopy and Imaging Systems"; chapter by Michael Sommerauer et al. "Optical Filters and Light Sources for FISH"). In combinatorial labeling, a computer is necessary, because at least one of the fluorochromes used has its emission maximum in a spectral region that is invisible to the human eye, while a computer is needed in ratio labeling because the slight color differences can only be resolved reliably by a computer program. Image acquisition itself can be performed using two different principles: (1) split spectra are acquired via a set of specific filters, as suggested by Speicher and coworkers [6] (M-FISH), or (2) complete emission spectra are acquired by an interferometer-based spectral imaging system, as recommended by Schröck and coworkers [7] (SKY); for more on optical filters, see chapter by Michael Sommerauer et al. ("Optical Filters and Light Sources for FISH").

Apart from human species, mFISH probe sets based on WCP probes have been published for the mouse (*Mus musculus f. domestica*) [7, 19], the rat (*Rattus norvegicus f. domestica*) [20], the chicken (*Gallus gallus f. domestica*) [21], the dog (*Canis lupus familiaris*) [22], and a beetle (*Nasonia* sp.) [23]. For further animal paints, see chapter by Fengtang Yang et al. ("Animal Probes and ZOO-FISH").

The protocol for M-FISH and SKY performed on human metaphase chromosome preparations is provided here. In both protocols, the 24 WCP probes are labeled with five different fluorochromes, and chromosomes are counterstained with DAPI (4,6-diamidino-2-phenylindole-2HCl).

2 Materials

Apart from standard cell biological and molecular cytogenetic equipment, including standard solutions (e.g., ethanol, methanol, formamide, formaldehyde, xylene, etc.), no more specialized items are required. The equipment needed for FISH itself is listed in chapter by Thomas Liehr et al. ("The Standard FISH Procedure").

2.1 Microscopy, Image Acquisition, and Evaluation Software

For standard M-FISH based on five fluorochromes and one counterstain, at the very least, a fluorescence microscope suitable for six filter sets is required. A charge-coupled device (CCD) camera connected to computer-based image acquisition and evaluation software is also needed; a number of providers are available on the market, like MetaSystems, Applied Imaging, and others.

For SKY, a microscope equipped with the SpectraCube provided by the Spectral Imaging Systems (ASI, Inc., Vista, CA, USA) and appropriate for two filter sets is needed. ASI is the only provider of this system, and they offer the complete system, including image acquisition, evaluation software, and SKY probes.

2.2 Chemicals and Other Materials

As previously mentioned for SKY, ASI provides corresponding ready-labeled probe sets with all 24 human chromosomes as probes; probes of the 22 murine chromosomes are also available

(*Mus musculus*). Probes for M-FISH can be purchased from, e.g., MetaSystems (Altlussheim, Germany).

3 Methods

When using commercial probes, we recommend that the manufacturer's instructions should be followed for M-FISH or SKY (*see* **Note 1**). In general, this corresponds to regular FISH protocol, as described in chapter by Thomas Liehr et al. ("The Standard FISH Procedure") (*see* **Note 2**).

In our laboratory we use homemade microdissection-based WCP probes for M-FISH [9]. They are labeled according to Fig. 1a. These probes are used successfully in clinical cases (Fig. 1b), in tumor cytogenetic cases (Fig. 1c), and in research (e.g., ZOO-FISH; Fig. 1d) (*see* **Note 3**). In terms of coworking, these M-FISH probes can be provided to other laboratories on request. We also use SKY, e.g., in studies on mouse chromosomes using the WCP probes of ASI [24] and also in human leukemia cases (Fig. 2) [25].

4 Notes

1. In general, probes suited to SKY can be evaluated without any problems in M-FISH, i.e., a six-filter microscope with the corresponding filter sets (FITC, SpectrumOrange, TexasRed, cyanine 5, cyanine 5.5, DAPI). The same holds true for M-FISH probe sets, which can be analyzed on a SKY system as long as they are not based on fluorochromes outside the frame covered by the SKY-1 filter (~450–850 nm). Thus, the M-FISH probe set provided by MetaSystems cannot be evaluated on a SKY system, for example, as it contains diethylaminocoumarin (DEAC) with an excitation maximum of ~430 nm.
2. Biotin may be detected by avidin-cyanine 5 here if not replaced by directly cyanine 5-labeled dUTP.
3. Good pseudocolors like those shown in Figs. 1 and 2 are not always possible to achieve in M-FISH or SKY. However, even when the hybridization quality is relatively poor, in most cases it is possible to arrive at results which then can be confirmed by one-, two-, or three-color FISH by applying the corresponding WCP probes; the latter should be done anyway. Moreover, there are multiple features in the M-FISH and SKY software that make it possible to create different pseudocolors, to analyze the different color channels separately, or to get the system to suggest which chromosome the hybridization pattern indicates (mainly in SKY).

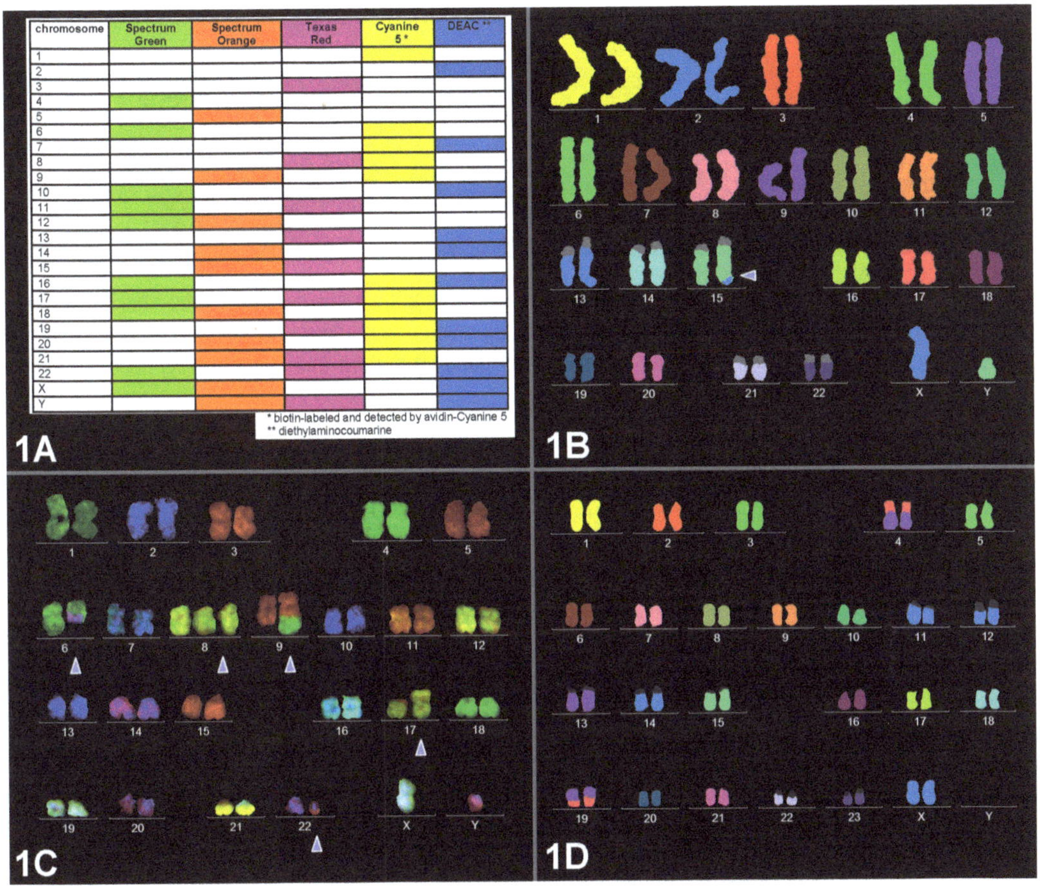

Fig. 1 (**a**) M-FISH labeling scheme as used in our laboratory for a homemade M-FISH set. (**b**) Prenatal case with a karyotype 46,XY,der(15)t(X;15)(q25;q26.3). The derivative chromosome 15 is denoted by an arrowhead. Here the M-FISH result is depicted in pseudocolors. The heterochromatic regions unlabeled by the M-FISH probe set are shown in gray pseudocolors. (**c**) Chronic myeloid leukemia case with a karyotype 47,XY,t(6;9;22)(q12;q34;q11),+8,i(17)(q10). Here the M-FISH result is depicted in a "real color" picture, i.e., an overlay of the different color channels. All derivatives are denoted by *arrowheads*. (**d**) ZOO-FISH: *Gorilla gorilla* chromosomes after M-FISH using human WCP probes. The corresponding homologous chromosomes can be identified by comparison with Fig. 1b. The evolutionarily conserved translocation t(5;17) and the distribution of human chromosome sequences on two gorilla chromosomes are easily recognizable. The heterochromatic regions unlabeled by the M-FISH probe set are shown in *gray* pseudocolors

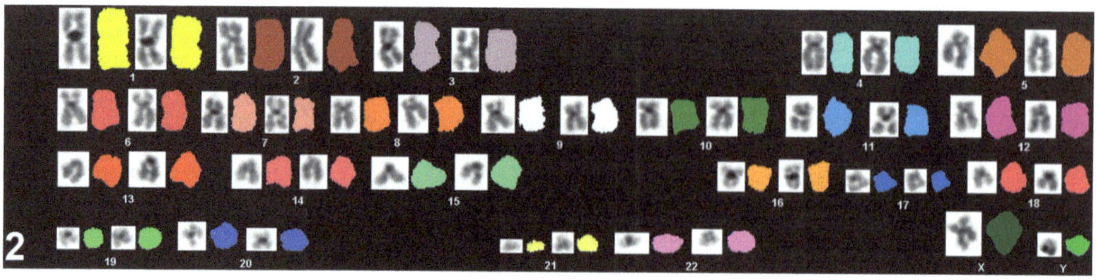

Fig. 2 SKY result for an acute lymphatic leukemia-derived metaphase spread. No cryptic aberrations were picked up here; note that SKY gave good, evaluable results even though the quality of the chromosomes was poor. One nice feature of SKY is that the inverted DAPI banding pattern and the SKY pseudocolors can be depicted side by side, as shown here

References

1. Claussen U, Michel S, Mühlig P et al (2002) Demystifying chromosome preparation and the implications for the concept of chromosome condensation during mitosis. Cytogenet Genome Res 98:136–146
2. Chang SS, Mark HF (1997) Emerging molecular cytogenetic technologies. Cytobios 90:7–22
3. Liehr T, Starke H, Weise A et al (2004) Multicolor FISH probe sets and their applications. Histol Histopathol 19:229–237
4. Nederlof PM, Robinson D, Abuknesha R et al (1989) Three-color fluorescence in situ hybridization for the simultaneous detection of multiple nucleic acid sequences. Cytometry 10:20–27
5. Yurov YB, Soloviev IV, Vorsanova SG et al (1996) High resolution multicolor fluorescence in situ hybridization using cyanine and fluorescein dyes: rapid chromosome identification by directly fluorescently labeled alphoid DNA probes. Hum Genet 97:390–398
6. Speicher MR, Gwyn Ballard S, Ward DC (1996) Karyotyping human chromosomes by combinatorial multi-fluor FISH. Nat Genet 12:368–375
7. Schröck E, du Manoir S, Veldman T et al (1996) Multicolor spectral karyotyping of human chromosomes. Science 273:494–497
8. Tanke HJ, De Haas RR, Sagner G et al (1998) Use of platinum coproporphyrin and delayed luminescence imaging to extend the number of targets FISH karyotyping. Cytometry 33:453–459
9. Senger G, Chudoba I, Plesch A (1998) Multicolor-FISH—the identification of chromosome aberrations by 24 colors. BIOforum 9:499–503
10. Tanke HJ, Wiegant J, van Gijlswijk RP et al (1999) New strategy for multi-colour fluorescence in situ hybridisation: COBRA: COmbined Binary RAtio labelling. Eur J Hum Genet 7:2–11
11. Azofeifa J, Fauth C, Kraus J et al (2000) An optimized probe set for the detection of small interchromosomal aberrations by use of 24-color FISH. Am J Hum Genet 66:1684–1688
12. Potapova TA, Unruh JR, Box AC et al (2015) Karyotyping human and mouse cells using probes from single-sorted chromosomes and open source software. Biotechniques 59:335–346
13. Aurich-Costa J, Vannier A, Gregoire E et al (2001) IPM-FISH, a new M-FISH approach using IRS-PCR painting probes: application to the analysis of seven human prostate cell lines. Genes Chromosomes Cancer 30:143–160
14. Szuhai K, Sandhaus E, Kolkman-Uljee SM et al (2001) A novel strategy for human papillomavirus detection and genotyping with SybrGreen and molecular beacon polymerase chain reaction. Am J Pathol 159:1651–1660
15. Tosi S, Giudici G, Rambaldi A et al (1999) Characterization of the human myeloid leukemia-derived cell line GF-D8 by multiplex fluorescence in situ hybridization, subtelomeric probes, and comparative genomic hybridization. Genes Chromosomes Cancer 24:213–221
16. Mrasek K, Heller A, Rubtsov N et al (2001) Reconstruction of the female *Gorilla gorilla* karyotype using 25-color FISH and multicolor banding (MCB). Cytogenet Cell Genet 93:242–248
17. Liehr T, Claussen U (2002) Multicolor-FISH approaches for the characterization of human

chromosomes in clinical genetics and tumor cytogenetics. Curr Genom 3:213–235

18. Liehr T (2016) Multicolor FISH homepage. http://ssmc-tl.com/mfish.html
19. Liyanage M, Coleman A, du Manoir S et al (1996) Multicolour spectral karyotyping of mouse chromosomes. Nat Genet 14:312–315
20. Buwe A, Steinlein C, Koehler MR et al (2003) Multicolor spectral karyotyping of rat chromosomes. Cytogenet Genome Res 103:163–168
21. Griffin DK, Haberman F, Masabanda J et al (1999) Micro- and macrochromosome paints generated by flow cytometry and microdissection: tools for mapping the chicken genome. Cytogenet Cell Genet 87:278–281
22. Milne SB, Hoather T, O'Brien PCM et al (2004) Karyotype of canine soft tissue sarcomas: a multi-colour, multi-species approach to canine chromosome painting. Chromosome Res 12:825–835
23. Rutten KB, Pietsch C, Olek K et al (2004) Chromosomal anchoring of linkage groups and identification of wing size QTL using markers and FISH probes derived from microdissected chromosomes in Nasonia (Pteromalidae: Hymenoptera). Cytogenet Genome Res 105:126–133
24. Karst C, Trifonov V, Romanenko SA et al (2006) Molecular cytogenetic characterization of the mouse cell line WMP2 by spectral karyotyping and multicolor banding applying murine probes. Int J Mol Med 17:209–213
25. Mkrtchyan H, Glaser M, Gross M et al (2006) Multicolor-FISH applied to resolve complex chromosomal changes in a case of T-ALL (FAB L2). Cytogenet Genome Res 114:270–273

FISH Banding Techniques

Thomas Liehr, Nadezda Kosyakova, and Anja Weise

Abstract

During the last decade, numerous chromosome banding techniques based on FISH were developed for the human and for the murine genome. Here we review FISH banding techniques, which were recently defined as "any kind of FISH-technique, which provide the possibility to characterize simultaneously several chromosomal subregions smaller than a chromosome arm" and are DNA specific. FISH banding methods are successfully applied in research and diagnostics.

Keywords Multicolor fluorescence in situ hybridization (mFISH), Human, Mouse, Multicolor banding (MCB/mBAND), Intrachromosomal aberrations, Submicroscopic alterations, Interphase, Nuclear architecture

1 Introduction

Conventional protein-based chromosome banding techniques have some profound technical restrictions, such as the fact that only (1) the black-and-white banding pattern combined with chromosome morphology, (2) changes within the normal pattern, (3) size variations in a chromosomal band or the chromosome itself, and (4) changes of the centromere index can be detected ([1]; chapter by Anja Weise and Thomas Liehr "Pre- and Postnatal Diagnostics and Research on Peripheral Blood, Bone Marrow, Chorion, Amniocytes, and Fibroblasts"). Thus, the origin of additional material in a structurally altered chromosome can often only be resolved by applying FISH (chapter by Thomas Liehr and Anja Weise "Back ground"). One straightforward approach of this type is to use multicolor FISH probe set(s), which enable each of the 24 different human chromosomes to be stained in different colors at the same

Thomas Liehr (ed.), *Fluorescence In Situ Hybridization (FISH)*, Springer Protocols Handbooks, DOI 10.1007/978-3-662-52959-1_26, © Springer-Verlag Berlin Heidelberg 2017

time using whole-chromosome painting (WCP) libraries (M-FISH [2] and SKY [3]) (chapter by Thomas Liehr and Nadezda Kosyakova "Multiplex FISH and Spectral Karyotyping").

However, FISH methods using WCP probes reach their limits when the exact localization of a chromosomal breakpoint is required or intrachromosomal aberrations are present, since these are not detectable and/or resolvable by M-FISH/SKY. Thus, different approaches were developed to overcome such limitations, and these can be grouped together under the term "FISH banding methods." As defined previously [4], FISH banding methods "are any kind of FISH technique, which provide the possibility to characterize simultaneously several chromosomal subregions smaller than a chromosome arm (excluding the short arms of the acrocentric chromosomes). FISH banding methods fitting that definition may have quite different characteristics, but share the ability to produce a DNA-specific chromosomal banding" ([4]; chapter by Thomas Liehr et al. "Bar-Coding Is Back").

As detailed below four different FISH banding approaches are currently available for the mouse (*Mus musculus f. domestica*) and nine for the human.

1.1 Murine FISH Banding Sets

Three whole-genome-directed murine YAC/BAC-based chromosome bar code sets (CBC) were developed as single-, dual-, or triple-color FISH approaches, respectively [5–7]. Only one approach for single-chromosome-directed FISH banding is available at present: microdissection-based multicolor banding (MCB or mBAND). This was developed for exclusively for chromosome 11 by a commercial provider [8] and was published to be available for all murine chromosomes in 2013 by our group [9].

1.2 Human FISH Banding Sets

For the human, four whole-genome-directed FISH banding probe sets have been developed: IPM-FISH [10], cross-species color banding (Rx-FISH) [11], somatic cell hybrid-based CBC [12], and multitude multicolor banding (mMCB) [13]. IPM-FISH is not performed anymore; the same holds true for Rx-FISH probe set, being previously commercially available from Cambio (Cambridge, UK) as "harlequin FISH." Also mMCB, a whole-genome-directed FISH banding set derived from single-chromosome-directed multicolor banding (MCB) (see below) [14, 15], is, like the somatic cell hybrid-based CBC [12], not commercially available at present.

Additionally, five single-chromosome-directed/chromosome-specific FISH banding probe sets have been published. In principle, two types of these chromosome-specific FISH banding sets can be distinguished: those based on microdissection-derived, chromosome-region specific probes and those based on locus-specific ones. Microdissection-based multicolor banding (MCB or mBAND) [14, 15], spectral color banding (SCAN) [16, 17], and M-FISH using chromosome-region-specific probes (CRP) [18] belong to the first group. YAC-/BAC-based multicolor banding (Y/B-MCB) and YAC-/BAC-based CBC (see also chapter by Thomas Liehr et al. “Bar-Coding Is Back”) belong to the second group [19–21]; for an overview, see [22].

1.3 FISH Banding Sets: Applications

FISH banding methods are applied for the characterization of marker and derivative chromosomes in clinical genetics and tumor cytogenetics. Studies to clarify the intranuclear structure have been done (chapter by Thomas Liehr and Nadezda Kosyakova “Three-Dimensional Interphase Analysis Enabled by Suspension FISH”), as well as analyses of cell lines. Characterizations of chromosomal changes after irradiation and during evolution (comparative cytogenetics) were also performed (for an overview, see [22]).

1.4 FISH Banding Sets: Advantages and Disadvantages

By definition, techniques like YAC-/BAC-based or somatic cell hybrid-based CBC and nonoverlapping microdissection libraries have the disadvantage that unstained and thus uninformative gaps are left along the chromosome. Such gaps can cause problems, as breakpoints within the unstained gaps cannot be determined exactly [23]. Conversely, techniques based on locus-specific probes (like BACs) would theoretically provide the ability to define chromosomal breakpoints very accurately. However, the coverage of the human genome by corresponding available non-chimeric clones appears to be too low to permit such a molecular cytogenetic approach [21]. All FISH banding probe sets developed and applied for the whole human genome provide the advantage of leaving no uninformative gaps in euchromatin. Noncommercial mMCB and MCB probe sets are currently the only ones that are anchored in the human genome sequence by array-CGH mapping [24]. The resolutions that can be achieved by the different FISH banding methods vary significantly. The Rx-FISH has a resolution of about 100–200 bands per human haploid human karyotype. The resolutions of SCAN and YAC-/BAC-based CBC are around 300–400 bands per haploid karyotype, while resolutions of between 400 and 800 bands per haploid karyotype have been described for (m)MCB [4]. Here we report the protocol for MCB/mBAND, as this is the most frequently used FISH banding approach.

2 Materials

Apart from standard cell biological and molecular cytogenetic equipment, including standard solutions (e.g., ethanol, methanol, formamide, formaldehyde, xylene, etc.), no more specialized items are required. The equipment needed for multicolor FISH is listed in chapter by Thomas Liehr et al. ("The Standard FISH Procedure" and chapter by Thomas Liehr and Nadezda Kosyakova "Multiplex FISH and Spectral Karyotyping").

Probes for mBAND can be purchased from MetaSystems (Altlussheim, Germany), as XCyte 1–22, X, and Y.

3 Methods

When using commercial probes, we recommend that the manufacturer's instructions should be followed for mBAND. In general, the regular FISH protocol is used, as described in chapter by Thomas Liehr et al. ("The Standard FISH Procedure") (*see* **Note 1**).

Examples of homemade FISH banding probe set (MCB) from clinical (Fig. 1) and tumor genetic diagnostic cases (Fig. 2), as well as from research (Fig. 3), are presented. FISH banding can be applied in postnatal (Fig. 1a, b), prenatal (Fig. 1c), and leukemia cases (Fig. 2). The mcb probe sets for *Mus musculus* are finished in the lab of the authors, and as an example murine chromosome 3 is shown in Fig. 3. In some cases, MCB is sufficient to characterize chromosomal breakpoints (Fig. 1b, c); in others it is not (Fig. 1a). In Fig. 2, one special advantage of mMCB is apparent, i.e., single-cell aberrations can be characterized in detail using this approach (*see* **Note 2**).

4 Notes

1. Biotin may be detected by avidin-Cyanine 5 here if not replaced by directly Cyanine 5-labeled dUTP.
2. Good pseudocolors like those shown in Figs. 1, 2, and 3 are not always achieved in mBAND/MCB. However, even when the hybridization quality is not particularly high, it is still possible to get results in most cases using the fluorochrome profiles and real colors for evaluation.

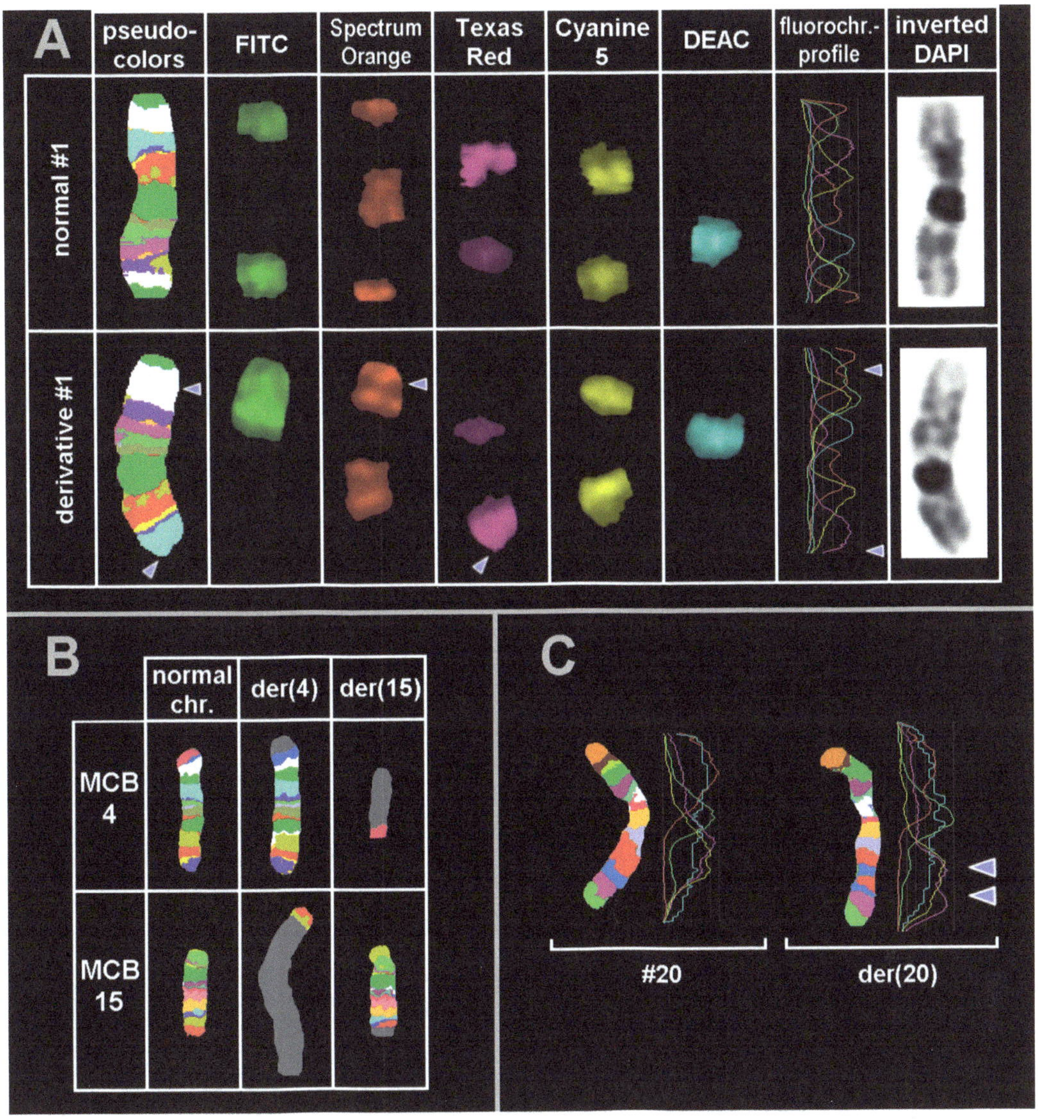

Fig. 1 Examples of the application of chromosome-specific MCB probe sets in humans. The clinical cases were kindly provided by Dr. Dilek Aktas (Ankara, Turkey), Dr. M. Stumm and Dr. R. Wegner (Berlin, Germany), and Dr. B. Schulze (Hannover, Germany). (**a**) A derivative chromosome 1 was detected in a mentally retarded child. MCB and application of subtelomeric probes (results not shown here) revealed that there was a der(1) (pter->p34.2::q43~44->p34.2). Thus, aside from an inversion, a deletion of 1q43~44 to 1qter is present in the derivative chromosome. The MCB results are depicted as pseudocolor banding (pseudocolors), and the different color channels of the five applied fluorochromes [fluorescein isothiocyanate (FITC), SpectrumOrange, TexasRed, Cyanine 5, and diethylaminocoumarin (DEAC)] plus fluorochrome profiles along each chromosome and inverted DAPI banding are shown. The breakpoints are marked as *arrowheads*. (**b**) A balanced translocation t(4;15) was characterized in a female studied cytogenetically due to infertility. MCB with the probe sets for the corresponding chromosomes refined the characterization of the breakpoints to 4p16.1 and 15q26.1. The results are depicted in pseudocolors. The pseudocolors were aligned to the corresponding chromosomal bands beforehand. *Chr* chromosome, *Der*, derivative chromosome. (**c**) In a prenatal case, a derivate 20 was detected by GTG banding. MCB using a chromosome 20-specific probe set revealed a molecular cytogenetically balanced inv(20)(q12q13.3). Breakpoints are marked by *arrowheads*

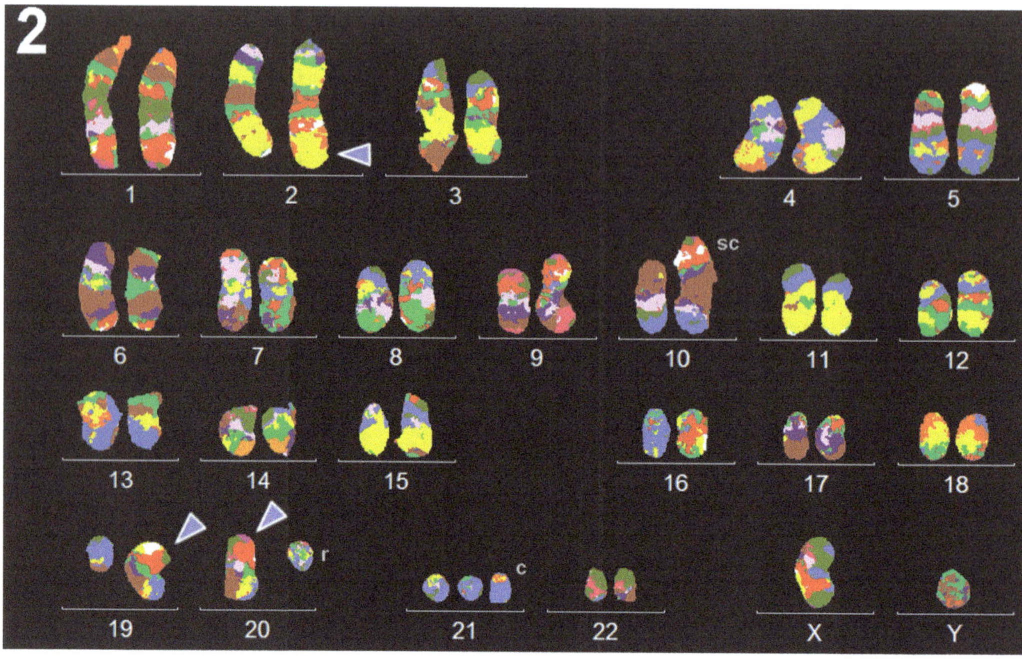

Fig. 2 Result of a multitude multicolor banding (mMCB) experiment on a patient with acute myeloid leukemia (AML) type M7 (pseudocolor image). The child had a constitutional trisomy 21 and other acquired chromosomal abnormalities. After application of mMCB, the karyotype could be described as follows: 47,XY,der(2)t(2;11)(q37.3;q12~13),der(19)t(1;19)(q31;p13.3),der(20)t(1;20)(q31;q12~13.1),r(20)(p11.2q12),+21. The derivative chromosomes are marked with *arrowheads*; the ring chromosome 20 is marked with an *r* and the constitutional additional chromosome 21 with a *c*. One chromosome 10 of this specific metaphase spread showed a single-cell aberration (sc), a der(10)t(1;10)(q41;p13). The cell suspension of bone marrow cells for this case was kindly provided by Dr. M.L. Macedo Silva (Rio de Janeiro, Brazil)

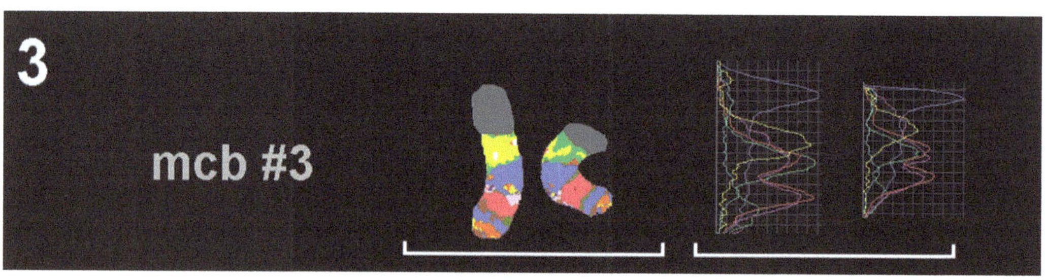

Fig. 3 Example of a mcb probe set for mouse chromosomes. A pseudocolor picture and fluorochrome profiles of the mcb for murine chromosome 3 on normal chromosomes of *Mus musculus* are depicted

References

1. Claussen U, Michel S, Mühlig P et al (2002) Demystifying chromosome preparation and the implications for the concept of chromosome condensation during mitosis. Cytogenet Genome Res 98:136–146
2. Speicher MR, Gwyn Ballard S, Ward DC (1996) Karyotyping human chromosomes by combinatorial multi-fluor FISH. Nat Genet 12:368–375

3. Schröck E, du Manoir S, Veldman T et al (1996) Multicolor spectral karyotyping of human chromosomes. Science 273:494–497
4. Liehr T, Heller A, Starke H et al (2002) FISH banding methods: applications in research and diagnostics. Expert Rev Mol Diagn 2:217–225
5. Liechty MC, Carpio CM, Aytay S et al (1999) Hybridization-based karyotyping of mouse chromosomes: hybridization-bands. Cytogenet Cell Genet 86:34–38
6. Liechty MC, Clase AC, Puschus KL et al (2000) Mouse linkage cytogenetics (L-C) probes. Cytogenet Cell Genet 88:163–167
7. Henegariu O, Dunai J, Chen XN et al (2001) A triple color FISH technique for mouse chromosome identification. Mamm Genome 12:462–465
8. Benedek K, Chudoba I, Klein G et al (2004) Rearrangements of the telomeric region of mouse chromosome 11 in Pre-B ABL/MYC cells revealed by mBANDing, spectral karyotyping, and fluorescence in-situ hybridization with a subtelomeric probe. Chromosome Res 12:777–785
9. Leibiger C, Kosyakova N, Mkrtchyan H et al (2013) First molecular cytogenetic high resolution characterization of the NIH 3T3 cell line by murine multicolor banding. J Histochem Cytochem 61:306–312
10. Aurich-Costa J, Vannier A, Gregoire E et al (2001) IPM-FISH, a new M-FISH approach using IRS-PCR painting probes: application to the analysis of seven human prostate cell lines. Genes Chromosomes Cancer 30:143–160
11. Müller S, Rocchi M, Ferguson-Smith MA et al (1997) Toward a multicolor chromosome bar code for the entire human karyotype by fluorescence in situ hybridization. Hum Genet 100:271–278
12. Müller S, O'Brien PC, Ferguson-Smith MA et al (1998) Cross-species colour segmenting: a novel tool in human karyotype analysis. Cytometry 33:445–452
13. Weise A, Heller A, Starke H et al (2003) Multitude multicolor chromosome banding (mMCB)—a comprehensive one-step multicolor FISH banding method. Cytogenet Genome Res 103:34–39
14. Chudoba I, Plesch A, Lörch T et al (1999) High resolution multicolor banding: a new technique for refined FISH analysis of human chromosomes. Cytogenet Cell Genet 84:156–160
15. Liehr T, Heller A, Starke H et al (2002) Microdissection based high resolution multicolor banding for all 24 human chromosomes. Int J Mol Med 9:335–339
16. Kakazu N, Ashihara E, Hada S et al (2001) Development of spectral colour banding in cytogenetic analysis. Lancet 357:529–530
17. Kakazu N, Bar-Am I, Hada S et al (2003) A new chromosome banding technique, spectral color banding (SCAN), for full characterization of chromosomal abnormalities. Genes Chromosomes Cancer 37:412–416
18. Tjia WM, Sham JS, Hu L et al (2005) Characterization of 3p, 5p, and 3q in two nasopharyngeal carcinoma cell lines, using region-specific multiplex fluorescence in situ hybridization probes. Cancer Genet Cytogenet 158:61–66
19. Lichter P, Tang CJ, Call K et al (1990) High-resolution mapping of human chromosome 11 by in situ hybridization with cosmid clones. Science 247:64–69
20. Lengauer C, Green ED, Cremer T (1992) Fluorescence in situ hybridization of YAC clones after Alu-PCR amplification. Genomics 13:826–828
21. Liehr T, Weise A, Heller A et al (2002) Multicolor chromosome banding (MCB) with YAC/BAC-based probes and region-specific microdissection DNA libraries. Cytogenet Genome Res 97:43–50
22. Liehr T (2016) Multicolor FISH homepage. http://ssmc-tl.com/mfish.html
23. Liehr T, Starke H, Weise A et al (2004) Multicolor FISH probe sets and their applications. Histol Histopathol 19:229–237
24. Weise A, Mrasek K, Fickelscher I et al (2008) Molecular definition of high resolution multicolor banding (MCB) probes—first within the human DNA-sequence anchored FISH-banding probe set. J Histochem Cytochem 56:487–493

cenM-FISH Approaches

Thomas Liehr, Anja Weise, and Nadezda Kosyakova

Abstract

Multicolor-FISH using whole-chromosome painting and FISH banding with partial-chromosome painting probes are for technical reasons not suited to characterize small centromeric and pericentromeric rearrangements. Thus, probe sets to characterize the different human centromeres, and others to enable studies of the centromere-near regions of human chromosomes were established. Such centromere-specific multicolor (cenM)- and subcentromere-specific (subcenM)-FISH techniques are presented in this chapter. How to set up and use these probe sets is reported.

Keywords Centromere, Centromere-specific multicolor-FISH (cenM-FISH), Subcentromere-specific multicolor-FISH (subcenM-FISH), Centromere-near imbalances, Small supernumerary marker chromosomes (sSMCs)

1 Introduction

24-color FISH using all human whole-chromosome libraries as probes has been termed M-FISH or SKY ([1, 2], chapter by Thomas Liehr and Nadezda Kosyakova "Multiplex FISH and Spectral Karyotyping"). These are useful techniques for the characterization of complex chromosomal aberrations [3]. Furthermore, FISH banding methods (chapter by Thomas Liehr et al. "FISH Banding Techniques") can be applied when 24-color FISH using WCP probes reaches its limits, i.e., when exact localization of a chromosomal breakpoint is required or intrachromosomal aberrations are present. However, for technical reasons, whole or partial chromosomal painting libraries are not informative about the centromeric and pericentric regions. This is on the one hand due to the blocking of repetitive sequences by COT1-DNA and on the other due to flaring effects of the painting probes (chapter by Vladimir Trifonov et al. "FISH with and Without COT1 DNA"). Thus, neither M-FISH/SKY nor FISH banding approaches are suitable for characterizing centromere-near euchromatic sequences. These restrictions particularly hamper analyses of the chromosomal origin

Thomas Liehr (ed.), *Fluorescence In Situ Hybridization (FISH)*, Springer Protocols Handbooks,
DOI 10.1007/978-3-662-52959-1_27,

and genetic content in small supernumerary marker chromosomes (sSMCs) [4, 5] and exact characterizations of centromere-near breakpoints. To overcome these restrictions, several centromere- and pericentromere-directed FISH approaches were recently developed. Centromere-specific multicolor-FISH (cenM-FISH [6]; CM-FISH [7]) is based on all of the available human centromere-specific DNA probes, which are labeled with five different fluorochromes and hybridized simultaneously. This approach allows the identification and characterization of all human centromeres by their individual colors in one single step. cenM-/CM-FISH is a very helpful technique, especially for the characterization of sSMC with no or very little euchromatin and/or small amounts of available chromosome suspension.

Two similar probe sets are available to characterize the short arms and the centromeric regions of the acrocentric chromosomes: acroM-FISH [8] and acrocenM-FISH [9]. These sets are helpful for determining the chromosomal origins of acrocentric-derived sSMCs and for distinguishing acrocentric p-arm polymorphism from cryptic translocations [4, 5, 9, 10]. Finally, subcentromere-specific multicolor-FISH (subcenM-FISH) [11] was developed to specifically stain near-centromere euchromatic material. No other probe set can resolve these regions, as they are either hidden by the flaring effect of the fluorescence-intense centromeric signals or underrepresented in other chromosome or chromosome region-specific probes. Further pericentric-oriented probe sets can be found elsewhere in this book (chapter by Thomas Liehr et al. "Bar-Coding Is Back"). Here we describe how cenM-FISH, acrocenM-FISH, and subcenM-FISH probe sets are constructed, and how they can be used to, e.g., characterize the genetic content of sSMCs in prenatal or postnatal diagnostics.

2 Materials

Apart from standard cell biological and molecular cytogenetic equipment, including standard solutions (e.g., ethanol, methanol, formamide, formaldehyde, xylene, etc.), no more specialized items are required. The equipment needed for multicolor-FISH is listed in chapter by Thomas Liehr et al. "The Standard FISH Procedure", and chapter by Thomas Liehr and Nadezda Kosyakova "Multiplex FISH and Spectral Karyotyping".

2.1 Chemicals and Solutions to Be Prepared

- Biotin nick translation kit (Cat. No. 11745824910, Roche, Basel, Switzerland)
- ChromaTide Texas Red-dUTP (Cat. No. C-7631, Invitrogen, Carlsbad, CA, USA)

- COT1-DNA (human) (Cat. No. 15279-001, Gibco BRL, Grand Island, NY, USA)
- Diethylaminocoumarin-5-dUTP (DEAC-dUTP) (Cat. No. NEL455, NENDuPont, Boston, MA, USA)
- Hybridization buffer: dissolve 2 g dextran sulfate in 10 ml 50 % deionized formamide/2 × SSC/50 mM phosphate buffer for 3 h at 70 °C. Aliquot and store at −20 °C.
- Nick translation kit (Cat. No. 11745808910, Roche)
- Phosphate buffer: prepare 0.5 M Na_2HPO_4 and 0.5 M NaH_2PO_4, mix these two solutions (1:1) to get pH 7.0, and then aliquot and store at −20 °C.
- SpectrumOrange-dUTP (Cat. No. 6J9415, Abbott, Abbott Park, IL, USA)
- SpectrumGreen-dUTP (Cat. No. 6J9410, Abbott)

3 Methods

3.1 Probe Preparation

3.1.1 cenM-FISH and acrocenM-FISH

1. Centromere-specific probes as well as a probe specific for the nucleolar organizing regions on acrocentric p-arms can be obtained on e-mail request from Prof. Dr. Mariano Rocchi, Bari, Italy (mail, rocchi@biologia.uniba.it); a p-arm-specific microdissection probe can be obtained (only labeled) on request by the last author of this chapter (mail, Thomas. Liehr@med.uni-jena.de).
2. Plasmid DNA, with centromere-specific DNA as inserts, is labeled by nick translation using commercially available kits according to the manufacturer's instructions. Labeling can be done according to the labeling schemes in Fig. 1a and b, or modified as required.
3. 1 μg of each probe can be labeled with one nick translation reaction; each labeling should be tested separately for its efficiency by hybridizing 10 ng of the probe on normal metaphase spreads.
4. All labeled probes are mixed and 50 aliquots are made in 1.5 ml microtubes.
5. Each aliquot is precipitated together with 10 μl of 1 μg/μl human COT1-DNA.
6. The pellets are vacuum dried and stored until use at −20 °C.

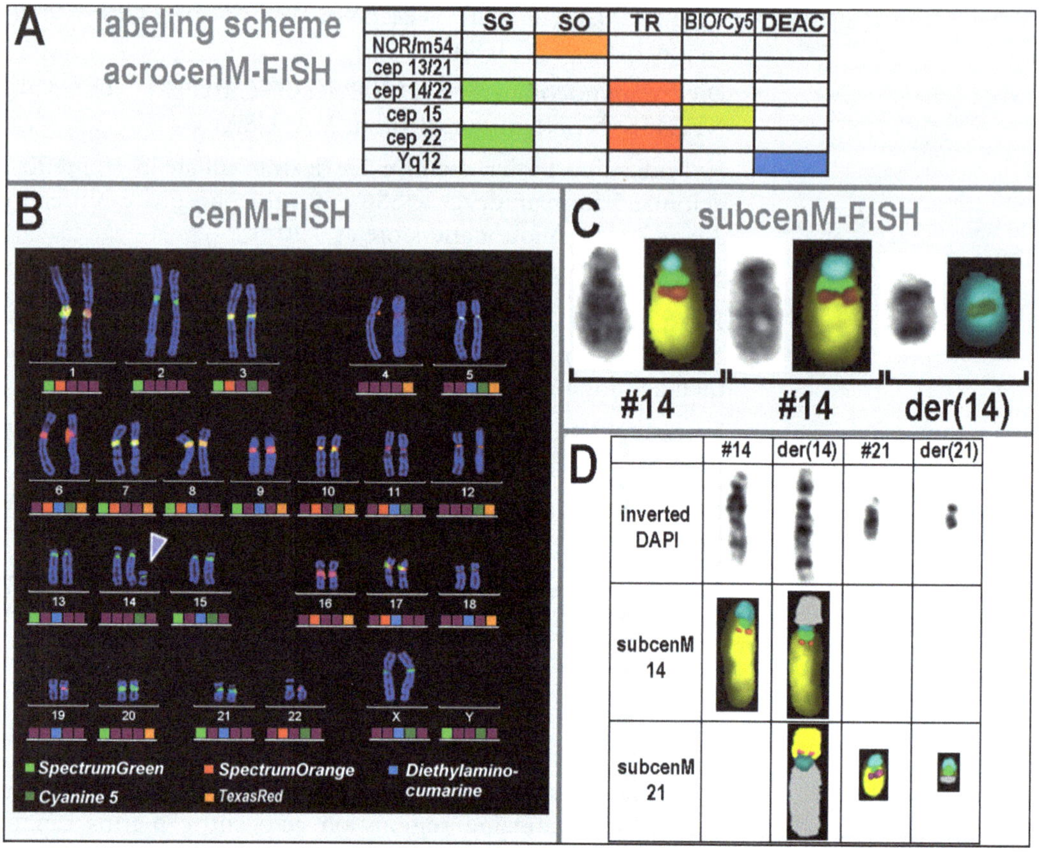

Fig. 1 (**a**) Labeling scheme for the acrocenM-FISH probe set. *BIO/Cy5* biotin detected by cyanine 5-avidin, *Cep* centromere-specific probe, *DEAC* diethylaminocoumarin, *m54* midi54, a microdissection-derived probe specific for the short arms of the acrocentric chromosomes, *NOR* NOR-specific probe, *SG* SpectrumGreen, *SO* SpectrumOrange, *TR*, TexasRed, *Yq12* probe specific for the heterochromatic satellite III region in Yq12. (**b**) The labeling scheme of the cenMFISH probe set is given below each chromosome; for the color code, see the *bottom* of the figure. An example of the characterization of a small supernumerary marker chromosome (sSMC) by cenM-FISH is given. The sSMC was derived from chromosome 14 (*arrowhead*). The case was kindly provided by Dr. Fuchs, Hamburg, Germany. (**c**) The presence of euchromatic material on the sSMC shown in **b** was checked by subcenM-FISH and excluded. For each chromosome 14 (#14) and the sSMC(14) [der(14)], an inverted DAPI and a multicolor-FISH depiction are shown. *Yellow*, partial-chromosome paint for 14q; *blue*, midi54, a microdissection-derived probe specific for the short arms of the acrocentric chromosomes; *green*, centromeric probe for chromosome 14 (and 22); *red*, BAC RP11-324B11 in 14q11.2. (**d**) The use of subcenM-FISH to narrow down two near-centromere chromosomal breakpoints in chromosomes 14 and 21 in a normal male with a karyotype of 46,XY,t(14;21). The breakpoints could be determined after the application of subcenM-FISH as 14p11.2 and 21q11.1. The labeling scheme for subcenM-FISH 14 is the same as in Fig. 1c; the labeling scheme for subcenM-FISH 21 is in concordance with that of chromosome 14; the BAC RP11-89H21 was used as the near-centromere probe in 21q11.2. The inverted DAPI banding shows how the derivative chromosomes looked after GTG banding. The regions unstained by the corresponding subcenM-FISH probe sets are pseudocolored in *gray*. The case was provided by Dr. Schmidtke and Dr. Pabst, Hannover, Germany

Table 1
Suggestions for subcenM-FISH suited to BAC probes

Region	Localization	Clone	Region	Localization	Clone
Xp	Xp11.21	RP11-570J18	10p	10p11.21	RP11-365P10
Xq	Xq12	RP11-J1065K8	10q	10q11.22	RP11-178A10
Yp	Yp11.2	RP11-115H13	11p	11p11.21	RP11-722K13
Yq	Yq11.21	RP11-71M14	11q	11q12	RP11-77M7
1p	1p12	RP11-130B18	12p	12p11.21	RP11-517B23
1q	1q21.1	RP11-35B4	12q	12q12	RP11-498B21
2p	2p11.2	RP11-316G9	13q	13q12.1	RP11-110K18
2q	2q11.2	RP11-708D7	14q	14q11.2	RP11-324B11
3p	3p12.1	RP11-91A15	15q	15q11.2	RP11-307C10
3q	3q12.1	RP11-21I16	16p	16p11.2	RP11-360L15
4p	4p12	RP11-100N21	16q	16q12.1	RP11-474B12
4q	4q12	RP11-535C7	17p	17p11.2	RP11-746M1
5p	5p12	RP11-19F12	17q	17q11.2	RP11-403E9
5q	5q11.2	RP11-160F8	18p	18p11.21	RP11-411B10
6p	6p11.2	RP1-61B2	18q	18q11.2	RP11-10G8
6q	6q12	RP11-387L5	19p	19p13.1	RP11-22G10
7p	7p11.2	RP11-10F11	19q	19q13.1	RP11-46J12
7q	7q11.21	RP11-3N2	20p	20p11.2	RP11-96L6
8p	8p11.21	RP11-503E24	20q	20q11.2	RP11-243J16
8q	8q11.21	RP13-116A	21q	21q11.2	RP11-89H21
9p/q	9p12	RP11-128P23	22q	22q11.21	RP11-172D7
9q	9q13	RP11-430C15			

7. Prior to use, the pellet is dissolved in 20 μl of the hybridization buffer, which is sufficient for one slide covered with a 24 × 60 mm coverslip.

3.1.2 subcenM-FISH

Available probes nearest to the centromere were selected from the human genomic sequence using corresponding databases (chapter by Thomas Liehr and Anja Weise "Background"). These were combined with homemade chromosome arm-specific microdissection-derived probes. The latter are not commercially available. However, an effective subcenM-FISH probe set can also be created that does not have these partial-chromosome painting probes. Table 1 suggests clones that can be used to create a three-color subcenM-FISH set consisting of a centromeric probe (available from Prof. Dr. Mariano Rocchi, see above, or commercially available) and near-centromere BAC probes (available from https://bacpac.chori.org/).

3.2 Slide Pretreatment

As described in chapter by Thomas Liehr et al. "The Standard FISH Procedure".

3.3 Fluorescence In Situ Hybridization (FISH)

As described in chapter by Thomas Liehr et al. "The Standard FISH Procedure".

3.4 Evaluation

See chapter by Thomas Liehr and Nadezda Kosyakova "Multiplex FISH and Spectral Karyotyping". No special evaluation software is available at present for the cenMFISH approaches described here (*see* **Note 1**).

CenM-FISH and acrocenM-FISH were applied successfully for the characterization of the chromosomal origins of ~1,600 cases with sSMC [4]. An example of an sSMC derived from chromosome 14 is given in Fig. 1b. This sSMC was proven to be heterochromatic by applying subcenM-FISH (Fig. 1c). SubcenM-FISH was applied to narrow down the sSMC breakpoints in ~ 800 cases [4].

Another application for which subcenM-FISH is extremely well suited is the characterization of near-centromere chromosomal breakpoints. In Fig. 1d, an example of this is given for a case of an infertile male with a karyotype of 46,XY,t(14;21) (p11.2;q11.1). About 150 similar cases have been studied by subcenM-FISH so far ([10] and unpublished data). AcrocenM-FISH (Fig. 1a) is a very useful approach for distinguishing short-arm variants of acrocentric chromosomes from cryptic translocations [10].

4 Notes

1. Good pseudocolors like those shown in the figures are not always achieved in cenM-FISH. However, even when the hybridization quality is relatively low, in most cases it is possible to get results which can then be confirmed by one-, two-, or three-color FISH by applying the corresponding centromeric probes. (The latter should be done anyway.)

References

1. Speicher MR, Gwyn Ballard S, Ward DC (1996) Karyotyping human chromosomes by combinatorial multi-fluor FISH. Nat Genet 12:368–375
2. Schröck E, du Manoir S, Veldman T et al (1996) Multicolor spectral karyotyping of human chromosomes. Science 273:494–497
3. Liehr T (2016) Multicolor FISH homepage. http://ssmc-tl.com/mfish.html

4. Liehr T (2016) sSMC homepage. http://ssmc-tl.com/sSMC.html
5. Liehr T (2012) Small supernumerary marker chromosomes (sSMC). A guide for human geneticists and clinicians, With contributions by UNIQUE (The Rare Chromosome Disorder Support Group). Springer, New York
6. Nietzel A, Rocchi M, Starke H et al (2001) A new multicolor-FISH approach for the characterization of marker chromosomes: Centromere-specific multicolor-FISH (cenM-FISH). Hum Genet 108:199–204
7. Henegariu O, Bray-Ward P, Artan S et al (2001) Small marker chromosome identification in metaphase and interphase using centromeric multiplex FISH (CM-FISH). Lab Invest 81:475–481
8. Langer S, Fauth C, Rocchi M et al (2001) AcroM fluorescent in situ hybridization analyses of marker chromosomes. Hum Genet 109:152–158
9. Trifonov V, Seidel J, Starke H et al (2003) Enlarged chromosome 13 p-arm hiding a cryptic partial trisomy 6p22.2-pter. Prenat Diagn 23:427–430
10. Starke H, Mrasek K, Liehr T (2005) 3 cases with enlarged acrocentric p-arms—2 cases with cryptic partial trisomies. J Histochem Cytochem 53:359–360
11. Starke H, Nietzel A, Weise A et al (2003) Small supernumerary marker chromosomes (SMCs): genotype-phenotype correlation and classification. Hum Genet 114:51–67

Heterochromatin-Directed mFISH (HCM-FISH)

Thomas Liehr, Nadezda Kosyakova, Anja Weise, and Ahmed B. Hamid Al-Rikabi

Abstract

Recently we presented a new multicolor FISH (mFISH) probe set, called heterochromatin-directed mFISH (HCM-FISH). It enables a one-step characterization of the large heterochromatic regions within the human genome and thus closes a gap in the up-to-now available mFISH probe sets. HCM-FISH covers the acrocentric short arms, the large pericentric regions of chromosomes 1, 9, 16, and 19, as well as the band Yq12.

Keywords Heterochromatin-directed mFISH (HCM-FISH), Acrocentric short arms, 1q12, 9q12, 16q11.2, 19p12, 19q12, NOR, 15p11.2, Yq12

1 Introduction

Neither 24-color FISH using all human whole-chromosome painting probes ([1, 2], chapter by Thomas Liehr and Nadezda Kosyakova "Multiplex FISH and Spectral Karyotyping"), FISH banding methods (chapter by Thomas Liehr et al. "FISH Banding Tec hniques"), nor centomeric multicolor FISH (mFISH) approaches (chapter by Thomas Liehr et al. "cenM-FISH Approaches") can be applied in case of rearrangements involving the large heterochromatic regions within the human genome, i.e., the acrocentric short arms; the large pericentric regions of chromosomes 1, 9, 16, and 19; as well as the band Yq12. Also array CGH (chapter by Eftychia Dimitriadou and Joris Vermeesch "Array CGH") does not cover these regions and thus is not suited to resolve weird cytogenetic results based in heterochromatic rearrangements (3) (Fig. 1). As there was no probe set available in the literature addressing all those regions in FISH simultaneously (4), we recently developed such as set and denominated it heterochromatin-directed mFISH (HCM-FISH) (3). Besides being useful in characterization of clinical cases including small supernumerary marker chromosomes (sSMC) (3) (Fig. 1), it can also be applied in tumor cytogenetics (unpublished

Thomas Liehr (ed.), *Fluorescence In Situ Hybridization (FISH)*, Springer Protocols Handbooks,
DOI 10.1007/978-3-662-52959-1_28, © Springer-Verlag Berlin Heidelberg 2017

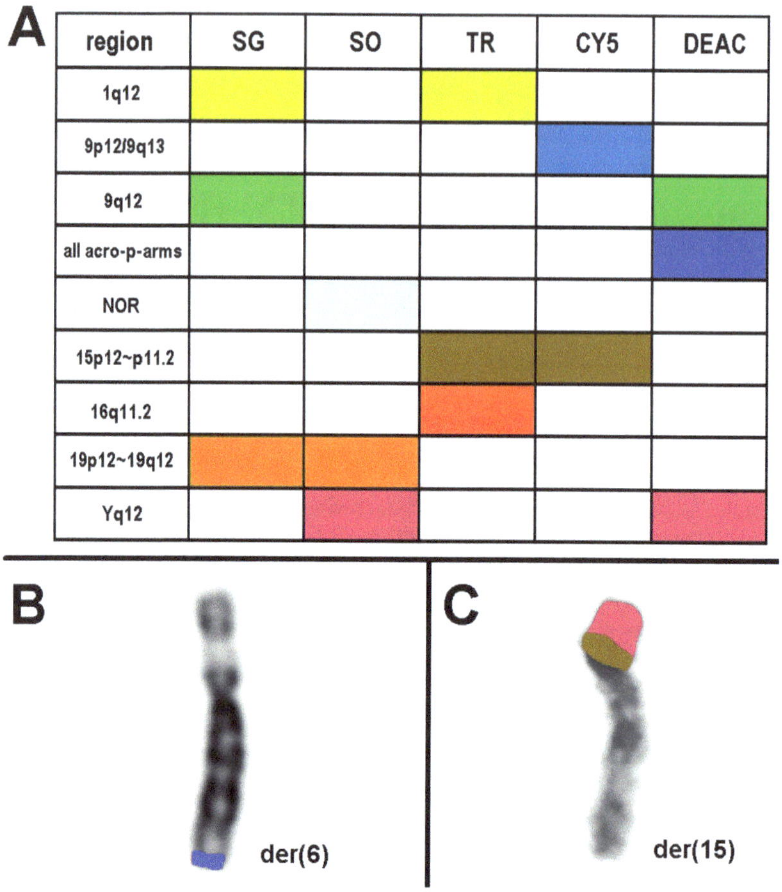

Fig. 1 (**a**) Label-scheme used for the HCM-FISH probe set. The pseudocolors used for the corresponding region-specific DNA-probes in (**b**) and (**c**) are used to indicate for the fluorochromes applied to generate the HCM-FISH probe set. Abbreviations: *acro-p-arms* short arms of all acrocentric human chromosomes; *CY5* cyanine 5; *DEAC* diethylaminocoumarin; *midi* microdissection derived; *NOR* nucleolus organizing region; *SG* SpectrumGreen; *SO* SpectrumOrange; *TR* Texas Red. (**b**) A derivative chromosome 6 turned out to be a der(6)t(6;13 or 14 or 15 or 21 or 22)(q27;p13) acc. to HCM-FISH. Case kindly provided by Dr. Petr Lonsky, Prague, Czech Republic. (**c**) The derivative chromosome 15 could be identified as a der(15)t(Y;15)(q12;p11.2) after HCM-FISH was applied. Case kindly provided by Dr. Raabe-Meyer, Hannover, Germany

data) and Zoo-FISH studies (chapter by Fengtang Yang et al. "Animal Probes and ZOO-FISH") as, e.g., in (5), being highly informative and time saving.

2 Materials

Apart from standard cell biological and molecular cytogenetic equipment, including standard solutions (e.g., ethanol, methanol, formamide, formaldehyde, xylene, etc.), no more specialized items are required. The equipment needed for multicolor FISH is listed in chapter by Thomas Liehr et al. "The Standard FISH Procedure" and chapter by Thomas Liehr and Nadezda Kosyakova "Multiplex FISH and Spectral Karyotyping".

3 Methods

3.1 HCM-FISH Probe Set

The HCM-FISH probe set can be obtained on request from the corresponding author of this chapter (Thomas.Liehr@med.uni-jena.de). As it is mainly based on homemade microdissection-derived probes (chapter by Nadezda Kosyakova et al. "FISH-Micro dissection"), they normally cannot be redone by other laboratories without corresponding equipment. The content and the labeling scheme are summarized in Fig. 1.

3.2 Slide Pretreatment

As described in chapter by Thomas Liehr et al. "The Standard FISH Procedure".

3.3 Fluorescence In Situ Hybridization (FISH)

As described in chapter by Thomas Liehr et al. "The Standard FISH Procedure".

3.4 Evaluation

See chapter by Thomas Liehr and Nadezda Kosyakova "Multiplex FISH and Spectral Karyotyping". No special evaluation software is available at present for the HCM-FISH described here.

References

1. Speicher MR, Gwyn Ballard S, Ward DC (1996) Karyotyping human chromosomes by combinatorial multi-fluor FISH. Nat Genet 12:368–375
2. Schröck E, du Manoir S, Veldman T et al (1996) Multicolor spectral karyotyping of human chromosomes. Science 273:494–497
3. Bucksch M, Ziegler M, Kosayakova N et al (2012) A new multicolor fluorescence in situ hybridization probe set directed against human heterochromatin: HCM-FISH. J Histochem Cytochem 60:530–536
4. Liehr T (2016) Multicolor FISH homepage. http://ssmc-tl.com/mfish.html
5. Sangpakdee W, Tanomtong A, Fan X et al (2016) Application of multicolor banding combined with heterochromatic and locus-specific probes identify evolutionary conserved breakpoints in Hylobates pileatus. Mol Cytogenet 9:17

Subtelomeric and/or Subcentromeric Probe Sets

Anja Weise and Thomas Liehr

Abstract

Besides microscopically visible chromosomal rearrangements, numerous cryptic chromosomal alterations were reported since the introduction of techniques like FISH, array-CGH, or MLPA. This holds especially true for dynamic regions in the subtelomere and subcentromere of any chromosome. To address these regions, locus-specific FISH probes are employed as single, chromosome-specific, or genome-wide probe sets. Here we present the chromosome-specific subtelomere-subcentromere multicolor FISH (subCTM) and the genome-wide subcentromere multicolor FISH (subcenM) probe sets, which are useful as screening tools in specific patient groups like infertile and mentally retarded but also in tumor cytogenetics and for evolutionary studies when applied as Zoo-FISH.

Keywords Subtelomeric region, Centromere-near region, Pericentromere, Subtelomere-subcentromere multicolor FISH (subCTM), Deletion, Duplication, Translocation, Inversion

1 Introduction

The extreme ends of all vertebrate chromosomes consist of non-coding, tandemly repeated hexanucleotide units TTAGGG (5′→3′direction) that are typically 5–15 kb long. Thus, an individual human telomere cannot be specifically stained using telomeric probes ([1], chapter by Gordana Joksic et al. "Telomere Length Measurement by FISH"; chapter by Nikolay Rubtsov and Natalya Zhdanova "The Replicative Detargeting FISH (ReD-FISH) Tech nique in Studies of Telomere Replication"). Proximal to the telomeres, the so-called subtelomeric regions start, which are (mostly) chromosome-specific and known for their enrichment of genes and segmental duplications that facilitate the formation of rearrangements [2]. Subtelomeres are often affected in mentally retarded individuals with normal banding cytogenetic outcomes [3, 4]. The detection rate is somewhere in the region of 2–10 %, depending on the method applied and the collective investigated [5]. Once subtelomeric alterations are found by molecular genetic approaches like qPCR, MAPH, MLPA, or array-CGH, they should be

Thomas Liehr (ed.), *Fluorescence In Situ Hybridization (FISH)*, Springer Protocols Handbooks,
DOI 10.1007/978-3-662-52959-1_29, © Springer-Verlag Berlin Heidelberg 2017

confirmed by FISH using locus-specific probes to define the cytogenetic location and type of aberration. Moreover, the parents of the corresponding patient should be studied, since polymorphic rearrangements in these regions have been repeatedly reported and can be transmitted through generations without any clinical signs [6]. When studying the subtelomeric regions in more detail, progress was made concerning the identification of single copy sequences that were located more and more distally. Thus, several so-called first- and second-generation probes are available for detecting subtelomeric rearrangements. Efforts were also made to develop subtelomeric mFISH probe sets [7–9].

Similarly, subcentromere-specific multicolor FISH (subcenM-FISH [10]) specifically labels a chromosomal region that no other FISH or mFISH probe set can characterize: the near-centromere euchromatic material (chapter by Thomas Liehr et al. "cenM-FISH Approaches"). This is due to the fact that these regions are either hidden by the flaring effect of the fluorescence-intense centromeric signals, or they are underrepresented in other chromosome or chromosome-region-specific probes. Subcentromeric or pericentromeric regions are, like the subtelomeric regions, enriched in segmental duplications that can lead to intra- and interchromosomal rearrangements [11, 12]. The clinical impact of subcentromeric rearrangements is neither well understood nor well studied, aside from some correlations with infertility and the role of subcentromeric imbalances caused by small supernumerary marker chromosomes (sSMC) [13].

Here we describe the composition and application of chromosome-specific combined subtelomeric and subcentromeric FISH probes in an approach called subCTM [14]. Moreover, a subcentromeric probe set directed at all chromosomes (ACM-FISH) is presented. For chromosome-specific subcenM-FISH, refer to chapter by Thomas Liehr et al. "cenM-FISH Approaches".

The application of the chromosome-specific subCTM probe set provides information about the dynamic subtelomere and subcentromere regions of the investigated chromosome, within one round of hybridization. If one probe is absent on one homologous chromosome, the other chromosomes should be checked for translocations before suggesting a deletion, especially for subtelomeric probes. Examples of subCTM sets on human chromosomes are depicted in Fig. 1. When signals are lacking on both chromosomes, the analysis should be repeated with the single copy probes to make sure that this is not due to methodological problems. This point is particularly relevant for, e.g., evolutionary studies, where changes in the subCTM pattern can affect both homologous chromosomes (Fig. 2). In the case of evolutionary studies, and also for expected interchromosomal rearrangements, we recommend the addition of a whole chromosome paint to the subCTM probe set for a better chromosome recognition.

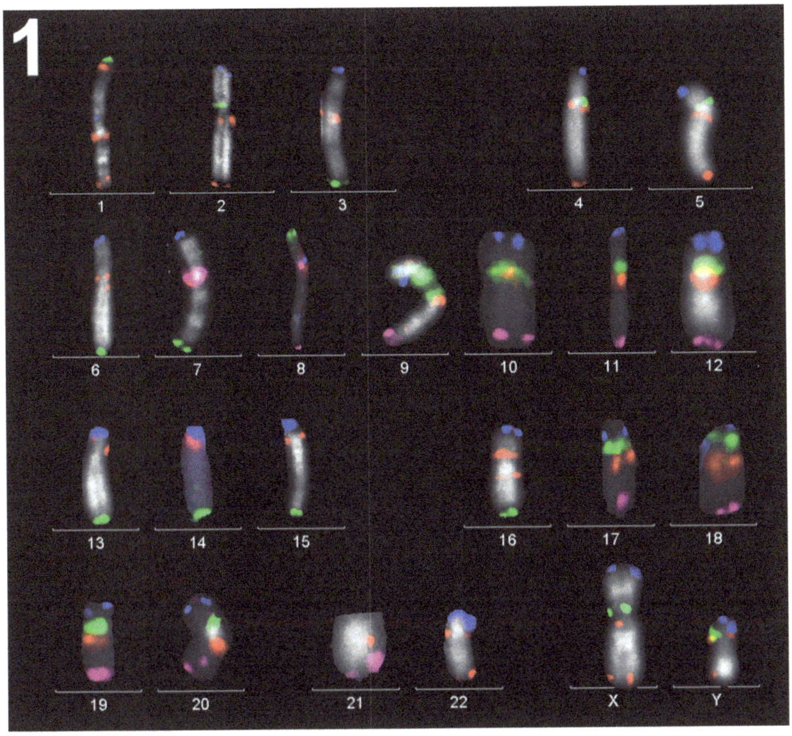

Fig. 1 subCTM for all human chromosomes. Composite picture from 24 single chromosome hybridizations

The genome-wide ACM-FISH probe set is useful when applied as a screening tool in order to find cryptic subcentromere rearrangements like deletions or inversions, which are easily missed in conventional banding cytogenetics. Inversions should be visible as a change in color between the short and the long arm of one chromosome. Deletions should be suggested after insertions on other chromosomes have been excluded and a single probe hybridization has been done for verification. Genome-wide ACM-FISH is set up for time-saving evaluations because only two fluorescent colors are needed.

2 Materials

Apart from the standard cell biological and molecular cytogenetic equipment, including standard solutions (e.g., ethanol, methanol, formamide, formaldehyde, etc.), the following more specialized items are needed. The equipment needed for FISH itself is listed in chapter by Thomas Liehr et al. "The Standard FISH Procedure"; for BAC probes see chapter by Thomas Liehr "Homemade Locus-Specific FISH Probes: Bacterial Artificial Chromosomes".

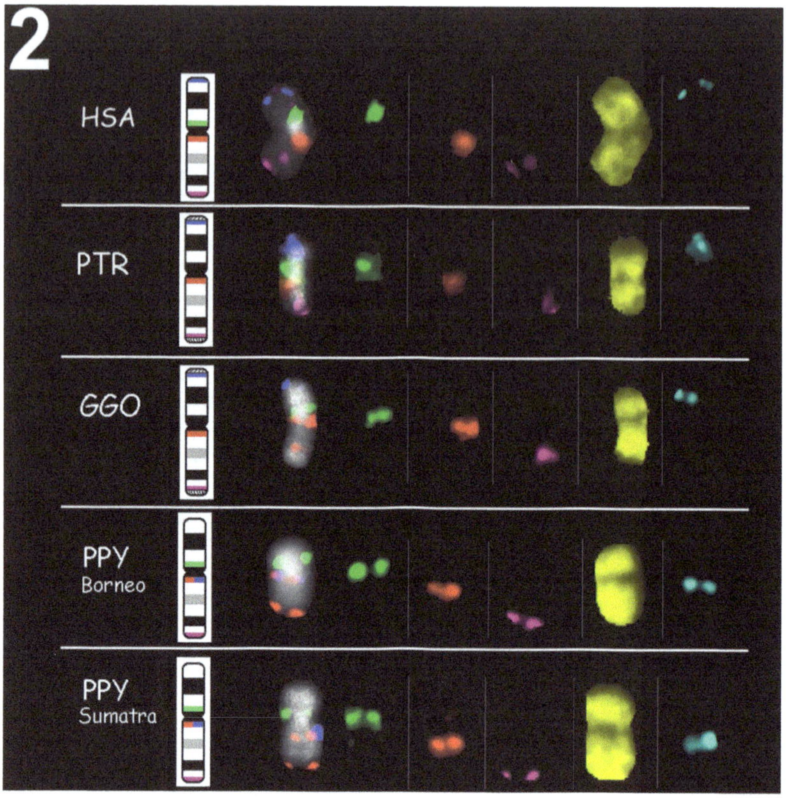

Fig. 2 subCTM hybridization for human chromosome 20 carried out in a Zoo-FISH study in great apes (HSA, *Homo sapiens*; PTR, *Pan troglodytes*; GGO, *Gorilla gorilla*; PPY, *Pongo pygmaeus*). The PPY chromosomes show the ancestral type of the homologous chromosome 20, where the human subtelomeric probe for the short arm is located in the pericentromeric region of the long arm. After splitting off GGO, two inversions appeared, leading to the signal pattern present in modern humans, gorillas, and chimpanzees. From *left* to *right*: species, ideogram with probe location, hybridization "mixcolor," subcentromere 20p (SpectrumGreen), subcentromere 20q (SpectrumOrange), subtelomere 20q (Texas Red), wcp (= whole chromosome paint) 20 (Cyanine 5), subtelomere 20p (DEAC, diethylaminocoumarin)

3 Methods

3.1 Chromosome-Specific subCTM-FISH

1. Chromosome-specific sets for metacentric and submetacentric chromosomes are combined as follows: two subtelomeric probes as commercially available [7] or selected from databases and ordered (chapter by Thomas Liehr "Homemade Locus-Specific FISH Probes: Bacterial Artificial Chromosomes") and two centromere-near probes (see Table 1 or chapter by Thomas Liehr et al. "cenM-FISH Approaches") are applied together. Optionally, a whole chromosome paint (wcp) or a centromeric probe (cep) can be added (*see* **Note 1**). All five probes are labeled accordingly with five different fluorochromes. Acrocentric chromosome sets consist only of one subtelomeric and one

Table 1
Suggested probe set for genome-wide ACM-FISH

chr	Cyto band	BAC	NCBI 36.2 [hg18] Start	Stop	Size	FISH cross hybs
1p	13.1	RP4-787H6	116.582.484	116.707.534	127.051	–
1q	21.1	RP11-441L11	145.337.817	145.51.131	175.494	–
2p	11.2	RP11-316G9	89.561.552	89.770.752	211.201	–
2q	11.1	RP11-11P22	95.110.968	95.267.371	158.404	–
3p	11.1	RP11-301H7	90.244.581	90.384.777	142.201	5q23.3
3q	11.2	RP11-631O4	95.307.753	95.569.618	263.876	3q27
4p	12	RP11-100N21	47.264.009	47.430.185	168.177	–
4q	11	RP11-365H22	52.354.875	52.530.859	177.985	–
5p	13.1	RP11-301A5	40.982.971	41.157.849	215.308	–
5q	11.2	RP11-289K10	52.686.877	52.776.534	161.812	–
6p	11.2	RP11-421P21	57.228.514	57.292.599	64.186	–
6q	12	RP11-387L5	68.415.186	68.562.704	147.619	–
7p	11.2	RP11-10F11	56.639.424	56.803.089	163.666	–
7q	11.21	RP11-144H20	61.606.122	61.791.403	187.282	–
8p	11.21	RP11-503E24	42.503.724	42.674.302	189.048	–
8q	11.21	RP13-116A4	48.320.010	48.368.352	48.347	–
9p	13.2	RP11-113O24	38.263.089	38.427.295	166.207	–
9q	21.11	RP11-430C15	70.642.632	70.758.000	115.469	–
10p	11.21	RP11-365P10	36.945.343	36.974.907	31.565	–
10q	11.21	RP11-92P6	43.174.613	43.219.888	45.276	–
11p	11.2	RP11-397M16	48.260.247	48.436.072	175.832	–
11q	11	RP11-100N3	56.238.248	56.397.619	159.372	–
12p	11.21	RP11-517B23	31.362.925	31.533.973	171.048	–
12q	12	RP11-498B21	39.833.150	39.900.092	67.150	–
13q	12.11	RP11-523H24	19.137.338	19.306.540	169.303	–
14q	11.2	RP11-14J7	20.057.964	20.172.932	162.209	–
15q	11.2	RP11-289D12	20.428.073	20.542.380	137.300	–
16p	11.2	RP11-360L15	28.873.631	29.083.631	210.000	–
16q	12.1	RP11-474B12	45.880.869	46.027.419	153.053	–
17p	11.2	RP11-746M1	20.824.144	21.055.082	235.968	–

(continued)

Table 1 (continued)

			NCBI 36.2 [hg18]			
chr	Cyto band	BAC	Start	Stop	Size	FISH cross hybs
17q	11.2	RP11-403E9	25.614.719	25.693.302	102.286	–
18p	11.21	RP11-411B10	13.997.946	14.143.336	154.589	–
18q	11.1	RP11-10G8	17.274.439	17.431.001	156.562	–
19p	12	RP11-22G10	22.970.881	23.120.518	139.887	–
19q	12	RP11-46I12	34.301.925	34.484.128	182.203	–
20p	11.21	RP11-96L6	25.465.310	25.522.324	57.115	–
20q	11.21	RP11-243J16	29.756.779	29.925.538	168.860	–
21q	11.2	RP11-89H21	14.850.742	15.000.742	150.000	–
22q	11.2	RP11-172D7	16.239.476	n.a.	n.a.	–
Xp	11.21	RP11-465B24	56.467.529	56.573.161	107.633	–
Xq	11.2	RP11-403E24	63.222.525	63.351.189	128.765	–
Yp	11.2	RP11-115H13	6.823.535	6.877.729	54.395	–
Yq	11.21	RP11-333E9	12.581.700	12.759.636	178.137	–

subcentromere probe of the long arm, each, and an optional wcp or cep. If available, a probe for the short arms of acrocentric chromosomes can be added (e.g., Midi54, [15]), and then a four-color FISH approach is done.

2. When using commercially available subtelomeric probes, follow the manufacturer's instructions and adjust the volume according to the other probes in the mix. All other probes for the chromosome set are mixed just before prehybridization (2–3 μl each, chapter by Thomas Liehr et al. "The Standard FISH Procedure") with 1–3 μg Cot 1 DNA to a final volume of 12 μl for one half slide.
3. After prehybridization, commercial and homemade probes are mixed together for the final hybridization (chapter by Thomas Liehr et al. "The Standard FISH Procedure").

3.2 Genome-Wide ACM-FISH

1. Available probes next to the centromere were selected from the human genomic sequence using databases, as introduced in chapter by Thomas Liehr and Anja Weise "Background" of this book (*see* **Notes 1** and **2**).
2. BAC clone DNA is isolated as described in chapter by Thomas Liehr "Homemade Locus-Specific FISH Probes: Bacterial

Artificial Chromosomes" and pooled into groups of four to five different loci (p- and q-arm-specific) (*see* **Note 3**).

3. After labeling the pooled centromere-near BAC DNA probes in two different fluorescence colors or haptens according to the chromosome arm (e.g., all centromere-near BACs in short arms in SpectrumGreen and all centromere-near BACs in the long arms in SpectrumOrange) by nick translation (chapter by Thomas Liehr et al. "The Standard FISH Procedure"), all probes are precipitated together and dissolved in the hybridization mix (chapter by Thomas Liehr et al. "The Standard FISH Procedure"). The FISH procedure is done as described in chapter by Thomas Liehr et al. "The Standard FISH Proce dure". Table 1 provides suggestions for clones that can be used to create a genome-wide two-color FISH ACM-FISH probe set (*see* **Note 4**).

3.3 Slide Pretreatment and Fluorescence In Situ Hybridization (FISH)

As described in chapter by Thomas Liehr et al. "The Standard FISH Procedure".

4 Notes

1. Especially when selecting probes from the pericentromeric regions for a multiplex approach, look out for cross-hybridizations caused by a high level of sequence homology in these regions. An example is given in Fig. 3 (genome-wide ACM-FISH), where the pericentromeric probe from chromosome 3p shows a cross-hybridization in the long arm of chromosome 5.
2. Locus-specific probes should be tested in single FISH experiments and then pooled into groups of four to five loci according to their FISH signal strength. Another strategy is to increase or reduce the amount of labeled probes for the genome-wide mix depending on the FISH signal strength in single FISH tests.
3. The problem with genome-wide sets where a number of different loci are applied in one hybridization is the risk of background and the failure of some of the probes (see chromosome 5p in Fig. 3). Therefore, a suggested rearrangement should always be confirmed by a second hybridization with the single probe or a set of probes that fail in the first round.
4. This approach should be used by experienced cytogeneticists, as the evaluation is finally done on a DAPI-stained karyotype that is comparable to GTG-stained chromosomes.

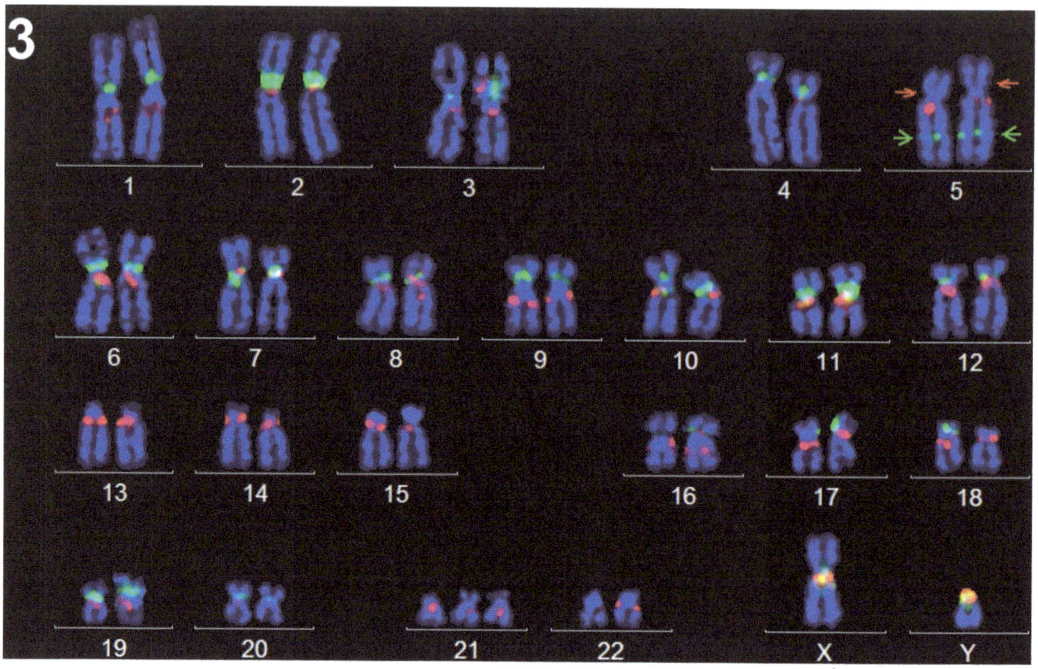

Fig. 3 Genome-wide ACM-FISH hybridization, where all p-arm probes are labeled in biotin/streptavidin FITC and all q-arm loci appear in digoxigenin/antidigoxigenin-rhodamine. This is an example of hybridization on a metaphase of a Down syndrome patient. Note the cross-hybridization of the 3p probe on the long arm of chromosome 5 (*green arrows*) and the failure of hybridization for the probe of the short arm of chromosome 5 (*red arrows*)

References

1. Moyzis RK, Buckingham JM, Cram LS et al (1988) A highly conserved repetitive DNA sequence, (TTAGGG)n, present at the telomeres of human chromosomes. Proc Natl Acad Sci USA 85:6622–6626
2. Linardopoulou EV, Williams EM, Fan Y et al (2005) Human subtelomeres are hot spots of interchromosomal recombination and segmental duplication. Nature 437:94–100
3. Knight SJ, Flint J (2000) Screening chromosome ends for learning disability. Br Med J 321:1240
4. Dawson AJ, Putnam S, Schultz J et al (2002) Cryptic chromosome rearrangements detected by subtelomere assay in patients with mental retardation and dysmorphic features. Clin Genet 62:488–494
5. Gardner RJM, Sutherland GR (2004) Chromosome abnormalities and genetic counseling, 3rd edn. Oxford University Press, Oxford
6. Balikova I, Menten B, de Ravel T et al (2007) Subtelomeric imbalances in phenotypically normal individuals. Hum Mutat 28:958–967
7. Knight SJ, Flint J (2000) Perfect endings: a review of subtelomeric probes and their use in clinical diagnosis. Med Genet 37:401–409
8. Brown J, Saracoglu K, Uhrig S et al (2001) Subtelomeric chromosome rearrangements are detected using an innovative 12-color FISH assay (M-TEL). Nat Med 7:497–501
9. Granzow M, Popp S, Keller M et al (2000) Multiplex FISH telomere integrity assay identifies an unbalanced cryptic translocation der (5)t(3;5)(q27;p15.3) in a family with three mentally retarded individuals. Hum Genet 107:51–57
10. Starke H, Nietzel A, Weise A et al (2003) Small supernumerary marker chromosomes (SMCs): genotype-phenotype correlation and classification. Hum Genet 114:51–67
11. Schmidt S, Claussen U, Liehr T et al (2005) Evolution versus constitution: differences in chromosomal inversion. Hum Genet 117:213–219
12. Mewborn SK, Lese Martin C, Ledbetter DH (2005) The dynamic nature and evolutionary

history of subtelomeric and pericentromeric regions. Cytogenet Genome Res 108:22–25
13. Hamid A, Weise A, Voigt M et al (2012) Clinical impact of proximal autosomal imbalances. Balkan J Med Genet 15:15–22
14. Gross M, Starke H, Trifonov V et al (2006) A molecular cytogenetic study of chromosome evolution in chimpanzee. Cytogenet Genome Res 112:67–75
15. Mrasek K, Heller A, Rubtsov N et al (2001) Reconstruction of the female Gorilla gorilla karyotype using 25-color FISH and multicolor banding (MCB). Cytogenet Cell Genet 93:242–248

Bar Coding Is Back

Thomas Liehr, Ahmed B. Hamid Al-Rikabi, and Anja Weise

Abstract

Chromosome bar codes (CBCs) based on locus-specific probes, i.e., bacterial artificial chromosome (BAC) clones, were popular in molecular cytogenetics for chromosome identification and characterization about 10 years ago. In the recent years, CBCs were introduced again but now meant for characterization of specific chromosomal subregions. Here we summarize a few such CBCs and highlight some points to be recognized when CBCs are set up.

Keywords Chromosome bar codes (CBCs), Locus-specific probes, Bacterial artificial chromosome (BAC) clones, Chromosomal subregions, Derivative chromosomes, Pericentric-ladder-FISH (PCL-FISH), Pericentromeric-critical region FISH (PeCR-FISH)

1 Introduction

As highlighted before (chapter by Thomas Liehr et al. "FISH Banding Techniques"), development of locus-specific probe-based chromosome bar codes (CBCs) is a rather old idea and available to some extent for human [1–5] and mouse [5–8]. Establishment of CBCs in human was meant for better breakpoint characterization in case of derivative chromosomes. CBCs developed for other species were primarily set up for characterization of hardly distinguishable chromosomes; studies in derivative chromosomes of the same or closely related species by these CBCs were possible but rarely done. CBCs in other species than human and mouse were published for the dog (*Canis lupus familiaris*) [9], a butterfly (*Bombyx mori*) [10], and some fish species (unpublished data). Also for plants CBCs are available, like *Arabidopsis* spec. [11], barley (*Hordeum vulgare*) [12], maize (*Zea mays*) [13], tomato (*Solanum lycopersicum*) [14, 15], or potato (*Solanum tuberosum*) [15].

However, most of these probe sets were just established and not much applied later on. Recently, new locus-specific probe-based CBCs in human were reported. In contrast to the previously

Thomas Liehr (ed.), *Fluorescence In Situ Hybridization (FISH)*, Springer Protocols Handbooks, DOI 10.1007/978-3-662-52959-1_30,

mentioned ones, they were developed to answer specific questions in connection with clinical problems. Such approaches are the parental origin determination FISH (pod-FISH) (chapter by Anja Weise and Thomas Liehr "Parental Origin Determination FISH: Pod-FISH"), subtelomeric and subcentromeric oriented FISH probe sets (chapter by Anja Weise and Thomas Liehr "Subtelomeric and/or Subcentromeric Probe Sets"), as well as probe sets implemented to study chromosomal breakpoints in the pericentric regions in detail [16–18].

Here we highlight the construction and application of CBCs oriented for characterization of specific chromosomal regions similar to the pericentric-ladder-FISH (PCL-FISH) probe set [17] or pericentromeric-critical region FISH (PeCR-FISH) probe set [18] as published recently by our group.

2 Materials

Apart from standard cell biological and molecular cytogenetic equipment, including standard solutions (e.g., ethanol, methanol, formamide, formaldehyde, xylene, etc.), no more specialized items are required. The equipment needed for multicolor FISH is listed in chapter by Thomas Liehr et al. "The Standard FISH Procedure" and chapter by Thomas Liehr and Nadezda Kosyakova "Multiplex FISH and Spectral Karyotyping".

3 Methods

3.1 Bar Coding Probe Sets/Selection of Probes

Pod-FISH (chapter by Anja Weise and Thomas Liehr "Parental Origin Determination FISH: Pod-FISH"), subtelomeric and subcentromeric oriented FISH probe sets (chapter by Anja Weise and Thomas Liehr "Subtelomeric and/or Subcentromeric Probe Sets"), as well as probe sets implemented to study chromosomal breakpoints in the pericentric regions in detail like PCL-FISH [17] or pericentromeric-critical region FISH (PeCR-FISH) [18] can be obtained on request from the first author of this chapter (mail: Thomas.Liehr@med.uni-jena.de).

Nowadays, CBCs based on locus-specific probes are constructed from bacterial artificial chromosome (BAC) clones. Selection of BACs is essential for a successful CBC. For sure, BAC probes are much more reliably mapped than previously used, much more instable yeast artificial chromosome (YAC) clones. However, before integration of BACs into a CBC, they need to be tested individually for correct location as well as for signal intensity. All BACs combined into one CBC need to be adjusted in terms of used amounts of DNA to obtain more or less comparable signal intensities. How to select and where to purchase BAC

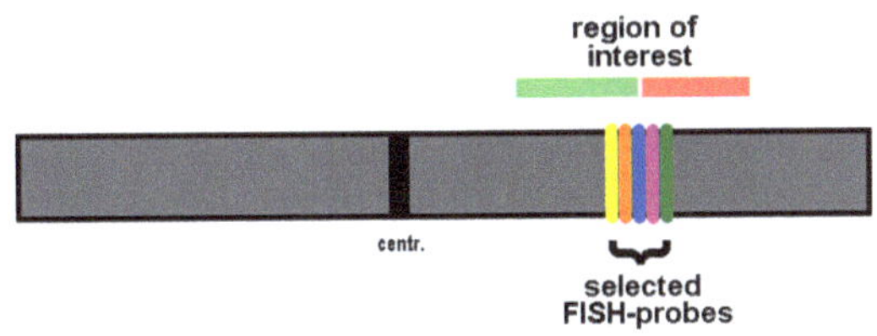

Fig. 1 Schematic depiction of how selection of a targeted chromosome-specific CBC probe set can be done: five consecutive BACs (FISH probes) are selected to span a region of interest

probes is described in chapter by Thomas Liehr "Homemade Locus-Specific FISH Probes: Bacterial Artificial Chromosomes". In Fig. 1 a possible way how to select BAC-probes is shown for a region of interest, e.g. a breakpoint.

3.2 Slide Pretreatment, Fluorescence In Situ Hybridization (FISH), and Evaluation

As described in chapter by Thomas Liehr et al. "The Standard FISH Procedure".

References

1. Lichter P, Tang CJ, Call K et al (1990) High-resolution mapping of human chromosome 11 by in situ hybridization with cosmid clones. Science 247:64–69
2. Lengauer C, Green ED, Cremer T (1992) Fluorescence in situ hybridization of YAC clones after Alu-PCR amplification. Genomics 13:826–828
3. Liehr T, Weise A, Heller A et al (2002) Multicolor chromosome banding (MCB) with YAC/BAC-based probes and region-specific microdissection DNA libraries. Cytogenet Genome Res 97:43–50
4. Speicher MR, Petersen S, Uhrig S et al (2000) Analysis of chromosomal alterations in non-small cell lung cancer by multiplex-FISH, comparative genomic hybridization, and multicolor bar coding. Lab Invest 80:1031–1041
5. Liehr T (2016) Multicolor FISH homepage. http://ssmc-tl.com/mfish.html
6. Liechty MC, Carpio CM, Aytay S et al (1999) Hybridization-based karyotyping of mouse chromosomes: hybridization-bands. Cytogenet Cell Genet 86:34–38
7. Liechty MC, Clase AC, Puschus KL et al (2000) Mouse linkage cytogenetics (L-C) probes. Cytogenet Cell Genet 88:163–167
8. Henegariu O, Dunai J, Chen XN et al (2001) A triple color FISH technique for mouse chromosome identification. Mamm Genome 12:462–465
9. Courtay-Cahen C, Griffiths LA et al (2007) Extensive coloured identification of dog chromosomes to support karyotype studies: the colour code. Cytogenet Genome Res 116:198–204
10. Shibata F, Sahara K, Naito Y et al (2009) Reprobing multicolor fish preparations in lepidopteran chromosome. Zoolog Sci 26:187–190
11. Berr A, Pecinka A, Meister A et al (2006) Chromosome arrangement and nuclear architecture but not centromeric sequences are conserved between Arabidopsis thaliana and Arabidopsis lyrata. Plant J 48:771–783
12. Karafiátová M, Bartoš J, Kopecký D et al (2013) Mapping nonrecombining regions in barley using multicolor FISH. Chromosome Res 21:739–751
13. Kato A, Lamb JC, Birchler JA (2004) Chromosome painting using repetitive DNA sequences as probes for somatic chromosome identification in maize. Proc Natl Acad Sci USA 101:13554–13559

14. Szinay D, Chang SB, Khrustaleva L et al (2008) High-resolution chromosome mapping of BACs using multi-colour FISH and pooled-BAC FISH as a backbone for sequencing tomato chromosome 6. Plant J 56:627–637
15. Tang X, Szinay D, Lang C et al (2008) Cross-species bacterial artificial chromosome-fluorescence in situ hybridization painting of the tomato and potato chromosome 6 reveals undescribed chromosomal rearrangements. Genetics 180:1319–1328
16. Castronovo C, Valtorta E, Crippa M et al (2013) Design and validation of a pericentromeric BAC clone set aimed at improving diagnosis and phenotype prediction of supernumerary marker chromosomes. Mol Cytogenet 6:45
17. Hamid AB, Kreskowski K, Weise A et al (2012) How to narrow down chromosomal breakpoints in small and large derivative chromosomes – a new probe set. J Appl Genet 53:259–269
18. Hamid AB, Pekova S, Fan X, et al. Small supernumerary marker chromosome may provide information on dosage-insensitive pericentric regions in human. Curr Genomics, in press

Fluorescence In Situ Hybridization onto DNA Fibres Generated Using Molecular Combing

Sandra Louzada, Jun Komatsu, and Fengtang Yang

Abstract

Molecular combing represents an advanced method for preparing extended DNA fibres. It enables uniform and parallel stretching of hundreds of DNA molecules on a glass surface, at an unprecedented resolution of two kilobase pairs per micrometre. When coupled with fluorescence in situ hybridization (FISH), molecular combing allows the direct visualisation of genomic structure and copy number variation, as well as the quantification of sizes of overlap and gap between sequence contigs in the genome assembly. Here we present a multicolor fibre-FISH protocol using DNA fibres prepared by molecular combing.

Keywords Molecular combing, DNA fibres, Fluorescence in situ hybridization (FISH), Fibre-FISH, Multicolor-FISH

1 Introduction

Fibre-FISH is a high-resolution fluorescence in situ hybridization (FISH) technique that uses extended chromatin and DNA fibres as hybridization targets for probes. DNA molecules are highly compacted in interphase nuclei (chapter by Thomas Liehr and Nadezda Kosyakova "Three-Dimensional Interphase Analysis Enabled by Suspension FISH"; chapter by Thomas Liehr and Sven Hauke "Interphase FISH in Diagnostics") and metaphase chromosomes (chapter by Anja Weise and Thomas Liehr "Pre- and Postnatal Diagnostics and Research on Peripheral Blood, Bone Marrow, Chorion, Amniocytes and Fibroblasts"). As a result, metaphase- and interphase FISH offer limited resolution in physical mapping of DNA markers, i.e. ~5–10 megabase pairs (Mbp) for the former and ~50 kilobase pairs (Kbp) for the latter. To increase the resolution of FISH, various methods for releasing chromatin and DNA fibres from interphase nuclei and metaphase chromosomes, via either chemical or mechanical treatment, were developed in the early 1990s [1–4], enabling a wide application of fibre-FISH in genomic studies. However, there is a major drawback in these methods, namely, the lack of

Thomas Liehr (ed.), *Fluorescence In Situ Hybridization (FISH)*, Springer Protocols Handbooks,
DOI 10.1007/978-3-662-52959-1_31,

consistency of the obtained fibres. In the same fibre preparation, there are chromatin fibres, single-molecule DNA fibres as well as partially lysed nuclei. The introduction of molecular combing, which enables a uniform stretching of hundreds of DNA molecules on a glass surface [5], represents an important advance in fibre-FISH. The main advantage of molecular-combing technology is its ability to comb high molecular weight genomic DNA. This procedure can produce a high density of fibres with most DNA fragments longer than several hundreds (200–700) Kbp [6]. During the combing process, DNA molecules are first bound, in a pH-dependent manner, to a chemically modified, hydrophobic surface (coverslips) and then slowly pulled at a constant speed (300 $\mu m\ s^{-1}$), with the receding meniscus promoting the uniform and parallel stretching of the anchored DNA molecules onto the glass surface [7]. The high consistency of DNA stretching (2 Kbp μm^{-1}) achieved by molecular-combing procedure also permits a direct conversion of the physical length of DNA fibres on the surface into DNA length in kilobase pairs (Kbp) [8]. When this technique is combined with FISH, it allows a direct visualisation of genomic structure and copy number variation, quantified analysis of the length, overlap and gap sizes in a given genome assembly as well as unambiguously defining the haplotypes in a diploid cell without pedigree information (e.g. [8–12]). Albeit molecular combing is a 20-year-old technology, the potential of molecular combing in the fields of genomics and comparative genomics remains to be fully explored.

Here we provide a multicolour fibre-FISH protocol using DNA fibres generated by molecular combing. The protocol is largely modified from a protocol that was developed originally by Genomic Vision (www.genomicvision.com). It includes two main parts: (1) the production of fibres by molecular combing and (2) sequential 3-color fibre-FISH using probes labelled by biotin-, digoxigenin-, and dinitrophenol-dUTPs.

2 Materials

2.1 Preparation of Fibres by Molecular Combing

- Gel plug mould and plastic plunger (Bio-Rad, Hertfordshire, UK)
- Cell counter

2.1.1 Equipment

- Molecular-combing system (MCS), including the soaking reservoir and the slide holders (Genomic Vision, Paris, France)
- Vinylsilane-coated coverslips (Genomic Vision)
- Glass engraver (Draper, Hampshire, UK)
- Fluorescence microscope equipped with filter specific for FITC (Zeiss Axio Imager, Zeiss, Oberkochen, Germany)
- Sample mixer with vertical rotation (HulaMixer, Thermo Fisher Scientific, MA, USA)
- Ceramic coverslip staining rack (Thomas Scientific, Swedesboro, NJ, USA)

2.1.2 Reagents

- 1 × PBS
- ACCUMAX™ cell dissociation solution (Sigma-Aldrich, MO, USA)
- 1 mM YOYO-1 (Thermo Fisher Scientific). Caution: This is an intercalating dye and a potential mutagen. Avoid skin contact.
- Trypan Blue Solution, 0.4 % (Thermo Fisher Scientific)
- β-Agarase enzyme (New England Biolabs, Ipswich, MA, USA)
- β-Agarase 10× reaction buffer (New England Biolabs)

2.1.3 Solutions to Be Prepared

- 1.2 % Low melting point (LMP) agarose solution: Add 0.12 g of NuSieve GTG agarose (Lonza, Basel, Switzerland) and 10 ml of 1 × PBS in a glass bottle and weigh the bottle. Heat in the microwave, and swirl several times until the agarose powder is dissolved completely. Weigh again and add ddH_2O to compensate any lost weight. Store at 4 °C in a falcon tube, up to 1 month.
- 0.5 M EDTA pH 8: Add 186.12 g of EDTA-2Na and 900 ml of ddH_2O. EDTA powder does not dissolve until pH increases. Bring the pH to 7 by adding NaOH pellets, and add 30 % NaOH solution to bring the pH to 8. Solution should now become clear. Top up with sterile distilled water to a final volume of 1 l and autoclave. Store at room temperature (RT). This solution is stable for several months.
- 10 % Sarkosyl: Add 1 g of sarkosyl and 10 ml of 0.5 M EDTA, pH 8. Store at RT.
- ESP solution: 0.5 M EDTA, 10 % sarkosyl, proteinase K (20 mg ml^{-1}) in a volume ratio of 8:1:1. Proteinase K should be added at the end. Prepare fresh before using.
- 1 M Tris, pH 8: Mix 88.8 g of Tris HCl, 53 g of Tris base and ddH_2O. Top up with ddH_2O to 1 l. Autoclave and store at RT.
- 1 × TE (10 mM Tris, 1 mM EDTA, pH 8.0): Mix 10 ml of 1 M Tris and 2 ml 0.5 M EDTA. Top up with ddH_2O to 1 l and autoclave. Store at RT.
- TE 40.2 (40 mM Tris, 2 mM EDTA pH 8.0): Mix 20 ml 1 M Tris and 2 ml 0.5 M EDTA. Top up with ddH_2O to 500 ml. Autoclave and store at RT.
- 0.5 M MES pH 5.5: Dissolve 9.76 g of MES in 80 ml of ddH_2O. Correct pH to 5 using NaOH pellets, and then use a 30 % NaOH solution to bring the pH to 5.5. Top up the volume with ddH_2O to 100 ml. Autoclave and store at 4 °C.

2.2 Probe Preparation

2.2.1 Equipment

- NanoDrop spectrophotometers (ND-8000, Thermo Fisher Scientific)
- Orbital incubator (SI50 Stuart Scientific, Bibby Scientific, Staffordshire, UK)

2.2.2 Reagents

- Unlabelled probe DNA (purified fosmid and or BAC DNA or long-range PCR product)
- GenomePlex® complete whole-genome amplification (WGA) kit (WGA2, Sigma-Aldrich)
- GenomePlex® reamplification kit (WGA3, Sigma-Aldrich)
- 100 mM individual dATP, dCTP, dGTP and dTTP (Thermo Fisher Scientific)
- Dig-11-dUTP (Jena Bioscience, Jena, Germany)
- Biotin-16-dUTP (Jena Bioscience)
- DNP-11-dUTP (Jena Bioscience)
- 50 mM MgCl2 (Bioline, London, UK)
- 10 U μl^{-1} DNase I recombinant (Roche, Basel, Switzerland)
- 10 × Incubation buffer DNase I (Roche)

2.2.3 Solutions to Be Prepared

- 0.5 M EDTA (see Sect. 2.1.3)

2.3 Fibre-FISH on Molecular-Combing Fibres

2.3.1 Equipment

- Epifluorescence microscope equipped with narrow band-pass filter sets for DAPI, FITC, Cy3 and Texas Red fluorescence. High numerical aperture 20× dry, and 40× oil, 63× oil, 100× oil objectives. Cooled CCD camera (Hamamatsu, ORCA-EA) and dedicated digital-imaging software (e.g. SmartCapture system from Digital Scientific, Cambridge, UK)
- Hot plate (CytoBrite, SciGene, CA, USA)
- Humid hybridization chamber
- Microscope slides: Microscope slides with frosted end (Thermo Fisher Scientific). The slides are cleaned by sonication in a 2 % Decon 90 solution for 10 min, then rinsed thoroughly in a large quantity of running tap water and distilled water. Store in 100 % ethanol and polish to dry with lint-free paper tissue.
- Gyratory rocker (GyroRocker, STR9, Bibby)

2.3.2 Reagents

- Blocking aid (Thermo Fisher Scientific)
- Ethanol (VWR, PA, USA)
- Formamide (Thermo Fisher Scientific). Caution: This is a hazardous substance, so it should always be handled in a fume hood.
- 20 × SSC buffer (Thermo Fisher Scientific)
- 1 μg μl^{-1} Blocking DNA (e.g. Human COT-1, Roche)
- DNA from salmon/herring sperm (e.g. Sigma-Aldrich)
- 3 M Sodium acetate, pH 5.2
- 1 × PBS
- Alkaline denature solution (0.5 M NaOH, 1.5 M NaCl, Sigma-Aldrich)
- Monoclonal anti-digoxin antibody produced in mouse (D8156, Sigma-Aldrich)
- Texas Red®-X goat anti-mouse, 2 mg ml^{-1} stock solution (T6390, Thermo Fisher Scientific)
- Anti-DNP unconjugated produced in rabbit, 2 mg ml^{-1} stock solution (SP-0603, Vector Laboratories, CA, USA)
- Donkey anti-rabbit IgG (H + L) secondary antibody, Alexa Fluor® 488 conjugate, 2 mg ml^{-1} stock solution (A-21206, Thermo Fisher Scientific)
- Streptavidin-Cy3 conjugate, 2 mg ml^{-1} stock solution (S6402, Sigma-Aldrich)
- Anti-streptavidin conjugated with CF543, 2 mg ml^{-1} stock solution [Customised antibody made in the lab by labelling anti-streptavidin unconjugated (SP-4000, Vector Laboratories) with CF™543 dye using the CF™ dye SE protein-labelling kit (Biotium, CA, USA)]
- SlowFade® gold antifade mountant (Thermo Fisher Scientific)

2.3.3 Solutions to Be Prepared

- Deionised formamide: For each 100 ml of formamide to be deionised, add 5 g of resin (MB-1 or MB-150, Sigma). Stir for 1 h using a magnetic stirrer, and then filter using a Buchner funnel and store at −20 °C.
- Hybridization buffer: Mix 100 μl of 20 × SSC, 50 μl of sarkosyl 10 %, 5 μl of NaCl 2 M, 50 μl of SDS 10 % and 295 μl of blocking aid. Make small aliquots and store at −20 °C. Before using the hybridization buffer, thaw it and add equivalent volume of deionised formamide (1:1).
- Post-hybridization stringent washing solution [50 % formamide/50 % 2 × SSC (v/v)]: Mix 50 ml formamide and 50 ml 2 × SSC.
- 2 × SSC: Mix 100 ml 20 × SSC and 900 ml ddH_2O.

- 2 × SSCT (with 0.05 % Tween 20): Mix 100 ml 20 × SSC, 900 ml ddH_2O and 500 μl Tween 20.
- Ethanol series in plastic pots (70 %/90 %/100 %).

3 Methods

3.1 Generation of DNA Fibres by Molecular Combing

The following protocol used in our lab was based on the experimental procedure for molecular-combing fibre preparation and hybridization provided by Genomic Vision, Paris, France. Please note that to facilitate the extraction of genomic DNA, a commercial kit (i.e. Fibre Prep DNA Extraction Kit) has been made available by Genomic Vision most recently, which can generate fibres with an average size of 300 Kbp.

For the preparation of fibres by molecular combing, genomic DNA from different origins can be used. We have successfully prepared the DNA solution and combed fibres from suspension/adherent cells, including human lymphoblastoid cell lines, pig fibroblast cell line, mouse spleen cells and chicken lymphocytes.

3.1.1 Outline of the Procedure

- From cell harvesting up to DNA fibres stretching on the coverslips, this procedure takes at least 3 days. At various points the products can be stored (Fig. 1).
- Cells are embedded in agarose plugs, which protect the genomic DNA from mechanical breaks.
- Proteins are digested, cell membranes solubilised and all residues removed by a series of washes with a buffered solution, preventing the genomic DNA from damages.
- High molecular weight (HMW) DNA is released into solution after enzyme digestion of the plugs.
- Attachment of DNA molecules to the surface of silanised hydrophobic coverslips, followed by uniform stretching.

3.1.2 Preparation

1. Both suspension and adherent cells can be used for this protocol. Generally, 1 plug should contain 1 million cells. Thus, the appropriate volume of medium, size and number of flasks should be taken into consideration beforehand.
2. Melt the 1.2 % LMT agarose at 68 °C for 10 min in a water bath and homogenise by inverting the tube. Lower the temperature to 50 °C and hold the agarose at this temperature.
3. Seal the base of the plug moulds with tape.

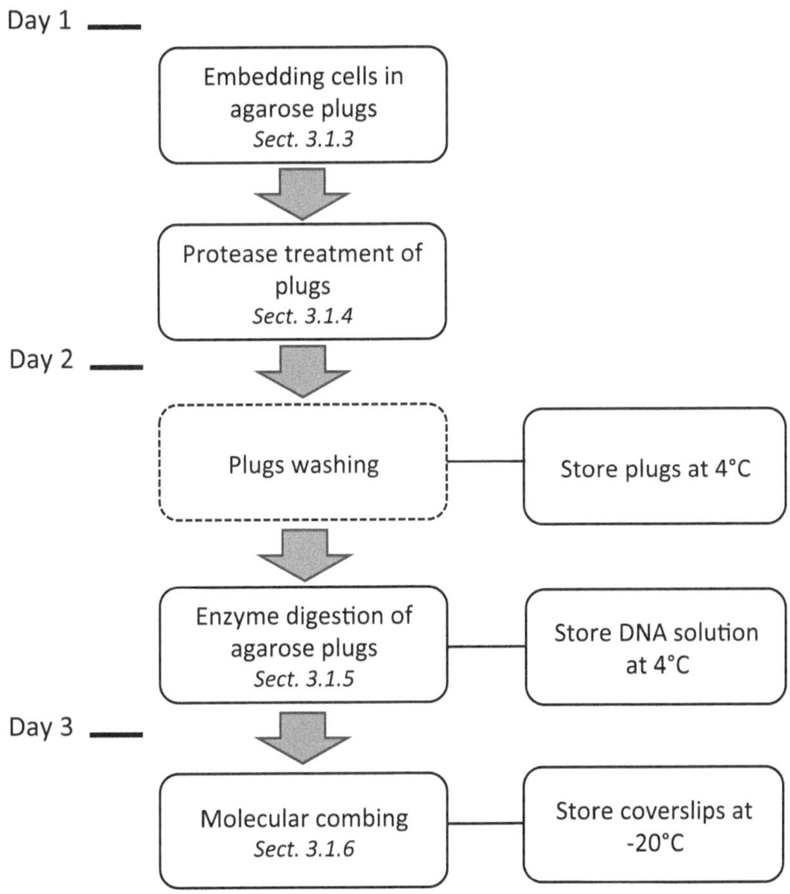

Fig. 1 Molecular-combing DNA fibre preparation. This workflow shows the main steps in the protocol. The preparation of the combed fibres will require at least 3 days

3.1.3 Embedding Cells in an Agarose Plug

1. Harvest cells as appropriate and centrifuge at 262× *g* for 5 min.
2. Remove the supernatant and resuspend the cells well in a mixture of 1 ml of Accumax and 1 ml of 1 × PBS, and incubate at 37 °C for 10 min in a water bath (*see* **Note 1**).
3. Centrifuge at 262× *g* for 5 min, remove the supernatant and add 10 ml of 1 × PBS.
4. Take 10 μl of the cell suspension, mix with the same amount of Trypan Blue Solution and load the suspension in the cell counter. Using an inverted microscope, count the number of cells.
5. Centrifuge the cell suspension at 262× *g* for 5 min, and gently remove the supernatant, as much as possible.
6. Add the appropriate volume of 1 × PBS, estimated from the result of the cell counting. 45 μl of 1 × PBS are needed for 1 plug, and each plug should contain 1 million cells (e.g. if the

cell count estimated that your culture has 5 million cells, this is sufficient for 5 plugs, 5 × 45 μl = 225 μl of 1 × PBS).

7. Homogenise well by pipetting 10× up and down. Ensure that no cell clumps are present.
8. Place the tube in the 50 °C water bath for 10 s to warm it up.
9. Without removing the tube from the water bath, add a volume of LMP agarose equal to the volume previously added for 1 × PBS (e.g. if you added 225 μl of PBS, add 225 μl of agarose).
10. Homogenise well by pipetting 10× up and down.
11. Quickly dispense the cell suspension into the DNA plug mould.
12. Put the plug mould in a sealed plastic box to prevent evaporation, and let the DNA plug set at 4 °C for 30 min.

3.1.4 ESP Treatment of Plugs

1. Prepare 250 μl of ESP solution for each plug to be treated. Use a 15 ml falcon tube when treating five or less plugs and a 50 ml falcon if more than five plugs are being treated simultaneously.
2. Push the plugs into the ESP solution using the plunger.
3. Place in the 50 °C water bath ensuring the tube is held vertically.
4. After 30 min, gently swirl the tubes to homogenise the solution.
5. Incubate overnight in a water bath at 50 °C (16–18 h).
6. Next day, carefully transfer the DNA plugs to a 15 ml falcon tube filled with 1 × TE, using a spatula. Maximum three plugs can be washed together in the same tube. Ensure that the tube is filled to the top and that no air bubbles remain in the tube, as these can damage the DNA plugs.
7. Wash the plugs for 1 h in a sample mixer with vertical rotation at 30 rpm.
8. Change the 1 × TE solution and wash again. Repeat 3× (total 4× washing).
9. DNA plugs can be stored in 0.5 M EDTA (maximum 10 plugs/5 ml 0.5 M EDTA) at 4 °C for several months.

3.1.5 Digestion of Agarose Plugs

1. If the plugs have been kept at 4 °C or if proceeding from Sect. 3.1.4 step 8, repeat the washing steps as in Sect. 3.1.4, steps 7–8 (*see* **Note 2**).
2. Carefully transfer each washed plug to a 2 ml round-bottom microtube, using a spatula.
3. Add 100 μl of TE 40.2 and 0.3 μl of YOYO-1, and verify that the plug is immersed in the solution. If not, tap gently on the bench until the plug drop in the bottom of the tube.

4. Leave at RT for 1 h, protected from light.
5. Pour the plug and solution out into plastic wrapping paper, and transfer the plug into a fresh 2 ml round-bottom microtube using a spatula. Tap the tube gently on the bench until the plug is at the bottom of the tube.
6. Place the tube in a 68 °C water bath for 10 min, in a fixed holder avoiding shaking (*see* **Note 3**).
7. Quickly transfer the tube into a 42 °C water bath and incubate for 5 min.
8. Add 10 μl of β-Agarase 10× reaction buffer prewarmed at 42 °C and then 3 μl of β-Agarase enzyme.
9. Incubate at 42 °C for 1 h, in the water bath, in a fixed holder avoiding agitation.
10. Add 1 ml of MES buffer prewarmed at 42 °C. Gently invert the tube and place it back in the water bath.
11. Leave at 42 °C overnight (16–18 h) in a fixed holder avoiding any agitation.
12. The following day, the high molecular weight DNA solution is ready to be combed, or it can be stored at 4 °C for few days (*see* **Note 4**).

3.1.6 Molecular Combing

1. If the solution has been stored at 4 °C, heat the solution to 42 °C before start.
2. Add 500 μl of MES buffer (prewarmed at 42 °C) to the combing reservoir.
3. Gently invert the tube with the DNA solution 2–3× to homogenise the solution.
4. Pour the DNA solution into the combing reservoir. Top up the reservoir with MES buffer (prewarmed at 42 °C) to within few mm from top.
5. Mark the coverslip (*see* **Note 5**) side that will be used in later experiments, using a glass engraver.
6. Insert the coverslip in the MCS slide holder slot and place it the MCS instrument. Ensure the writing part is at the top.
7. Place the reservoir in the MCS instrument and initialise the instrument. The coverslip will be dipped and incubated in the DNA solution and will be pulled at a constant speed of 300 μm s^{-1}.
8. Check the coverslips under a fluorescence microscope equipped with FITC filter, using the 20× objective. The coverslip should have its entire surface covered in long-range fibres (see Fig. 2). If the coverslips show very bright fibre bundles, gently pour the DNA solution back to the tube and repeat step 3–8 (*see* **Note 6**).
9. Bake the coverslips at 68 °C for 4 h.

2

Fig. 2 Human genomic DNA fibres obtained by molecular combing and stained with YOYO-1. The stretched and parallel fibres cover the entire surface of the coverslip. The image was captured with 20 ×, NA 0.8 objective, showing DNA fibres spanning hundreds of kilobases. The scale bar represents 50 μm (≈100 Kbp)

10. Coverslips are ready to use after baking, or they can be stored at −20 °C in a sealed container for several months.
11. DNA solution can be stored at 4 °C (*see* **Note 4**).

3.2 Fibre-FISH

3.2.1 Probe Generation from BAC Clones, Fosmid Clones and PCR Amplification

For fibre-FISH, we have been using probes derived from genomic DNA clones (BACs and fosmids) or probes generated by PCR amplification (1.5 Kbp or above). When selecting the probe combinations, pay attention to the size and complexity of the target region you want to visualise. Long molecules are prone to break; to reduce artefacts associated with broken fibres, it is desirable to include references (e.g. fosmid clones) marking the two ends of the target region, particularly when the target region is above 300 Kbp.

3.2.2 Primary WGA Amplification

Before the probe labelling, the DNA needs to be amplified using the GenomePlex® Complete Whole Genome Amplification Kit (WGA2, Sigma-Aldrich).

1. Start from 10 ng of template DNA.
2. Check the concentration and purity of the DNA using NanoDrop Spectrophotometers.

3. Perform WGA amplification by strictly following the manufacturer's instructions (see https://www.sigmaaldrich.com/content/dam/sigma-aldrich/docs/Sigma/Bulletin/wga2bul.pdf).
4. Run 2 μl of the amplified products on a 1 % agarose gel to verify the yield and size.

3.2.3 Labelling Probes Using WGA Reamplification Kit

We label the probes using a modified protocol based on the GenomePlex® Reamplification Kit (WGA3, Sigma-Aldrich). This kit comes with 10× Amplification Master Mix without the dNTP mixture (provided in separate). We prepare homemade tailored dNTP mixtures that contain different concentrations of dTTP to allow efficient incorporation of specific dUTP.

1. Calculate the number of probes to be labelled, and prepare a master mix for each reaction. For each 25 μl reaction, use 0.5–1 μl of WGA amplified DNA.
2. For labelling with biotin-, dig-, DNP-dUTPs, combine the following solutions:

25 μl Reaction Master Mix for Biotin-, Dig-dUTPs (Per Sample)	
dH_2O	15.7 μl
WGA3 10× amplification master mix	2.5 μl
10 × dNTP with 70 % dTTP (dATP, dCTP, dGTP 2 mM each, dTTP 1.4 mM)	2.5 μl
1 mM labelled dUTP	1.5 μl
$MgCl_2$ (50 mM)	0.5 μl
WGA3 DNA polymerase	1.8 μl
WGA2 amplified DNA	0.5 μl

25 μl Reaction Master Mix for DNP-dUTP (Per Sample)	
dH_2O	15.7 μl
WGA3 10× amplification master mix	2.5 μl
10 × dNTP with 80 % dTTP (dATP, dCTP, dGTP 2 mM each, dTTP 1.6 mM)	2.5 μl
1 mM labelled dUTP	1.0 μl
$MgCl_2$ (50 mM)	0.5 μl
WGA3 DNA polymerase	1.8 μl
WGA2 amplified DNA	1.0 μl

3. Place the PCR reactions in the thermocycler, and start the following programme:
   (a) 94 °C, 3 min
   (b) 94 °C, 15 s
   (c) 65 °C, 5 min
   (d) Repeat steps 2–3, 14 cycles (for labelling with DNP-dUTPs, increase the number of cycles to 25 cycles).
   (e) 4 °C. Pause.
4. Run 2 μl of each sample on a 1 % agarose gel to check for the presence and size of the product (see Fig. 3a).

3.2.4 Further Optimisation of the Probe Size by DNase I Digestion

In general the DNA fragment sizes of WGA-labelled probes are far too large to be used directly in FISH, and they need to be cut down to a size range between 200 and 500 bp using DNase I.

1. Dilute the DNase I enzyme in 10× incubation buffer in the proportion of 1:15.
2. Add 2 μl of diluted enzyme per 25 μl of labelling reaction, and incubate in the thermocycler at 15 °C for 25 min (*see* **Note 7**).
3. After the incubation, quickly return the tubes to ice.
4. Run 2 μl of each sample on a 1 % agarose gel to check if the product sizes range between 200 and 500 bp (see Fig. 3b) (*see* **Note 8**).
5. If the probe size still too large, return the tubes to the thermocycler at 15 °C for another 10–15 min (*see* **Note 7**).
6. Repeat steps 3 and 4.

3

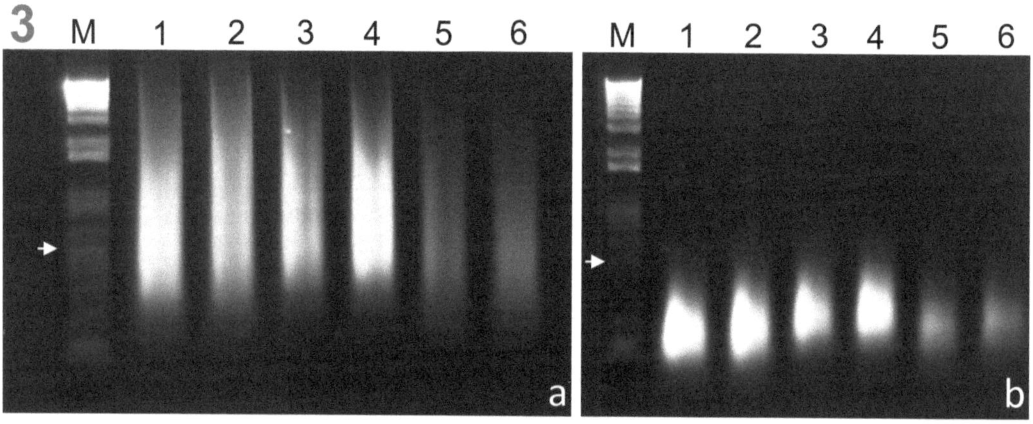

Fig. 3 Agarose gel electrophoresis of products labelled using modified WGA3 kit (**a**) and probes after further fragmentation with DNase I (**b**) M: 1 Kbp DNA ladder (Thermo Fisher Scientific). *Lane 1*: Human salivary amylase 1 gene PCR product labelled with biotin. *Lane 2*: G248P800725F7 fosmid clone labelled with biotin. *Lane 3*: Human pancreatic *amylase 2A* gene PCR product labelled with digoxigenin. *Lane 4*: G248P89536F11 fosmid clone labelled with digoxigenin. *Lane 5*: Human pancreatic *amylase 2B* gene PCR product labelled with DNP. *Lane 6*: G248P8836H11 fosmid clone labelled with DNP. The *arrow* indicates the 500 bp band

7. Add 1 μl of 0.5 M EDTA to each tube, and incubate at 65 °C for 10 min to stop the enzyme digestion when the desired probe size range is achieved.
8. Store at −20 °C until required.
9. Before being used in fibre-FISH, all probes should be subject to quality control by hybridising them onto interphase/metaphase control samples. This will allow to verify the probe quality and if they map in the expected position.

3.2.5 Fibre-FISH

3.2.5.1 Ethanol Precipitation of Probes

1. Prepare the probe mix in a 1.5 ml tube by adding the following reagents:

	1 coverslip
DNA labelled probe	5 μl each
1 μg μl^{-1} blocking DNA	5 μl × number probes
Salmon/herring sperm DNA	2 μl
3 M sodium acetate, pH 5.2	1/10 final volume
Ice-cold 100 % ethanol	2.0–2.5 × final volume

2. If the volume of the probes, plus salmon/herring sperm DNA and blocking DNA, is less than 50 μl, top up with ddH_20 or 1 mM TE.
3. Mix thoroughly by vortexing, and incubate at −20 °C for 2 h or at −70 °C for 30 min.
4. Calculate the amount of hybridization buffer needed. Thaw the hybridization buffer and add the appropriate amount of formamide (1:1). Use 22 μl of (hybridization buffer + formamide) mix for each coverslip (e.g. for hybridising one coverslip, mix 11 μl of hybridization buffer with 11 μl of formamide). Warm up at 37 °C for at least 20 min.
5. Centrifuge the probe mix in a precooled microcentrifuge at 13,000× g at 4 °C for 30 min.
6. Discard the supernatant and invert the tube in a paper towel to drain for a few seconds.
7. Wash the pellet with 500 μl of chilled 70 % ethanol, and centrifuge at 13,000× g at 4 °C for 10 min.
8. Remove the supernatant using a P1000 pipette.
9. Re-spin at 13,000× g for 1 min. Remove the remaining supernatant with a P10 pipette tip while avoiding touching the pellet.
10. Air-dry the pellet.

11. Resuspend the probe in the preheated hybridization buffer by vigorous vortexing or pipetting.
12. Spin the tube briefly to collect the probe mixture in the bottom of the tube, and incubate at 65–72 °C for 10 min to ensure the pelleted probes are fully resuspended (*see* **Note 9**).
13. Keep at 37 °C until needed.

3.2.5.2 Coverslips Pretreatments

Handle all combed coverslips with care as they can easily break. Always hold the coverslip by the lower corner; make sure the engraved side contacts with the probe mixture when setting up hybridization.

1. If the coverslips have been kept at −20 °C, remove them from the freezer and allow them to reach RT. If they have been baked, start from the next step.
2. Put the coverslips in a staining rack, and dip the rack successively in plastic pots containing an ethanol series (70, 90, 100 % ethanol) for 3 min incubation in each pot.
3. Carefully drain the ethanol from the coverslips on tissue and air-dry the coverslips at RT.
4. *Optional step*: chemical ageing can be performed in order to promote a better fixation of the fibres to the coverslip.
   - Warm up 100 % ethanol at 65 °C in a plastic pot.
   - Incubate the coverslips for 30–45 s.
   - Air-dry the coverslips at RT.

3.2.5.3 Denaturation and Hybridization

Co-denature

1. Pipette 20 μl of probe mix into a clean microscope glass slide.
2. Carefully lower the coverslip over the probe (with the engraved side facing down). Ignore bubbles underneath the coverslip and do not apply pressure to the coverslip as this will lead to probe overflow.
3. Place the slides on a hot plate at 75 °C for 10 min.
4. Transfer the slides to a humid chamber and incubate at 37 °C overnight.

Alkaline Denature

1. Place the coverslips in a ceramic staining rack, and dip it in the alkaline denature solution for 10 min.
2. Transfer the coverslips to 1 × PBS and wash three times, 5 min each.
3. Transfer the coverslips to plastic pots containing an ethanol series (70, 90, 100 % ethanol), and incubate 3 min in each pot.
4. Carefully drain the ethanol from coverslips on tissue and air-dry the coverslips at RT.

5. Denature the probe mix at 65 °C for 10 min.
6. Pipette 20 μl of the probe mix into a clean glass slide.
7. Carefully lower the coverslip over the probe (with the engraved side facing down).
8. Transfer the slides to a humid chamber and incubate at 37 °C overnight.

3.2.5.4 Post-hybridization Washes

During the post-hybridization washes, cover all plastic pots with foil.

1. Prepare three plastic pots with formamide/2 × SSC (1:1) solution and three plastic pots with 2 × SSC prewarmed at 25 °C.
2. Remove the coverslips from the humid chamber, and carefully transfer them into a staining rack immersed in 2 × SSC at RT.
3. Transfer the rack into the first pot containing formamide/2 × SSC (1:1) solution, and then incubate for 5 min on a gyratory rocker plate at 5 rpm at RT.
4. Repeat this wash two more times.
5. Transfer the rack into the pot containing the 2 × SSC solution, and incubate for 5 min on a gyratory rocker plate at 5 rpm at RT
6. Repeat this wash two more times.

3.2.5.5 Signal Detection and Amplification

The following steps are required for probes labelled with biotin, digoxigenin and DNP. Here we show an example of antibody combination for the simultaneous detection of three indirectly labelled probes. Other combinations can be used, as long as the selected antibodies do not cross-react.

1. Prepare the first layer for antibody detection. The antibodies should be diluted in blocking aid and prepared as shown in the table below. Prepare 20 μl of antibody mix for each coverslip.
2. Pipette 20 μl of antibody mix on a clean glass slide and gently overlay the coverslip with the engraved part facing down.
3. Incubate in a humid chamber at 37 °C for 20 min.
4. Carefully remove the coverslips and place them in a staining rack immersed in 2 × SSCT.
5. Wash for 3 min in 2 × SSCT on the gyratory rocker plate at 5 rpm at RT.
6. Repeat this wash two more times.
7. Prepare the antibody mix for the second layer (follow table below).

8. Pipette 20 μl of antibody mix on a clean glass slide and gently overlay the coverslip with the engraved part facing down.
9. Incubate in a humid chamber at 37 °C for 20 min.
10. Repeat steps 4–6.
11. Wash in 1 × PBS for 3 min.

Detection and Amplification of Three Color Probe

1st layer detection		*2nd layer detection*	
Dilution	*Antibody*	*Dilution*	*Antibody*
1:100	Mouse anti-digoxin	1:100	Goat anti-mouse Texas Red®
1:100	Rabbit anti-DNP	1:100	Donkey anti-rabbit Alexa Fluor®488
1:100	Streptavidin Cy3	1:100	Anti-streptavidin CF™543

12. Put a drop of antifade mountant in a clean microscope slide.
13. Remove a coverslip from the rack and drain the excess of 1 × PBS on a tissue.
14. Carefully lower the coverslip (engraved side down) onto the mounting solution.
15. Blot the excess of mounting solution by placing the slide over tissue.
16. Seal with nail varnish and store at 4 °C.

3.2.5.6 Image Capture

1. Manually scan the coverslip for desired hybridization patterns using a dual pass filter for FITC and Texas Red, under a 40× oil objective.
2. Capture FISH signals from FITC, Cy3 and Texas Red using SmartCapture software (Digital Scientific) via a cooled CCD camera (Hamamatsu, ORCA-EA). Use a 40× oil objective for long-range hybridization patterns, while 63× oil or 100× oil objective for shorter ones (*see* **Note 10**; Fig. 4).

4 Notes

1. Any clump of cells must be removed before embedding the cells in the agarose. After the treatment with Accumax, check if the cell suspension is homogeneous and no cell clumps remain.
2. The washing steps are very important for removing all residues that can cause DNA breaks and lower the fibre quality. More washes do no harm, only lead to an improvement in the combing preparation.

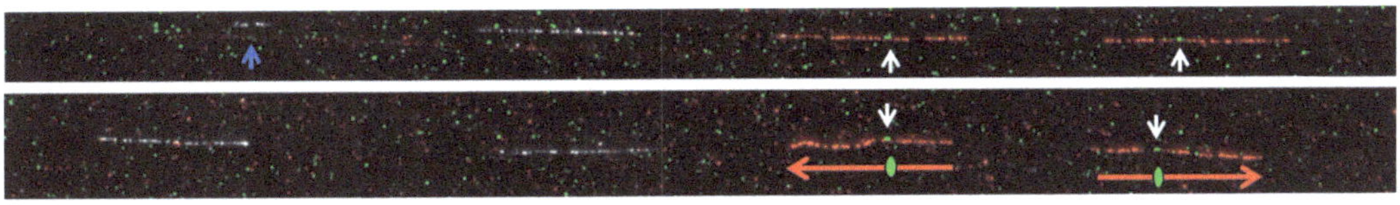

Fig. 4 Pig fibres prepared by molecular combing and hybridised with Y-specific probes. The *white* and *red signals* correspond to fosmid clones (~40 Kbp) carrying the genes *CUL4B* and *SRY*, respectively; the *small green signal* (*white arrows*) corresponds to a *SRY* cDNA probe (~1.5 Kbp). These results show two copies of the genes *CUL4B* and *SRY*, indicating that these genes are duplicated in the pig Y chromosome. *SRY* is duplicated but in inverted orientation, forming a mini palindrome (schematic representation with *red arrows*, for more details see 11). The use of fibres obtained by molecular combing allows the estimate of the distance between the *CUL4B* and *SRY* fosmids to be ~30 Kbp. The estimated total length of these fibres is ~280 Kbp. These long-range fibres are difficult to maintain intact and are prone to break (*blue arrow*)

3. From this step onwards, the DNA will no longer be protected from shearing, and it should be manipulated with care.
4. The DNA solution can be stored at 4 °C for a short period of time depending on the thoroughness of washing (*see* **Note 2**). Prolonged storage may result in the degradation of DNA fibres. For better results, always use freshly made DNA solution. If the combed fibres are good, use the solution to prepare as many coverslips as you need, and store the coverslips instead of the DNA solution.
5. Be aware that the coverslips coating can deteriorate in contact with air, so their shelf life is reduced to 6 months after opening the box. The coverslip degradation is indicated by a reduction in fibre density or contrast (fibres visualisation is reduced due to intense green background with YOYO-1 staining). Manufacturers recommend to store the coverslips in their original box, well closed, and to put the box in the original aluminium bag, well closed with the zip, at RT, in a dark place.
6. Quality control of the combed fibres often reveals the following problems:
   - Short-length DNA fibres: Insufficient washing of the plugs can cause this. It is recommended to use a new plug and increase the washing times.
   - The presence of fibre bundles: This can result from deficient homogenisation of the DNA solution. In this case gently return the DNA solution to the tube and gently invert (maximum 8 times; it is recommended to invert 2 times for the first time, then comb a new coverslip and verify the fibre quality and repeat this step if necessary).
   - Low fibre density: This could be due to coverslip problem (*see* **Note 5**) or low cell count. Try to comb again using a coverslip from a newly opened box or from a

different batch. If no improvement, combine the DNA solution from two plugs in the same reservoir before combing.

7. The amount of enzyme and the incubation time may have to be adjusted and will depend on the size of the products after labelling. If the size is still too large after the first incubation, the tube can be incubated for further minutes and more enzyme can also be added.
8. It is very important to cut the probes into the correct size range. If the probe size is too large, it will result in nonspecific bright spots due to the deposit of large DNA fragments on the coverslip. If the probe is too small, it will result in low-intensity hybridization signals.
9. The full resuspension of the probes constitutes a critical step. If the probes are not fully dissolved, it may lead to a lot of nonspecific spots, together with relatively dim hybridization signals, which will make the subsequent image capture process difficult.
10. An automated laser scanning and image capture system, Fiber-Vision® automated system, is available from Genomic Vision (France).

Acknowledgements

We would like to thank Genomic Vision for kindly providing the protocols and for their constant availability and help in troubleshooting. We also thank our team members from the Molecular Cytogenetics Core Facility at the Wellcome Trust Sanger Institute, Beiyuan Fu and Ruby Banerjee as well as the former team member Elizabeth Langley, for the contributions in the optimisation of these protocols. The Molecular Cytogenetics Core Facility at the Wellcome Trust Sanger Institute is funded by the Wellcome Trust (grant number WT098051).

References

1. Heng HHQ, Squire J, Tsui LC (1992) High resolution mapping of mammalian genes by in situ hybridization to free chromatin. Proc Natl Acad Sci USA 89:9509–9513
2. Fidlerová H, Senger G, Kost M et al (1994) Two simple procedures for releasing chromatin from routinely fixed cells for fluorescence in situ hybridization. Cytogenet Cell Genet 65:203–205
3. Heiskanen M, Peltonen L, Palotie A (1996) Visual mapping by high resolution FISH. TIG 12:379–382
4. Heng HH, Shi XM (1997) From free chromatin analysis to high resolution fiber FISH. Cell Res 7:119–124
5. Bensimon A, Simon A, Chiffaudel A et al (1994) Alignment and sensitive detection of DNA by a moving interface. Science 265:2096–2098
6. Lebofsky R, Bensimon A (2003) Single DNA molecule analysis: applications of molecular combing. Brief Funct Genomic Proteomic 1:385–396
7. Bensimon D, Simon AJ, Croquette VV et al (1995) Stretching DNA with a receding

meniscus: experiments and models. Phys Rev Lett 74:4754–4757

8. Michalet X, Ekong R, Fougerousse F et al (1997) Dynamic molecular combing: stretching the whole human genome for high-resolution studies. Science 277:1518–1523
9. Polley S, Louzada S, Forni D et al (2015) Evolution of the rapidly mutating human salivary agglutinin gene (DMBT1) and population subsistence strategy. Proc Natl Acad Sci USA 112:5105–5110
10. Skinner BM, Lachani K, Sargent CA et al (2015) Expansion of the HSFY gene family in pig lineages: HSFY expansion in suids. BMC Genomics 16:442
11. Skinner BM, Sargent CA, Churcher C et al (2016) The pig X and Y Chromosomes: structure, sequence, and evolution. Genome Res 26:130–139
12. Carpenter D, Dhar S, Mitchell LM, Fu B, Tyson J, Shwan NA, Yang F, Thomas MG, Armour JA (2015) Obesity, starch digestion and amylase: association between copy number variants at human salivary (AMY1) and pancreatic (AMY2) amylase genes. Hum Mol Genet 24:3472–3480

Parental Origin Determination FISH: Pod-FISH

Anja Weise and Thomas Liehr

Abstract

Except for individuals with variations in the pericentric heterochromatic chromosomal regions (including acrocentric short arms), a distinction of homologue chromosomes on a single-cell level is not possible. Due to this limitation, various questions of scientific and diagnostic relevance could not be studied by now. Based on copy number variations (CNV) spanning up to several megabasepair of DNA, we developed a molecular cytogenetic approach for an interindividual differentiation of homologue chromosomes, the so-called parental origin determination FISH (pod-FISH) technique. For this, all human chromosomes were covered with CNV-spanning BAC-probes in one- up to five-color chromosome-specific pod-FISH sets. With this approach to study the parental origin of individual human chromosomes on a single-cell level, new horizons for diagnostics and basic research were opened.

Keywords Copy number variations (CNV), Parental origin determination FISH (pod-FISH), Single cell level, Uniparental disomy

1 Introduction

The DNA sequences of any two human individuals differ by at least 0.1 %. The most common variations are single-nucleotide polymorphisms (SNP), which appear every thousand base pairs on average and are located in or outside coding regions [1]. Others include small insertion-deletion polymorphisms (INDEL) and noncoding polymorphisms, like mini- and microsatellites [2]. Although these variations can be found as alleles on homologous chromosomes, they cannot be used to distinguish them at a cytogenetic level (chapter by Anja Weise and Thomas Liehr "Pre- and Postnatal Diagnostics and Research on Peripheral Blood, Bone Marrow, Chorion, Amniocytes, and Fibroblasts"). Only molecular genetic methods are currently available for differentiating between homologous sequences. Starting from a mixture of many different single cells, DNA is extracted and analyzed by approaches like microsatellite analysis [3] or methylation-sensitive PCR [4]. On the other hand, conventional banding cytogenetics enables differentiation with respect to maternal or paternal origin of the

Thomas Liehr (ed.), *Fluorescence In Situ Hybridization (FISH)*, Springer Protocols Handbooks,
DOI 10.1007/978-3-662-52959-1_32,

chromosomes at a single-cell level, but only for a specific subset of human chromosomes and in exceptional cases [5]. Such cases may occur due to size variations of the heterochromatic regions of chromosomes 1, 9, 16, and Y or of the short arms of acrocentric chromosomes, inversion polymorphisms, or even less frequently, if different dimensions of centromeric heterochromatin (cen-, cen+, or cenh variants) are observed [6]. In the absence of such microscopically visible heteromorphisms, cytogenetic discrimination between homologous chromosomes is impossible. Consequently, a variety of scientifically and diagnostically important questions could not be answered due to that technical limitation.

Our understanding of human genome variations was significantly changed and extended by the discovery of a new kind of polymorphism affecting the copy number of euchromatic regions, ranging in size from ten up to several hundreds of thousands of base pairs [7, 8]. These so-called copy number variations (CNV) were found by DNA microarray technology (chapter by Eftychia Dimitriadou and Joris Vermeesch "Array CGH") and include hundreds of previously undetected structural variants in the human genome, like deletions, gains, and inversions. Up to February 2016, 491,894 CNVs and 1,745 polymorphic inversions have been reported (e.g., [7–11]), which are detailed, described, and collected in the database of genomic variants (http://dgv.tcag.ca/dgv/app/home).

Regarding the huge size of these structural variants, now it is possible to connect the DNA polymorphisms at a molecular genetic level with microscopically visible homologous chromosomes, making it possible to distinguish between all homologous chromosomes using a special fluorescence in situ hybridization (FISH) technique: "parental origin determination FISH" (pod-FISH) [12].

In conventional FISH, where signals for locus-specific probes on homologous chromosomes are expected to be equal in size and intensity, discordant signal patterns or cross-hybridizations are normal side effects (chapter by Thomas Liehr and Sven Hauke "Inter phase FISH in Diagnostics"). Pod-FISH takes special advantage of such variations in order to get discriminable signal patterns along the chromosome. Not all of the BAC clones used for polymorphic regions will show a distinctive pattern in one individual; this depends on the frequency of the polymorphism in the population. In our studies [12, 13], up to 29 % of the tested BAC clones showed a signal difference in one test person. Even if a signal difference is not directly visible by eye, it could be measured by software approaches like SCION (https://en.wikiversity.org/wiki/Scion_Image).

Once a distinctive signal pattern has been found in one individual, the parents should be tested with the same pod-FISH set in order to find clues about the chromosome segregation [13].

Alternatively, this can be used to differentiate between different cell lines, e.g., in cases of maternal contamination in prenatal diagnostics, or in follow-up studies after bone marrow transplantations. Differences from cell to cell within a normal, healthy individual can also be found when looking for single cells. This can be expected to some extent when searching for partial/complete disomies caused by mitotic recombination. On the other hand, this could also be induced by inconsistent FISH procedures (see below).

2 Materials

Apart from the standard cell biological and molecular cytogenetic equipment, including standard solutions (e.g., ethanol, methanol, formamide, formaldehyde, etc.), the following more specialized items are needed. The equipment needed for FISH itself is listed in the chapter by Thomas Liehr et al. ("The Standard FISH Procedure"); for BAC-probes, see the chapter by Thomas Liehr ("Homemade Locus-Specific FISH Probes: Bacterial Artificial Chromosomes").

3 Methods

3.1 Selection and Ordering of BACs

Chromosomes or regions of interest for pod-FISH studies can be selected from the database of genomic variants (http://projects.tcag.ca/variation/, chapter by Thomas Liehr and Anja Weise "Background"). In order to find BAC clones as FISH probes from corresponding regions, we recommend selecting regions of interest that are over 150 kb in size. Moreover, the CNV should have been reported by more than one person or study, and a loss should have been described, as a signal deletion on one homologous chromosome is easier to evaluate than a direct duplication caused by a copy number increase in the investigated region. A list of 225 BAC clones for the initial genome-wide pod-FISH is given in [12]. Once selected, there are several sources that can be used to order BAC clones (chapter by Thomas Liehr "Homemade Locus-Specific FISH Probes: Bacterial Artificial Chromosomes").

3.2 Creating Pod-FISH Sets

Depending on the fluorescence microscope filter sets available, BACs for the polymorphic regions of one chromosome (for example) can be labeled in different colors and applied in one hybridization step (*see* **Note 1**). An example is given for chromosome 16 in a five-color pod-FISH approach. When working with more than five BAC clones at the same time, it appears to be more convenient to split the probe sets and do two successive hybridizations. This approach has worked well, especially for several loci on long chromosomes, where we created chromosome-arm-specific pod-FISH sets. After labeling different BAC DNAs with different fluorescence colors or haptens, e.g., by DOP-PCR or nick translation (chapter by Thomas Liehr "Homemade Locus-Specific FISH Probes: Bacterial Artificial Chromosomes"), the probes are precipitated together and solvated in dextran sulfate.

3.3 Slide Pretreatment and Fluorescence In Situ Hybridization (FISH)

As described in the chapter by Thomas Liehr et al. ("The Standard FISH Procedure") (*see* **Note 2**).

3.4 Analysis of Pod-FISH Results

Pod-FISH probes can be evaluated in several ways: (1) by eye using a fluorescence microscope (Fig. 1), (2) by analyzing fluorescence profiles with appropriate software (Fig. 1), and (3) by measuring the signal intensity and area with software that has previously been shown to be suitable for measuring FISH signal intensity, such as the freely available software SCION (https://en.wikiversity.org/wiki/Scion_Image) (*see* **Note 3**) [12].

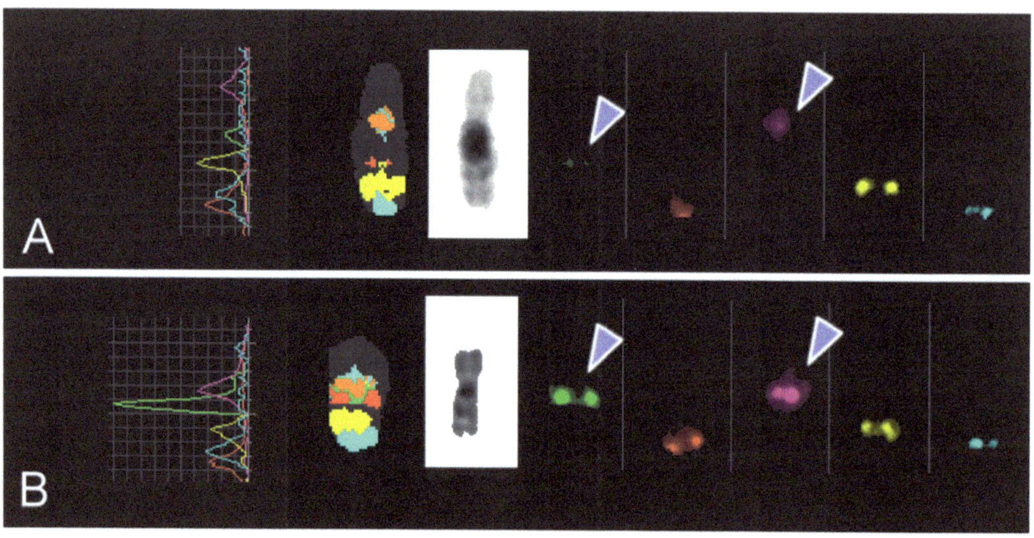

Fig. 1 Example of a five-color simultaneous BAC hybridization for chromosome 16. Here, a special FISH BAC technique is applied: so-called parental origin determination FISH (pod-FISH [12]). In pod-FISH, only BAC clones that are located in regions that are polymorphic for copy number are employed in order to make them visible by FISH and then use them to distinguish between homologous chromosomes (**a** and **b**). In this example, polymorphisms are visible for the Spectrum Green and Texas Red signals (*arrows*). From *left* to *right*: fluorescence profile, pseudo color image, inverted DAPI, BACs labeled in Spectrum Green, Spectrum Orange, Texas Red, Cy5, and DEAC

4 Notes

1. Always be careful when mixing clones for a pod-FISH set and test them beforehand in order to check whether the signals are very bright, normal, or weak. For pod-FISH sets, adjust the concentration of the BACs used by adding more labeled DNA for the weaker BACs and less for the brighter ones.
2. The FISH procedure for this approach has to be highly standardized and done by experienced lab staff to exclude the possibility that signal intensity differences are induced by fluctuations in technique. Concerning the specific limitations of the FISH method, we recommend testing this beforehand in the lab using nonpolymorphic clones and commercially available locus-specific probes in order to measure the signal differences between homologous chromosomes. The mean value of the signal difference will give you an idea of the laboratory-specific cutoff to be applied when measuring real signal differences caused by copy number variations.
3. Although single-cell analysis can be done by pod-FISH, the results are more reliable when done on several cells and then averaged, because of the abovementioned technical limitations. Also, it should be mentioned that software that can directly compare a whole pod-FISH set hybridization is not available at present, and so the analysis is still time consuming.

References

1. Lee C (2005) Vive la difference! Nat Genet 37:660–661
2. Schlötterer C (2004) The evolution of molecular markers—just a matter of fashion? Nat Rev Genet 5:63–69
3. Liehr T, Nietzel A, Starke H et al (2003) Characterization of small marker chromosomes (SMC) by recently developed molecular cytogenetic approaches. J Assoc Genet Technol 29:5–10
4. Nietzel A, Albrecht B, Starke H et al (2003) Partial hexasomy 15pter->15q13 including SNRPN and D15S10: first molecular cytogenetically proven case report. J Med Genet 40: e28
5. Müller H, Klinger HP, Glasser M (1975) Chromosome polymorphism in a human newborn population II: potentials of polymorphic chromosome variants for characterizing the idiogram of an individual. Cytogenet Cell Genet 15:239–255
6. Liehr T (2014) Benign & pathological chromosomal imbalances, 1st edition microscopic and submicroscopic copy number variations (CNVs) in genetics and counseling. Academic Press, New York
7. Iafrate AJ, Feuk L, Rivera MN et al (2004) Detection of large-scale variation in the human genome. Nat Genet 36:949–951
8. Sebat J, Lakshmi B, Troge J et al (2004) Large-scale copy number polymorphism in the human genome. Science 23:525–528
9. Shaw-Smith C, Redon R, Rickman L et al (2004) Microarray based comparative genomic hybridisation (array-CGH) detects submicroscopic chromosomal deletions and duplications in patients with learning disability/mental retardation and dysmorphic features. J Med Genet 41:241–248
10. Sharp AJ, Locke DP, McGrath SD et al (2005) Segmental duplications and copy-number variation in the human genome. Am J Hum Genet 77:78–88
11. Tuzun E, Sharp AJ, Bailey JA et al (2005) Fine-scale structural variation of the human genome. Nat Genet 37:727–732

12. Weise A, Gross M, Mrasek K et al (2008) Parental-origin-determination fluorescence in situ hybridization distinguishes homologous human chromosomes on a single-cell level. Int J Mol Med 21:189–200

13. Weise A, Bhatt S, Piaszinski K et al (2016) Chromosomes in a genome-wise order: evidence for metaphase architecture. Mol Cytogenet 9:36

Simultaneous Fluorescence Immunostaining and FISH

Christine J. Ye, Guo Liu, and Henry H.Q. Heng

Abstract

DNA–protein in situ codetection is a powerful method to study chromosomal/nuclear structure, function, and behavior along with their variations. Following the success of various large-scale genomic projects, the combination of fluorescence immunostaining and FISH may potentially become the method of choice to validate the molecular observations. In this chapter, different protocols are described for the preparation of mitotic and meiotic chromosomal/nuclear slides, as well as DNA–protein in situ codetection. The application of these methods will play an important role in the post-sequencing era.

Keywords DNA–protein codetection, Fluorescence immunostaining, FISH, Mitotic and meiotic chromosomes, SKY, Tissue sections

1 Introduction

The remarkable success of various large-scale omics technologies has unexpectedly highlighted the importance of molecular cytogenetic/cytogenomic analyses (chapter by Eftychia Dimitriadou and Joris Vermeesch "Array CGH"; chapter by Jiří Štika and Oldřich Mazal "Sequencing of Microdissection Derived FISH-Probes"). Based on the genome theory that the karyotype (rather than individual genes) defines genome context and determines a biological system by providing the system inheritance [1, 2], it becomes urgent to refocus on the characterization of the overall structure, function, and behavior of an entire genome as well as the heterogeneity of a given cell population rather than solely focus on one specific region of the genome based on the average profile (chapter by Thomas Liehr and Nadezda Kosyakova "Three-Dimensional Interphase Analysis Enabled by Suspension FISH"). The beauty of molecular cytogenetics lies in its power to bridge molecular characterization and morphological visualization of a specified genome, as well as the capability to profile single cell and its population [3, 4]. Among various molecular cytogenetic-based technologies, FISH represents the method of choice. In the past

Thomas Liehr (ed.), *Fluorescence In Situ Hybridization (FISH)*, Springer Protocols Handbooks,
DOI 10.1007/978-3-662-52959-1_33, © Springer-Verlag Berlin Heidelberg 2017

three decades, the continued optimization of FISH technologies has revolutionized the field of chromosome and genome research ([5–8]; chapter by Anja Weise and Thomas Liehr "Parental Origin Determination FISH: Pod-FISH"; chapter by Thomas Liehr et al. "Bar-Coding Is Back"). Some of these key improvements include sensitive small probe detection on banded chromosomes ([9, 10]; chapter by Thomas Liehr "FISH on Insect Cells Transfected with Heterologous DNA"), high-resolution fiber FISH ([11, 12]; chapter by Sandra Louzada et al. "Fluorescence In Situ Hybridization onto DNA Fibers Generated Using Molecular Combing"), RNA FISH ([13]; chapter by Bin Ma and Naoko Tanese "RNA Imaging in Living Cells"; chapter by Bin Ma and Naoko Tanese "RNA Imaging in Living Cells"), multicolor FISH and SKY ([14–16]; chapter by Thomas Liehr and Nadezda Kosyakova "Multiplex FISH and Spectral Karyotyping"), multicolor DNA–protein in situ codetection ([10, 17–19]; chapter by Bin Ma and Naoko Tanese "RNA-Directed FISH and Immunostaining"; chapter by Tiphaine Aguirre-Lavin and Nathalie Beaujean "Three-Dimensional Immuno-Fluorescence In Situ Hybridization in Preimplantation Mouse"), three-dimensional image analysis ([20, 21]; chapter by Thomas Liehr and Nadezda Kosyakova "Three-Dimensional Interphase Analysis Enabled by Suspension FISH"), Q-FISH ([22]; chapter by Gordana Joksic et al. "Telomere Length Measurement by FISH"), and FISH in living cells ([23]; chapter by Bin Ma and Naoko Tanese "RNA Imaging in Living Cells"). In this chapter, the combined multicolor FISH/SKY and immunostaining method (one form of DNA–protein in situ codetection) and the associated detailed protocols will be described and briefly discussed.

1.1 Development of Methodologies

The application of FISH methodologies was historically focused on gene and physical mapping. The detection of chromosome-associated proteins was less frequently applied. Fluorescence immunostaining, on the other hand, has long been used to study chromosomal and nuclear proteins and to identify specific cell types by detecting cell membrane antigens. Examples of these applications include the use of anti-5-methylcytosine antibody to illustrate the heterochromatic region on human (chapter by Anna Pendina et al. "Immunofluorescent Staining for Cytosine Modifications Like 5-Methylcytosine and Its Oxidative Derivatives and FISH") and chimpanzee chromosomes [24], as well as use of autoantibodies isolated from patients with autoimmune diseases to highlight multi-component structures such as nucleoli and centromeres ([25]; chapter by Elisabeth Klein and Thomas Liehr "CENP-Antibodies Used Additionally to FISH"). As chromosomes are complex DNA–protein structures, it was logical to combine the detection methods of FISH and immunostaining to investigate the DNA–protein interaction in situ. Early applications of simultaneous DNA–protein codetection techniques were limited to the colocalization of centromeric DNA sequences and the signals of anti-centromere sera (CREST) along chromosomes [26–29] and the simultaneous detection of cell membrane antigens as well as specific DNA targets within the same cell [30–32]. A new development for DNA–protein codetection was successfully demonstrated by the visualization of the structure and function of meiotic chromosomes [10]. Targeted DNA fragments painted by FISH signals and detected along the protein core of mouse meiotic prophase chromosomes were highlighted with the immunostaining of anti-core antibodies. This application of DNA–protein codetection opened the door to the study of chromatin loop structures as well as the meiotic process. This straightforward but powerful approach was soon used by many research groups [33–43]. The application of these technologies has led to many significant observations, including to demonstrate that the size of meiotic chromatin loops is related to their position along the chromosome core [3, 44], that a discrete X-recognition element can distinguish the X chromosome from autosomes to recruit the dosage compensation complex [45], that active genes can share sites of ongoing transcription [46], and that genes on the active X chromosome are more significantly associated with promyelocytic leukemia protein (PML) bodies than with their silenced homolog on the inactive X chromosome [47].

Multicolor FISH and immunostaining methodologies were introduced by combining spectral karyotyping (SKY) and synaptonemal complex (SC) protein detection ([18, 19, 48, 49]; chapter by Maria Bonet Oliver "Sperms, Spermatocytes and Oocytes"). This method was developed to precisely identify and measure each meiotic prophase chromosome. SC proteins were detected using antibodies to synaptonemal complex proteins SYCP1 and SYCP3 tagged with either fluoresce in isothiocyanate (FITC), rhodamine, or gold particle-conjugated secondary antibodies. Antibody detection of SC proteins was followed by labeling of chromosomes with SKY. This study revealed an inconsistency in the size of mitotic and meiotic chromosome lengths that allowed for a systematic analysis of the relationship between the length of meiotic chromosomes, GC content, and the genetic recombination rate ([48, 50], Heng unpublished observation). Soon after our initial demonstration of SKY and immunostaining as a powerful DNA–protein codetection technique, additional studies using SKY or M-FISH technologies combined with SC detection on human and mouse meiotic chromosomes were reported by other investigators [43, 51–54]. Multicolor interphase detection coupled with immunophenotyping has also proven to be a useful tool for clinical research [55]. In recent years, increased efforts have been focusing on the characterization of DNA–RNA–protein complexes, and integrated set of protocols are developed to visualize specific RNAs, DNAs, proteins, and histone modifications in single cells [56].

1.2 Significance and Implications

DNA–protein in situ codetection will play an increasingly important role in the post-genome era. According to the genome centric concept (genome theory), more studies are required to characterize the genome system defined by overall genomic context rather than individual genes [1]. For decades, the attention of molecular biology has focused more on genes and pathways than the genome system. Current molecular knowledge is mostly generated by averaging biochemical analyses without systematical integration of genomic topology. Chromosomes are not simply the vehicles of genes; more importantly, they are the key genomic coding system that defines the blueprints of cells [1, 2, 57]. Accordingly, techniques that can be applied to fill the gaps between molecular mechanism and cellular structure, between in vitro analysis and in vivo systems, between individual cells and cell populations, between genetic network and genomic topology, and between a snapshot perspective and the entire biological process over time are invaluable to the future of genomic research. Karyotypic variations including these seemingly random non-clonal chromosomal aberrations (or NCCAs) are important, and molecular cytogenetic analyses will continue to play an important role in future genomics [2, 58]. Simultaneous fluorescence immunostaining and FISH will play an increasingly important role in bridging these gaps [3, 38, 48, 50]. With these techniques, we consider this an important finding from our group: Genetic heterogeneity (detected at the genome level as NCCAs) is not "genetic noise," but a key adaptive feature of dynamic biological systems, including normal development, cancer progression, and organismal evolution [58]. Therefore, the use of these technologies will be proven to be ever more important in monitoring genetic heterogeneity through the tracking of chromosomal aberration patterns at the levels of both individual cells and cell populations. Particularly pertinent to clinical cancer research, an important application of these technologies is the ability to monitor genome level aberrations to correspond to the changes in gene expression profiles. For example, the correlation between gene amplification and oncoprotein expression of both *C-MYC* and *HER-2* in invasive breast cancer cases has been studied using these techniques [59–61]. These codetection principles are extremely valuable to many areas of clinical and basic research, as they are well suited to the simultaneous detection of polyploidy, aneuploidy, gene amplification, and protein content in the same cell. In particular, knowing (1) that various types of genome chaos frequently occurred during different physiological and pathological conditions [62, 63], (2) altered genome (such as various types of NCCAs) are coupled with elevated transcriptome dynamics [64], and (3) gene expression is closely related to the topological associated domains of the given genome [65], the DNA–protein in situ codetection method is therefore essential to validate large-scale genomic data systematically.

Here we describe four protocols for the induction/preparation of mitotic chromosomes/nuclei and one protocol for the preparation of meiotic chromosomes, which are then combined with methods of sequential protein and DNA detection. Optimal conditions for these, producing the strongest signals for FISH and immunostaining, sometimes conflict with each other. If both FISH and immunostaining work well separately, then the codetection should also work well. Technical compromise is sometimes needed to optimize signals of both DNA and protein. Also below we show some of our results obtained in special applications.

1.3 DNA–Protein Codetection to Monitor Mitotic Chromosomes/Nuclei

DNA–protein in situ codetection can be applied to various investigations that analyze the structure, function, and behavior of chromatin, chromosomes, or nuclei. During our characterization of defective mitotic figures (DMFs), for example, SMC proteins and SKY detection were simultaneously applied to study the potential order of chromosome condensation. The techniques detailed here were applied to discover a differential association of the structural maintenance of chromosome (SMC2) protein, characterized by negative SMC2 staining in the uncondensed region. Sequential SKY was used to assign a chromosome number to the defective chromosome. Similarly, these codetection methods facilitated the characterization of a novel form of mitotic cell death termed "chromosome fragmentation" and genome chaos [62, 66]. In these studies, the mechanisms underlying chromosome fragmentation were compared with other known forms of cell death by labeling cells and chromosome-associated proteins with specific antibodies along known cell-death pathways.

Recently, the formation and biological consequence of various NCCAs have undergone expanded investigation [67]. For example, somatic evolution of chaotic genomes has been linked to a new type of cell division that generates a cluster of small cells with positive markers of stem cell. Interestingly, each of these small cells contains partial genome illustrated by SKY analyses (Fig. 1a). The DNA–protein codetection method will certainly play an important role to further characterize the function of these cells.

1.4 Meiotic Chromosomes

FISH immunostaining codetection is a powerful method of studying meiotic chromosome structure and function. In addition to providing an experimental system to trace chromosomal pairing behavior during various stages, this method has revealed some unique features of meiotic chromosomes. Data from a number of transgenic mice with the same DNA inserts but different integration sites illustrated that various loop sizes along different regions of chromosomes (such as the telomere region) were different [44] (Fig. 1b). These patterns observed from transgenic experiments have also been confirmed by endogenous sequences [48, 50]. SKY immunostaining codetection also provides investigators with an elegant method of studying the order of pairing among chromosomes.

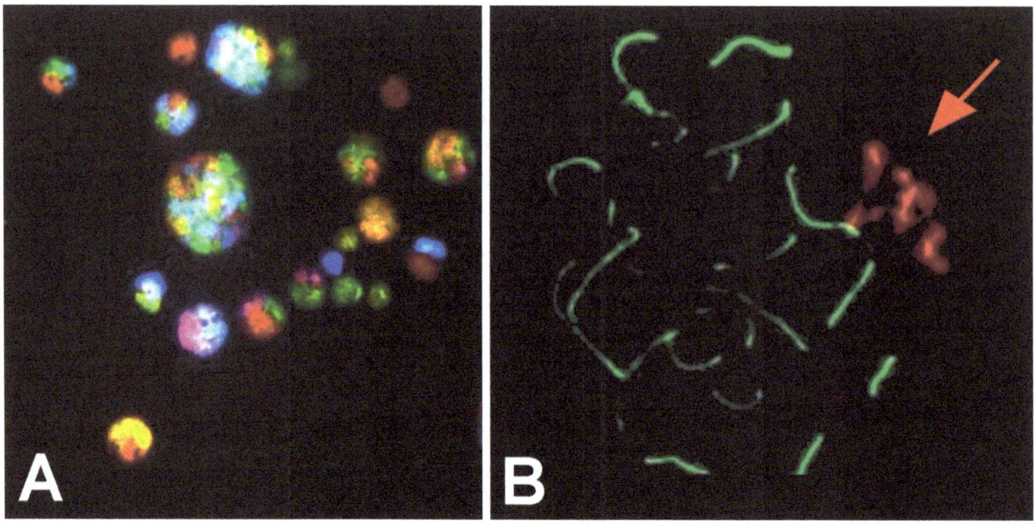

Fig. 1 (**a**) SKY image of a group of small cells with partial genomes following cytogenetic preparation. All these cells are generated from a single cell with chaotic genome through one abnormal nuclear division. The chromosomal composition of each small cell varies, from a single chromosome to 15 chromosomes in this specific figure (reproduced from [68]). Only a small portion of these cells display positive stem cell markers (unpublished observation). (**b**) Example of the use of FISH and protein codetection to study the transgenic insertion of mouse meiotic chromosome. Protein cores are visualized *green* with FITC-conjugated secondary antibody attached to antisynaptonemal complex antibody. The red signal (indicated by the *red arrow*) represents the insert of human DNA detected by FISH

1.5 Tissue Sections

Using tissue sections to detect specific genomic changes within a defined cell type is of importance to clinical diagnosis. Many recent publications have used these methodologies to visualize important features of normal and abnormal cells. In particular, when monitoring cancer progression and/or studying the heterogeneity of tumors, there is often a need to correlate genome level variation, such as the level of gene amplification, with the level of proteins detected. These techniques are frequently used in transplantation experiments as well as in posttransplant detection of foreign cells within a host's tissue.

1.6 Conclusions and Future Perspectives

Simultaneous fluorescence immunostaining and DNA detection with SKY or FISH represent a powerful tool to study the structure, function, and behavior of chromatin domains, individual chromosomes, genomes, and genome variations within cell populations, especially in the setting of genome level heterogeneity [19, 48–50, 64, 67, 68]. In comparison with other molecular methods that do not provide morphological characteristics, the direct visualization techniques described in the previous sections are superior for several reasons. First, these techniques can provide information that addresses the issue of genomic heterogeneity within cell populations. Additionally, these techniques can identify unique distribution associations and localization patterns of specific DNA–protein complexes within the nuclei or chromosomes of a single cell. For example, it is known that gene expression can be highly variable within a population of cells. In a similar vein, we have observed that transcription factors were limited to specific regions on a chromosome instead of being evenly distributed (Heng et al., unpublished observations). Similarly, we have also observed viral infections of a cell population that is remarkably heterogeneous, with only a small portion of the cells being infected at any given time [19]. Taken together, these observations call attention to the limitations of various in vitro assays that require DNA isolation and destruction of the genome/chromosome or cellular context. Despite the enormous implications, we would like to discuss a few aspects of the applications of this method based on our own research experiences. Detailed characterization of the higher-order structure of the chromosome remains one of the major challenges in molecular cytogenetics research. Following several decades of extensive research, there is no generally accepted model for high-order structure for both mitotic and meiotic chromosomes. DNA--protein in situ visualization methods will help to establish these models. We have recently suggested a model of meiotic chromosome loop organization based on the findings that loop size correlates well with AT-GC chromosome content and that the telomere regions display smaller loops ([44, 50]; Heng et al., unpublished results).

The study of gene expression represents an extremely common theme in current biological research. During the past two decades, a great deal of effort has been made to characterize the promoter regions, enhancer regions, and some protein-binding motifs within regulatory regions. Recently, increased attention has focused on higher-order chromatin-based regulation of gene expression. These include the study of chromatin loop dynamics [48, 50] and the identification of regulatory DNA, RNA, and protein complexes [69]. Additionally, the examination of genome variations such as gene duplication, chromosome duplication, or chromosome translocation on gene expression profiles reflects a new area of higher-order chromosome research which integrates multiple levels of gene regulation. In particular, the rapid development of various chromosome conformation capture technologies has generated a large amount of data to illustrate the quantitative chromosomal/protein interaction, which will require in situ validation. It is also timely to connect genome level alterations, such as translocations and aneuploidy, to pattern changes of topological context of chromatin [70, 71].

Another trend in the use of these methods is the examination of the time course of chromosomal events as they relate to other known biological processes. For example, these techniques can be used to relate chromosomal events to different phases of the cell cycle, to trace chromosome pairing in meiosis [10, 44, 50], and to illustrate the karyotypic pattern of evolution during cancer progression [58]. SKY-FISH codetection provides the ability to monitor the changes of specific loci within the framework of an altered genome. This can be achieved by using codetection methodologies to correlate specific protein markers with local gene and whole-genome level changes.

Several challenges need to be addressed to further improve upon the technologies presented in this chapter. One area that should be targeted for future development is the improvement of protein detection techniques for increased sensitivity. The development of multicolor fluorescent immunostaining methodologies for different target proteins is currently underway [72] and should provide valuable information on the relationships among different proteins, chromosome structure, and function. With more sensitive multiple color protein and DNA codetection, it is anticipated that detailed DNA–protein interactions will be traced. This will allow for the analysis of transcription factor binding order to the promoter region using high-resolution fiber FISH methods [3, 11, 50]. Future projects that push the current limits of these technologies are the combination of DNA–protein codetection with the three-dimensional study of chromosome structure and the application of this technology within live cells. Efforts to advance powerful combinations of these technologies will prove to be worthwhile, as its applications are far reaching in the study of chromosome dynamics and their relationship to structural and regulatory proteins. Most encouragingly, novel combinations and increased sensitivity of DNA–RNA–protein codetections have been developed, which will certainly increase the power of the technology in this field.

2 Materials

Apart from standard cell biological and molecular cytogenetic equipment, including standard solutions (e.g., ethanol, methanol, formamide, formaldehyde, etc.), the following specialized reagents and equipment are needed.

2.1 Mitotic Chromosome Preparation

2.1.1 Induction of Chromosomal/Nuclear Variations

- Doxorubicin
- Mitomycin C

2.1.2 Direct Preparation of Cells Growing on Slides

- Fixative solution (ratio of methanol to acetic acid 3:1, for convenience, we refer to it as Carnoy's fixative)
- Hypotonic solution (0.4 % KCl)
- PBS-azide solution pH 7.4: 1 mM ethylene glycol tetraacetic acid (EGTA) (38 mg); 0.01 % sodium azide (NaN_3) (1 ml);

150 mM NaCl (877 mg); 10 mM sodium phosphate ($NaPO_4$) (10 ml of 100 mM, pH 7.4); distilled water to 100 ml

- Sterilized slides or coverslips

2.1.3 Slide Preparation by Cytospin® Centrifugation

- 0.05 M borate buffer stock solution: 1.91 % (w/v) sodium borate in distilled water, adjust pH to 9.2 with 0.5 M NaOH. Prepare a working solution of 0.01 M by diluting stock 1:4 with distilled water.
- CSK solution: 10 mM pipes, pH 7.8; 100 mM NaCl; 0.3 M sucrose; 3 mM $MgCl_2$; 0.5 % Triton® X-100.
- Cytospin® and Cytofunnel®.
- Hypotonic solution (0.4 % KCl).
- 1 % paraformaldehyde solution: 1 % (w/v) paraformaldehyde (1 g); 0.5 M NaOH (10 ml); 30 ml phenol red; distilled water to 100 ml. Bring to 55 °C on a hot plate in a fume hood. Shut off the heat, place the Erlenmeyer flask in a container of cold water on the hot plate, and continue stirring until all powder has dissolved. The temperature should not go higher than 60 °C. If the pH acidifies, add 0.01 M borate buffer dropwise. Filter, cool to room temperature (RT), and adjust the final pH to 8.2 with borate buffer using paper pH indicators. *Caution:* Paraformaldehyde is harmful if ingested and can be absorbed through the skin. The fine powder is easily dispersed through the air.
- PBS-azide (pH 7.4) (see Sect. 2.1.2).
- Phenol red indicator for pH monitoring: 0.5 % phenol red in distilled water. Filter and store at RT indefinitely; (alkaline pH = purple-red color, acidic pH = yellow).
- 0.4 % Photo-flo® 200 (Kodak) in distilled water. Add 30 ml of phenol red indicator per 100 ml of solution. Using the paper pH indicator, adjust the pH to 8.0 with borate buffer.

2.1.4 Slide Preparation from Sample of 3:1 Fixative

- Bovine serum albumin (BSA) (10 %): 10 g BSA; add prewarmed to 37 °C distilled water and vortex to mix well. Leave at RT until dissolved.
- Chromosome swelling buffer: 1 ml TWEEN® 20 (10 %), 1 ml BSA (10 %) with 100 ml TEEN buffer.
- PBS-azide solution: see Sect. 2.1.2.
- TEEN buffer: 1 mM triethanolamine HCl, pH 8.5 (186 mg), 0.2 mM ethylene dinitrilotetraacetic acid (EDTA) (75 mg), 25 mM sodium chloride (NaCl), (1461 mg), distilled water to 1000 ml.
- TWEEN® 20 solution (10 %): 10 ml TWEEN® 20, distilled water to 100 ml.

2.2 Surface Spreading of Testicular Material

- 0.05 M borate buffer stock solution: see Sect. 2.1.3.
- Five 10 ml coplin jars, a 10 ml pipettor and tips, wide-bore plastic transfer pipettes.
- Dissection tray, scissors, fine forceps, 50 ml beaker, dental wax, single edge razor, microcentrifuge tubes, centrifuge.
- Minimum essential medium (MEM) with Hanks salts, without L-glutamine: Purchased ready to use from supplier or made from powder (10× concentrate). Adjust pH to 7.3 with 0.05 M borate buffer.
- 1 % paraformaldehyde solution (see Sect. 2.1.3).
- Phenol red for pH monitoring (see Sect. 2.1.3).
- 0.4 % Photo-flo® 200 (Kodak) in distilled water (see Sect 2.1.3).
- 60 mg ml^{-1} sodium dodecyl sulfate (SDS) stock solution with 30 ml of phenol red indicator per 100 ml of solution. Adjust pH to 8.2 with borate buffer. Store at RT. Depending on the degree of the chromatin dispersion desired, use from 0 to 0.06 % SDS in the first paraformaldehyde fixation. *Caution:* SDS is harmful if ingested or inhaled and irritates the eyes and skin. The fine powder is easily dispersed through the air.
- Spreading (hypotonic) solution: 0.5 % NaCl in distilled water; adjust the pH to 8.0 with borate buffer.
- Windex®.

2.3 Treatment of Tissue Sections

- 1 % paraformaldehyde (see Sect. 2.1.3)
- Ethanol
- PBS-azide (see Sect. 2.1.2)
- Pepsin solution: 0.005 % pepsin in 0.01 M HCl
- PK solution: 20 mg ml^{-1} Proteinase K in 0.1 M of TRIS HCl and 0.05 M EDTA
- Slide warmer 60–70 °C
- Xylene

2.4 Immunostaining

- ADB (antibody dilution buffer) solution: 10 % goat serum; 3 % bovine serum albumin (BSA). Prepare BSA solution using pre-warmed (37 °C) PBS. Vortex well and leave at RT until dissolved.
- Triton® X-100. You will add 1 % v/v to the second of three washes.
- Triton® X-100 (0.05 %) in PBS.
- Small humid chamber such as a Plexiglas box with a support for holding the slides.
- Kodak Photo-flo® 200 (1 %) in PBS.
- Wash buffer (10 % ADB in PBS).

2.5 FISH: Fluorescence In Situ Hybridization Detection

2.5.1 DNA Probe Labeling and Purification

- Biotin labeling kit (Roche Diagnostics, Basel, Switzerland)
- Equilibrium buffer: Use TE buffer, pH 7.5
- 3 M NaAc
- Nick column
- Salmon sperm DNA (100–500 bp fragments obtained by sonicating)

2.5.2 Hybridization and Detection

- Antifade solution. ProLong Antifade (Thermo Fisher Scientific).
- Avidin-FITC (fluorescein isothiocyanate): 2 mg ml^{-1} (stock solution). FITC detection working solution: 5 ml of avidin-FITC stock solution to 2 ml of detection solution. Store in the dark at 4 °C. Good for up to 6 months.
- Biotinylated goat anti-avidin antibody: 500 mg ml^{-1} (stock solution). Aliquots (50 ml each) can be kept at −20 °C. Working solution: 5 ml of anti-avidin antibody stock solution to 0.5 ml of detection solution.
- Cot-1 DNA (Thermo Fisher Scientific).
- DAPI (Sigma): 0.2 mg ml^{-1} of stock solution in H_2O. Store in the dark at 4 °C.
- DAPI/Antifade: 0.2 μg ml^{-1}.
- Denaturation solution: 70 % deionized formamide (Sigma) in 2 × SSC (saline sodium citrate) (20 × SSC stock solution: 3 M NaCl, 300 mM Na citrate).
- Detection solution: 1 % BSA and 0.05 % TWEEN® 20 in 4 × SSC. Store at 4 °C.
- Detection washing solution: 0.05 % TWEEN® 20 in 4 × SSC.
- Hybridization solution I (for use with genomic DNA probes): 50 % deionized formamide (Sigma); 10 % dextran sulfate in 2 × SSC.
- Hybridization solution II (for use with repetitive DNA probes): 65 % formamide; 10 % dextran sulfate in 2 × SSC.
- 25 ml plastic slide mailers (Surgipath, Richmond, IL, USA).
- Plastic slide chamber (slide holder) (VWR CanLab, Mississauga, ON, Canada).
- RNase A (Sigma).
- Hybridization washing solution A (for non-repetitive DNA probes): 50 % formamide in 2 × SSC.

- Hybridization washing solution B (for repetitive DNA clones): 65 % formamide in 2 × SSC.
- Water baths at 37, 42, 70, and 75 °C; 37 °C incubator.

2.6 SKY: Spectral Karyotyping

- Antifade DAPI reagent (see Sect. 2.5 or can be purchased from Applied Spectral Imaging, Vista, CA, USA).
- Prepare BSA solution using prewarmed (37 °C) PBS. Vortex well and leave at RT until dissolved.
- Pretreated chromosome slides (following the immunostaining step).
- Cy5 staining reagent: 5 μl of anti-digoxin (Sigma, D8156); 5 μl of Cy5 streptavidin (1 mg ml^{-1}, Amersham, Little Chalfont, UK; PA45001); 1 ml of 4 × SSC.
- Cy5.5 staining reagent: 5 μl of Cy5.5 sheep anti-mouse (1 mg ml^{-1}) (Rockland, Gilbertsville, PA, USA; 610–113–121); 1 ml of 4 × SSC. For long-term storage of the diluted Cy5 and Cy5.5 needed, add 1 % of BSA fraction V (Roche 735078 or for USA only: Roche 100062) to the 4 × SSC solution.
- SKYPaints™ (painting probes in hybridization buffer can be purchased from Applied Spectral Imaging or ASI).
- 20 × SSC (prepare 1 × SSC, 2 × SSC, 4 × SSC).
- Washing solution I: 50 % formamide in 2 × SSC.
- Washing solution II: 1 × SSC.
- Washing solution III: 0.1 % TWEEN® 20 in 4 × SSC.

3 Method

3.1 Mitotic Chromosomes and Interphase Nuclei Preparation

3.1.1 Induction of Chromosomal/Nuclear Variations

The following two treatment options can be integrated into the following protocols for slide preparation, when necessary.

(a) Induction of genome chaos in cell lines: When cells reach 50–70 % confluence, treat them with doxorubicin (Fisher) at 0.5–2.0 μg ml^{-1} or mitomycin C (Fisher) at 1–5 μg ml^{-1} for 2 h and briefly wash them in PBS. Treated cells are re-cultured with fresh medium and harvested 24–48 h following the drug treatment.

(b) Induction of defective mitotic figures (DMFs): When cells reach 50–70 % confluence, treat cells with doxorubicin

($0.5–2\ \mu g\ ml^{-1}$) for 2 h and then harvest cells following a brief colcemid treatment.

3.1.2 Direct Preparation of Cells Growing on Slides

1. Plate cells onto sterilized slides or coverslips within a culture dish. Let them grow for 2 days. To accumulate mitotic cells, treat the cells with colcemid for 2–4 h ($0.1\ \mu g\ ml^{-1}$). The colcemid treatment time may be extended depending on the mitotic rate (*see* **Note 1**).
2. Remove culture medium by aspiration. Pipette hypotonic solution onto the slide to cover the area for 10–20 min at 37 °C.
3. Add an equal amount of Carnoy's fixative for 2 min. Remove fixative by aspiration. Add new fixative to cover the area. Incubate for an additional 5 min (*see* **Notes 2** and **3**).
4. Remove fixative and dry slide by gentle aspiration or under a slow stream of air.
5. As soon as the surface is dry, rehydrate the slide by applying PBS-azide for 10 min.
6. Air-dry. The slide can be used immediately for immunostaining (see Sect. 3.4) or stored at −20 °C.

3.1.3 Cytospin® Centrifugation for Cell Preparation on Glass

The Cytospin® allows thin-layer preparations to be made from cells in a liquid matrix, including cells in suspension, or cells harvested from culture flasks or dishes.

(a) Harvest cells by spinning at 800 × *g* at RT for 10 min.
(b) Wash cells with 10 ml of 1 x PBS-azide.
(c) Resuspend cells in 10 ml of isotonic CSK solution on ice for 15 min (*see* **Note 4**).
(d) Apply 40 μl of resuspended cells into the Cytofunnel®; spread by centrifugation with the Cytospin® at 800 rpm for ~5 min (depending on the cell concentration, 1.0×10^4 cells per ml).
(e) Immediately put the slide into paraformaldehyde solution (1 % paraformaldehyde, 10 μl of 0.5 M HCl, 300 μl of 0.04 % phenol red indicator) for 3–5 min.
(f) Transfer the slide to Photo-flo® solution; incubate on ice for 3 min. Then incubate for 3 min at RT.
(g) Air-dry the slide. It can be used immediately for antibody detection (see Sect. 3.4) or stored at −20 °C (*see* **Note 5**).

3.1.4 Slide Preparation from Samples Stored Under Routine Cytogenetic Conditions (Carnoy's Fixative)

The protocol below describes the preparation of cells/chromosomes from specimens already treated with fixative. These procedures are typically used when reexamining past specimens or when assessing cases available from others. Materials stored in fixative can be treated as follows:

1. Drop a Carnoy fixed cell preparation onto an ice-cold glass slide.
2. Closely watch the fixative solution evaporating, but do not let the solution completely dry out. This is a critical step. When the cells just begin to appear from the fixative solution during evaporation on the slide, quickly proceed with step 3 (*see* **Note 6**).
3. Immediately transfer the slide into 1 × PBS-azide buffer in a coplin jar and incubate for 15–20 min.
4. Gently transfer the slides to chromosome swelling buffer for 2 × 10 min.
5. Air-dry. Use immediately for antibody detection (see Sect. 3.4) or store at −20 °C for future use. Use phase-contrast microscopy to examine the quality of the spread before antibody detection or storage.

3.2 Preparation of Meiotic Chromosomes

3.2.1 Slide Preparation

1. Wash the slides with a glass cleaner like Windex® just prior to use.
2. Rinse in hot water and then rinse in distilled water.
3. Rub dry with a lint-free wipe such as a Kimwipe®.
4. To reuse the slides for future experiments, wash with detergent, sonicate in a bleach/detergent solution, rinse and dry as in steps 1–3, and store. The more the slide is reused, the better the material adheres to the glass surface.

3.2.2 Preparation of Tissue

1. Remove the testes of a relatively young male (about 25-day-old for rats, mice, or hamsters, where there will be few spermatozoa).
2. Remove all fat from the testes.
3. Using a transfer pipette, run MEM over the testis. Blot off the excess MEM with a lint-free wipe.
4. Hold the testis with forceps. Using a razor or scalpel, cut open the side of the testes in the location with the fewest blood vessels.
5. Extrude the seminiferous tubules into a drop of MEM on dental wax. Do not allow the outer casing of the testes to touch the MEM.
6. Pick up the tubule with clean forceps and run about 3–5 ml of fresh MEM over the bundle. Drain on a lint-free wipe, and then place the tubules in a fresh drop of MEM on the dental wax.
7. Cut the tubules several times with a new, grease-free blade.

8. Squeeze the tubules with clean, grease-free forceps to release the spermatocytes from the tubules.
9. Transfer the cell suspension to a 1.5 ml microcentrifuge tube.
10. Fill the tube with MEM and draw the suspension up and down through a wide-bore plastic transfer pipette to separate the cells. Allow to stand for 1 min until all of the tubules have settled.
11. Transfer the supernatant to a clean 1.5 ml microcentrifuge tube. Add 1 ml of fresh MEM and centrifuge for 5 min at $160\times g$.
12. Pour off the supernatant and gently resuspend the cells in the residual MEM by tapping the side of the tube.

3.2.3 Surface Spreading

1. Fill a small Petri dish with 0.5 % hypotonic NaCl solution until the surface of the liquid is convex.
2. Gently tap the cell suspension to mix and draw up 5 μl with a pipette.
3. Wipe the pipette tip clean with a lint-free wipe and carefully expel the cell suspension such that a drop hangs from the pipette tip.
4. Touch the lower edge of the drop to the convex surface top to allow the cells to spread.
5. Allow the cells to stabilize for 10 s, and then carefully lower a slide onto the surface to pick up the cells.
6. Let the slide sit for 10 s.
7. Peel the slide off the NaCl carefully with a rolling motion. Begin by lifting along one long edge and then the rest.
8. Place the slide in a coplin jar with paraformaldehyde and SDS, if required, for 3 min (the degree of chromosome spreading can be adjusted by varying the concentration of SDS in the first paraformaldehyde fixation from 0 to 0.06 %; the more SDS used, the greater the spreading).
9. Transfer the slide to a second coplin jar containing only paraformaldehyde for an additional 3 min.
10. Wash 3 × 1 min each in Photo-flo® solution and then air-dry.
11. While the slides are in the fixative and washing solution, additional spreads can be made: Discard the used hypotonic solution and rinse the spreading dish in soapy water, hot water, and distilled H_2O. Add fresh hypotonic solution and spread the next 5 ml by repeating steps 1–9.
12. The slides can be used for antibody detection when dried or stored at −20 °C (*see* **Note 7**).

3.3 Treatment of Tissue Sections

1. Slide deparaffinization (tissue section with a thickness of 4 μm is preferred): Briefly melt the wax using a slide warmer (68 °C). As soon as the wax is melted, immediately place slides in the fresh xylene 3 × 5–10 min each (processed in a ventilated hood).
2. Rehydrate the slide by transferring through a series of fresh ethanol solutions: 100 % ethanol, 2 × 3 min each; 95 % ethanol, 2 × 3 min each; 80 % ethanol, 1 × 3 min; 50 % ethanol, 1 × 3 min.
3. Rinse with distilled water, PBS, two changes, 3 min, each. Once the sections have been rehydrated, do not allow them to dry.
4. Denature in 0.01 M HCl for 15 min.
5. Rinse in fresh PBS for 3 × 3 min.
6. Digest in prewarmed PK solution at 50 °C for 30–60 min.
7. Rinse in fresh PBS 3 × 3 min.
8. *Optional step*: Digest in pepsin solution at 37 °C for 5–30 min.
9. Rinse in fresh PBS 3 × 3 min.
10. Fix in 1 % post-fixation solution (PFA) for 10 min at RT.
11. Rinse in PBS for 5 min.
12. Dehydrate the slide in ethanol of increasing concentrations as follows: 70 % ethanol, 2 min; 90 % ethanol, 2 min; 100 % ethanol, 2 min.
13. Air-dry and proceed with the immunostaining procedure (see Sect. 3.4) or store at −20 °C.

3.4 Immunostaining

1. Wash the slide in wash buffer (10 % ADB) for 3 × 10 min each, adding 1 % Triton® X-100 to the second wash. The buffer can be mixed by leaving the stir bar in the wash jar.
2. Remove the slide from the jar and briefly drain off some of the buffer (do not let the slide dry).
3. Dilute the primary antibody in ADB, add this onto the slide, and cover with Parafilm®.
4. Incubate the slide in a humidified container at 37 °C for 1 h or overnight at RT.
5. Repeat slide wash steps 1 and 2.

   All the following steps should be performed in the dark room!
6. Repeat steps 3 and 4 to dilute, apply, and incubate with the secondary antibody.
7. Wash the slide in PBS buffer with 1 % Photo-flo® 3 × 10 min each, adding add 1 % Triton® X-100 to the second wash, and air-dry the slide.
8. Proceed with FISH or SKY after checking the quality of the antibody detection (*see* **Note 8**).

3.4.1 Antibodies Used in Detection

Both biotin-labeled and digoxigenin-labeled probes can be detected by different fluorescent molecules (e.g., FITC, Rhodamine, Texas Red, etc.), and therefore different combinations can be used for detection depending on availability. However, antibodies chosen from different species should be used with caution to avoid cross-reactions between antibodies.

3.5 FISH: Fluorescence In Situ Hybridization

3.5.1 Slide Preparation

See immunostaining: see Sect. 3.4 (*see* **Notes 9** and **11**).

3.5.2 DNA Probe Labeling

1. DNA probes can be labeled by different methods (nick translation and random primers) and different molecules indirectly (e.g., biotin, digoxigenin) or directly (e.g., FITC, Rhodamine, SpectrumOrange), with either in-house kits or commercial kits.

3.5.3 Labeled DNA Purification

1. Add 3 ml of equilibrium buffer to wash the Nick column (*see* **Note 10**).
2. Load the DNA sample and allow the column to dry.
3. Add 4 μl of equilibrium buffer to the column and allow the column to dry.
4. Place a new tube under the column.
5. Add another 400 μl of equilibrium buffer to the column and collect the purified DNA sample in a new tube.
6. Add 6 μl of ssDNA, 40 μl of 3 M NaAc, and 880 μl of 100 % ethanol for DNA precipitation.
7. Wash the DNA with 70 % cold ethanol and then allow to air-dry.
8. Add 10 μl of 10 mM Tris HCl buffer.

3.5.4 Probe Denaturation

1. Denature the probe at 75 °C for 10 min in hybridization solution (using either hybridization solution I or II, according to the type of probe used) (*see* **Note 12**).
2. Incubate in a water bath at 37 °C for 10 min.

3.5.5 RNase A Treatment of Slides (Optional Step)

1. Incubate slides in the 25 ml jars containing RNase A (100 μg ml^{-1} in 2 × SSC) at 37 °C for 1 h.
2. Wash the slides in 2 × SSC for 2 min.
3. Dehydrate the slide in increasing concentrations of ethanol as follows: 75 % ethanol, 2 min; 90 % ethanol, 2 min; 100 % ethanol, 2 min; air-dry slide.

3.5.6 Slide Denaturation

1. Denature the slides in a prewarmed fresh denaturation solution at 70 °C for 1–2 min.

2. Quickly remove the slide and dehydrate the slide in increasing concentrations of ethanol as follows: 75 % ethanol, 2 min; 90 % ethanol, 2 min; 100 % ethanol, 2 min; allow the slide to air-dry.

3.5.7 Hybridization and Post-hybridization Washing

1. Load the denatured probe onto the denatured slide.
2. Cover the slide with a coverslip and seal with rubber cement.
3. Allow for hybridization by placing the slide in a humidified container at 37 °C overnight.
4. After overnight incubation, peel off the rubber cement and remove the coverslip.
5. Wash the slide in prewarmed washing solution at 42 °C, 3 × 5 min each (using either solution A or B, according to the type of probe used).
6. Wash slide in 2 × SSC at 42 °C, 3 × 5 min.

3.5.8 Detection and Amplification

1. Briefly drain the slide and add avidin-FITC.
2. Place the coverslip onto the slide and incubate at 37 °C for 30 min.
3. Remove the coverslip and wash the slide in detection washing solution at RT for 3 × 5 min.
4. Briefly drain the slide and if amplification is required, continue with the following amplification steps; otherwise, go directly to the rinse step in 2 × SSC.
5. Briefly drain the slide and add anti-avidin and then place the coverslip onto the slide and incubate at 37 °C for 30 min.
6. Remove the coverslip and wash the slide in the detection washing solution at RT 3 × 5 min.
7. Briefly drain the slide and add avidin-FITC and then place the coverslip onto the slide and incubate at 37 °C for 30 min.
8. Remove the coverslip and wash the slide in the detection washing solution at RT 3 × 5 min.
9. Rinse in 2 × SSC and then air-dry the slide.

3.5.9 Counterstaining

1. Counterstain the slide with DAPI/antifade, and cover with a coverslip.
2. The slide is now ready for observation or for storage at −20 °C.

3.5.10 Signal Observation

1. FISH signals can be observed with fluorescent microscopy.
2. Images can be captured by a charge-coupled device (CCD) camera and analyzed.

   DNA–protein codetection can be achieved using various commercially available software packages for imaging (*see* **Notes 13** and **14**).

3.6 SKY: Spectral Karyotype Analysis

3.6.1 Chromosome Denaturation

1. Heat 40 ml of denaturation solution to 72 °C (±2 °C) in a glass coplin jar. Place slides in the solution for 60–90 s.
2. Immediately place slides in cold 70 %, 80 %, 100 % ethanol, 2 min each, and air-dry.

3.6.2 Probe Denaturation and Hybridization

1. Briefly centrifuge the content of the spectral karyotyping reagent (vial no. 1 supplied by ASI).
2. Mix the contents of the vial well, including the red precipitation, by pupating up and down for several times. Take 25 μl for each slide, put in an Eppendorf tube, and denature the probe by incubation at 80 °C in a water bath for 7 min.
3. Transfer the tube into a 37 °C water bath for 30–60 min.
4. Add 20 μl of denatured probes to the denatured chromosome preparation (where previous antibody detection was performed), and place a 24 × 24 mm coverslip over the probe mix. Seal the edges with rubber cement.
5. Transfer the slides to a humidified chamber at 37 °C for 24–36 h.

3.6.3 Detection

1. Remove the slides from the hybridization chamber and carefully remove the rubber cement.
2. Put slides into a coplin jar of washing solution I.
3. Wash the slides twice in washing solution II (1 × SSC) at 45 °C for 5 min each.
4. Dip the slides in washing solution III (4 × SSC/0.1 % TWEEN® 20) at 45 °C for 2 min each.
5. Tilt the slides and allow the fluid to drain. Add 80 μl of Cy5 staining reagent. Place a plastic coverslip (24 × 60 mm) over the top, and incubate for 40 min at 37 °C.
6. Wash slides three times in washing solution III at 45 °C for 2 min each.
7. Apply 80 μl of Cy5.5 staining reagent and cover with a plastic coverslip. Incubate at 37 °C for 40 min.
8. Repeat step 6.
9. Tilt slides and allow fluid to drain. Apply 20 μl of antifade DAPI reagent; carefully place a cover glass (24 × 60 mm) over the top and remove air bubbles.

3.6.4 Image Analysis

For imaging SKY combined with protein immunostaining, the protein (such as SC) image needs to be acquired using the SKY filter (not the DAPI filter). The second "DAPI" image needs to be captured as well (*see* **Note 15**).

4 Notes

1. The plastic culture flask with cells can be cut into pieces to replace the use of slides or coverslips.
2. The Carnoy's fixative can be replaced by paraformaldehyde solution.
3. In general, 3:1 (methanol: acetic acid) fixation preserves chromosome morphology and is therefore ideal for FISH or SKY detection. In DNA–protein codetection, paraformaldehyde fixation is preferred, as it maintains the protein in its native conformation and facilitates antibody detection. The type and time of fixatives can be adjusted according to the specific situation.
4. If it is required to break down the cell membrane, 0.4 % KCl hypotonic solution can be used.
5. Use phase-contrast microscopy to examine the quality of the spread before antibody detection or storage.
6. If the slide is transferred too early into PBS solution, more material will be lost. If the slide is transferred too late, it will reduce antibody detection.
7. Use phase-contrast microscopy to examine the quality of the spread before antibody detection or storage. Nuclei and chromosome cores should be visible. Upon finishing the immunostaining step for protein detection, the slides can directly be used for DNA detection by FISH or SKY. The slides can also be stored at −20 °C, and then used for DNA detection later.
8. Weak signals in protein detection can be caused by various factors, including low protein abundance, inappropriate fixation conditions, low quality of antibody, excess dilution of antibody, insufficient incubation time, incorrect incubation temperature, and incorrect washing conditions. Like many biological assays, optimization depends on a balance between sensitivity and specificity or signal to noise. For example, reducing the amount of antibody will increase the specificity but could reduce the intensity of the signal from the protein of interest. Antibody incubation and washing conditions should be optimized based on the same principle. In general, alterations in antibody concentration, incubation time, and washing conditions should be designed based on the level of background fluorescence of the slide. For low-abundance proteins with weak antibody detection signals, immunostaining images can be captured just after completion of the immunostaining procedures. The position of the captured

images can be manually (with slide finder) or digitally recorded and used later for image co-registration. As the denaturation and detection steps of the FISH protocol will reduce the immunostaining signal, protein immunostaining image capture prior to FISH or SKY detection is recommended.

9. Images can be taken prior to DNA detection in the next step.
10. For purifying labeled probes, QIAquick Nucleotide Removal Kit (Qiagen, Inc., Venlo, The Netherlands) can be used as an alternative choice.
11. The order of protein and DNA detection depends on the aims underlying each individual experiment. Since the antigen–antibody complex is much more tightly bound than the DNA hybridization complex, the antibody detection of protein should be performed prior to FISH detection. The antibody signal will survive the harsh denaturation steps of the FISH protocol. The antibody signal from meiotic core proteins is particularly resistant to FISH denaturation steps because of the high protein density in the meiotic chromosome core. In the case of weak protein and strong FISH signals, it is also possible to perform these protocols in reverse order.
12. As with any fluorescently detected slide, avoid exposure to light whenever possible.
13. Weak signals from FISH can be caused by small probe size, low quality of the DNA probe, and insufficient denaturation of the slides. For DNA–protein codetection, larger probes (cosmids or BACs) will give an improved signal. Slide denaturation time varies depending on the nature and age of the slide. Usually, freshly prepared slides require less denaturation time. Distorted chromosome morphology can be caused by drying the slides too quickly during preparation, over-denaturation of the slides, or incomplete drying of slides prior to immersion in denaturation solution.
14. When studying chromosomes using DNA–protein codetection, the cytoplasm can interfere with the results. To remove the cytoplasm, pepsin treatment can be used prior to immunostaining. Treat the slides with pepsin (10–30 μg ml^{-1}) in 0.01 M HCl, at 37 °C for 3–5 min, rinse in PBS, air-dry, and check the results using phase microscope. If there is not enough digestion, treat the slides again with pepsin. However, if membrane protein marker is to be used for the DNA–protein codetection, the cytoplasm should not be digested. Different hypotonic solutions, chromosomal swelling buffers, and spreading solutions can be used in an exchangeable fashion to increase the accessibility of DNA probes and antibodies for chromosomal/nuclear detection. The treatment time can also be adjusted.
15. For SKY-protein codetection, optimal results are obtained for both DNA and protein when the protein signal is not so strong

as to interfere with the color of SKY chromosomal identification. Regardless of color, a protein signal that is excessively bright causes a color shift in the SKY signal. This problem can be resolved by using multiple filters to record the signals for protein and DNA separately. The choice of color for secondary antibody labeling is an important consideration for codetection involving SKY. FITC labeling prior to SKY detection produces better results than labeling with rhodamine.
By using combinations of various fluorophores and antibodies generated in different host species (rabbit, mouse, etc.), multiple color detection of various targets can easily be achieved. For example, simultaneous detection of multiple DNA targets can be achieved with direct labeling and detection of different color combinations. If rehybridization with probes tagged with new fluorophores is to be done, it is recommended that direct labeling be used. This is because it is much more difficult to remove the signal of indirectly labeled probes by denaturation due to the strength of the antibody interactions. Thus, DNA detection can be performed multiple times by changing the probes for each an additional hybridization.

Acknowledgments

This work was partially supported by the start-up fund for Christine J. Ye from the Department of Internal Medicine, Division of Hematology/Oncology, University of Michigan. Due to space limitations, we regret that we are unable to cite all notable references deserving acknowledgement.

References

1. Heng HH (2009) The genome-centric concept: resynthesis of evolutionary theory. Bioessays 31:512–525
2. Heng HH (2015) Debating cancer: the paradox in cancer research. World Scientific, Singapore
3. Heng HH, Spyropoulos B, Moens PB (1997) FISH technology in chromosome and genome research. Bioessays 19:75–84
4. Abdallah BY, Horne SD, Stevens JB et al (2013) Single cell heterogeneity: why unstable genomes are incompatible with average profiles. Cell Cycle 12:3640–3649
5. Liehr T, Claussen U (2002) Current developments in human molecular cytogenetic techniques. Curr Mol Med 2:283–297
6. Beatty B, Heng HH (2004) Gene mapping by fluorescence in situ hybridization. In: Meyer R (ed) Encyclopedia of molecular cell biology and molecular medicine, vol 5. Wiley-VCH, Weinheim, pp 137–171
7. Speicher MR, Carter NP (2005) The new cytogenetics: blurring the boundaries with molecular biology. Nat Rev Genet 6:782–792
8. Ferguson-Smith MA (2015) History and evolution of cytogenetics. Mol Cytogenet 8:19
9. Heng HH, Tsui L-C (1993) Modes of DAPI banding and simultaneous in situ hybridization. Chromosoma 102:325–332
10. Heng HH, Tsui LC, Moens PB (1994) Organization of heterologous DNA inserts on the mouse meiotic chromosome core. Chromosoma 103:401–407
11. Heng HH, Squire J, Tsui LC (1992) High-resolution mapping of mammalian genes by in

situ hybridization to free chromatin. Proc Natl Acad Sci USA 89:9509–9513
12. Parra I, Windle B (1993) High resolution visual mapping of stretched DNA by fluorescent hybridization. Nat Genet 5:17–21
13. Long RM, Elliott DJ, Stutz F et al (1995) Spatial consequences of defective processing of specific yeast mRNAs revealed by fluorescent in situ hybridization. RNA (New York, NY) 1:1071–1078
14. Ried T, Baldini A, Rand TC et al (1992) Simultaneous visualization of seven different DNA probes by in situ hybridization using combinatorial fluorescence and digital imaging microscopy. Proc Natl Acad Sci USA 89:1388–1392
15. Speicher MR, Gwyn Ballard S, Ward DC (1996) Karyotyping human chromosomes by combinatorial multi-fluor FISH. Nat Genet 12:368–375
16. Schröck E, du Manoir S, Veldman T et al (1996) Multicolor spectral karyotyping of human chromosomes. Science 273:494–497
17. Heng HH, Krawetz SK, Lu W et al (2001) Redefining the chromatin loop domain. Cytogenet Cell Genet 93:155–161
18. Ye CJ, Lu W, Liu G et al (2001) The combination of SKY and specific loci detection with FISH or Immunostaining. Cytogenet Cell Genet 93:195–202
19. Ye CJ, Stevens JB, Liu G et al (2006) Combined multicolor-FISH and immunostaining. Cytogenet Genome Res 114:227–234
20. Cremer T, Cremer C (2001) Chromosome territories, nuclear architecture and gene regulation in mammalian cells. Nat Rev Genet 2:292–301
21. Parada L, Misteli T (2002) Chromosome positioning in the interphase nucleus. Trends Cell Biol 12:425–432
22. Poon SS, Lansdorp PM (2001) Quantitative fluorescence in situ hybridization (Q-FISH). Curr Protoc Cell Biol Unit 18(4)
23. Vargas DY, Raj A, Marras SAE et al (2005) Mechanism of mRNA transport in the nucleus. Proc Natl Acad Sci USA 102:17008–17013
24. Schnedl W, Dev VG, Tantravahi R et al (1975) 5-Methylcytosine in heterochromatic regions of chromosomes: chimpanzee and gorilla compared to the human. Chromosoma 52:59–66
25. Tan EM (1983) autoantibodies to nuclear antigens (ANA): their immunobiology and medicine. Adv Immunol 33:167–240
26. Masumoto H, Sugimoto K, Okazaki T (1989) Alphoid satellite DNA is tightly associated with centromere antigens in human chromosomes throughout the cell cycle. Exp Cell Res 181:181–196
27. Haaf T, Ward DC (1994) Structural analysis of alpha-satellite DNA and centromere proteins using extended chromatin and chromosomes. Hum Mol Genet 3:697–709
28. Haaf T, Warburton PE, Willard HF (1992) Integration of human alpha-satellite DNA into simian chromosomes: centromere protein binding and disruption of normal chromosome segregation. Cell 70:681–696
29. Page SL, Earnshaw WC, Choo KHA et al (1995) Further evidence that CENP-C is a necessary component of active centromeres: studies of a dic(X;15) with simultaneous immunofluorescence and FISH. Hum Mol Genet 4:289–294
30. Van den Berg H, Vossen JM, Langlois van den Bergh R et al (1991) Detection of Y chromosome by in situ hybridization in combination with membrane antigens by two-color immunofluorescence. Lab Invest 64:623–628
31. Weber-Matthiesen K, Deerberg J, Muller-Hermelink A et al (1993) Rapid immunophenotypic characterization of chromosomally aberrant cells by the new FICTION method. Cytogenet Cell Genet 63:123–125
32. Baurmann H, Cherif D, Berger R (1993) Interphase cytogenetics by fluorescent in situ hybridization (FISH) for characterization of monosomy-7-associated myeloid disorders. Leukemia 7:384391
33. Hunt P, LeMaire R, Embury P et al (1995) Analysis of chromosome behavior in intact mammalian oocytes: monitoring the segregation of a univalent chromosome during female meiosis. Hum Mol Genet 4:2007–2012
34. Barlow AL, Hulten MA (1996) Combined immunocytogenetic and molecular cytogenetic analysis of meiosis I human spermatocytes. Chromosome Res 4:562–573
35. Scherthan H, Weich S, Schwegler H et al (1996) Centromere and telomere movements during early meiotic prophase of mouse and man are associated with the onset of chromosome pairing. J Cell Biol 134:1109–1125
36. Heng HH, Shi XM, Tsui L-C, et al (1998) Visualization of mitotic and meiotic chromatin loop domains. Presented at the International Genetic conference, Beijing
37. Fukagawa T, Pendon C, Morris J et al (1999) CENP-C is necessary but not sufficient to induce formation of a functional centromere. EMBO J 18:4196–4209
38. Heng HH, Spyropoulos B, Moens P (2000) DNA-protein in situ covisualization for

chromosomal analysis. Methods Mol Biol 123:15–27

39. Hiatt EN, Kentner EK, Dawe RK (2002) Independently regulated neocentromere activity of two classes of tandem repeat arrays. Plant Cell 14:407420
40. Trelles-Sticken E, Loidl J, Scherthan H (2003) Increased ploidy and KAR3 and SIR3 disruption alter the dynamics of meiotic chromosomes and telomeres. J Cell Sci 116:2431–2442
41. Hudson DF, Vagnarelli P, Gassmann R et al (2003) Condensin is required for nonhistone protein assembly and structural integrity of vertebrate mitotic chromosomes. Dev Cell 5:323336
42. Craig JM, Earle E, Canham P et al (2003) Analysis of mammalian proteins involved in chromatin modification reveals new metaphase centromeric proteins and distinct chromosomal distribution patterns. Hum Mol Genet 12:3109–3121
43. Hassold T, Judis L, Chan ER et al (2004) Cytological studies of meiotic recombination in human males. Cytogenet Genome Res 107:249–255
44. Heng HH, Chamberlain J, Shi XM et al (1996) Regulation of meiotic chromatin loop size by chromosomal position. Proc Natl Acad Sci USA 93:2795–2800
45. Csankovszki G, McDonel P, Meyer BJ (2004) Recruitment and spreading of the C. elegans dosage compensation complex along X chromosomes. Science 303:1182–1185
46. Osborne CS, Chakalova L, Brown KE et al (2004) Active genes dynamically colocalize to shared sites of ongoing transcription. Nat Genet 36:1065–1071
47. Wang J, Shiels C, Sasieni P et al (2004) Promyelocytic leukemia nuclear bodies associate with transcriptionally active genomic regions. J Cell Biol 164:515–526
48. Heng HH, Goetze S, Ye CJ et al (2004) Chromatin loops are selectively anchored using scaffold/matrix-attachment regions. J Cell Sci 117:999–1008
49. Ye CJ, Lawrenson L, Liu G, et al (2009) Simultaneous fluorescence immunostaining and FISH. In: Liehr T (ed) Fluorescence in situ hybridization (FISH) – Application. Springer, pp 193-216
50. Heng HH, Stevens J, Yang F, et al (2004) Packaging of meiotic chromosomes correlated to GC-content, loop size, and recombination rates. Annual Meeting of American Society of Human Genetics, Toronto, Program No. 161
51. Lynn A, Koehler KE, Judis L et al (2002) Covariation of synaptonemal complex length and mammalian meiotic exchange rates. Science 296:2222–2225
52. Froenicke L, Anderson LK, Wienberg J et al (2002) Male mouse recombination maps for each autosome identified by chromosome painting. Am J Hum Genet 1:13531368
53. Tease C, Hartshorne GM, Hulten MA (2002) Patterns of meiotic recombination in human fetal oocytes. Am J Hum Genet 70:1469–1479
54. Sun F, Oliver-Bonet M, Liehr T et al (2004) Human male recombination maps for individual chromosomes. Am J Hum Genet 74:521–531
55. Martin-Subero JI, Chudoba I, Harder L et al (2002) Multicolor-FICTION: Expanding the possibilities of combined morphologic, immunophenotypic, and genetic single cell analyses. Am J Pathol 161:413–420
56. Hinten M, Maclary E, Gayen S et al (2016) Visualizing long noncoding RNAs on chromatin. Methods Mol Biol 1402:147–164
57. Heng HH, Liu G, Stevens JB et al (2011) Decoding the genome beyond sequencing: the new phase of genomic research. Genomics 98:242–252
58. Heng HH, Stevens JB, Liu G et al (2006) Stochastic cancer progression driven by non-clonal chromosome aberrations. J Cell Physiol 208:461–472
59. Selim AG, El-Ayat G, Naase M et al (2002) C-myc oncoprotein expression and gene amplification in apocrine metaplasia and apocrine change within sclerosing adenosis of the breast. Breast 11:466–472
60. Lewis JT, Ketterling RP, Halling KC et al (2005) Analysis of intratumoral heterogeneity and amplification status in breast carcinomas with equivocal (2+) HER-2. Am J Clin Pathol 124:273–281
61. Lottner C, Schwarz S, Diermeier SH, Hartmann A et al (2005) Simultaneous detection of HER2/neu gene amplification and protein overexpression in paraffin-embedded breast cancer. J Pathol 205:577–584
62. Liu G, Stevens JB, Horne SD, Abdallah BY et al (2014) Genome chaos: survival strategy during crisis. Cell Cycle 13:528–537
63. Horne SD, Ye CJ, Heng HH (2015) Chromosomal instability (CIN) in cancer. eLS 1–9
64. Stevens JB, Liu G, Abdallah BY et al (2014) Unstable genomes elevate transcriptome dynamics. Int J Cancer 134:2074–2087
65. Heng HH (2017) The genomic landscape of cancer. In: Ujvari B, Roche B, Thomas F (eds)

Ecology and evolution of cancer. Elsevier (in press)

66. Stevens JB, Liu G, Bremer SW et al (2007) Mitotic cell death by chromosome fragmentation. Cancer Res 67:7686–7694
67. Heng HH, Regan SM, Liu G et al (2016) Why is it crucial to analyze non clonal chromosomal aberrations or NCCAs? Mol Cytogenet 9:15
68. Heng HH, Liu G, Stevens JB et al (2013) Karyotype heterogeneity and unclassified chromosomal abnormalities. Cytogenet Genome Res 139:144–157
69. Azzalin CM, Reichenbach P, Khoriauli L et al (2007) Telomeric repeat containing RNA and RNA surveillance factors at mammalian chromosome ends. Science 318:798–801
70. Jin F, Li Y, Dixon JR et al (2013) A high-resolution map of the three-dimensional chromatin interactome in human cells. Nature 503:290–294
71. Horne SD, Pollick SA, Heng HH (2015) Evolutionary mechanism unifies the hallmarks of cancer. Int J Cancer 136:2012–2021
72. Buchwalow IB, Minin EA, Boecker W (2005) A multicolor fluorescence immunostaining technique for simultaneous antigen targeting. Acta Histochem 107:143–148

RNA-Directed FISH and Immunostaining

Bin Ma and Naoko Tanese

Abstract

RNA plays a critical role in the health/maintenance of cells and misregulation of RNA contributes to the development of many disorders. Fluorescence in situ hybridization (FISH) is a useful tool for detecting the location of RNA and its targeting in intact cellular and tissue environment. The combination of FISH and immunofluorescence staining (IFS) presents a powerful method for visualizing spatial relationships or interactions between mRNA and proteins, or for localizing mRNA in certain cell types, while preserving the anatomical structure of the cell or tissue. Although seemingly straightforward, FISH/IFS of mRNA and proteins can be difficult to perform simultaneously on the same specimen often generating variable results. Here, we describe a combined method of multicolor FISH/IFS and explore various factors that influence the outcomes of protein and mRNA detection in detail.

Keywords RNA, mRNA, Fluorescence in situ hybridization (FISH), Immunofluorescence staining (IFS)

1 Introduction

Proper functioning of RNA is crucial to the health/maintenance of cells, and RNA misregulation plays an important role in the development of many disorders [1–3]. Therefore, the study of RNA has been a critical area of biomedical research and has found numerous applications in diagnostics and treatment of some diseases [4, 5]. To obtain better insights into the various functions of RNA, each transcript needs to be examined from biogenesis to decay (chapter by Bin Ma and Naoko Tanese "RNA Imaging in Living Cells"). Fluorescence in situ hybridization (FISH) is a powerful tool that enables us to determine the location of RNA and its targeting while preserving intact cellular and tissue environment (chapter by Thomas Liehr and Anja Weise "Background"). Advances in RNA FISH have allowed investigators to detect individual RNA molecules in single cells, enabling accurate and highly sensitive quantification of gene expression [6–10]. In addition, multicolor FISH permits detection of multiple transcripts inside the cells or tissue ([11, 12]; Part 4).

Thomas Liehr (ed.), *Fluorescence In Situ Hybridization (FISH)*, Springer Protocols Handbooks,
DOI 10.1007/978-3-662-52959-1_34,

Immunostaining is an extremely useful means of identifying specific cell populations expressing specific proteins (i.e., the antigen, chapter by Christine Ye et al. "Simultaneous Fluorescence Immunostaining and FISH"; chapter by Anna Pendina et al. "Immunofluorescent Staining for Cytosine Modifications Like 5-Methylcytosine and Its Oxidative Derivatives and FISH"; chapter by Elisabeth Klein and Thomas Liehr "CENP-Antibodies Used Additionally to FISH"; chapter by Tiphaine Aguirre-Lavin and Nathalie Beaujean "Three-Dimensional Immuno-Fluorescence In Situ Hybridization in Preimplantation Mouse"). The combination of FISH and immunofluorescence staining (IFS) presents an impressive method for visualizing spatial relationships or interactions between mRNA and proteins, and for localizing sites of mRNA expression within a certain cell type, while preserving the anatomical structure of the cell or tissue [7, 13]. Although the concept of combining these two approaches seems obvious, the specific materials and conditions used during IFS and RNA FISH make it difficult to perform these two procedures simultaneously on the same specimen. For example, immunodetection signals are often weak following FISH because of rigorous sample treatments during the FISH procedure. We have optimized experimental conditions of FISH/immunostaining and created a robust protocol to detect proteins and mRNA in a single cell at a high resolution. In our previous studies, we have successfully applied this protocol to identify proteins and mRNAs in both primary neuronal cultures and brain sections [6, 7]. In this chapter, we describe this protocol and discuss various factors that influence the outcomes of protein and mRNA detection in detail (Fig. 1).

2 Materials

2.1 For FISH Probe Labeling

- Digoxigenin (DIG)-nick translation mix (Cat. No.: 11745816910, Roche Diagnostics)
- Plasmid DNA carrying a cDNA of interest and a corresponding empty vector for negative control
- Illustra ProbeQuant™ G-50 micro columns (Cat. No.: 28-9034-08, GE Healthcare)

2.2 For Pretreatment

- 24-well flat-bottom Corning® Costar® cell culture plates (Cat. No.: CLS3524, Sigma).
- Thermanox coverslips for 24-well multi-dish 13 mm diameter (Cat. No.: 10252961, Fisher Scientific) for cell culture.
- Coverslip Best circular, 13 mm diameter, 0.08 to 0.12 mm thick (Cat. No.: 12392128, Fisher Scientific) for cryosections and paraffin sections.

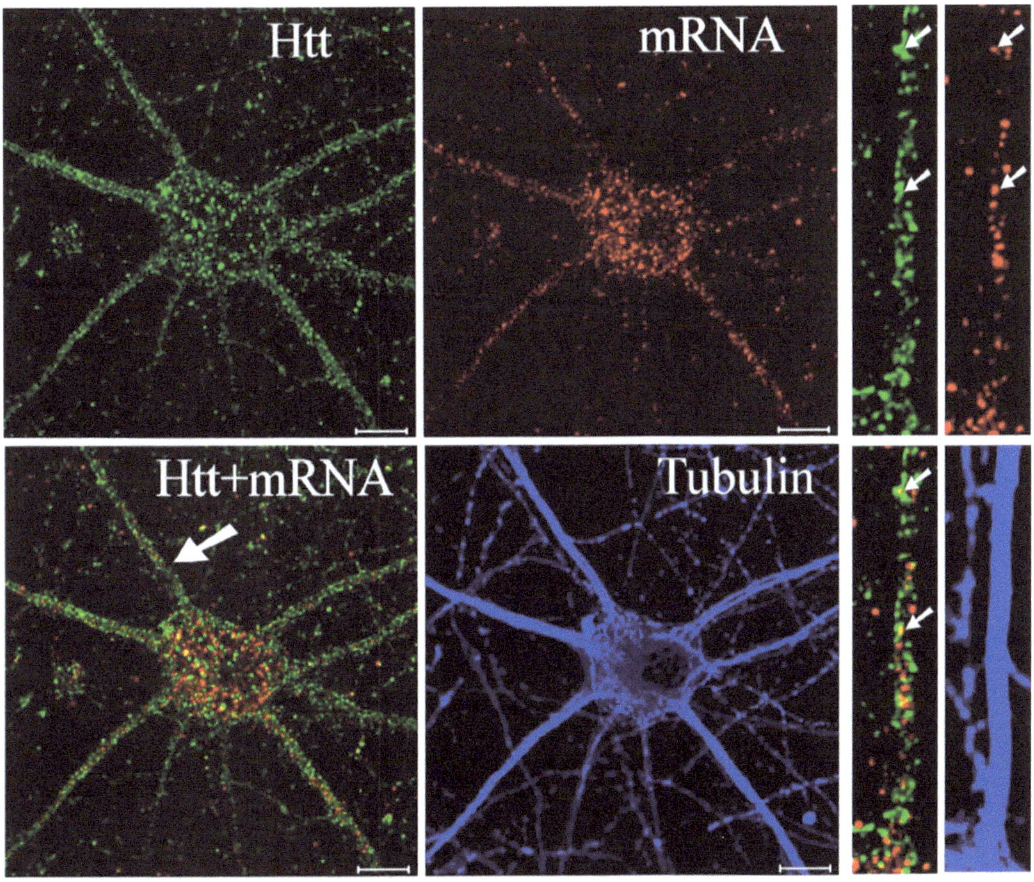

Fig. 1 Colocalization of Huntingtin (Htt) protein and β-actin mRNA in day in vitro 9 (DIV9) rat cortical neurons [7]. Htt, β-actin mRNA, and α-tubulin are shown in *green*, *red*, and *blue*, respectively. The *right panels* are high-magnification views of a single dendrite (pointed with a *white arrow* in the *left panels*). The *arrows* indicate the colocalization of Htt and β-actin mRNA. *Scale bar*: 10.0 μm

- Poly-L-lysine (PLL) solution, 0.1 % (w/v) in H_2O (Cat. No.: P8920, Sigma). Dilute (1:10) to make 0.01 % working solution for coating coverslips.
- RNase-free (diethylpyrocarbonate (DEPC, Cat. No.: D5758, Sigma)-treated) water. Add 1 ml DEPC to 1 l deionized water. Stir overnight at room temperature (RT). Properly autoclave the solution the next day to hydrolyze and inactivate DEPC. Use DEPC-treated water for preparation of all solutions used in FISH and immunostaining.
- 10 × PBS: for 1 l combine 80.0 g NaCl, 2.0 g KCl, 14.4 g Na_2HPO_4, 2.4 g KH_2PO_4, and 800 ml DEPC-treated water. Adjust pH to 7.4 and add DEPC-treated water to 1 l. Autoclave and store at RT.
- 1 × PBS: dilute 10 × PBS with DEPC-treated water to make 1 × PBS.

- 4 % paraformaldehyde (PFA, toxic!—needs to be discarded after use as hazardous waste): dilute 20 % PFA (Cat. No.: 15713, Electron Microscopy Sciences, Hatfield, PA, USA) with 1 × PBS.
- 0.25 % Triton X-100: dilute Triton X-100 (Cat. No.: T8787, for molecular biology, Sigma) with 1 × PBS.

2.3 For FISH Procedure Itself

Solutions and equipment needed for FISH itself are listed in chapter by Thomas Liehr et al. "The Standard FISH Procedure". Besides you need the following:

- 25 % dextran sulfate sodium. Dilute 50 % dextran sulfate (Cat. No.: D8906, Sigma) with DEPC-treated water.
- 200 mM ribonucleoside vanadyl complex (RVC, Cat. No.: R3380, Sigma). Aliquot and store at −20 °C.
- 1 M sodium phosphate buffer, pH 7.0. Combine 57.7 ml of 1 M Na_2HPO_4 and 42.3 ml of 1 M NaH_2PO_4 to make 100 ml solution. Adjust pH to 7.0 with Na_2HPO_4 or NaH_2PO_4. Autoclave and store at RT.
- Single-stranded salmon sperm DNA (ssDNA, 10 mg/ml, Cat. No.: D7656, Sigma).
- Yeast tRNA (Cat. No.: R5636-1ML, Sigma, used as carrier).
- BSA (20 mg/ml). Dissolve 0.4 g bovine serum albumin (BSA, IgG-free, protease-free, Cat. No.: 001-000-161, Jackson ImmunoResearch, West Grove, PA, USA) in DEPC-treated water to make 20 ml solution. Aliquot and store at −20 °C.
- Hybridization buffer. Combine 400 μl 25 % dextran sulfate, 200 μl BSA (20 mg/ml), 100 μl 20 × SSC, 100 μl 200 mM RVC, 10 μl 1 M sodium phosphate buffer pH 7.0, and 190 μl DEPC-treated distilled H_2O to make 1 ml hybridization buffer. Mix well by vortexing. Spin at 4 °C (12,000 RCF) for 5 min. Keep on ice until use.
- Formamide, Deionized (Cat. No.: S4117, Millipore; discard formamide solution as hazardous waste).
- Plastic containers (clean and RNase-free) and Parafilm for lining the bottom of the container.

2.4 For Immunostaining

- BSA (see Sect. 2.3)
- Goat serum (Cat. No.: G9023-10 ml, Sigma)
- Blocking buffer (2 % goat serum, 1 % BSA in PBS)
- Chicken anti-digoxigenin affinity-purified (Cat. No.: CDIG-65A, Immunology Consultants Laboratory, Atlanta, GA, USA) or another similar anti-DIG antibody

- Primary antibodies for detection of proteins
- Nonimmune IgG to control for specificity of immunostaining
- Goat anti-chicken IgY (IgG) DyLight 549 or similar secondary antibodies
- Secondary antibodies for protein detection
- Fluorescence mounting medium (DAKO, Carpinteria, CA, USA) or equivalent mounting medium (e.g., Fluoromount G from SouthernBiotech, Birmingham, AL, USA)

3 Methods

3.1 FISH Probes

Ensure all solutions, containers, and instruments are RNase-free for FISH and IFS. Purchase reagents of analytical grade. Diligently follow all waste disposal regulations when disposing waste materials.

3.1.1 Commercially Available Probes

Commercially available FISH probes and corresponding control probes (~50 nucleotides in length) are offered as fluorescence- or hapten-labeled probe DNA (*see* **Note 1**). Perform a BLAST search to assure specificity to target mRNA. Optimal probes have high G/C content (>50 %) and low self-complementarity. DNAStar or other comparable software should be used for probe design. Typically, two to four DIG-labeled oligonucleotides are combined at an equal molar ratio for use in FISH.

3.1.2 Homemade Probes

1. DIG-labeled DNA probes are generated using a DIG-nick translation mix according to the manufacturer's instructions. 1.0 μg plasmid DNA (described above) is used to generate probes. Plasmids without any insert are used as controls, which should give no FISH signals. 1.0 μg template DNA yields ~50 μl probe (*see* **Note 2**).
2. Illustra ProbeQuant™ G-50 Micro Columns are used to purify probes according to the manufacturer's instructions.

3.2 Pretreatment

1. (a) Grow cells on 13 mm diameter round glass coverslips (Thermanox) placed in 24-well culture plates. Before FISH, wash the cells 2 × 5 min with PBS and fix with 4 % PFA for 10 min at RT; (b) rinse coverslips (Best circular) briefly with 70 % ethanol to clean them and then coat the coverslips with 0.01 % PLL for 10 min. Air-dry and store in a clean box at RT until use. Cut 5–10 μm cryosections from tissues to be examined by using a cryotome (Leica) and mount cryosections on pretreated coverslips. Fix the sections with 4 % PFA for 10 min at RT; (c) cut 3–5 μm paraffin-embedded sections and mount on pretreated coverslips [described in (b)]. Deparaffinize the

sections by using a standard protocol for immunohistochemistry/immunofluorescence staining of paraffin-embedded sections.

2. Wash the coverslips for 3 × 5 min with PBS.
3. Permeabilize cells for 5 min with 0.25 % Triton X-100 in 1 × PBS. Wash the cells with PBS for 2 × 5 min (*see* **Note 3**).
4. Incubate coverslips with 1 × SSC for 10 min.

3.3 FISH

The volume of solutions indicated below applies to one 13 mm diameter coverslip in one well of a 24-well culture plate for FISH and IFS:

1. For each coverslip, prepare 20 μl 80 % formamide/1 × SSC solution containing 1 μg each of ssDNA and tRNA as carriers. Heat this suspension for 5 min at 85 °C and place on ice. Add 20 μl hybridization buffer per coverslip to the carrier/formamide suspension to form prehybridization mix.
2. Spot 40 μl pre-hybridization mix on Parafilm placed inside a small plastic container and put a coverslip over each drop with the cell side facing down. Incubate the box at 37 °C in a humid chamber for 1 h. To prevent cell/tissue from drying out, place wet tissue paper inside the box with a tight lid to make the chamber humid.
3. For each coverslip, prepare 20 μl 80 % formamide/1 × SSC solutions containing 20 ng oligonucleotide probes, 1 μg each of ssDNA and tRNA. For nick-translated DNA probes, use 1 μl probe per coverslip (*see* **Note 4**).
4. Heat the mixture at 85 °C for 5 min in a dry bath and place on ice for 5 min. After the suspension has cooled, add 20 μl hybridization buffer per coverslip to generate probe hybridization solution.
5. Spot 40 μl of probe hybridization solution on the Parafilm inside a small plastic container, and place a coverslip over each drop with the cell side facing down. Incubate the box at 37 °C in a humid chamber for 5 h.
6. After hybridization, pick up the coverslips and place them in a 24-well culture plate.
7. Wash coverslips with 40 % formamide/1 × SSC for 30 min at 37 °C with gentle shaking, followed by 3 × 10 min washes at RT with 1 × SSC with gentle shaking on an orbital shaker (*see* **Note 5**).
8. Wash coverslips in 1 × PBS for 10 min.

3.4 Immunostaining

1. Incubate coverslips for 30 min in blocking buffer (*see* **Note 6**).
2. Dilute primary antibody (including anti-DIG antibody) in 150 μl blocking buffer and incubate coverslips with these

antibodies at RT for 1 h. Wash for 3 × 5 min in PBS/0.02 % Triton X-100 (*see* **Note 7**).

3. Dilute secondary antibody in 150 μl blocking buffer and incubate coverslips with secondary antibodies at RT for 45 min. Wash 3 × 5 min in PBS/0.02 % Triton X-100 (*see* **Note 8**).
4. Add 150 μl DAPI (1:1000 dilution in PBS) working solution to the well and incubate 15 min at RT. Wash 2 × 5 min in PBS/0.02 % Triton X-100.
5. Apply 20 μl Fluorescence Mounting Medium (DAKO) onto a pre-cleaned microscope slide. Put coverslips upside down on the mounting medium on the slide.
6. Allow coverslips to dry on a bench at RT (protected from light). Note that one can add preservative/antifade reagents to the mounting medium to extend the lifetime of the signal generated from the fluorophore. Slides may be kept in the dark at 4 °C.

4 Notes

1. (a) Increasing the number of DIG (or another hapten) molecules incorporated into oligonucleotide probes will improve the FISH signal; e.g., each GreenStar* DIG hyper-labeled probe (GeneDetect, Sarasota, FL, USA) has ten DIG molecules at its 3′ end. (b) The amount of probe used for hybridization should be determined by the user for individual applications. (c) We recommend using DIG (or another hapten)-labeled probes and amplify the signal using anti-DIG (or another hapten) antibodies. Directly labeled probes are not recommended. (d) Avoid the use of biotin-labeled probes since endogenous biotin will interfere with the results (even after a biotin-blocking step is applied), especially for the tissue or cells containing high amounts of biotin.
2. (a) Control FISH with a nick-translated probe (generated from a control plasmid) or a sense probe should give no FISH signal. (b) For nick translation, probes of 200–400 bp in length are typically used for hybridization. Probe length should be checked on a DNA gel before application. If a desired probe length or amount is not achieved, long incubation time may be needed to improve the yield of the probe. (c) The use of coding sequence only, instead of the entire plasmid or of shorter sequences for generating probes, is likely to improve probe specificity. Linear DNA is labeled more efficiently than circular or supercoiled DNA.
3. 0.25 % Triton X-100 can be used to permeabilize all membranes of the cell. 0.02 % saponin can also be used to

permeabilize outer cell membrane while leaving membranes of internal organelles intact. If saponin is used for permeabilization, it should also be included in solutions used for subsequent incubations and wash steps replacing Triton X-100.

4. (a) A mock control FISH should give no signals. (b) Poly-dT probe allows confirmation of the quality of mRNA in the sample; it will serve as a positive control. (c) 20 ng of probe DNA is recommended for hybridization. If 2–4 types of probes are used, combine equal amounts of each probe to make 20 ng probe mix for hybridization. The amount of probe to be used should be determined by the user for individual applications.
5. Sufficient washing will help to reduce the background of FISH. However, too long or too rigorous washes are not recommended, since cells or tissue may be washed off the coverslips.
6. Use high-quality normal goat serum and IgG-free/protease-free BSA in blocking buffer. If using secondary antibodies generated from other animals, use the serum from the same species for blocking.
7. (a) Use highly sensitive polyclonal anti-DIG antibodies for the detection of DIG. In this protocol, a polyclonal chicken anti-DIG antibody is used. (b) Use a highly specific primary antibody for the detection of target proteins. (c) If paraffin sections are used, certain antigen retrieval methods (e.g., protease-induced epitope retrieval, heat-induced epitope retrieval) might be needed to unmask the antigen epitopes. (d) Each antibody should be titrated for each application.
8. (a) Each secondary antibody should be titrated prior to each application; no FISH/IFS signal should be detected when primary antibodies are omitted. (b) Secondary antibodies used for detection in three channels should have clearly separate fluorescence spectra. (c) Use fluorescent dyes that generate bright fluorescence and low background/noise. Alexa and DyLight serial dye-conjugated secondary antibodies are suitable for these applications. (d) Background and noise can be further reduced by using the $F(ab')_2$ fragment (instead of whole antibody molecule) as secondary antibodies. (e) Pre-absorbed (by serum from the species from which cells and tissues are obtained) secondary antibody will also reduce the background and noise. For example, for FISH and immunostaining of mouse cells or tissue, use commercially available secondary antibodies that are pre-absorbed by mouse serum. Alternatively, mouse serum (1:1000) can be added to the secondary antibody solution that will be subsequently added to the cell/tissue after 10 min incubation at RT. Alexa Fluor® 488 AffiniPure $F(ab')_2$ Fragment Goat Anti-Rabbit IgG (H + L) (min X Hu, Ms, Rat Sr Prot, Cat. No.: 111-546-144, Jackson

ImmunoResearch) is a good antibody that can fulfill the requirements mentioned in (c–e). This antibody has been absorbed by human, mouse, and rat serum, and therefore it has minimal cross-reactivity to the serum proteins (including IgG) from these species.

References

1. Mortimer SA, Kidwell MA, Doudna JA (2014) Insights into RNA structure and function from genome-wide studies. Nat Rev Genet 15:469–479
2. Ibrahim F, Nakaya T, Mourelatos Z (2012) RNA dysregulation in diseases of motor neurons. Annu Rev Pathol 7:323–352
3. Li Y, Wang X (2016) Role of long noncoding RNAs in malignant disease. Mol Med Rep 13:1463–1469
4. Wang J, Chen J, Sen S (2016) MicroRNA as biomarkers and diagnostics. J Cell Physiol 231:25–30
5. Mansoori B, Sandoghchian Shotorbani S, Baradaran B (2014) RNA interference and its role in cancer therapy. Adv Pharm Bull 4:313–321
6. Ma B, Savas JN, Chao MV et al (2012) Quantitative analysis of BDNF/TrkB protein and mRNA in cortical and striatal neurons using α-tubulin as a normalization factor. Cytometry A 81:704–717
7. Ma B, Savas JN, Yu MS et al (2011) Huntingtin facilitates β-actin mRNA transport in dendrites of cortical neurons. Sci Rep 1:140
8. Zimmerman SG, Peters NC, Altaras AE et al (2013) Optimized RNA ISH, RNA FISH and protein-RNA double labeling (IF/FISH) in Drosophila ovaries. Nat Protoc 8:2158–2179
9. Kwon S (2013) Single-molecule fluorescence *in situ* hybridization: quantitative imaging of single RNA molecules. BMB Rep 46:65–72
10. Shaffer SM, Wu MT, Levesque MJ et al (2013) Turbo FISH: a method for rapid single molecule RNA FISH. PLoS One 8, e75120
11. Renwick N, Cekan P, Masry PA et al (2013) Multicolor microRNA FISH effectively differentiates tumor types. J Clin Invest 123:2694–2702
12. Lea R, Bonev B, Dubaissi E et al (2012) Multicolor fluorescent *in situ* mRNA hybridization (FISH) on whole mounts and sections. Methods Mol Biol 917:431–444
13. Kochan J, Wawro M, Kasza A (2015) Simultaneous detection of mRNA and protein in single cells using immunofluorescence-combined single-molecule RNA FISH. Biotechniques 59:209–212

Immunofluorescence Staining for Cytosine Modifications Like 5-Methylcytosine and Its Oxidative Derivatives and FISH

Anna A. Pendina, Olga A. Efimova, Andrei V. Tikhonov, Olga G. Chiryaeva, Irina D. Fedorova, Alla S. Koltsova, Mikhail I. Krapivin, Sergey E. Parfenyev, Tatyana V. Kuznetzova, and Vladislav S. Baranov

Abstract

In this chapter, the protocol of combined immunofluorescence detection of 5-methylcytosine and its oxidative derivatives and FISH is given. This approach can be applied on fixed preparations of human chromosomes and nuclei. Human PHA-stimulated adult and fetal lymphocytes, uncultured tissues comprising spontaneously dividing cells, such as chorion, embryonic and testicular tissues (prepared using "direct" technique), oocytes, zygotes and blastomeres of preimplantation human embryos, and non-cultured uterine leiomyoma samples may be used. The combined immunostaining for DNA methylation allows simultaneous assessment of the whole-genome methylation pattern as well as of genomic subunits, thus indicating the functional status of nuclei with different karyotypes; it is also suited for homologous or structurally abnormal chromosomes characterizable by FISH. Combined immunostaining/FISH is an indispensable method for investigation of both programmed and abnormal de novo alterations of DNA methylation.

Keywords DNA methylation, 5-methylcytosine, 5-hydroxymethylcytosine, FISH, Marker chromosomes, Nuclei, Human karyotype, Immunofluorescence staining, Direct chromosome preparation

1 Introduction

In 1974, Miller and coworkers published in *Nature* their immunofluorescence-based studies of DNA methylation patterns in murine and human metaphase chromosomes, reporting heterochromatin-specific location of 5-methylcytosine (5mC) residues [1]. Their results initiated a new direction of research, focusing not only on methylated DNA extracted from cells but also on study of DNA methylation in situ, directly on metaphase chromosomes. With the discovery of DNA methylation as a mechanism of cell memory [2, 3] and as a major epigenetic mechanism regulating gene activity and chromatin structure ([4]; chapter by Tiphaine

Thomas Liehr (ed.), *Fluorescence In Situ Hybridization (FISH)*, Springer Protocols Handbooks,
DOI 10.1007/978-3-662-52959-1_35, © Springer-Verlag Berlin Heidelberg 2017

Aguirre-Lavin and Nathalie Beaujean "Three-Dimensional Immuno-Fluorescence In Situ Hybridization in Preimplantation Mouse"), in situ immunofluorescence detection of antibodies against 5-methylcytosine in metaphase chromosomes became of specific interest [5–7].

Barbin et al. [6] have optimized the procedure of 5mC immunodetection by modifying the DNA denaturation protocol. This allowed mapping of 5mC-rich bands not only in constitutive heterochromatin but also along chromosome arms [6, 7]. Thus the authors observed heavily labeled bands both in heterochromatic regions and T-bands, while weaker labeled bands corresponded to R-bands of human metaphase chromosomes [6]. R- (T-) and G-bands have different nucleotide composition, including CpG dinucleotides which are sites of possible cytosine methylation. CpG dinucleotides have the highest density in T-banding, and less intense staining features in G-banding. R-bands are enriched with CpG dinucleotides to a different extent [8, 9]. Overall, CpG enrichment correlates with gene density in chromosomal subregions [10].

The methylated regions in human chromosomes provide a combined R/C-like banding pattern. Methylated DNA preferentially accumulates in R- and C-bands and to a much less extent in G-bands. This banding is persistent in metaphases from PHA-stimulated, cultured adult and fetal lymphocytes, as well as in the chorion, placenta, and embryonic liver, lung, kidney, and brain cells [11, 12]. Notably, 5mC pattern forms distinct borders which correspond to known chromosome band boundaries, making each chromosome easily recognizable by its DNA methylation pattern; the latter is called MeC banding. The transitions of methylated to non-methylated DNA at the R/G-band boundaries may contribute to a marking of structural/functional genomic "units."

The intensity of immunofluorescence varies strongly among MeC-positive bands. Sharp contrast in MeC immunostaining of chromosome bands allows identification of homologous chromosomes by their landmarks. Enrichment of chromosome band in CpG dinucleotides is not the only determinant of the DNA methylation intensity. Several highly GC-rich R-bands demonstrate weak methylation like human bands 6q15, 6q21, 6q23, 9p13, 9q22, 9q32, 10q24, 13q22, 15q15, and 15q24 [11]. Another notable feature of MeC banding is the developmental stage specificity: bands 1q42, 2q23, 2q31, 2q33, 3q21, 3q25, 5q13, 6p11.2, 7p13, 7p15, 12p13, 12q13, 13q22, 14q13, 15q11.2-13, 15q15, 16q22, 17q21, 18q11.2, 20q11.2, 21q11.2, and 22q11.2 show different strengths of methylation in lymphocytes from adult and fetal blood [13]. These facts advocate for the functional significance of DNA methylation pattern and, thus, allow categorizing MeC staining along with replication banding as the functional chromosome banding.

Functional significance of chromosome methylation pattern is especially obvious in early development as chromosomes of human preimplantation embryos lack MeC banding, while R-, G-, and C-banding patterns can be easily detected. At 2-cell stage, chromosomes demonstrate asymmetric labeling for 5mC with bright fluorescence in one chromatid and pale fluorescence in the other one. At subsequent cleavages, along with asymmetric ones, chromosomes with equally weak methylation in both chromatids appear. These chromosome methylation patterns are explained by genome-wide replication-dependent (passive) loss of 5mC during cleavage divisions. The almost total erasure of the DNA methylation patterns during the first cleavages is followed by remethylation at the 8-cell/blastocyst stage. At 8-cell stage, with gradual loss of blastomere totipotency, MeC banding could be still found in some metaphases; however, by the blastocyst stage, MeC banding predominates [14]. DNA methylation reprogramming also comprises active (replication-independent) mechanisms. Active demethylation is mediated by ten-eleven translocation (TET) oxygenases, which consequently oxidize 5mC to 5-hydroxymethylcytosine (5hmC), 5-formylcytosine (5fC), and 5-carboxylcytosine (5caC). Subsequent replication-dependent loss of 5mC oxidative derivatives also results in asymmetrical staining of chromatids [15].

In immunofluorescence studies of human preimplantation embryos, the quality of chromosome spreads and the small sample amount cause major problems. Meanwhile, identification of chromosomes is of particular importance for comparison of DNA methylation and demethylation patterns in homologues of different parental origins. Fluorescence in situ hybridization (FISH) is the basic technique to identify chromosomes and specific chromosome or genome regions. When combined with immunofluorescence staining of the same preparation, it allows simultaneous identification of homologous chromosomes and their DNA methylation/demethylation patterns. The key step of both techniques is DNA denaturation; when FISH and immunodetection of 5mC are subsequently performed, DNA denaturation is required only once. As FISH and immunostaining cannot be done at the same target site, it is reasonable to use centromeric DNA probes for persistently non-methylated centromeric regions and thus reserve the chromosome arms as a target of immunocytochemistry.

Immunodetection of 5mC can be necessary to assess structural and/or functional status of derivative chromosomes or small supernumerary marker chromosomes. However, DNA denaturation has an adverse effect on the quality of chromosome banding, making such aforementioned derivative chromosomes difficult to identify. Using FISH on the same metaphases prior to immunostaining for

DNA methylation helps to circumvent this problem (own unpublished data in collaboration with the editor of this book).

The approach of the combined FISH and immunofluorescence staining for DNA methylation/demethylation is also informative when studying interphase nuclei. Samples from some tissues may comprise several cell clones differing by their ploidy, presence of aneuploidy, and/or structural chromosome rearrangements. When studying DNA methylation/demethylation in non-cultured and nondividing cells, FISH allows identification of karyotypically different nuclei within the same preparation (Fig. 1). Thus, combined FISH/immunostaining technique may also be informative when studying malignant and nonmalignant tumors or testicular tissue samples, which comprise genetically heterogeneous cells.

In conclusion, immunofluorescence detection of 5mC and its oxidative is an essential technique to simultaneously assess the whole-genome methylation/demethylation pattern and thus its functional status and/or the functional status of genomic subunits. This approach allows also the analysis of very small and heterogeneous samples. As it is single cell directed, it also permits the evaluation of methylation/demethylation profiles in individual cells or embryos and even in parental sets of chromosomes [16]. Here, we present the protocol of combined FISH/immunostaining for DNA methylation which can be applied on preparations of

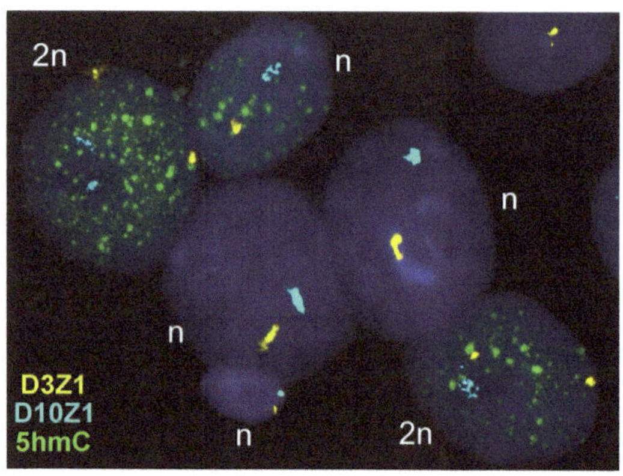

Fig. 1 Haploid (n) and diploid (2n) nuclei from human spermatogenic and in situ fixed nurse cells are depicted. Preparations were hybridized with DNA probes specific for centromere 3 (3p11.1-q11.1 = D3Z1) and centromere 10 (10p11.1-q11.1 = D10Z1) and immunostained with primary antibodies against 5-hydroxymethylcytosine (5hmC). The latter were detected by secondary antibodies conjugated with Alexa 488 in *green*. The counterstain for the nuclei used was 4′,6-diamidino-2-phenylindole (DAPI, *dark blue*), and the picture was taken by a Leica Ocular HC Plan 10×/25, objective NHCX Pl 100×/1.30-0.60. Nuclei exhibit distinct 5hmC patterns: peripheral one haploid and two diploid nuclei are heavily hydroxymethylated, whereas in the central three haploid nuclei 5hmC is totally absent

human chromosomes and nuclei derived from PHA-stimulated lymphocytes and non-cultured embryonic/extraembryonic cells, blastomeres of preimplantation embryos, germ cells, and cells from uterine leiomyomas.

2 Materials

Apart from standard cell biological and molecular cytogenetic equipment, including standard solutions (e.g., ethanol, methanol, formamide, formaldehyde, etc.), more specialized items required are listed below. Solutions and equipment needed for FISH itself are listed in chapter by Thomas Liehr et al. "The Standard FISH Procedure".

2.1 Culture and Slide Preparation

- Collagenase from *Clostridium histolyticum* (Sigma-Aldrich)
- Colchicine (Sigma-Aldrich)
- Eagle's medium with L-glutamine (BioloT, Russia)
- Fetal calf serum (PanEco)
- Gentamicin (40 mg/ml)
- ISM1 medium (Origio)
- L-Glutamine (Serva, Heidelberg, Germany)
- Phytohemagglutinin-P (Gibco)
- Pronase from *Streptomyces griseus* (Sigma-Aldrich)
- RPMI 1640 medium with glutamine (PanEco)

2.2 Immunofluorescence Staining with Antibodies Against 5mC and 5hmC

- Antibodies against 5hmC (rabbit polyclonal, Active Motif, 39769, Carlsbad, CA, USA)
- Antibodies against 5mC (mouse monoclonal, Eurogentec, BI-MECY-0100, Seraing, Belgium)
- Bovine serum albumin (BSA) (Sigma-Aldrich)
- DAPI-containing Vectashield antifade H-1200 (Vector Laboratories, Burlingame, CA, USA)
- Goat anti-rabbit Alexa Fluor 488 antibodies (Life Technologies, A-11008)
- Goat anti-mouse Alexa Fluor 555 antibodies (Life Technologies, A-21424)
- Goat anti-mouse Cy3 (GE Healthcare Life Sciences, PA43002, Uppsala, Sweden) antibodies
- Tween 20 (Sigma-Aldrich)

3 Methods

3.1 Culture and Slide Preparation of Human Peripheral and Cord Blood Lymphocytes

1. Add 0.6 ml of blood supplemented with heparin to a 25 cm flask containing 9 ml of RPMI 1640 medium, 1 ml of fetal calf serum, phytohemagglutinin in a concentration recommended by manufacturer, and gentamicin (see also chapter by Anja Weise and Thomas Liehr "Pre- and Postnatal Diagnostics and Research on Peripheral Blood, Bone Marrow, Chorion, Amniocytes, and Fibroblasts"). Culture lymphocytes for 72 h at 37 °C in a closed system.
2. Add colchicine at a final concentration 8.0 μg/ml 40 min prior to the end of culturing.
3. Transfer flask content to two tubes.
4. Centrifuge for 10 min at 1,000 rpm. Remove supernatant leaving 0.3–0.5 ml above the pellet, resuspend the pellet, and add 5 ml of hypotonic solution (0.55 % KCl); pipette and incubate for 25–30 min at room temperature (RT).
5. Add 0.5 ml of freshly prepared fixative (methanol/glacial acetic acid, 3:1), shake carefully by hand, and repeat step 4.
6. Add to the pellet 5 ml of freshly prepared cold fixative, shake briefly, and incubate for 30 min at −20 °C.
7. Change fixative 3 times with at least 15 min at −20 °C fixation each time and repeat step 4.
8. Make preparations of metaphase spreads and interphase nuclei by dropping suspension from a 30–50 cm distance on cold wet slides. Age slides at +55 °C overnight before further use under Sect. 3.5.

3.2 "Direct" Technique of Slide Preparation from Uncultured Human Tissues

The method was originally suggested by us [17, 18] and later modified as outlined below. The technique is applicable to tissues which contain dividing cells, such as chorionic villi, embryonic tissues, testicular tissue, etc.:

1. Divide tissue samples with two jeweler forceps into small fragments and place into vials (5–10 mg per vial) in 5 ml of hypotonic solution (0.9 % sodium citrate) supplemented with colchicine at a final concentration 2.5 μg/ml. Incubate 60–80 min at RT.
2. Replace 2.5 ml of hypotonic solution with 2.5 ml of freshly prepared fixative (methanol/glacial acetic acid, 3:1). Incubate 25–35 min at RT.
3. Replace all the solution in vials with cold fixative. Incubate for at least 90 min at +4 °C.
4. To make preparations, add approximately 2.5 ml of distilled water (RT) in the vials. 2–5 min later, extract the tissue

fragments from the vials by jeweler forceps and dry on filter paper. Put tissue fragments on prewarmed slides into drops of freshly prepared 60 % acetic acid and macerate for 30–90 s. Control the saturation of acetic acid with cells under the stereomicroscope. Discard tissue fragments from the slides with forceps and spread drops of 60 % acetic acid saturated with cells on the slides. Fix preparation with 2–3 drops of freshly prepared fixative (methanol/glacial acetic acid, 3:1). Age slides overnight at +55 °C before further use under Sect. 3.5.

3.3 Chromosome and Nuclei Preparation from Human Oocytes, Zygotes, and Blastomeres of Preimplantation Embryos

The method was originally suggested by Tarkowski [19] and modified by Dyban [20]. Control all steps under stereomicroscope:

1. Select unfertilized oocytes, triploid zygotes, or preimplantation embryos unsuitable for transfer in the uterus and put each one in ISM1 medium.
2. Add to medium 0.1 % colchicine solution 2–24 h prior to slide preparation: 2–4 h for oocytes and zygotes, 2–6 h for 2–8-cell embryos, and 6–24 h for morula-blastocyst stage.
3. Put each sample in a drop of 0.9 % sodium citrate, supplemented with 1 % protease; incubate up to 5 min at RT. Control when zona pellucida disappears under stereomicroscope.
4. Transfer each sample in freshly prepared ice-cold (−20 °C) of fixative (methanol + glacial acetic acid, 3:1) and incubate for 30 s at minimum.
5. Transfer each sample on a marked area of a slide and dry out a little. Drop 2.5 μl of freshly prepared ice-cold (−20 °C) of fixative (methanol/glacial acetic acid, 3:1). Age slides overnight at +55 °C before further use under Sect. 3.5.

3.4 Slide Preparation from Human Uterine Leiomyoma Samples

1. Transfer tissue sample in Eagle's medium (MEM) immediately after surgery. Put sample to a Petri dish with 1 × PBS and separate cells with scissors into 2–3 mm fragments.
2. Transfer fragments to 15 ml tubes with 5 ml 1 × PBS and centrifuge for 5 min at 100×*g*. Discard all supernatant.
3. Add 5 ml of the Eagle's medium (MEM), 100 μl of collagenase (final concentration 200 U/ml), and incubate in slanting position for 90–120 min at 37 °C.
4. Add 5 ml of 1 × PBS and pipette fragments up and down. Centrifuge for 5 min at 1200 rpm. Remove supernatant leaving 0.3–0.5 ml; resuspend the pellet.
5. Repeat step 4.
6. Add 2.5 ml of 0.55 % KCl and 2.5 ml of 0.9 % sodium citrate; incubate for 25 min at 37 °C.
7. Add 30 μl of fixative (methanol/glacial acetic acid, 1:1) and shake carefully by hand.

8. Centrifuge for 10 min at 1,200 rpm. Remove supernatant leaving 0.3–0.5 ml and resuspend the pellet.
9. Add 5 ml of fixative, resuspend, incubate for 30 min at +4 °C, and repeat step 8.
10. Repeat step 9.
11. Make preparations of interphase nuclei by dropping suspension from a 30–50 cm on cold wet slides. Age slides overnight at +55 °C before further use under Sect. 3.5.

3.5 FISH and Immunodetection of 5-Methylcytosine and 5-Hydroxymethylcytosine

3.5.1 Slide Pretreatment

1. Age slides overnight at +55 °C.
2. Dehydrate slides in ethanol series (70 %, 80 %, and 95 %) for 3 min each. Air-dry at RT.
3. Wash slides in 4 × SSC for 3 min at +37 °C.
4. Add 50 μl of 10 % pepsin stock solution and 50 μl of 2 M HCl to 50 ml of distilled water, and incubate 3–20 min depending on the amount of cells and amount of cytoplasm on preparation.
5. Wash slides in 4 × SSC for 3 min at 37 °C.
6. Fix in 2.5 % paraformaldehyde 2 × SSC solution for 10 min at RT.
7. Wash slides in two changes of fresh 4 × SSC for 3 min at 37 °C.
8. Drop slides in dH_2O for 1 min at RT.
9. Dehydrate slides in ethanol series (70 %, 80 %, and 95 %) for 3 min each and air-dry at RT.

3.5.2 Probe Preparation

Briefly vortex and spin down the probe vial before use and let it adjust to RT before use.

3.5.3 Denaturation

1. Add 1 μl of probe or probe mix on the marked place (Ø 1 cm) on the slide. Put coverslip Ø 1 cm and remove air bubbles by gently pushing on coverslip. Seal with rubber cement.
2. Denature sample and probe on a Thermobrite for 10 min at 78 °C.

3.5.4 Hybridization

Incubate 4–72 h at 37 °C in a Thermobrite or in a humidified chamber.

3.5.5 Post-hybridization Wash

1. Carefully remove rubber cement with forceps and wash slides until coverslips slide off in 4 × SSC, supplemented with Tween 20 (0.5 %) at 37 °C.

2. Wash slides in two changes of 4 × SSC for 3 min at 37 °C.
3. Wash slides in dH_2O for 1 min at RT.
4. Dehydrate slides in ethanol series (70 %, 80 %, and 95 %) for 3 min each. Air-dry at RT.

3.5.6 Immunofluorescence Staining with Antibodies Against 5mC and 5hmC

1. Add to each slide 100 μl of blocking solution; cover with a 24 × 60 mm coverslip. Incubate for 40 min in a humidified chamber at +37 °C. Do not allow slides to dry out.
2. Add to each slide 100 μl of primary antibody against 5mC (or primary antibody mixture, e.g., mouse anti-5mC antibody and rabbit anti-5hmC antibody) diluted in blocking solution in a concentration recommended by the manufacturer. Cover with a 24 × 60 mm coverslip and incubate for 60–90 min in a humidified chamber at RT or overnight at +4 °C.
3. Carefully remove coverslips and wash slides in three changes of 1 × PBS supplemented with Tween 20 (0.5 %) for 3 min in a shaking bath at 37 °C. Do not allow slides to dry out.
4. Add to each slide 100 μl of secondary antibody (or secondary antibody mixture, e.g., goat anti-mouse Alexa 555 and goat anti-rabbit Alexa 488) diluted in blocking solution in a concentration recommended by the manufacturer. Cover with 24 × 60 mm coverslip and incubate for 60–90 min in a humidified chamber at +37 °C.
5. Carefully remove coverslips and wash slides in three changes of 1 × PBS supplemented with Tween 20 (0.5 %) for 3 min in shaking bath at 37 °C.
6. Rinse slides in 1 × PBS and dH_2O and air-dry at RT.
7. Dehydrate slides in ethanol series (70 %, 80 %, and 95 %) for 3 min each. Air-dry at RT.
8. Add to each slide 15 μl of DAPI-containing Vectashield antifade; cover with 24 × 60 mm coverslip. Keep in the dark for 10–15 min for DAPI staining. Proceed with microscopy.

Acknowledgments

This work was supported by the Russian Scientific Foundation (the protocols concerning immunofluorescence study of human uterine leiomyomas, grant number 14-15-00737) and by the Russian Foundation of Basic Research (the protocols concerning immunofluorescence study of preimplantation embryos and uncultured embryonic tissues, grant number 16-04-01438_a; the protocols concerning immunofluorescence study of lymphocytes and germ cells, grant number 16-34-60107_mol_a_dk). O.A.E. and A.V.T. are grantees of RF President Scholarship.

References

1. Miller OJ, Schnedl W, Allen J et al (1974) 5-Methylcytosine localized in mammalian constitutive heterochromatin. Nature 251:636–637
2. Holliday R, Pugh JE (1975) DNA modification mechanisms and gene activity during development. Science 186:226–232
3. Riggs AD (1975) X-inactivation, differentiation and DNA methylation. Cytogenet Cell Genet 14:9–25
4. Razin A, Riggs AD (1980) DNA methylation and gene function. Science 210:604–610
5. Lubit BW, Pham TD, Miller OJ et al (1976) Localization of 5-methylcytosine in human metaphase chromosomes by immunoelectron microscopy. Cell 9:503–509
6. Barbin A, Montpellier C, Kokalj-Vokac N et al (1994) New sites of methylcytosine-rich DNA detected on metaphase chromosomes. Hum Genet 94:684–692
7. Montpellier C, Burgeois CA, Kokalj-Vokac N et al (1994) Detection of methylcytosine-rich heterochromatin on banded chromosomes. Application to cells with various status of DNA methylation. Cancer Genet Cytogenet 78:87–93
8. Rodionov AV (1985) Genetic activity of the G- and R-band DNA of human mitotic chromosomes. Genetika 21:2057–2065
9. Bicmore WA, Sumner AT (1989) Mammalian chromosome banding – an expression of genome organization. Trends Genet 5:144–148
10. Zoubak S, Clay O, Bernardi G (1996) The gene distribution of the human genome. Gene 174:95–102
11. Pendina AA, Efimova OA, Kaminskaia AN et al (2005) Immunocytochemical analysis of human metaphase chromosome methylation status. Tsitologiia 47:731–737, in Russian
12. Baranov VS, Pendina AA, Kuznetsova TV et al (2005) Peculiarities of metaphase chromosome methylation pattern in preimplantation human embryos. Tsitologiia 47:723–730, in Russian
13. Efimova OA, Pendina AA, Tikhonov AV et al (2015) A comparative immunocytochemical analysis of DNA methylation patterns in human metaphase chromosomes of adult and fetuses. Mol Med 3:17–21, in Russian
14. Pendina AA, Efimova OA, Fedorova ID et al (2011) DNA methylation patterns of metaphase chromosomes in human preimplantation embryos. Cytogenet Genome Res 132:1–7
15. Efimova OA, Pendina AA, Tikhonov AV et al (2015) Chromosome hydroxymethylation patterns in human zygotes and cleavage-stage embryos. Reproduction 149:223–233
16. Santos F, Dean W (2006) Using immunofluorescence to observe methylation changes in mammalian preimplantation embryos. Methods Mol Biol 325:129–137
17. Baranov VS (1989) A method of shaking-blotting – a simple and reliable means for obtaining direct chromosomal preparations from chorionic biopsies. Tsitologiia 31:251–253
18. Baranov VS, Lebedev VM, Poleev AV et al (1990) Fast direct method of obtaining metaphase and prometaphase chromosomes from chorion biopsy cells and human embryos during the 1st trimester of pregnancy. Biull Eksp Biol Med 110:196–198
19. Tarkowski AK (1966) An air-drying method for chromosome preparations from mouse eggs. Cytogenetics 5:394–400
20. Dyban AP, De Sutter P, Verlinsky Y (1993) Preimplantation cytogenetics analysis. In: Verlinsky Y, Kuliev AM (eds) Preimplantation diagnosis of genetic diseases: a new technique in assisted reproduction. Wiley-Liss, New York

CENP Antibodies Used Additionally to FISH

Elisabeth Klein and Thomas Liehr

Abstract

Chromosomal analyses in most instances can be sufficiently performed by banding cytogenetics or FISH; even though being performed much less frequently, immunohistochemistry on chromosomes is also an important option here. Consequently this kind of approach was also included in this book dedicated mainly to FISH approaches. Here the visualization of centromeric proteins (CENP) on chromosomes is described. CENP-B is present at both active and inactive centromeres, while CENP-C can only be found at active ones. Here a protocol is reported, being suited to distinguish mitotically active from inactive centromeres, especially important when analyzing dicentric human chromosomes.

Keywords Dicentric human chromosomes, Immunohistochemistry, Centromeric proteins (CENP), CENP-B, CENP-C, Active centromere, Inactive centromere

1 Introduction

Cytogenetic banding and FISH including array-comparative genomic hybridization are approaches coming to our minds when thinking about chromosomal analyses (see all other chapters of this book apart from chapter by Anna Pendina et al. "Immunofluorescence Staining for Cytosine Modifications Like 5-Methylcytosine and Its Oxidative Derivatives and FISH"). Immunohistochemistry being performed on chromosomes is much less frequently applied. This is due to the fact that Carnoy's fixative tends to alter or destroy protein structures; still this is the major ingredient used during chromosome preparation and storage of cytogenetically worked-up and fixed cells (chapter by Anja Weise and Thomas Liehr "Pre- and Postnatal Diagnostics and Research on Peripheral Blood, Bone Marrow, Chorion, Amniocytes, and Fibroblasts"). Nonetheless, DNA methylation patterns ([1]; chapter by Anna Pendina et al. "Immunofluorescence Staining for Cytosine Modifications Like 5-Methylcytosine and Its Oxidative Derivatives and FISH"), as well as centromeric proteins (as described in this chapter), can be visualized on chromosomes themselves [2, 3].

Thomas Liehr (ed.), *Fluorescence In Situ Hybridization (FISH)*, Springer Protocols Handbooks, DOI 10.1007/978-3-662-52959-1_36, © Springer-Verlag Berlin Heidelberg 2017

As we recently summarized, "anticentromeric antibodies were identified in the sera of patients with the calcinosis, Raynaud's syndrome, esophageal dysmotility, sclerodactyly, and telangiectasia (CREST) variety of scleroderma. These sera recognize both centromeres in normal and dicentric chromosomes except for the Y chromosome. Different proteins were recognized by the CREST sera and their location in the centromere determined: centromere protein (CENP)-A is a centromere-specific histone similar to H3, CENP-B is distributed in the centromere region beneath the kinetochores where it binds to a recognition sequence in human alpha-satellite DNA, CENP-C is a component of the inner kinetochore plate" [3]. Whereas CENP-B is present at both active and inactive centromeres, CENP-C can only be found at active centromeres.

The protocol we adapted here from [2] is meant for distinguishing mitotically active from inactive centromeres by application of CENP-B- and CENP-C-specific antibodies. This approach can be extremely helpful in the analyses of dicentric human chromosomes.

2 Materials

Cell suspension from peripheral blood cells or amniotic cells in Carnoy's fixative (chapter by Anja Weise and Thomas Liehr "Pre- and Postnatal Diagnostics and Research on Peripheral Blood, Bone Marrow, Chorion, Amniocytes, and Fibroblasts") is used for analyses.

2.1 Chemicals and Solutions to Be Prepared

- Antifade Vectashield (Cat. No.: H1000, Vector Laboratories/Biozol; store at +4 °C).
- DAPI (4,6-diamidino-2-phenylindole.2HCl) stock solution (Cat. No.: 124653, Merck; store at −20 °C).
- DAPI solution: dissolve 1.5 µl of 1 M DAPI stock solution in 1 ml antifade Vectashield (store at +4 °C, can be used at least for 3 months).
- Na-EDTA 0.5 M (e.g., Merck; store at −20 °C).
- 1 × PBS-azide solution: 10 mM $NaPO_4$ (pH 7.4), 0.15 M NaCl, 1 mM EGTA (e.g. Roth), 0.01 % NaN_3 (*see* **Note 1**).
- 1 × TEEN solution: 1 mM triethanolamine/HCl (pH 8.5), 0.2 mM Na-EDTA, and 25 mM NaCl.
- Triethanolamine stock solution (e.g., Fluka): Tris((hydroxymethyl)-aminomethan) (e.g., Roth), 0.1 % Triton X-100 (e.g., NeoLab) + 0.1 % BSA.
- 1 × KB washing buffer: 10 mM Tris: HCl (pH 7.7), 0.15 M NaCl, 0.1 % BSA.

3 Methods

3.1 Immunohistochemistry on Chromosomes

Immunohistochemical tests are done on Carnoy's fixed cells [2]. A rabbit polyclonal antibody against CENP-B (Abcam, Cambridge, UK) was used to label all centromeres (dilution 1:50). The specific staining of the active centromeres is performed using an anti-CENP-C antibody guinea pig serum (1:100) [4]. Both antibodies are simultaneously applied. FITC-labeled goat anti-rabbit IgG and CyTM3-conjugated AffiniPure goat anti-guinea pig IgG (Dianova, Hamburg, Germany) are applied as secondary antibodies. Chromosome banding is achieved by DAPI counterstaining (4′6-diamidino-2-phenylindole, Sigma).

3.2 Slide Preparation and Immunohistochemistry

1. Drop the cells in Carnoy's fixative on a clean slide as described in chapter by Anja Weise and Thomas Liehr "Pre- and Postnatal Diagnostics and Research on Peripheral Blood, Bone Marrow, Chorion, Amniocytes, and Fibroblasts" following the air-drying method.
2. To avoid unnecessary damage of proteins associated with the chromosomes, the slides with metaphase spreads are dried immediately at 39 °C on a heating plate.
3. As soon as the slides are dry, transfer them in 100 ml PBS-azide solution (Coplin jar) at room temperature (RT) and incubate for 5 min. This step is important to rehydrate the slides and if necessary to perform "antigen retrieval" (*see* **Note 2**).
4. Wash the slides three times for 1 min, each in 1 × TEEN solution at RT (*see* **Note 3**).
5. Place 100 μl of the primary antibody solution on the slide(s) and cover carefully with a 24 × 60 mm elastic coverslip; elastic coverslips prevent damage of chromosomal structures compared to those from glass (*see* **Note 4**).
6. Incubate the slides in a humid chamber for 40 min at 37 °C (*see* **Note 5**).
7. Remove coverslip and transfer slides into a Coplin jar filled with 1 × KB washing buffer at RT; incubate for 2 min on a shaker.
8. Exchange 1 × KB washing buffer and wash for 5 min on a shaker.
9. Exchange 1 × KB washing buffer and wash for 3 min on a shaker.
10. Remove slide from the Coplin jar and add 100 μl of the secondary antibody solution on the slide(s), cover them carefully with a 24 × 60 mm elastic coverslip, and incubate again in a humid chamber for 40 min at 37 °C (*see* **Note 6**).

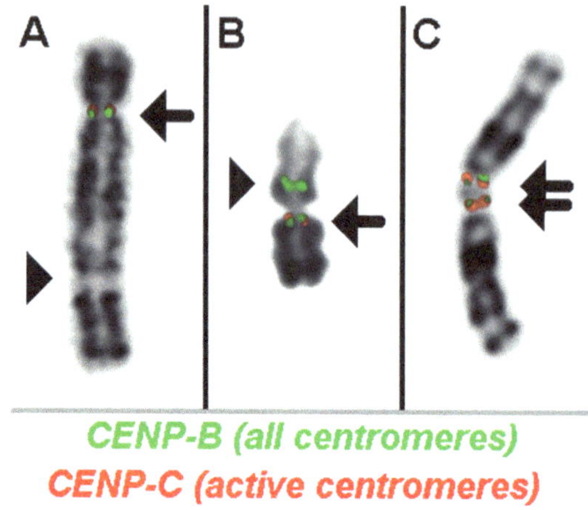

Fig. 1 Results of immunohistochemistry highlighting CENP-B (*green*, staining all centromeres apart from Y-chromosome) and CENP-C (*red*, staining only mitotically active centromeres) in one normal and two different dicentric human chromosomes; centromeres are highlighted by an *arrow*. (**a**) Derivative chromosome 5 with one active centromere only, thus being labeled in *green* and *red*. After cytogenetics a der(5)t(5;18)(q35;q10) was suggested; however only one centromeric structure was detectable by immunostaining. We know five more such cases; the significance of such a finding needs to be determined. (**b**) Dicentric chromosome dic(18;22)(p11.32;p12~13) with only chromosome 18-derived centromere being active. (**c**) Dicentric chromosome der(13;14)(13qter->13p11.2::14p11.2->14qter) with two active centromeres

11. Wash the slides two times in 1 × KB washing buffer for 2 min at RT.
12. Counterstain the slides with 20 μl of DAPI solution, cover with a 24 × 60 mm glass coverslip, and evaluate the results under a fluorescence microscope.

3.3 Evaluation

As described in chapter by Thomas Liehr et al. “The Standard FISH Procedure” for FISH slides. An example on how a result may look like is given in Fig. 1.

4 Notes

1. The PBS-azide washing solution contains NaN_3. However, this agent is toxic and according to our experience not necessary for successful experiments, at least with the antibodies used here.
2. Antigen retrieval is necessary especially after proteins in Carnoy’s fixative were “stressed.”
3. During the whole immunohistochemistry procedure until the last washing steps, it is important to avoid drying out the slide surface. Otherwise background can show up during evaluation.

4. The incubation time has to be adjusted for each primary antibody.
5. Elastic coverslip may be purchased; instead, e.g., Parafilm® may be cut into 24 × 60 mm pieces and applied instead.
6. The fluorochrome-coupled antibody detects the primary antibodies and leads to the indirect detection of the protein in question.

References

1. Barski A, Cuddapah S, Cui K et al (2007) High-resolution profiling of histone methylations in the human genome. Cell 129: 823–837
2. Earnshaw WC III, Ratrie R, Stetten G (1989) Visualization of centromere proteins CENP-B and CENP-C on a stable dicentric chromosome in cytological spreads. Chromosoma 98:1–12
3. Ewers E, Yoda K, Hamid AB et al (2010) Centromere activity in dicentric small supernumerary marker chromosomes. Chromosome Res 18:555–562
4. Ando S, Yang H, Nozaki N et al (2002) CENP-A, -B, and -C chromatin complex that contains the I-type alpha-satellite array constitutes the prekinetochore in HeLa cells. Mol Cell Biol 22:2229–2241

Part V

Interphase FISH

Interphase FISH in Diagnostics

Thomas Liehr and Sven Hauke

Abstract

Interphase FISH (iFISH) is applied in many fields of molecular cytogenetics. Besides in research applications for nuclear architecture studies, iFISH is performed in routine diagnostics of clinical and tumor cytogenetics. Here we give an overview on probes and tissues suited for iFISH applications.

Keywords Interphase FISH (iFISH), Centromeric satellite probes, Locus-specific probes, Heteromorphisms, Double minutes, Homogeneously staining regions, Microdeletion, Duplication, Translocation

1 Introduction

Interphase FISH (iFISH) is widely applied in pre- and postnatal clinical genetics, leukemia, and lymphoma diagnostics as well as in defining subclasses in growing numbers of solid tumors [1, 2]. Each kind of FISH probe (chapter by Thomas Liehr "Classification of FISH Probes") can be applied in research for nuclear architecture studies (chapter by Thomas Liehr and Nadezda Kosyakova "Three-Dimensional Interphase Analysis Enabled by Suspension FISH"). However, probes suited for routine iFISH are exclusively locus-specific ones and centromeric satellite probes. While locus-specific probes normally do not have the ability to produce confusing additional signals, centromeric satellite probes can do due to the presence of rare heteromorphisms [3]. Probes used are summarized in (chapter by Thomas Liehr "Commercial FISH Probes"). iFISH can be done in numerous tissues as outlined in Part III of this book (see also Table 1).

The major goal of iFISH is to achieve a quick result and/or to make nondividing tissues accessible to cytogenetic studies. iFISH probes can be used to detect numerical chromosomal aberrations and chromosomal rearrangements—how this can be achieved is summarized in Fig. 1. Also iFISH may be used for detection of

Thomas Liehr (ed.), *Fluorescence In Situ Hybridization (FISH)*, Springer Protocols Handbooks,
DOI 10.1007/978-3-662-52959-1_37,

Table 1
Summary of tissues/material suited for iFISH

Material/Tissue	Reference
Blood (including EDTA treated and umbilical cord blood)	Chapter by Anja Weise and Thomas Liehr "Pre- and Postnatal Diagnostics and Research on Peripheral Blood, Bone Marrow, Chorion, Amniocytes, and Fibroblasts"
Bone marrow	Chapter by Anja Weise and Thomas Liehr "Pre- and Postnatal Diagnostics and Research on Peripheral Blood, Bone Marrow, Chorion, Amniocytes, and Fibroblasts"
Amnion	Chapter by Anja Weise and Thomas Liehr "Pre- and Postnatal Diagnostics and Research on Peripheral Blood, Bone Marrow, Chorion, Amniocytes, and Fibroblasts" Chapter by Anja Weise et al. "FISH in Uncultivated Amniocytes"
Chorion	Chapter by Anja Weise and Thomas Liehr "Pre- and Postnatal Diagnostics and Research on Peripheral Blood, Bone Marrow, Chorion, Amniocytes, and Fibroblasts"
Fibroblasts	Chapter by Anja Weise and Thomas Liehr "Pre- and Postnatal Diagnostics and Research on Peripheral Blood, Bone Marrow, Chorion, Amniocytes, and Fibroblasts"
Smears	Chapter by Eyad Alhourani et al. "Tumorcytogenetic Diagnostics and Research on Blood and Bone Marrow Smears or Effusions"
Effusions	Chapter by Eyad Alhourani et al. "Tumorcytogenetic Diagnostics and Research on Blood and Bone Marrow Smears or Effusions"
Skin	Chapter by Thomas Liehr and Nadezda Kosyakova "Characterization of Mosaicism in Different Easy-to-Acquire Body Tissues such as Buccal Smears, Skin Abrasions, Hair Root Cells, or Urine"
Buccal mucosa	Chapter by Thomas Liehr and Nadezda Kosyakova "Characterization of Mosaicism in Different Easy-to-Acquire Body Tissues such as Buccal Smears, Skin Abrasions, Hair Root Cells, or Urine"
Hair root cells	Chapter by Thomas Liehr and Nadezda Kosyakova "Characterization of Mosaicism in Different Easy-to-Acquire Body Tissues such as Buccal Smears, Skin Abrasions, Hair Root Cells, or Urine"
Urine	Chapter by Thomas Liehr and Nadezda Kosyakova "Characterization of Mosaicism in Different Easy-to-Acquire Body Tissues such as Buccal Smears, Skin Abrasions, Hair Root Cells, or Urine"
Cryofixed tissue	Chapter by Thomas Liehr "Characterization of Archived Formalin-Fixed/Paraffin-Embedded or Cryofixed Tissue, Including Nucleus Extraction"
Formalin-fixed paraffin-embedded tissue	Chapter by Thomas Liehr "Characterization of Archived Formalin-Fixed/Paraffin-Embedded or Cryofixed Tissue, Including Nucleus Extraction"

(continued)

Table 1
(continued)

Material/Tissue	Reference
Sperm	Chapter by Maria Bonet Oliver "Sperms, Spermatocytes and Oocytes"
Oocytes	Chapter by Maria Bonet Oliver "Sperms, Spermatocytes and Oocytes"
Extended DNA	Chapter by Sandra Louzada et al. "Fluorescence In Situ Hybridization onto DNA Fibers Generated Using Molecular Combing"
Micronuclei	Chapter by Galina Hovhannisyan et al. "Micronucleus FISH"
Insect cells transfected with heterologous DNA	Chapter by Thomas Liehr "FISH on Insect Cells Transfected with Heterologous DNA"

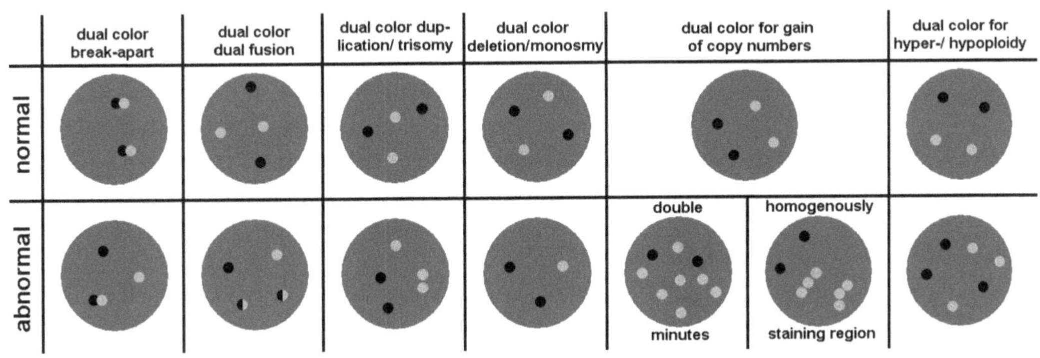

Fig. 1 Schematic depiction of how locus-specific (or in parts) centromeric probes can be combined in commercially available probe sets; the signal distribution as observed in a normal interphase cell is shown in the *upper row*, the abnormal situation in the *lower row*. All examples can be found as constitutional or malignant cell-associated aberrations—only double-minute and homogeneously staining regions are restricted to tumor cells, as they are the cytogenetic equivalents of oncogene amplification

mosaicism (chapter by Ivan Iourov et al. "Interphase FISH for Detection of Chromosomal Mosaicism").

2 Materials

Apart from standard cell biological and molecular cytogenetic equipment, including standard solutions (e.g., ethanol, methanol, formamide, formaldehyde, xylene, etc.), no more specialized items are required. The equipment needed for multicolor FISH is listed in chapter by Thomas Liehr et al. ("The Standard FISH Proce dure"); chapter by Thomas Liehr and Nadezda Kosyakova ("Multi plex FISH and Spectral Karyotyping").

3 Methods

3.1 Slide Pretreatment, Fluorescence In Situ Hybridization (FISH)

As described in chapter by Thomas Liehr et al. "The Standard FISH Procedure"; chapter by Thomas Liehr "Characterization of Archived Formalin-Fixed/Paraffin-Embedded or Cryofixed Tissue, Including Nucleus Extraction".

Variations:

- Pretreatment and denaturation time might be longer than in standard protocol FISH as no metaphase structure has to be preserved.

3.2 Evaluation

As the analysis of iFISH results lacks the possibility to combine signal localization and structural information of metaphase chromosomes, evaluation of FISH signals in interphase nuclei requires some specific prerequisites.

The main evaluation principle of probes for the detection of structural aberrations in iFISH is the presence or absence of fusion signals (see Fig. 1). Fusion signals or signals located in close proximity can occur either due to the hybridization of two differently labeled probes to two regions of one DNA stretch with only a small distance to each other or they can be artifacts caused by signals located in different focus layers of a nucleus but on the same spot. Even when focusing through the different layers, differentiating true fusion signals from coincidentally colocalizing signals might be impossible. Therefore, iFISH always requires establishing cutoff levels first by determining the number of nuclei showing coincidentally colocalizing signals with the probe of interest in a large number of normal nuclei [4].

Additionally, decondensation of DNA stretches could result in fusion signals revealing a small gap between both signals. As a rule of thumb, for most break-apart probes, a signal separation of at least two signal diameters indicates a true break-apart event, while smaller signal separations are most likely caused by decondensation. Nevertheless, signal separation is, among others, affected by the target that is addressed, the probe that is used, and the kind of rearrangement that is detected (with, e.g., small inversions leading to just small inter-signal gaps). Thus, systematic validation studies have to be performed before using probes for diagnostic purposes [5–8].

The same holds true for probes addressing copy number alterations. While, e.g., *MDM2* amplifications are characterized by large amplification clusters, *EGFR* amplifications frequently just appear like signals of increased size and *ERBB2* is known to reveal distinct extra-signals up to large amplification clusters. Additionally, nuclei displaying aberrant signal patterns are distributed either homogenously or heterogeneously in tumor sections and might display different patterns of positivity for one marker, e.g., isolated tumor nuclei with high-level cluster amplification in comparison to homogeneously distributed tumor nuclei with microcluster amplification [9].

For additional remarks considering the evaluation of iFISH results using tissue sections, please refer to chapter by Thomas Liehr "Characterization of Archived Formalin-Fixed/Paraffin-Embedded or Cryofixed Tissue, Including Nucleus Extraction".

References

1. Liehr T, Weise A, Hamid AB et al (2013) Multicolor FISH methods in current clinical diagnostics. Expert Rev Mol Diagn 13: 251–255
2. Liehr T, Othman MA, Rittscher K et al (2015) The current state of molecular cytogenetics in cancer diagnosis. Expert Rev Mol Diagn 15:517–526
3. Liehr T (2014) Benign & pathological chromosomal imbalances. In: Microscopic and submicroscopic copy number variations (CNVs) in genetics and counseling, 1st edn. Academic Press, New York
4. Friedrichs N, Kriegl L, Poremba C et al (2006) Pitfalls in the detection of t(11;22) translocation by fluorescence in situ hybridization and RT-PCR: a single-blinded study. Diagn Mol Pathol 15:83–89
5. Reddy KS, Tripodi J (2008) Contemplations on preclinical validation of fluorescence in situ hybridization probe assay for paraffin-embedded tissues in hematologic disorders. Cancer Genet Cytogenet 183:1–5
6. Watson MS, Buchanan PD, Cohen MM et al (2000) Technical and clinical assessment of fluorescence in situ hybridization: an ACMG/ASHG position statement. I. Technical considerations. Genet Med 2:356–361
7. Wiktor AE, Van Dyke DL, Stupca PJ et al (2006) Preclinical validation of fluorescence in situ hybridization assays for clinical practice. Genet Med 8:16–23
8. Wolff DJ, Bagg A, Cooley LD et al (2007) Guidance for fluorescence in situ hybridization testing in hematologic disorders. J Mol Diagn 9:134–143
9. Schildhaus HU, Heukamp LC, Merkelbach-Bruse S et al (2012) Definition of a fluorescence in-situ hybridization score identifies high- and low-level FGFR1 amplification types in squamous cell lung cancer. Mod Pathol 25 (11):1473–1480

Interphase FISH for Detection of Chromosomal Mosaicism

Ivan Y. Iourov, Svetlana G. Vorsanova, and Yuri B. Yurov

Abstract

Interphase fluorescence in situ hybridization (iFISH) allows qualitative and quantitative detection of chromosomal DNA at all stages of the cell cycle and at molecular resolutions. Accordingly, this methodology offers an opportunity to address chromosome numbers and structures in all the human tissues. In this light, a variety of iFISH techniques have been acknowledged as indispensable for studying intercellular genomic variation or somatic chromosomal mosaicism. Here, an interphase FISH protocol for the detection of intercellular genomic variations and low-level chromosomal mosaicism in somatic tissues is described. Additionally, technical issues influencing the results of iFISH are considered.

Keywords Interphase fluorescence in situ hybridization (iFISH), Evaluation, Mosaicism, Molecular resolution

1 Introduction

Interphase cytogenetics is essentially based on single-cell molecular cytogenetic techniques enabling the analysis of chromosomal DNA at molecular resolutions and at all stages of the cell cycle. Among the latter, one of the most widely applied molecular cytogenetic approaches is interphase fluorescence in situ hybridization (iFISH), representing an umbrella term for all the techniques (or probe sets) used for detecting chromosomal DNA in interphase nuclei ([1–14]; chapter by Thomas Liehr and Sven Hauke "Interphase FISH in Diagnostics"; chapter by Thomas Liehr and Nadezda Kosyakova "Three-Dimensional Interphase Analysis Enabled by Suspension FISH").

There is a variety of iFISH protocols used to address intercellular chromosomal variations, the spontaneous (background) level of chromosomal mutations, or somatic chromosomal mosaicism (chapter by Anja Weise and Thomas Liehr "Pre- and Postnatal Diagnostics and Research on Peripheral Blood, Bone Marrow, Chorion, Amniocytes, and Fibroblasts"; chapter by Thomas Liehr and Monika Ziegler "Application of FISH to Previously

Thomas Liehr (ed.), *Fluorescence In Situ Hybridization (FISH)*, Springer Protocols Handbooks,
DOI 10.1007/978-3-662-52959-1_38, © Springer-Verlag Berlin Heidelberg 2017

GTG-Banded and/or Embedded Cytogenetic Slides"; chapter by Anja Weise et al. "FISH in Uncultivated Amniocytes"; chapter by Eyad Alhourani et al. "Tumorcytogenetic Diagnostics and Research on Blood and Bone Marrow Smears or Effusions"; chapter by Thomas Liehr and Nadezda Kosyakova "Characterization of Mosaicism in Different Easy-to-Acquire Body Tissues such as Buc cal Smears, Skin Abrasions, Hair Root Cells or Urine"; chapter by Thomas Liehr "Characterization of Archived Formalin-Fixed/Par affin-Embedded or Cryofixed Tissue, Including Nucleus Extrac tion"; chapter by Maria Bonet Oliver "Sperms, Spermatocytes and Oocytes"). Despite a number of technical limitations (i.e., limited number of targets in a single iFISH analysis), these approaches are found applicable for uncovering changes in chromosome numbers (aneuploidy or polyploidy) or structural chromosome rearrangements in interphase [2, 4–6]. Furthermore, an increase in efficiency of iFISH can be achieved through multiple target analysis by a multicolor interphase FISH protocol [11, 12, 15–17]. In addition, iFISH can be combined with other molecular/cellular approaches aimed at studying cellular, subcellular, or extracellular structures (i.e., immuno-FISH, FISH combined with micronucleus test) ([11, 18–21]; chapter by Christine Ye et al. "Simultaneous Fluorescence Immunostaining and FISH"; chapter by Bin Ma and Naoko Tanese "RNA-Directed FISH and Immunostaining"; chapter by Tiphaine Aguirre-Lavin and Nathalie Beaujean "Three-Dimensional Immuno-Fluorescence In Situ Hybridization in Preimplantation Mouse"; chapter by Galina Hovhannisyan et al. "Micronucleus FISH").

The development and introduction of new molecular cytogenetic techniques have made apparent the significant contribution of somatic mosaicism to biodiversity and disease during the last decade [1, 3, 5, 8, 22–34]. Consequently, somatic chromosomal mosaicism and chromosome instability have become a focus for current biomedical research. Somatic chromosomal mosaicism is common in humans. It is abundantly found in almost all fetal human tissues at all stages of prenatal development [i.e., preimplantation embryos, spontaneous abortions, normal fetal tissues (prenatal diagnosis/induced abortions), extraembryonic tissues] ([1, 3, 22, 24, 25, 33, 35, 36]; chapter by Anja Weise and Thomas Liehr "Pre- and Postnatal Diagnostics and Research on Peripheral Blood, Bone Marrow, Chorion, Amniocytes, and Fibroblasts"; chapter by Thomas Liehr and Nadezda Kosyakova "Characterization of Mosai cism in Different Easy-to-Acquire Body Tissues such as Buccal Smears, Skin Abrasions, Hair Root Cells or Urine"). Postnatally, somatic chromosomal mosaicism is found in the normal human tissues, manifesting as stochastic (background) aneuploidy or (more rarely) as structural chromosome aberrations [3, 5, 19, 24, 26–28, 30, 32, 34, 37–39, 41, 42]. Chromosome instability resulting in karyotypic changes diversification among somatic cells (an effect equal to chromosomal mosaicism) is the main mechanism for cancer ([43–45]; chapter by Eyad Alhourani et al. "Tumorcytogenetic Diagnostics and Research on Blood and Bone Marrow Smears or Effusions"). Additionally, somatic chromosomal mosaicism hallmarks such processes as aging [24, 28, 38, 42, 46] and failure of genome stability maintenance [29, 38, 40, 45, 47–49]. Moreover, iFISH is able to help in shedding light on mechanisms of chromosome mosaicism and instability formation [29, 40, 49, 50]. Finally, there is a wide spectrum of diseases, which are associated with increased rates of somatic chromosomal mosaicism. In this instance, chromosomal mosaicism can directly result in the phenotypic manifestations or can be an important element in the pathogenic cascade of a disease [7, 8, 18, 19, 27, 44, 48, 50]. To be more precise, somatic chromosomal mosaicism has been shown to play a role in pathogenesis of Alzheimer's disease [19, 42, 47, 48, 51, 52], autism [23, 53], autoimmune diseases [54], chromosome instability syndromes [18, 19], and schizophrenia [37, 55]. In total, one can conclude that biomedical studies require a technical solution for uncovering somatic chromosomal mosaicism.

Technically, apart from combinations of FISH with other cellular/molecular assays, iFISH procedure does not imply any specific tools, reagents, or conditions comparing to the other FISH protocols. However, due to specificity and variability of chromosome organization and behavior in interphase nuclei (i.e., chromosomal DNA replication, chromosome associations, etc.), an analysis of iFISH results has its own peculiarities ([4, 56–58]; chapter by Thomas Liehr and Sven Hauke "Interphase FISH in Diagnostics"). To overpass these problems, an elegant technique allowing multicolor banding of an interphase chromosome pair—interphase chromosome-specific multicolor banding (ICS-MCB)—has been proposed ([11, 12, 39, 53, 59]; chapter by Thomas Liehr et al. "FISH Banding Techniques"). Using microdissection-derived DNA probes (chapter by Fengtang Yang et al. "Generation of Paint Probes from Flow-Sorted and Microdissected Chromosomes") for multicolor chromosome banding (chapter by Nadezda Kosyakova et al. "FISH-Microdissection"), the entire banded interphase chromosome can be detected in an interphase nucleus.

Overall, the detection of somatic chromosomal mosaicism (intercellular genomic variations) is usually a process of detecting rare events. Since cells with abnormal chromosomal content are detectable in almost all the human tissues, there is a difficulty on the definition of a sample as mosaic. Fortunately, suggesting that deviations in signal appearance are more appropriate for mosaicism benchmarks, recommendations for the detection of chromosomal mosaicism or somatic genomic variations have been proposed (for more details, see [5, 26]). Somatic mosaicism has been repeatedly noted to make a contribution to human intercellular diversity and diseases [3, 5, 7, 18, 19, 22–24, 26–31, 40, 41, 45, 47, 51, 52, 55]. Still, our knowledge about these phenomena is certainly incomplete.

Here, an iFISH protocol for high-resolution single-cell scoring of somatic chromosomal mosaicism (including low-level mosaicism) and solutions for iFISH technical problems are presented.

Outline of the Procedure

1. Cell suspension preparation (see as well [15, 60])
   a. Suspension preparation
   b. Slide pretreatment
   c. Quality control
2. iFISH (see as well [3, 4, 9, 15, 39, 58])
   a. Denaturation
   b. Hybridization
   c. Detection
3. Microscopy (see as well [9, 15, 18, 22, 23, 25, 37, 55, 57–59, 61])
   a. Visual/digital analysis
   b. Quantitative iFISH

2 Materials

In addition to the standard molecular cytogenetic equipment, including standard solutions (e.g., ethanol, methanol, formamide, formaldehyde), the following items are needed. The equipment needed for iFISH itself is listed in chapter by Thomas Liehr et al. "The Standard FISH Procedure".

2.1 Solutions to Be Prepared

- Carnoy's fixative: methanol/glacial acetic acid 3:1, freshly prepared, at 4 °C
- DAPI-antifade solution: 300 nM DAPI (4,6-diamidino-2-phenylindole)

3 Protocol

3.1 Cell Suspension Preparation, Slide Pretreatment, and Quality Control

1. Rinse a tissue sample (~3 mm^3) with 0.9 % NaCl and transfer to a homogenizer glass tube.
2. Homogenize tissue by rotating the Teflon pestle.
3. Add 2 ml of PBS and continue to homogenize until a liquid-like substance is present (some tissues do not need to be homogenized; in this case, skip the first three steps here).

4. Transfer the homogenized tissue to a plastic or glass tube, add 1 ml of 60 % glacial acetic acid, and incubate for 3–5 min at room temperature (RT).
5. Add 9 ml of Carnoy's fixative (−20 °C) and centrifuge at 1,000 g (5 min, RT).
6. Discard supernatant, add ~9 ml Carnoy's fixative (−20 °C), and spin down at 1,000 g (8 min, RT).
7. Repeat step 6 at least 3 times.
8. Put the suspension into a 2 ml tube; the suspension can be stored for a long period (up to 1 year) at −20 °C.
9. For further use in FISH, put 50–100 ml of suspension obtained as described before on a microscope slide and air-dry for 15–20 min (RT).
10. Slide pretreatment is performed as described in chapter by Thomas Liehr et al. "The Standard FISH Procedure".
11. For quality control do the following: drop 10–15 ml of cell suspension onto a microscope slide and air-dry.
12. Look into the light microscope using phase contrast. Depending on the distribution of the nuclei, the suspension can be further diluted or concentrated. If the distribution of nuclei is satisfactory, skip the following steps.
    a. Suspension characterized by a low nucleus distribution: centrifuge (2500 g, 7 min) and decrease the volume in the tube twofold. Then mix by inverting the cap and repeat step 11.
    b. Suspension characterized by a high nucleus distribution: add 0.3–0.7 ml of fixative mixture and repeat step 11.
13. One can use the slides in an iFISH assay after the FISH quality control procedure (not recommended).

3.2 iFISH

1. Put 5 μl of a DNA probe (*see* **Note 1**) on the pretreated slide and cover the liquid with an 18 × 18 mm coverslip.
2. Put the slide on a hot plate at 72–76 °C for 5–7 min.
3. Transfer it into a humid chamber at 37 °C overnight.
4. Remove the coverslip by putting distilled water on its edges.
5. Wash the slide in 50 % formamide in 2 × SSC at 42 °C for 10 min.
6. Exchange the washing solution for 2 × SSC/Tween20 at 42 °C and leave it for 15 min.
7. For directly labeled probes, continue with step 8. If the probes contain modified nucleotides with ligands (indirect labeling), additional procedures after step 6 are needed. For an example see chapter by Thomas Liehr et al. "The Standard FISH Procedure".

8. Add 24 μl of DAPI-antifade solution and cover the slide with a coverslip.
9. Proceed to microscopic analysis.

3.3 Microscopic Analysis

The microscopic analysis cannot be arranged as a step-by-step protocol. However, several points additional to those presented in chapter by Ivan Iourov ("Microscopy and Imaging Systems") can be indicated in this chapter's part, as well. These are generally related to applications of QFISH ([57, 62]; chapter by Gordana Joksic et al. "Telomere Length Measurement by FISH") and ICS-MCB ([39, 59, 61]; chapter by Thomas Liehr and Nadezda Kosyakova "Three-Dimensional Interphase Analysis Enabled by Suspen sion FISH"). In iFISH, the former is used to differ between signal association and the loss of a chromosome/chromosomal region, whereas the latter allows the analysis of interphase chromosomes in their entirety (*see* **Note 1**). Currently, there are softwares which have specified options/macros for quantification of iFISH signals from each digital image. For more details, one can address protocols in [57, 62].

3.4 Applications and Evaluation

3.4.1 Chromosome Enumeration Probes and QFISH

iFISH results using chromosome enumeration DNA probes, which paint pericentromeric satellite DNAs, appear as distinct spots corresponding to these chromosomal loci and numbers. Accordingly, one can at least determine chromosome numbers (or, more precisely, numbers of chromosomal loci) in a nucleus (Figs. 1a–i). Multiprobe iFISH performed using more than five differently labeled chromosome-specific probes usually requires a digital analysis of each nucleus for making multicolor images showing simultaneously the FISH results. Visual analysis will be exceedingly complex in these given circumstances. Since signals produced by a specific fluorochrome are observed with specific filter, autofluorescence particles can be differentiated from FISH signals because autofluorescence is seen at all the fluorescent microscope filters [9, 60].

For a successful iFISH analysis, the following two recommendations are to be considered: [1] scoring of intact undamaged nuclei (nuclei of more or less regular form specific for a given cell population) and [2] registration of all the signal appearance and patterns in a nucleus (i.e., one signal, associated signals, two signals, three signals, and so on) [3, 5, 8, 9]. Simultaneous use of several probes (i.e., multiprobe iFISH assays) requires that all nuclear signals exhibiting deviant signal appearance or pattern should be analyzed digitally. To solve the problem of signal associations, which are observed in almost all tissues studied by iFISH-based molecular cytogenetic techniques [2, 8, 18, 19, 22, 23, 39, 55, 57], one should apply QFISH. The intensity of signals is proportional to the amount of DNA in a chromosomal region. Therefore, the

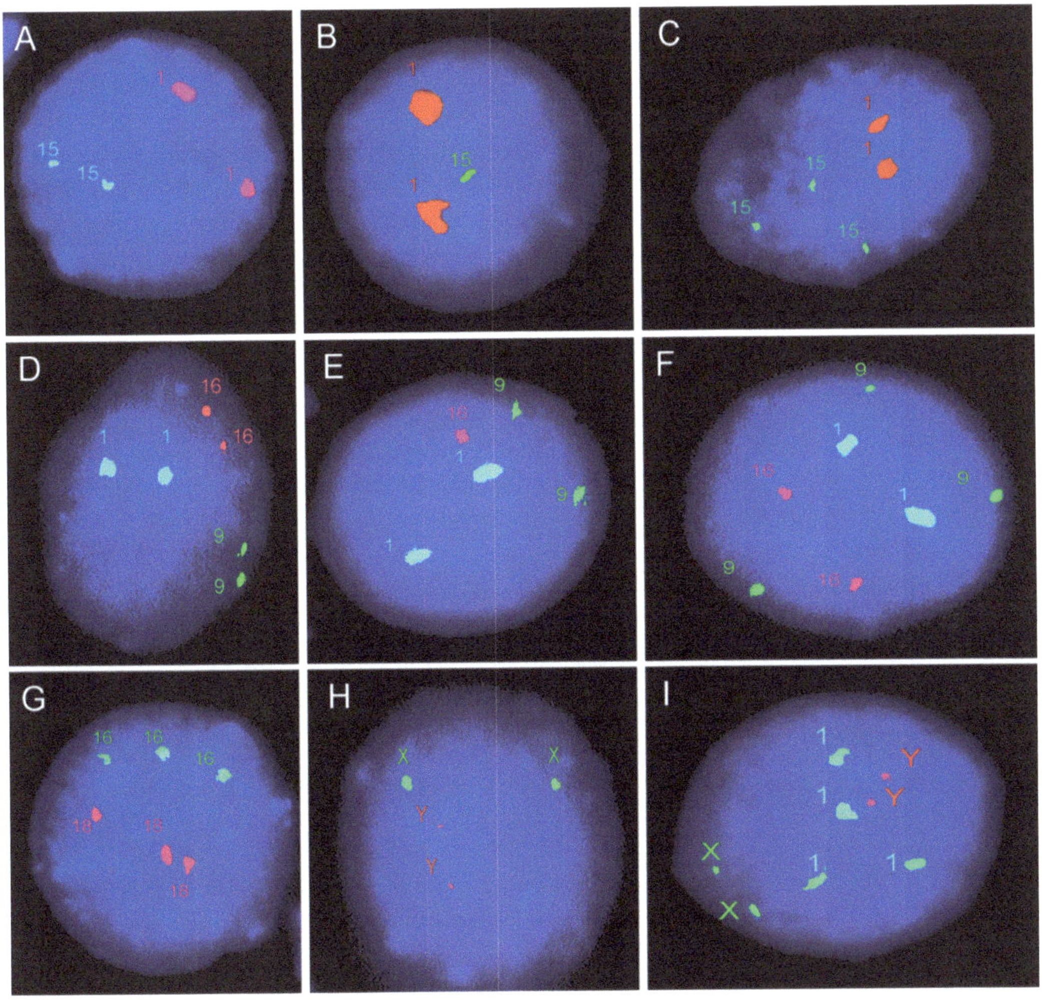

Fig. 1 Two- and three-color interphase fluorescence in situ hybridization (iFISH) with centromeric DNA probes: (**a**) Normal diploid nucleus with two signals for chromosome 1 and chromosome 15. (**b**) Monosomic nucleus with two signals for chromosome 1 and one signal for chromosome 15. (**c**) Trisomic nucleus with two signals for chromosome 1 and three signals for chromosome 15. (**d**) Normal diploid nucleus with two signals for chromosome 1, chromosome 9, and chromosome 16. (**e**) Monosomic nucleus with two signals for chromosome 1 and chromosome 9 and one signal for chromosome 16. (**f**) Trisomic nucleus with two signals for chromosome 1 and chromosome 16 and three signals for chromosome 9. (**g**) Triploid nucleus with three signals for chromosome 16 and chromosome 18. (**h**) Tetraploid nucleus with two signals for chromosome X and chromosome Y. (**i**) Tetraploid nucleus with two signals for chromosome X and chromosome Y and four signals for chromosome 1. (Copyright © [2]; licensee BioMed Central Ltd. This is an Open Access article distributed under the terms of the Creative Commons Attribution License (http:/creativecommons.org/licenses/by/2.0)

association of two signals in a diploid nucleus should be nearly twice more intensive as to the intensity of an unassociated signal [57]. The problem of the replication of chromosomal DNA [56] can be generally solved by a number of recommendations. These are generally related to determination of two distinct signals as those separated by a distance more than the diameter of a signal; since the latter is not always the case, another recommendation to solve the problem of replicated signals is related to uncovering a tiny fluorescent line connecting the two spots of a signal, hallmarking DNA replication [9]. Another problem affecting iFISH is related to chromosomal heteromorphisms. Although it is a rare problem, additional cross-hybridization signals on nonhomologous chromosomes or a lack of a signal on the second homologous chromosome can be observed in some individuals [58, 63]. To solve this one, QIFH can help to succeed [9, 16, 57]. Finally, the essential nontechnical disadvantage of iFISH originates from the possibility of another chromosomal unbalance (unbalanced translocations, supernumerary derivative chromosomes, etc.) to produce signal patterns featuring aneuploidy or polyploidy [4, 8, 18, 64]. This problem can be solved by ICS-MCB.

3.4.2 FISH Banding

ICS-MCB is a high-resolution FISH-based technique for the simultaneous visualization of several chromosome regions in interphase [39, 59, 61]. Examples of ICS-MCB are depicted in Figs. 2a–f. The essential part of ICS-MCB is the digital analysis of FISH results for defining chromosomal axis and aligning the signals for each different chromosome regions. The problem of interpreting results of ICS-MCB is associated with high levels of background fluorescence within the nuclear volume. To solve this problem, one can apply imaging software with threshold options for intensity diminishing within the nuclear area. This procedure allows obtaining a set of signals corresponding to chromosomal regions only. To exclude false-positive results arising from chromosomal associations, it is better to apply QFISH [59, 61].

4 Notes

1. There is a risk for unproductive use of DNA probes in such a sophisticate procedure as scoring rare events (low-level mosaicism) in unique samples with limited potential for repeating the cell suspension preparation. Therefore, the need to select appropriate slides to be prepared for scoring these rare events exists. This problem is solved using quality control [60]. Denaturation, hybridization, and detection are not usually associated with commonly encountered problems. Still, a number of DNA probes require to be hybridized for a longer time period (two or even more days) than chromosome enumeration probes. The existence of published guidelines for identifying chromosomal mosaicism in somatic cells simplifies the task for defining samples as mosaic, pseudomosaic, chromosomally

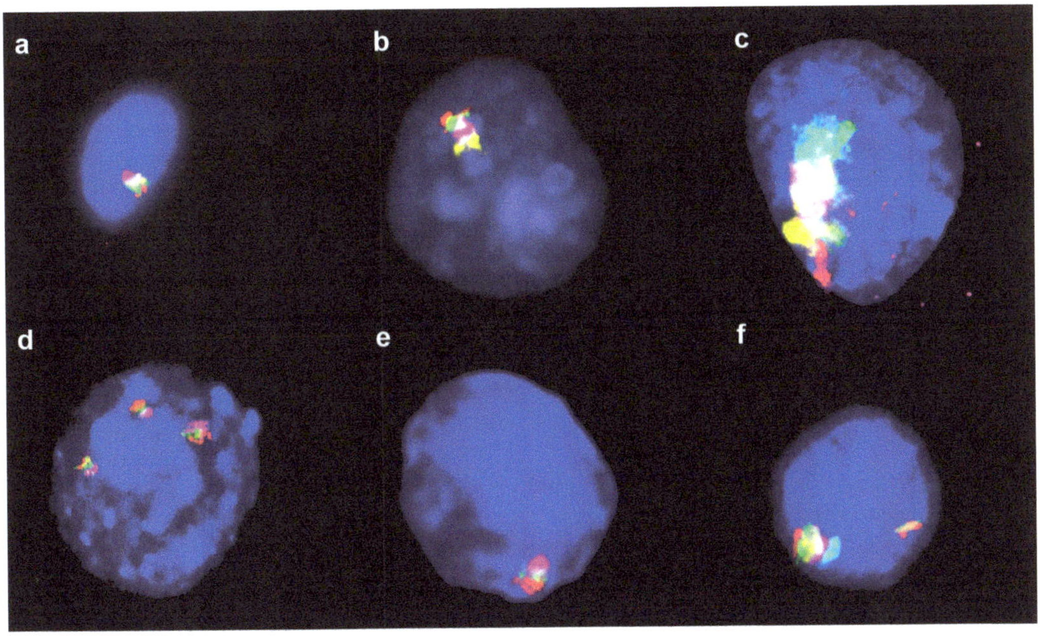

Fig. 2 ICS-MCB generated on nuclei of the developing and adult human brain: (**a**) Loss of chromosome 18 (monosomy) in a cell isolated from telencephalic regions of the fetal brain. (**b**) Loss of chromosome 16 (monosomy) in a cell isolated from the cerebral cortex of the normal human brain. (**c**) Loss of chromosome 1 (monosomy) in a cell isolated from the cerebral cortex of the schizophrenia brain. (**d**) Gain of chromosome 21 (trisomy) in a cell isolated from the cerebral cortex of the Alzheimer's disease brain. (**e**) Loss of chromosome 21 (monosomy) in a cell isolated from the cerebellum of the ataxia-telangiectasia brain. (**f**) Chromosome instability in the cerebellum of the ataxia-telangiectasia brain manifesting as the presence of a rearranged chromosome 14 order(14)(14pter->14q12:) (from [59], reproduced with permission of Springer in the format reuse in a book/textbook via Copyright Clearance Center)

instable, etc. Generally, nuclei with unusual shapes compared to other ones or overlapping nuclei are to be excluded from the analysis and are not to be scored. As shown previously applications of multiprobe iFISH with QFISH and ICS-MCB are only rarely uninterpretable [2, 4, 8, 9, 18, 19, 22, 23, 26, 39, 50, 51, 55, 59, 64].

Acknowledgments

The chapter is dedicated to Ilia V. Soloviev. The authors are supported by a grant from the Russian Science Foundation (project #14-35-00060).

References

1. Vorsanova SG, Kolotii AD, Iourov IY et al (2005) Evidence for high frequency of chromosomal mosaicism in spontaneous abortions revealed by interphase FISH analysis. J Histochem Cytochem 53:375–380

2. Vorsanova SG, Yurov YB, Iourov IY (2010) Human interphase chromosomes: a review of available molecular cytogenetic technologies. Mol Cytogenet 3:1
3. Yurov YB, Iourov IY, Monakhov VV et al (2005) The variation of aneuploidy frequency in the developing and adult human brain revealed by an interphase FISH study. J Histochem Cytochem 53:385–390
4. Yurov YB, Vorsanova SG, Iourov IY (eds) (2013) Human interphase chromosomes: biomedical aspects. Springer Science and Business Media, New York
5. Iourov IY, Vorsanova SG, Yurov YB (2006) Chromosomal variation in mammalian neuronal cells: known facts and attractive hypotheses. Int Rev Cytol 249:143–191
6. Iourov IY, Vorsanova SG, Yurov YB (2008) Recent patents on molecular cytogenetics. Recent Pat DNA Gene Seq 2:6–15
7. Iourov IY, Vorsanova SG, Yurov YB (2012) Single cell genomics of the brain: focus on neuronal diversity and neuropsychiatric diseases. Curr Genomics 13:477–488
8. Iourov IY, Vorsanova SG, Yurov YB (2013) Interphase chromosomes of the human brain: the biological and clinical meaning of neural aneuploidy. In: Yurov YB, Vorsanova SG, Iourov IY (eds) Human interphase chromosomes: biomedical aspects. Springer, New York, pp 53–83
9. Iourov IY, Vorsanova SG, Yurov YB (2016) Detection of nuclear DNA by interphase fluorescence in situ hybridization. Encyclopedia of Analytical Chemistry: 1–12
10. Arendt T, Mosch B, Morawski M (2009) Neuronal aneuploidy in health and disease: a cytomic approach to understand the molecular individuality of neurons. Int J Mol Sci 10: 1609–1627
11. Liehr T (ed) (2009) Fluorescence in situ hybridization (FISH)—application guide: Springer protocols. Springer, Heidelberg
12. Liehr T, Weise A, Hamid AB et al (2013) Multicolor FISH methods in current clinical diagnostics. Expert Rev Mol Diagn 13:251–255
13. Riegel M (2014) Human molecular cytogenetics: from cells to nucleotides. Genet Mol Biol 37:S194–S209
14. Bakker B, van den Bos H, Lansdorp PM et al (2015) How to count chromosomes in a cell: an overview of current and novel technologies. Bioessays 37:570–577
15. Yurov YB, Soloviev IV, Vorsanova SG et al (1996) High resolution fluorescence in situ hybridization using cyanine and fluorescein dyes: ultra-rapid chromosome detection by directly fluorescently labeled alphoid DNA probes. Hum Genet 97:390–398
16. Iurov II, Vorsanova SG, Solov'ev IV et al (2011) Original molecular cytogenetic approach to determining spontaneous chromosomal mutations in the interphase cells to evaluate the mutagenic activity of environmental factors. Gig Sanit 5:90–94
17. Potapova TA, Unruh JR, Box AC et al (2015) Karyotyping human and mouse cells using probes from single-sorted chromosomes and open source software. Biotechniques 59: 335–336
18. Iourov IY, Vorsanova SG, Liehr T et al (2009) Increased chromosome instability dramatically disrupts neural genome integrity and mediates cerebellar degeneration in the ataxia-telangiectasia brain. Hum Mol Genet 18: 2656–2669
19. Iourov IY, Vorsanova SG, Liehr T et al (2009) Aneuploidy in the normal, Alzheimer's disease and ataxia-telangiectasia brain: differential expression and pathological meaning. Neurobiol Dis 34:212–220
20. Devadhasan JP, Kim S, An J (2011) Fish-on-a-chip: a sensitive detection microfluidic system for Alzheimer's disease. J Biomed Sci 18:33
21. Hovhannisyan G, Aroutiounian R, Liehr T (2012) Chromosomal composition of micronuclei in human leukocytes exposed to mitomycin C. J Histochem Cytochem 60:316–322
22. Yurov YB, Iourov IY, Vorsanova SG et al (2007) Aneuploidy and confined chromosomal mosaicism in the developing human brain. PLoS One 2, e558
23. Yurov YB, Vorsanova SG, Iourov IY et al (2007) Unexplained autism is frequently associated with low-level mosaic aneuploidy. J Med Genet 44:521–525
24. Yurov YB, Vorsanova SG, Iourov IY (2010) Ontogenetic variation of the human genome. Curr Genomics 11:420–425
25. Vorsanova SG, Iourov IY, Kolotii AD et al (2010) Chromosomal mosaicism in spontaneous abortions: analysis of 650 cases. Rus J Genet 46:1197–1200
26. Vorsanova SG, Yurov YB, Soloviev IV et al (2010) Molecular cytogenetic diagnosis and somatic genome variations. Curr Genomics 11:440–446
27. Iourov IY, Vorsanova SG, Yurov YB (2008) Chromosomal mosaicism goes global. Mol Cytogenet 1:26
28. Iourov IY, Vorsanova SG, Yurov YB (2010) Somatic genome variations in health and disease. Curr Genomics 11:387–396

29. Iourov IY, Vorsanova SG, Yurov YB (2013) Somatic cell genomics of brain disorders: a new opportunity to clarify genetic-environmental interactions. Cytogenet Genome Res 139: 181–188
30. Rodríguez-Santiago B, Malats N et al (2010) Mosaic uniparental disomies and aneuploidies as large structural variants of the human genome. Am J Hum Genet 87:129–138
31. Bushman DM, Chun J (2013) The genomically mosaic brain: aneuploidy and more in neural diversity and disease. Semin Cell Dev Biol 24:357–369
32. Hultén MA, Jonasson J, Iwarsson E et al (2013) Trisomy 21 mosaicism: we may all have a touch of down syndrome. Cytogenet Genome Res 139:189–192
33. Taylor TH, Gitlin SA, Patrick JL et al (2014) The origin, mechanisms, incidence and clinical consequences of chromosomal mosaicism in humans. Hum Reprod Update 20:571–581
34. Campbell IM, Shaw CA, Stankiewicz P et al (2015) Somatic mosaicism: implications for disease and transmission genetics. Trends Genet 31:382–392
35. Liehr T, Ziegler M (2005) Rapid prenatal diagnostics in the interphase nucleus: procedure and cut-off rates. J Histochem Cytochem 53:289–291
36. Weier JF, Ferlatte C, Weier HU (2010) Somatic genomic variations in extra-embryonic tissues. Curr Genomics 11:402–408
37. Yurov YB, Vostrikov VM, Vorsanova SG et al (2001) Multicolor fluorescent in situ hybridization on post-mortem brain in schizophrenia as an approach for identification of low-level chromosomal aneuploidy in neuropsychiatric diseases. Brain Dev 23:S186–S190
38. Yurov YB, Vorsanova SG, Iourov IY (2009) GIN'n'CIN hypothesis of brain aging: deciphering the role of somatic genetic instabilities and neural aneuploidy during ontogeny. Mol Cytogenet 2:23
39. Iourov IY, Liehr T, Vorsanova SG et al (2006) Visualization of interphase chromosomes in postmitotic cells of the human brain by multicolour banding (MCB). Chromosome Res 14:223–229
40. Iourov IY, Vorsanova SG, Zelenova MA et al (2015) Genomic copy number variation affecting genes involved in the cell cycle pathway: implications for somatic mosaicism. Int J Genomics 2015:757680
41. Devalle S, Sartore RC, Paulsen BS et al (2012) Implications of aneuploidy for stem cell biology and brain therapeutics. Front Cell Neurosci 6:36
42. Andriani GA, Vijg J, Montagna C (2016) Mechanisms and consequences of aneuploidy and chromosome instability in the aging brain. Mech Ageing Dev. pii: S0047-6374(16)30028-8. doi: 10.1016/j.mad.2016.03.007
43. Das K, Tan P (2013) Molecular cytogenetics: recent developments and applications in cancer. Clin Genet 84:315–325
44. Heng HH, Bremer SW, Stevens JB et al (2013) Chromosomal instability (CIN): what it is and why it is crucial to cancer evolution. Cancer Metastasis Rev 32:325–340
45. Heng HH, Liu G, Stevens JB et al (2013) Karyotype heterogeneity and unclassified chromosomal abnormalities. Cytogenet Genome Res 139:144–157
46. Faggioli F, Wang T, Vijg J et al (2012) Chromosome-specific accumulation of aneuploidy in the aging mouse brain. Hum Mol Genet 21:5246–5253
47. Arendt T, Brückner MK, Lösche A (2015) Regional mosaic genomic heterogeneity in the elderly and in Alzheimer's disease as a correlate of neuronal vulnerability. Acta Neuropathol 130:501–510
48. Yurov YB, Vorsanova SG, Iourov IY (2011) The DNA replication stress hypothesis of Alzheimer's disease. Sci World J 11:2602–2612
49. Iourov IY, Vorsanova SG, Yurov YB (2014) In silico molecular cytogenetics: a bioinformatic approach to prioritization of candidate genes and copy number variations for basic and clinical genome research. Mol Cytogenet 7:98
50. Iourov IY, Vorsanova SG, Demidova IA et al (2015) (2015) 5p13.3p13.2 duplication associated with developmental delay, congenital malformations and chromosome instability manifested as low-level aneuploidy. Springerplus 4:616
51. Yurov YB, Vorsanova SG, Liehr T et al (2014) X chromosome aneuploidy in the Alzheimer's disease brain. Mol Cytogenet 7:20
52. Hou Y, Song H, Croteau DL, Akbari M, Bohr VA (2016). Genome instability in Alzheimer disease. Mech Ageing Dev. pii: S0047-6374(16)30047-1. doi: 10.1016/j.mad.2016.04.005
53. Iourov IY, Yurov YB, Vorsanova SG (2007) Mosaic X chromosome aneuploidy can help to explain the male-to-female ratio in autism. Med Hypotheses 70:456
54. Svyryd Y, Hernández-Molina G, Vargas F et al (2012) X chromosome monosomy in primary and overlapping autoimmune diseases. Autoimmun Rev 11:301–304
55. Yurov YB, Iourov IY, Vorsanova SG et al (2008) The schizophrenia brain exhibits low-

level aneuploidy involving chromosome 1. Schizophr Res 98:139–147
56. Soloviev IV, Yurov YB, Vorsanova SG et al (1995) Prenatal diagnosis of trisomy 21 using interphase fluorescence in situ hybridization of postreplicated cells with site-specific cosmid and cosmid contig probes. Prenat Diagn 15:237–248
57. Iourov IY, Soloviev IV, Vorsanova SG et al (2005) An approach for quantitative assessment of fluorescence in situ hybridization (FISH) signals for applied human molecular cytogenetics. J Histochem Cytochem 53: 401–408
58. Vorsanova SG, Iourov IY, Beresheva AK et al (2005) Non-disjunction of chromosome 21, alphoid DNA variation, and sociogenetic features of Down syndrome. Tsitol Genet 39: 30–36
59. Iourov IY, Liehr T, Vorsanova SG et al (2013) Interphase chromosome-specific multicolor banding. In: Yurov YB, Vorsanova SG, Iourov IY (eds) Human interphase chromosomes: biomedical aspects. Springer, New York, pp 161–169
60. Iourov IY, Vorsanova SG, Pellestor F et al (2006) Brain tissue preparations for chromosomal PRINS labeling. Methods Mol Biol 334:123–132
61. Iourov IY, Liehr T, Vorsanova SG et al (2007) Interphase chromosome-specific multicolor banding (ICS-MCB): a new tool for analysis of interphase chromosomes in their integrity. Biomol Eng 24:415–417
62. Weise A, Gross M, Hinreiner S et al (2010) POD-FISH: a new technique for parental origin determination based on copy number variation polymorphism. Methods Mol Biol 659:291–298
63. Liehr T (2016) Cytogenetically visible copy number variations (CG-CNVs) in banding and molecular cytogenetics of human; about heteromorphisms and euchromatic variants. Mol Cytogenet 9:5
64. Iourov IY, Vorsanova SG, Liehr T et al (2008) Dynamic mosaicism manifesting as loss, gain and rearrangement of an isodicentric Y chromosome in a male child with growth retardation and abnormal external genitalia. Cytogenet Genome Res 121:302–306

Comet-FISH

Galina Hovhannisyan and Rouben Aroutiounian

Abstract

The comet assay was developed as a sensitive method for detecting DNA damage and repair at the level of individual cells. The combination of comet assay with fluorescence in situ hybridization (FISH) or comet-FISH technique permits to localize DNA breaks within the specific DNA sequences. The position of FISH signals within the comet head or tail indicates intact or damaged DNA in genome region selected by the probe, respectively. Comet-FISH offers unique possibility to detect on the same specimen the total DNA damage and evaluate the damage of specific regions of genome as well.

Keywords DNA damage, Comet assay, Comet-FISH

1 Introduction

Single-cell gel electrophoresis, or comet assay, is a rapid, sensitive, and relatively simple method of visualization and measurement of DNA damage and repair on the level of the whole genome of single cells [1]. Comet assay was first described by Ostling and Johanson [2] to detect DNA damage induced by radiation. Since its development, the alkaline version [1], which includes DNA denaturation under alkaline conditions, has become the most used and is recommended for a broad spectrum of detected DNA damage (single and double breaks and alkali-label sites). In this technique cells are embedded in agarose on a microscope slide. After the gel has solidified, the cells are lysed to remove most of histones and to breakdown membranes. Subsequent DNA unwinding and electrophoresis under alkaline conditions permit to distinguish damaged from undamaged DNA. Undamaged DNA is too large to move in the electric field and therefore remains in the place of the former nucleus. This part of the comet is termed the comet's head. The damaged (fragmented) DNA is separated electrophoretically and generates the so-called comet tail. The relative amount of DNA in the comet tail indicates DNA break frequency. After staining with a suitable dye, the relative intensity of tail DNA fluorescence is

Thomas Liehr (ed.), *Fluorescence In Situ Hybridization (FISH)*, Springer Protocols Handbooks,
DOI 10.1007/978-3-662-52959-1_39,

measured as an index of DNA break frequency [3]. Comet analysis is performed using image analysis software. This assay is now widely used in genotoxicity testing, human biomonitoring, and ecogenotoxicology [4].

The comet assay combined with fluorescence in situ hybridization (FISH) is a powerful technique for simultaneous detection of overall DNA damage and DNA breaks in specific regions of the genome within single cells [5]. Comet-FISH was first introduced in 1997 [6] to study the spatial distribution of specific chromosome sequences and chromatin fibers in comet cells. Subsequently, it was used to study damage in a number of specific genes or other loci of the genome [7–10]. The presence of the signal in comet tail indicates damage to the DNA sequence of interest (see Fig. 1).

Mladinic et al. [11] proposed simultaneous temperature denaturation for both slide and probe instead of separated chemical denaturation of slides and temperature denaturation of probe [5] and achieved an increase of hybridization rate by 16 %. Last developments in the methodology of comet-FISH include the design and generation of fluorescent probes [12]. In general, the probes should be no longer than ~300 nucleotides (single or double stranded) to be able to penetrate the gel in which the target genomic DNA is embedded, they should be sequence specific, and their signal should be detectable and distinct from the background fluorescence and the dye used to stain the DNA.

In short, comet-FISH technique consists of the following steps:

- Preparation of the comet assay slides
- Performing the lysis, denaturation, and electrophoresis
- Performing the fluorescent in situ hybridization on comet assay slides

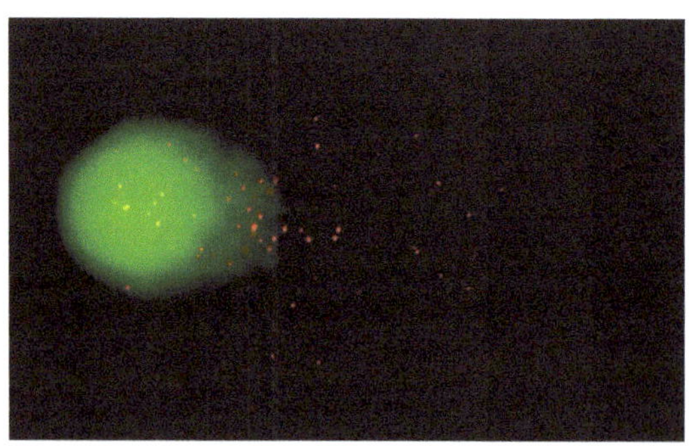

Fig. 1 Comet-FISH in bleomycin-treated human leukocyte using telomeric PNA probe (*red*) for the detection of telomeric repeat sequences and SYBR Green for staining the total DNA. Telomeric signals can be found in comet head and tail

– Analysis of slides (evaluation of primary DNA damage and/or specific gene damage)

Here we describe comet-FISH analyses of blood samples after addition of a mutagen.

2 Materials

2.1 Comet Assay

Equipment

- Comet assay tank for single-cell gel electrophoresis (SCGE) (Cat. No.: EL1920, Alpha Laboratories)
- Comet Assay IV analysis system (Perceptive Instruments, Suffolk, UK)
- Coverslips sized 24 × 60 mm (Cat. No.: C9056-1CS, Sigma-Aldrich, St. Louis, MO 63103, USA)
- Glass slides (frosted) (Cat. No.: CLS294875X25-72EA, Sigma-Aldrich, St. Louis, MO 63103, USA)

Chemicals

- Agarose BioReagent, for molecular biology, low EEO (normal melting) (Cat. No.: A9539, Sigma-Aldrich, St. Louis, MO 63103, USA).
- Normal melting agarose solution: Dissolve 1 mg of agarose in 100 ml PBS by heating in a microwave and keep fluid in a water bath.
- Agarose, low gelling temperature (Cat. No.: A9414, Sigma-Aldrich, St. Louis, MO 63103, USA).
- Agarose, low gelling temperature solution: Dissolve 0.9 mg of agarose in 100 ml PBS by heating in a microwave and keep fluid in a water bath.
- Ethylenediaminetetraacetic acid disodium salt dihydrate (Na_2EDTA) (Cat. No.: E5134, Sigma-Aldrich, St. Louis, MO 63103, USA).
- PBS 1 × (phosphate-buffered saline, Cat. No.: L1825, Biochrom; store at room temperature = RT).
- Sodium chloride (NaCl) (Cat. No.:S3014, Sigma-Aldrich, St. Louis, MO 63103, USA).
- Sodium hydroxide (NaOH) (Cat. No.:S5881, Sigma-Aldrich, St. Louis, MO 63103, USA).
- SYBR® Green I nucleic acid gel stain 10,000 × in DMSO (Cat. No.:S9430, Sigma-Aldrich, St. Louis, MO 63103, USA).
- SYBR Green solution: (1:10,000) (30 μl per slide).

- Triton™ X-100 (Cat. No.: T8787, Sigma-Aldrich, St. Louis, MO 63103, USA).
- Trizma® base (Cat. No.:T1503, Sigma-Aldrich, St. Louis, MO 63103, USA).
- Lysis solution: 10 mM Trizma® base, pH 7.5, 100 mM Na_2EDTA, 2.5 M NaCl, 1 % Triton X-100, pH 10; make fresh as required.
- Electrophoresis solution: 1 mM Na_2EDTA, 300 mM NaOH, pH 13; store at 4 °C.
- Neutralization buffer: 0.08 M Trizma® base, pH 7.2.

2.2 FISH

Besides things mentioned in chapter by Thomas Liehr et al. "The Standard FISH Procedure", the following things are needed:

- Telomere PNA FISH Kit/Cy3 (Cat. No.: K5326, Dako, Denmark) comprising
  - Wash solution 50 ×
  - Telomere PNA probe
  - TBS—Tris-buffered saline
- Other FISH probes

3 Methods

3.1 Human Blood Sample Preparation and Treatment

Immediately after sampling, add mutagen (e.g., mitomycin C or bleomycin) to heparinized human blood and incubate for 1–3 h at 37 °C.

3.2 Comet Assay (Acc. To (1))

1. Cover frosted microscope slides with ground layer of 1.0 % normal melting point agarose in PBS.
2. Keep the slides for at least 24 h at 37 °C to solidify agarose.
3. Distribute 100 μl of cell/agarose suspension (containing 10 μl of whole blood with 90 μl 0.9 % low gelling temperature agarose in PBS preheated to 37 °C) onto the microscope slide (*see* **Note 1**).
4. Cover slides with glass coverslips rapidly for uniform distribution of cell/agarose suspension on the slide surface.
5. Keep the slides at 4 °C to solidify agarose.
6. Remove coverslips (*see* **Note 2**) and immerse slides in cold lysis solution for at least 60 min at 4 °C. Slides can be stored overnight in lysis solution.
7. Place slides in an electrophoresis tank containing alkaline electrophoresis buffer for DNA unwinding for 20 min at 4 °C.

8. Switch on the current and carry out electrophoresis at 1.25 V/cm, 300 mA for 25 min at 4 °C.
9. Remove the slides from the electrophoresis tank and wash once with neutralization buffer for 20 min at RT.

3.3 Comet-FISH

1. Before hybridization store slides for at least 3 days in absolute ethanol at 4 °C for dehydration.
2. Rehydrate slides in H_2O for 15 min at RT.
3. Denature DNA by incubation of slides in 0.5 M NaOH for 25 min at RT.
4. Dehydrate slides immediately in an ethanol series (75, 80, and 95 %, 5 min each) at RT.
5. Air-dry slides until all ethanol evaporated.
6. Denature PNA probes (Telomere PNA FISH Kit/Cy3, Dako Cytomation, Denmark) by preheating to 80 °C in water bath for 3 min.
7. Apply 10 μl denatured probe to an area of approximately 20 × 20 mm (*see* **Note 3**).
8. Cover the gel with glass coverslip and place the slides in a humidified chamber at 37 °C overnight.
9. The next day, place the slides for 30 min at RT.
10. Place the slides in 1 × rinse solution supplied with the PNA probes, in order to facilitate removal of the coverslips.
11. Put the slides into a staining jar containing prewarmed post-hybridization washing solution (from the PNA Kit) at 65 °C in water bath, without agitation for 2.5 min.
12. Cool the slides in cold 1x phosphate-buffered detergent (PBD).
13. Counterstain the slides with SYBR Green (1:10,000) including 50 % antifade (30 μl per slide), cover with a coverslip 24 × 60 mm, and evaluate the results under fluorescence microscope.

3.4 Evaluation

1. For comet image analysis, apply comet analysis software, e.g., Comet Assay IV analysis system (Perceptive Instruments, Suffolk, UK) (*see* **Note 4**). For comet-FISH image analysis, record the position of the FISH signals in the comet head or comet tail. Cell numbers scored for comet-FISH ranged from 50 to 100 cells/slide.

4 Notes

1. Avoid the use of the LMA agarose at a temperature higher than 37 °C to prevent cell damage.
2. Carefully remove the coverslips to prevent gel rupture.
3. Avoid using damaged gels (with breaks and air bubbles) for hybridization.
4. Before applying the comet-FISH, evaluation of the effect of hybridization procedure on comet measurements is recommended. Record the positions of several comets on the evaluated slides and relocate the same comets after the hybridization process. Compare comet parameters before and after hybridization. Consider the differences in the assessment of total DNA damage.

References

1. Singh NP, McCoy MT, Tice RR et al (1988) A simple technique for the quantitation of low levels of DNA damage in individual cells. Exp Cell Res 175:184–191
2. Ostling O, Johanson KJ (1984) Microelectrophoretic study of radiation-induced DNA damages in individual mammalian cells. Biochem Biophys Res Commun 123:291–298
3. Rapp A, Bock C, Dittmar H et al (2000) UV-A breakage sensitivity of human chromosomes as measured by COMET-FISH depends on gene density and not on the chromosome size. J Photochem Photobiol B 56:109–117
4. Azqueta A, Collins AR (2013) The essential comet assay: a comprehensive guide to measuring DNA damage and repair. Arch Toxicol 87:949–968
5. Glei M, Schlörmann W (2014) Analysis of DNA damage and repair by comet fluorescence in situ hybridization (Comet-FISH). Methods Mol Biol 1094:39–48
6. Santos SJ, Singh NP, Natarajan AT (1997) Fluorescence in situ hybridization with comets. Exp Cell Res 232:407–411
7. Arutyunyan R, Gebhart E, Hovhannisyan G et al (2004) Comet-FISH using peptide nucleic acid probes detects telomeric repeats in DNA damaged by bleomycin and mitomycin C proportional to general DNA damage. Mutagenesis 19:403–408
8. Hovhannisyan G, Rapp A, Arutyunyan R et al (2005) Comet-assay in combination with PNA-FISH detects mutagen-induced DNA damage and specific repeat sequences in the damaged DNA of transformed cells. Int J Mol Med 15:437–442
9. Glei M, Hovhannisyan G, Pool-Zobel BL (2009) Use of Comet-FISH in the study of DNA damage and repair: review. Mutat Res 681:33–43
10. McKenna DJ, Doherty BA, Downes CS et al (2012) Use of the comet-FISH assay to compare DNA damage and repair in p53 and hTERT genes following ionizing radiation. PLoS One 7, e49364
11. Mladinic M, Zeljezic D (2014) Modification of comet-FISH technique by using temperature instead of chemical denaturation. Methods 1:162–167
12. Spivak G (2015) New developments in comet-FISH. Mutagenesis 30:5–9

Micronucleus FISH

Galina Hovhannisyan, Tigran Harutyunyan, and Thomas Liehr

Abstract

Micronuclei (MN) originate from chromosomal fragments or whole chromosomes that fail to be incorporated into daughter nuclei. MN frequency has been extensively used as a biomarker to measure rates of chromosomal damage. By MN test combined with fluorescence in situ hybridization (FISH), the chromosomal contents of the MN can be characterized. The application of FISH probes allows to distinguish MN originating either from chromosome loss or breakage and to determine the involvement of specific chromosomes and chromosome fragments in MN formation. Understanding the MN origin and content using FISH is essential for the proper use of this cytogenetic endpoint.

Keywords Chromosome damage, Micronuclei, Micronucleus FISH

1 Introduction

The micronucleus (MN) assay is one of the best validated cytogenetic techniques for evaluating chromosomal damage [1]. MN are small bodies that arise in dividing cells from acentric chromosomal fragments and whole chromosomes lagging behind in anaphase and not included in the daughter nuclei [2]. In comparison with chromosomal aberrations, the scoring of MN is simpler and less time-consuming. MN were first described by Howell and Jolly in the late 1800s and early 1900s as Feulgen-positive nuclear bodies in human reticulocytes, representing chromosomes separated from the mitotic spindle [3]. In the early 1970s, the term micronucleus test was suggested for the first time by Boller and Schmidt [4] and Heddle [5] who showed that this assay provided a simple method to detect the genotoxic potential of mutagens in bone marrow erythrocytes after in vivo exposure of animals. A few years later, Countryman and Heddle [6] studied MN in human lymphocytes. The cytokinesis-block micronucleus (CBMN) assay, based on cytokinesis inhibition by cytochalasin B, has been developed by Fenech and Morley in 1985 [7]. This approach allows evaluating the MN frequency in once-divided cells accumulated in the binucleated

Thomas Liehr (ed.), *Fluorescence In Situ Hybridization (FISH)*, Springer Protocols Handbooks,
DOI 10.1007/978-3-662-52959-1_40, © Springer-Verlag Berlin Heidelberg 2017

stage. Now MN test is widely applied in genetic toxicology testing, radiation biodosimetry, and population biomonitoring [1].

However, by the conventional MN test, it is impossible to describe the chromosomal content of MN. Attempts to overcome this limitation of the MN test have included in situ hybridization with DNA probes. The MN assay combined with fluorescence in situ hybridization (FISH) allows to characterize the occurrence of material of different chromosomes in MN (see Fig. 1). It also permits to discriminate between aneugenic and clastogenic effects ([8, 9]; chapter by Galina Hovhannisyan and Rouben Aroutiounian "Comet-FISH"). This approach was first introduced by Becker et al. [10] with application of centromere-specific DNA probe (p82H) in human lymphocytes. The involvement of chromosomes in MN has been studied by centromeric (chapter by Thomas Liehr et al. "cenM-FISH Approaches") and telomeric DNA probes ([11]; chapter by Gordana Joksic et al. "Telomere Length Measurement by FISH"), chromosome painting ([12]; chapter by Thomas Liehr et al. "Two- to Three-Color FISH"), spectral karyotyping ([13]; chapter by Thomas Liehr and Nadezda Kosyakova "Multiplex FISH and Spectral Karyotyping"), and multicolor FISH ([14]; chapter by Thomas Liehr and Nadezda Kosyakova "Multiplex FISH and Spectral Karyotyping").

FISH analysis of MN is based on the achievements of interphase FISH (chapter by Thomas Liehr and Sven Hauke "Inter phase FISH in Diagnostics"; chapter by Ivan Iourov et al. "Interphase FISH for Detection of Chromosomal Mosaicism"). A major condition of the quantitative accuracy of the MN assay is integrity of cell membrane and preservation of the cytoplasm during the cell harvesting, while interphase FISH technique allows the destruction of cellular membrane [9]. Commercial (chapter by Thomas Liehr "Commercial FISH Probes") and homemade

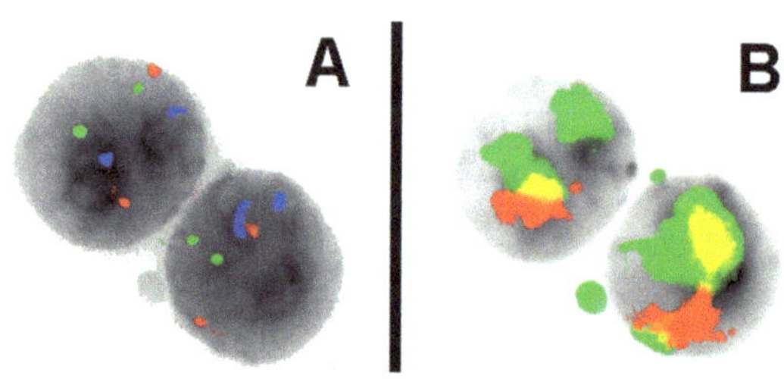

Fig. 1 Bleomycin-induced micronuclei (MN) in human leukocyte evaluated by a three-color FISH probe set consisting of centromeric probes (cep) for chromosomes 9 (*SpectrumAqua*) and 16 (*SpectrumGreen*) as well as a probe for 1q12 (*SpectrumOrange*). In the second round of hybridization, the same nuclei/MN was hybridized with whole-chromosome paints (wcp) for chromosomes 1 (*SpectrumOrange*), 9 (Cy5), and 16 (*SpectrumGreen*). In (A) there are no cep signals in MN and in (B) MN contains wcp signal for chromosome 1

FISH probes (chapter by Thomas Liehr "Homemade Locus-Specific FISH Probes: Bacterial Artificial Chromosomes") for selective painting of specific DNA sequences and software's for image analysis are also suitable for description of MN composition.

In short, micronucleus FISH technique consists of the following steps:

- Preparation of the MN slides
- Performing the fluorescent in situ hybridization on MN slides
- Assessment of localization of FISH signals in main nuclei or MN

2 Materials

2.1 Human Whole-Blood Culture

- RPMI 1640 medium with L-glutamine (Cat. No.: 72400–021, Gibco BRL)
- Fetal bovine serum (Can. No.: S0113, Biochrom)
- Penicillin/streptomycin (Cat. No.: A 2212, Seromed, Berlin, Germany)
- Phytohemagglutinin (Cat. No.: M 5030, Biochrom)

2.2 MN Test

- Cytochalasin B (Cat. No.: C6762, Sigma)
- Cytochalasin B stock solution in ethanol (0.6 mg/ml; store at −20 °C)
- KCl (Cat. No.: 1.04936.1000, Merck, Darmstadt, Germany)
- Hypotonic solution: 0.075 M KCl, freshly prepared and used at 4 °C
- Methanol (Cat. No.: 1.060092500, Merck)
- Glacial acetic acid (Cat. No.: 1.00063.2500, Merck)
- Carnoy's fixative: methanol/glacial acetic acid 3:1, freshly prepared and used at 4 °C

2.3 FISH

In chapter by Thomas Liehr et al. "The Standard FISH Procedure", material needed for FISH is listed.

3 Methods

3.1 Cell Culture

1. Add 1 ml of heparinized human whole blood to 9 ml of cell culture medium RPMI 1640 supplemented with 10 % fetal bovine serum (FBS), 1 % antibiotics (penicillin and streptomycin), and 10 μg/ml phytohemagglutinin (PHA) and incubate at 37 °C.

2. After 22 h growth, add mutagen (e.g., mitomycin C or bleomycin) to the cell culture.
3. After 44 h growth, remove treatment medium; add fresh medium with cytochalasin B at a final concentration of 3 μg/ml to inhibit cytoplasmic division.
4. After 72 h growth, transfer the culture to 15 ml tube.
5. Centrifuge the solution at room temperature (RT) for 5 min at 1,000 rounds per minute (rpm) and discard the supernatant with glass pipette.
6. Resuspend the pellet in cold hypotonic solution of 0.075 M KCl (4 °C) and incubate at RT for 3 min.
7. Repeat step 5.
8. Resuspend the pellet in 10 ml of Carnoy's fixative (4 °C) and incubate at RT for 20 min.
9. Repeat step 5.
10. Resuspend the pellet in 5 ml of fixative (4 °C) and incubate at RT for 10 min.
11. Repeat step 5.
12. Resuspend the pellet in 1 ml of fixative (4 °C).
13. Drop the cell suspension onto the glass slide and air-dry at least 2 h.
14. Freeze the slides at −20 °C until in situ hybridization.

3.2 Micronucleus FISH

Perform FISH according to standard procedure (see the chapter by Thomas Liehr et al. "The Standard FISH Procedure") with minor modifications (*see* **Notes 1–4**). Commercially available as well as homemade and/or self-labeled probes can be used.

4 Notes

1. In slide pretreatment procedure after step 1 (dehydrate slide in an ethanol series and air-dry), add additional postfixation step (see the chapter by Thomas Liehr et al. "The Standard FISH Procedure", step 6 in Sect. 2.2 Slide Pretreatment).
2. Incubate slides for denaturation on a warming plate for 5–10 min (instead of 2–4 min) at 75 °C (see the chapter by Thomas Liehr et al. "The Standard FISH Procedure", step 2 in Sect. 2.3 Fluorescence In Situ Hybridization (FISH)).
3. Incubate slides for hybridization 36 h at 37 °C in a humid chamber (see the chapter by Thomas Liehr et al. "The Standard FISH Procedure", step 6 in Sect. 2.3 Fluorescence In Situ Hybridization (FISH)).

4. Postwash the slides 5 min in 1 × SSC solution at 60 °C (instead of 62–64 °C) without agitation (instead of gentle agitation) (see the chapter by Thomas Liehr et al. "The Standard FISH Procedure", step 9 in Sect. 2.3 Fluorescence In Situ Hybridization (FISH)).

References

1. Bolognesi C, Fenech M (2013) Micronucleus assay in human cells: lymphocytes and buccal cells. Methods Mol Biol 1044:191–207
2. Kirsch-Volders M, Plas G, Elhajouji A et al (2011) The in vitro MN assay in 2011: origin and fate, biological significance, protocols, high throughput methodologies and toxicological relevance. Arch Toxicol 85:873–899
3. Sabharwal R, Verma P, Syed MA et al (2015) Emergence of micronuclei as a genomic biomarker. Indian J Med Paediatr Oncol 36: 212–218
4. Boller K, Schmid W (1970) Chemical mutagenesis in mammals: the Chinese hamster bone marrow as an in vivo test system. Hematological findings after treatment with trenimon. Humangenetik 11:35–54
5. Heddle JA (1973) A rapid in vivo test for chromosomal damage. Mutat Res 18:187–190
6. Countryman PI, Heddle JA (1976) The production of micronuclei from chromosome aberrations in irradiated cultures of human lymphocytes. Mutat Res 41:321–332
7. Fenech J, Morley AA (1985) Measurement of micronuclei in lymphocytes. Mutat Res 147: 29–36
8. Norppa H, Falck GC (2003) What do human micronuclei contain? Mutagenesis 18:221–233
9. Hovhannisyan GG (2010) Fluorescence in situ hybridization in combination with the comet assay and micronucleus test in genetic toxicology. Mol Cytogenet 3:17
10. Becker P, Scherthan H, Zankl H (1990) Use of a centromere-specific DNA probe (p82H) in nonisotopic in situ hybridization for classification of micronuclei. Genes Chromosomes Cancer 2:59–62
11. Attia SM (2009) Use of centromeric and telomeric DNA probes in situ hybridization for differentiation of micronuclei induced by lomefloxacin. Environ Mol Mutagen 50: 394–403
12. Hovhannisyan G, Aroutiounian R, Liehr T (2012) Chromosomal composition of micronuclei in human leukocytes exposed to mitomycin C. J Histochem Cytochem 60:316–322
13. Leach NT, Jackson-Cook C (2001) The application of spectral karyotyping (SKY) and fluorescent in situ hybridization (FISH) technology to determine the chromosomal content(s) of micronuclei. Mutat Res 495: 11–19
14. Balajee AS, Bertucci A, Taveras M et al (2014) Multicolour FISH analysis of ionising radiation induced micronucleus formation in human lymphocytes. Mutagenesis 29:447–455

Three-Dimensional Interphase Analysis Enabled by Suspension FISH

Thomas Liehr and Nadezda Kosyakova

Abstract

Here an established method is described for performing the entire FISH procedure in suspension instead of on a slide, as usual. This so-called suspension FISH (S-FISH) opened new possibilities for the analysis of shape internal structure of the human interphase nucleus in different tissues starting. The procedure is described and some applications of this approach are presented.

Keywords Suspension FISH (S-FISH), Nuclear architecture, Three-dimensional structure, Interphase nucleus, Clinical questions, Evolution genetics

1 Introduction

FISH on human meta- and interphase chromosomes (chapter by Thomas Liehr et al. "The Standard FISH Procedure") is a well-established technique in clinical (chapter by Anja Weise and Thomas Liehr "Pre- and Postnatal Diagnostics and Research on Peripheral Blood, Bone Marrow, Chorion, Amniocytes, and Fibroblasts"; chapter by Thomas Liehr and Nadezda Kosyakova "Characterization of Mosaicism in Different Easy-to-Acquire Body Tissues such as Buccal Smears, Skin Abrasions, Hair Root Cells or Urine") and tumor cytogenetics (chapter by Eyad Alhourani et al. "Tumor Cytogenetic Diagnostics and Research on Blood and Bone Marrow Smears or Effusions"; chapter by Thomas Liehr "Characterization of Archived Formalin-Fixed/Paraffin-Embed ded or Cryofixed Tissue, Including Nucleus Extraction") as well as in studies of evolutional (chapter by Fengtang Yang et al. "Ani mal Probes and ZOO-FISH") and interphase architecture [1]. However, almost all such FISH studies are based on the air-drying procedure of chromosome preparation; in other words, after hypotonic treatment and fixing the cells in methanol/acetic acid, they are spread on the slide surface and air-dried (chapter by Anja Weise

Thomas Liehr (ed.), *Fluorescence In Situ Hybridization (FISH)*, Springer Protocols Handbooks,
DOI 10.1007/978-3-662-52959-1_41, © Springer-Verlag Berlin Heidelberg 2017

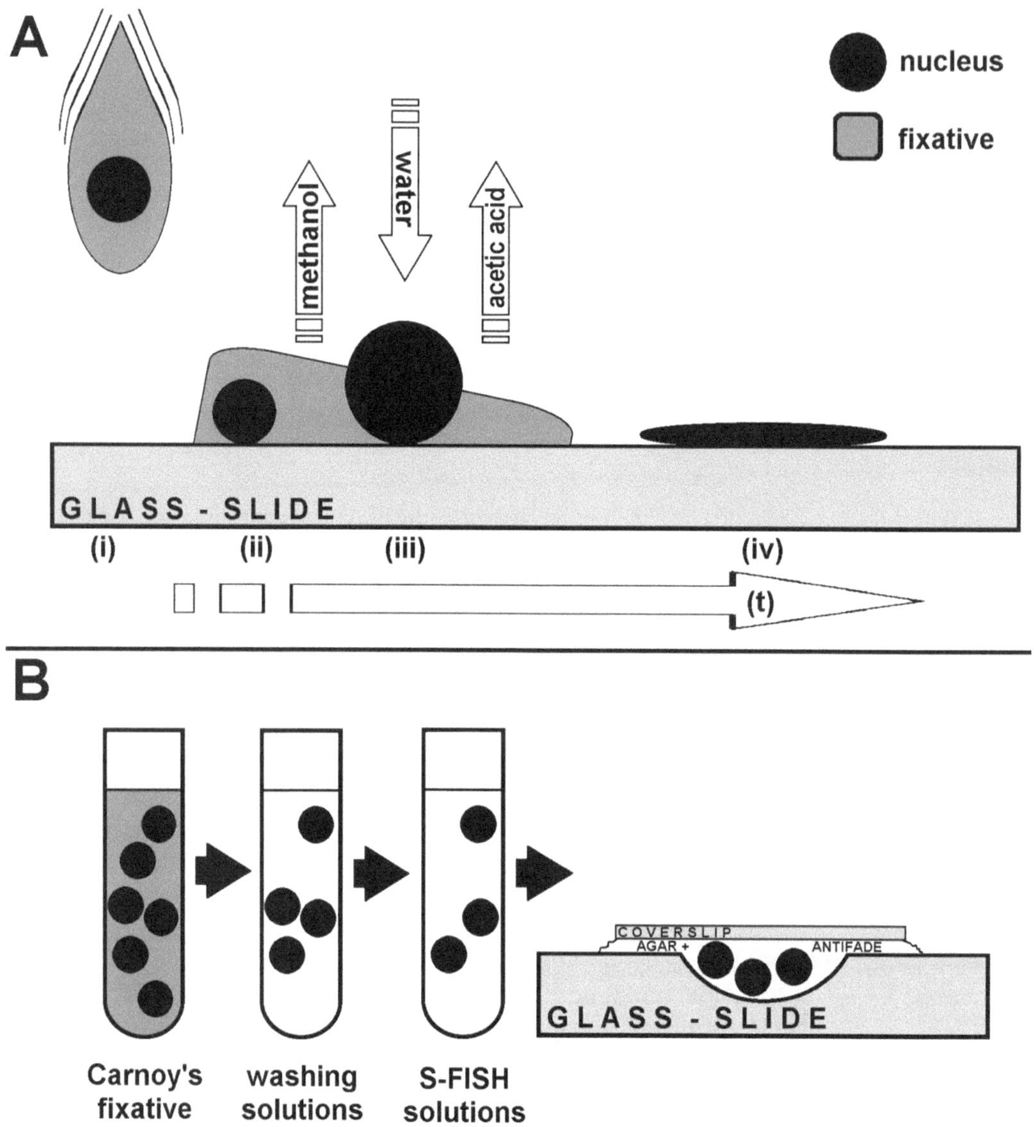

Fig. 1 (**a**) Schematic drawing of what happens to the interphase nuclei (*black symbols*) in cell suspension (*dark gray*) during the air-drying method. After dropping it onto the glass slide (i), the nucleus is attached to the slide surface (ii). The methanol then evaporates, water is acquired from the air due to the hydrophilic nature of the remaining acetic acid, and the nucleus swells (iii). Finally, the acetic acid evaporates and the nucleus flattens to a pancake-like structure (iv). Note that the nucleus is much bigger than it was originally! (**b**) S-FISH avoids this flattening and artificial swelling of the nuclei. The whole procedure is performed in suspension. The details of the protocol are described in the text

and Thomas Liehr "Pre- and Postnatal Diagnostics and Research on Peripheral Blood, Bone Marrow, Chorion, Amniocytes, and Fibroblasts"). This procedure leads to well-spread metaphases on the slide surface if the air is humid enough and to flattening of the originally spherical interphase nuclei (Fig. 1a). This air-drying procedure is well suited to nearly all FISH approaches; however, when the interphase architecture is being studied [2–4], the flattening and swelling of the nuclei (Fig. 1a) may lead to questionable results.

Several years ago, we developed an approach where the whole FISH procedure is performed in cell suspension and the nuclei are finally placed on a polished concave slide in the final step of the procedure (Fig. 1b), just before evaluation using suspension FISH (S-FISH; Fig. 2). Using this approach, it is possible to perform three-dimensional (3D) analyses on totally spherical interphase nuclei [5–14] or even on three-dimensional metaphases [5]. Using S-FISH, it is possible to perform one- or multicolor FISH experiments. In summary, three-dimensional (3D) analysis of the interphase architecture can be performed using S-FISH in samples derived from human (normal or malignant tissue) or animals (Fig. 3); plant cells were not tested yet in S-FISH.

2 Materials

Apart from the standard cell biological and molecular cytogenetic equipment, including typical solutions (e.g., methanol, formaldehyde, etc.), the following more specialized items are needed. The equipment and chemicals needed for FISH and multicolor FISH are listed in chapter by Thomas Liehr et al. "The Standard FISH Procedure" and chapter by Thomas Liehr and Nadezda Kosyakova "Multiplex FISH and Spectral Karyotyping".

2.1 Equipment

To evaluate the results, hardware and software that can perform three-dimensional image acquisition (create image stacks) and evaluation are required. A relatively affordable possibility is to use a fluorescence microscope with motorized X-/Y-/Z-axes and the Cell-P software (from Olympus).

2.2 Solutions to Be Prepared

As described in chapter by Thomas Liehr et al. "The Standard FISH Procedure".

Variations:

- DAPI solution: dissolve 2 μl of DAPI stock solution in 2 ml Vectashield antifade.
- 0.5 % DAPI–Vectashield gel: add 250 mg agarose to 25 ml 0.9 % NaCl. To make a 1 % agarose gel, incubate for 1 min at 600 W in a microwave. Add 2 ml of this suspension to 2 ml Vectashield antifade and mix on a shaker.
- Hybridization buffer: dissolve 2 g dextran sulfate in 10 ml 50 % deionized formamide/2x SSC/50 mM phosphate buffer for 3 h at 70 °C. Aliquot and store at −20 °C.

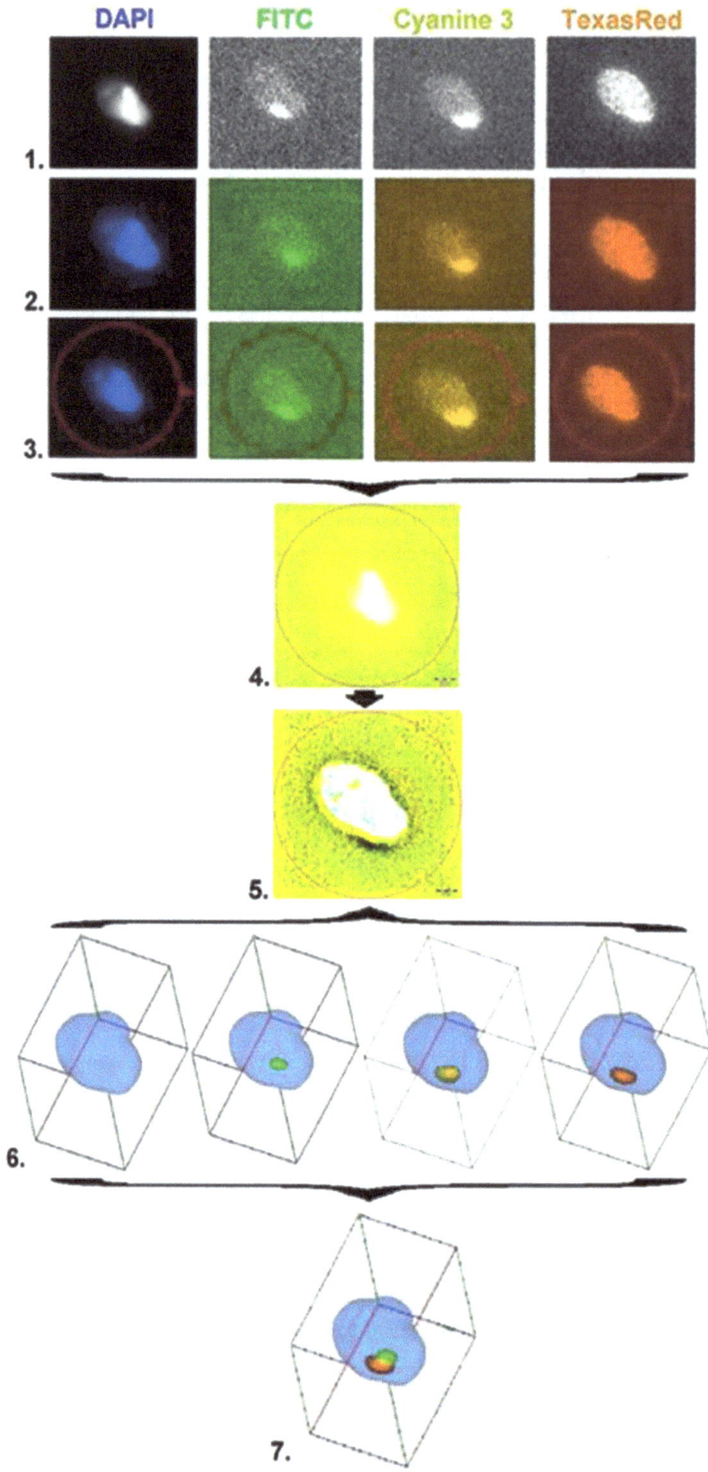

Fig. 2 Flowchart showing how 3D analysis of S-FISH results is achieved using the Cell-P software (from Olympus). The details of the analytical process are described in the text

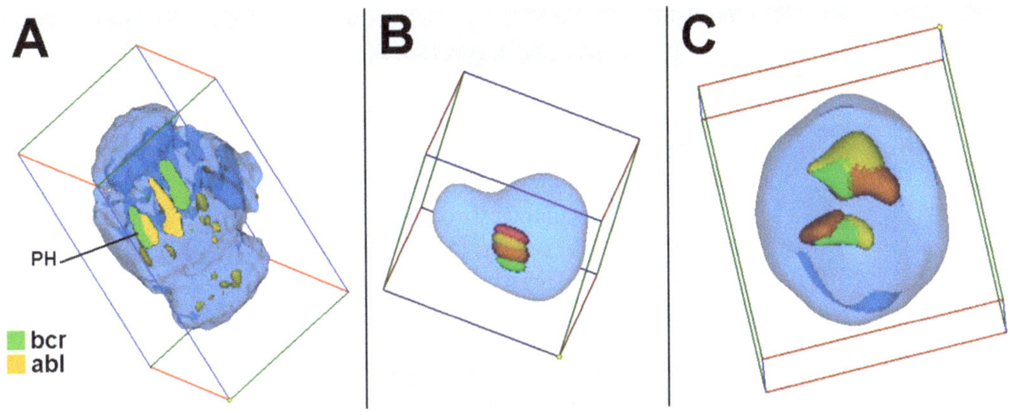

Fig. 3 (**a**) 3-D S-FISH results after applying commercially available probes (Abbott, Vysis) for the genes bcr (*green*) and abl (*yellow*) on a bone marrow nucleus taken from a patient with chronic myelogenous leukemia (CML). Besides two single signals for each gene region, a typical Philadelphia fusion signal translocation (PH) can be detected. Some *yellow* and *green* background noise is also visible (**b**) Multicolor banding (MCB) upon applying the probe set for human chromosome 10 in human sperm; for more details on MCB, see chapter by Thomas Liehr et al. "FISH Banding Techniques" (**c**) MCB probe set for human chromosome 22 after an S-FISH experiment in a B lymphocyte of *Gorilla gorilla* (for more on ZOO-FISH, see chapter by Fengtang Yang et al. "Animal Probes and ZOO-FISH")

- Pepsin solution: mix 950 µl distilled water with 50 µl 0.2 N HCl and place in a water bath at 37 °C. 10 min before application, add 5 µl of pepsin stock solution.

3 Methods

An S-FISH protocol using commercially available, directly labeled probes is described here (*see* **Note 1**).

3.1 Preparation of the Cell Pellet for S-FISH

1. Cytogenetic pellets of any tissue in Carnoy's fixative (chapter by Anja Weise and Thomas Liehr "Pre- and Postnatal Diagnostics and Research on Peripheral Blood, Bone Marrow, Chorion, Amniocytes, and Fibroblasts") can be used (*see* **Note 2**).
2. Pellet the cells by centrifugation for 10 min at 1,500 rpm and at 4 °C, and discard the supernatant carefully with a micropipette (*see* **Note 3**).
3. Resuspend in 500 µl methanol and incubate for 2 min and repeat step 2.
4. Wash for 3 min with 500 µl 0.9 % NaCl and repeat step 2.
5. Add 500 µl of pepsin solution and place at 37 °C for 5 min; then repeat step 2.

6. Add 500 μl 0.9 % NaCl solution, incubate for 1–2 min at room temperature (RT), and repeat step 2. About 50 μl of suspension should be left in the tube.

3.2 S-FISH Procedure

1. Dissolve a concentration three times that applied in a normal FISH experiment (or recommended by the provider) of commercially available, directly labeled probe in 25 μl hybridization buffer. Also use at least 5–50 μg of COT1 DNA to block undesired background (*see* **Note 4**).
2. Denature at 95 °C for 5 min and prehybridize at 37 °C for 30–60 min.
3. Denature 50 μl of the suspension from above at 95 °C for 5 min, pellet the cells by centrifugation for 10 min at 1500 rpm and at 4 °C, and then discard 20 μl of supernatant; finally, add the prehybridized probe from step 2 (*see* **Note 5**).
4. Incubate for ~12–16 h (overnight) at 37 °C.

3.3 S-FISH Postwashing

1. Perform the first postwashing step in suspension as follows: add 500 μl of 0.4 × SSC (68 °C) and incubate for 2 min at this temperature.
2. Pellet the cells by centrifugation for 10 min at 1,500 rpm and discard the supernatant carefully using a micropipette.
3. Perform the second postwashing step: add 500 μl 4x SSC at RT and incubate for 2 min (RT); repeat step 2.
4. Add 150 μl of DAPI solution (RT) and incubate for 10 min. Then add 500 μl 0.9 % NaCl (RT) and repeat step 2.
5. Resuspend in 50 μl 0.5 % DAPI–Vectashield gel (microwave beforehand to make it fluid) and transfer immediately onto a 15 μl well slide. Cover with a coverslip. After the gel has set, the slide is ready for microscopic inspection.

3.4 Three-Dimensional Analysis

As an example, the Cell-P software (from Olympus) can be used for three-dimensional analysis of the results. The application of the Cell-P software for 3D-FISH analysis is detailed here; see also Fig. 2:

1. Capture the figures for each color channel used (defined by the number of fluorochromes plus the counterstain), and acquire an image stack that shows the same specimen area at different focal planes.
2. As a black-and-white CCD camera is used, define the fluorescence pseudocolor for each channel to be used in further analysis.
3. Define the regions of interest.
4. Combine the image stack into a raw data image.

5. Use the 3D deconvolution feature of the Cell-P software.
6. Obtain the 3D image using the voxel viewer feature of the Cell-P software.
7. Combine all of the colors and images for interpretation into a 3D projection.

4 Notes

1. Indirectly labeled probes can also be used, although in our hands secondary detection has given a worse signal-to-background ratio than directly labeled probes. Thus, we normally try to avoid indirectly labeled probes.
2. In principle, a pellet of any tissue in Carnoy's fixative can be used for S-FISH experiments. However, for as-yet unknown reasons, multicolor FISH studies have sometimes given better results when chromosome/interphase preparations prepared without using colchicine are employed.
3. Due to the significant loss (30–70 %) of interphase cells during the preparation for and the procedure of S-FISH, make sure that the cell pellet is sufficient.
4. COT1 DNA should be used in excess. In order to avoid knocking the hybridization buffer out of balance, COT1 can be aliquoted into 0.5 or 1 ml reaction cups and lyophilized. The hybridization buffer with probe can be added to this lyophilized DNA.
5. FISH hybridization can also be improved by microwave treatment, as described in chapter by Anja Weise and Thomas Liehr "Microwave Treatment for Better FISH Results in a Shorter Time".

References

1. Liehr T (2016) Multicolor FISH homepage http://ssmc-tl.com/mfish.html
2. Chevret E, Volpi EV, Sheer D (2000) Mini review: form and function in the human interphase chromosome. Cytogenet Cell Genet 90:13–21
3. Cremer T, Cremer C (2001) Chromosome territories, nuclear architecture and gene regulation in mammalian cells. Nat Rev Genet 2:292–301
4. Fraser P, Bickmore W (2007) Nuclear organization of the genome and the potential for gene regulation. Nature 447:413–417
5. Steinhaeuser U, Starke H, Nietzel A et al (2002) Suspension (S)-FISH, a new technique for interphase nuclei. J Histochem Cytochem 50:1697–1698
6. Iourov IY, Liehr T, Vorsanova SG et al (2006) Visualization of interphase chromosomes in postmitotic cells of the human brain by multicolour banding (MCB). Chromosome Res 14:223–229
7. Iourov IY, Liehr T, Vorsanova SG et al (2007) Interphase chromosome-specific multicolor banding (ICS-MCB): a new tool for analysis of interphase chromosomes in their integrity. Biomol Eng 24:415–417
8. Manvelyan M, Hunstig F, Mrasek K et al (2008) Position of chromosomes 18, 19, 21 and 22 in 3D-preserved interphase nuclei of

human and gorilla and white hand gibbon. Mol Cytogenet 1:9

9. Manvelyan M, Hunstig F, Bhatt S et al (2008) Chromosome distribution in human sperm: a 3D multicolor banding-study. Mol Cytogenet 1:25
10. Manvelyan M, Kempf P, Weise A et al (2009) Preferred co-localization of chromosome 8 and 21 in myeloid bone marrow cells detected by three dimensional molecular cytogenetics. Int J Mol Med 24:335–341
11. Klein E, Manvelyan M, Simonyan I et al (2012) Centromeric association of small supernumerary marker chromosomes with their sister-chromosomes detected by three dimensional molecular cytogenetics. Mol Cytogenet 5:15
12. Othman MAK, Lier A, Junker S et al (2012) Does positioning of chromosomes 8 and 21 in interphase drive t(8;21) in acute myelogenous leukemia? BioDiscovery 4:4
13. Roediger J, Hessenkemper W, Bartsch S et al (2014) Supraphysiological androgen levels induce cellular senescence in human prostate cancer cells through the Src-Akt pathway. Mol Cancer 13:214
14. Karamysheva T, Kosyakova N, Guediche N et al (2015) Small supernumerary marker chromosomes and the nuclear architecture of sperm: a study in a fertile and an infertile brother. Syst Biol Reprod Med 61:32–36

Part VI

Applications of FISH in Zoology, Botany and Microbiology

Animal Probes and ZOO-FISH

Fengtang Yang and Alexander S. Graphodatsky

Abstract

The invention of cross-species chromosome painting (Zoo-FISH) represents the most significant technical breakthrough in animal cytogenetics after the introduction of chromosomal banding techniques in late 1960 and the early 1970s. This made it possible to compare the karyotypes of virtually any two vertebrate species that diverged up to 100 million years ago. With the availability of paint probes for more and more vertebrate species, Zoo-FISH has made a far-reaching impact on animal comparative cytogenetics, leading to the birth of the new cytogenetics—cytogenomics. Here we present two detailed protocols for cross-species chromosome painting.

Keywords Cross species, Chromosome painting, Zoo-FISH, Cytogenetics, Evolution, Animal probes, Mammalia, Insects, Fishes, Reptiles, Amphibia

1 Introduction

Zoo-FISH (fluorescence in situ hybridization) [1], also known as cross-species chromosome painting or comparative chromosome painting, essentially describes the use of whole chromosome- or chromosomal arm- or region-specific painting probes to delimit homologous segments (chromosome or chromosomal segments with evolutionarily conserved synteny) in other species by means of FISH. As most molecular cytogenetic approaches, the technique of chromosome painting was introduced firstly to human cytogenetics in 1988 [2]. Wienberg and colleagues [3] were among the pioneers who introduced chromosomal painting to the field of comparative cytogenetics of primates. They established the first genome-wide chromosome maps between human and a Japanese macaque. In the beginning, painting probes were derived from libraries of human chromosome-specific DNA clones. Limited by the availability of painting probes and technical difficulties in comparing distantly related species, early cross-species chromosome painting experiments used primarily apes and Old World monkeys [3, 4]. Scherthan and colleagues [1] were among the first who

Thomas Liehr (ed.), *Fluorescence In Situ Hybridization (FISH)*, Springer Protocols Handbooks, DOI 10.1007/978-3-662-52959-1_42, © Springer-Verlag Berlin Heidelberg 2017

demonstrated the feasibility of comparing such distantly related species as those from different orders. Meanwhile, the invention of degenerate oligonucleotide-primed PCR (DOP-PCR) ([5], chapter by Nadezda Kosyakova et al. "FISH-Microdissection"), coupled with chromosomal sorting by flow cytometry (chapter by Fengtang Yang et al. "Generation of Paint Probes from Flow-Sorted and Microdissected Chromosomes"), made it possible to generate painting probes for any given vertebrate species [6–8] and to carry out multidirectional cross-species chromosome painting [9]. The whole set of human chromosome-specific painting probes derived from DOP-PCR were made commercially available around 1992 (chapter by Thomas Liehr "Commercial FISH Probes"), and the whole set of chromosome painting probes for mouse became available in late 1995 [7]. Human chromosome-specific painting probes have been the main workhorses in the field of molecular comparative cytogenetics among eutherian mammals due to the wide availability of commercial paints. The ever-increasing availability of probes from more than 100 mammalian species covering the major branches of mammalian tree has revolutionized comparative cytogenetics. Notably, genome-wide comparative chromosome maps between humans and representative species of almost all 18 extant eutherian orders have been established [8]. Such a task is beyond the capacity of chromosomal banding-based cytogenetic comparison. Our laboratories have had the opportunity to being onboard the boat of "Zoo-FISHing" over the past 20 or so years. In comparison with protocols of such intraspecies chromosomal painting as human onto human, mouse onto mouse, etc., the three most critical modifications in Zoo-FISH protocols include (1) an increment in the concentration of probes and (2) in hybridization time and (3) improved accessibility of the target chromosomal DNA by enzymatic digestion of unwanted cytoplasm background cover, debris, and chromosomal proteins. Here we present two Zoo-FISH protocols that have been widely used in our laboratories and hope they will be useful to the community of animal comparative cytogenetics.

2 Materials

2.1 Simultaneous DAPI Banding and Multicolor Cross-Species Chromosome Painting

2.1.1 Equipment

- Modern epifluorescence microscopes with specific filters for DAPI, FITC, Cy3, and Cy5 fluorescence, CCD camera, and dedicated digital imaging software that allows digital enhancement of reversed DAPI-banding (e.g., SmartCapture system from Digital Scientific, Cambridge, UK; CytoVision from Leica Biosystems, Wetzlar, Germany; and In Situ Imaging System from Metasystems, Altlussheim, Germany)

- Refrigerated microcentrifuge (e.g., R5417R, Eppendorf, Hamburg, Germany)

2.1.2 Chemicals

- Animal painting probes derived from DOP-PCR amplification of flow-sorted chromosomes and microdissected chromosomes (chapter by Fengtang Yang et al. "Generation of Paint Probes from Flow-Sorted and Microdissected Chromosomes").
- Antibodies (Cy3-avidin or Cy3-streptavidin, Cy5-avidin or Cy5-streptavidin, monoclonal mouse anti-digoxigenin, FITC-conjugated goat anti-mouse IgG, rabbit anti-FITC IgG, FITC-conjugated goat anti-rabbit IgG, Sigma-Aldrich).
- Formamide (BDH, AnalaR, Prod 103264R, >99.5 % purity)
- 50 × Denhardt's solution (Sigma-Aldrich D2532, for molecular biology, liquid, a 1 % solution of BSA, Ficoll® and PVP)
- Dextran sulfate (Sigma-Aldrich D8906, average molecular weight 500,000)
- Pre-cleaned microscope slides (*see* **Note 1**): 76 × 26 mm microscopic slides with grounded edges twin frosted end (Menzel-Gläser, Braunschweig, Germany)
- Pepsin (Sigma-Aldrich, P6887)
- Coverslip sealant (Fixogum rubber cement, Marabu, Tamm, Germany, or CytoBond from SciGene, Sunnyvale, CA, USA)

2.1.3 Solutions to Be Prepared

- Denaturation solution [1] = 70 % formamide/2 × SSC (v/v): 70 ml formamide + 30 ml 2 × SSC. Store at 4 °C refrigerator after use, discard after 2 months.
- Ethanol series in Coplin jars (10-slide capacity) or Hellendahl jars (16-slide capacity) (70 %/90 % 90 %/100 %).
- Fixative: methanol/acetic acid (v/v) 3:1. Make fresh before use.
- Hybridization buffers (stored at −20 °C).

	Hyb-P (Hyb55)	Hyb-Z (Hyb50)
Formamide (deionized)	25 ml	25 ml
50 % Dextran sulfate	10 ml	10 ml
20 × SSC (filtered)	5 ml	5 ml
0.5 mM phosphate buffer pH7 (2.3 ml 0.5 M Na_2HPO_4 + 1.7 ml 0.5 NaH_2PO_4)	4 ml	4 ml

50 × Denhardt's solution (Sigma)	1 ml	1 ml
0.5 mM EDTA (optional)	50 μl	50 μl
dH_2O	–	5 ml

- 2 μg ml^{-1} hybridization blocking DNA in hybridization buffer: 1,000 μg of salmon/herring sperm DNA or Cot-1 DNA, precipitate down in ethanol and then resuspend in Hyb-Z (for more on ethanol precipitation, see "Ethanol Precipitation of DNA probes in Sect. 3.1.2).
- 1 % stock pepsin solution (store at −20 °C): 1 g pepsin (Sigma P6887) + 99 ml dH_2O. Aliquot into 1.5 ml Eppendorf tubes and store at −20 °C.
- 0.01 % working pepsin solution: 1 ml 1 % stock pepsin solution + 99 ml 10 mM HCl.
- Post-hybridization stringent washing solution **A** = 50 % formamide/50 % 2 × SSC (v/v): 50 ml formamide + 50 ml 2 × SSC.
- Post-hybridization stringent washing solution **B** = 50 % formamide/50 % 1 × SSC (v/v): 50 ml formamide + 25 ml 2 × SSC + 25 ml dH_2O. Keep at 4 °C refrigerator after use, discard after 2 months.
- UltraPure 20 × SSC buffer (Invitrogen, Carlsbad, CA, USA).
- 2 × SSC: 100 ml 20 × SSC + 900 ml dH_2O.
- 1 × SSC: 50 ml 20 × SSC + 950 ml dH_2O.
- 4 × SSCT solution: 200 ml 20 × SSC + 800 ml dH_2O + 500 μl Tween 20.

2.2 Sequential G-Banding and Zoo-FISH

2.2.1 Equipment

- Humid (moisture) chamber: Petri dish with filter paper at the bottom damped with a small amount of 2 × SSC.

2.2.2 Chemicals

- Formamide: for best results use only freshly deionized formamide. Otherwise follow deionization protocols described. For each 100 ml of formamide to be deionized, add 5 g of resin (MB-1 or MB-150, Sigma-Aldrich). Stir for 1 h using a magnetic stirrer. Then filter using a Buchner funnel, and store at −20 °C freezer.
- Trypsin: 0.25 % sterile trypsin solution.

2.2.3 Solutions to Be Prepared

- RNase A solution: 2 × SSC with 0.1 mg ml^{-1} RNase A (Sigma, DNase free, to remove DNase warm the stock solution (10 mg ml^{-1}) for 10 min at 96 °C).

- FITC-avidin solution: make a 1:500 dilution of FITC-avidin stock (1 mg ml^{-1}) in blocking solution. Spin in a microcentrifuge at 12,000× *g* for 5 min; transfer the supernatant into a fresh tube.
- Biotinylated anti-avidin solution: make a 1:200 dilution of biotinylated goat anti-avidin IgG stock (0.5 mg ml^{-1}) with blocking solution. Centrifuge the solution for 5 min. Both FITC-avidin and biotinylated anti-avidin solutions can be kept in a refrigerator in the dark for no more than 1 day.
- Blocking solution: 3 % of dry milk (fat-free) or of blocking reagent dissolved in 4 × SSCT. Dissolve the milk using shaker. Centrifuge the solution for 5 min and take the solution without pellet and pellicle and put in another tube.
- DABCO solution: 0.233 g of 1,4-diazobicyclo[2,2]octane dissolve in 10 ml of solution containing 90 % glycerol, 100 mM Tris-HCl pH 8.0.
- Denaturation solution: 70 % formamide in 2 × SSC, pH 7.0. For 100 ml combine: 10 ml 20 × SSC (pH = 7.0) + 70 ml formamide + 20 ml dH_2O.
- Stock Giemsa solution (2 g of Giemsa-dye (lyophilized, MERCK) dissolved in mixture (250 ml of glycerol, Serva, analytical grade and 250 ml of methanol and remove unresolved particles by filtering).
- Working Giemsa solution: take 2 ml of stock solution; add distilled water up to 50 ml and 1 ml of 0.1 % sodium carbonate in H_2O.
- Hybridization buffer: combine 400 μl of 100 % formamide, 200 μl of 50 % dextran sulfate, 100 μl of 20 × SSC, and 50 μl of deionized water. Note the concentration of formamide can vary from 40 % for distantly related species to 50 % for closely related species.

3 Methods

3.1 Simultaneous DAPI Banding and Multicolor Cross-Species Chromosome Painting

For achieving high-quality metaphase spreads, see also **Note 2**

3.1.1 Slide Preparation

Metaphase Spreads

Scheme 1

1. Take out the metaphase preparation from −20 °C freezer and place on ice for 10 min (*see* **Note 3**).

2. Thoroughly resuspend the cells by flicking the tube several times or by gentle pipetting with a pasture pipette.
3. Prepare a test slide by applying 10 μl of metaphase suspension onto a dry and clean slide (*see* **Note 1**), and allow the slide to dry in the air.
4. Evaluate the quality of metaphase chromosome spreading on the slide under a phase-contrast microscope using a 20× objective (for spreading, cytoplasm, cell density, etc.). If underspreading occurs, add a drop of 3:1 or 2:1 fixative immediately after placing the suspension on the slide. Avoid using over-concentrated samples; dilute the suspension with cold (3:1) fixative if necessary. If the cells are too sparse, spin down the cells and resuspend in a smaller volume of freshly made fixative. The temperature and humidity of the laboratory are critical for controlling the spreading of metaphase chromosomes. Ideally drop the slides inside a hood with both controlled temperature (20 °C) and humidity (50–55 %).

Scheme 2

Alternatively, controlled spreading can be achieved using a tray floating inside a water bath with covering lid (for more information see [10]).

1. Place the slide inside a 1–2 cm (height) × 10 cm (width) × 20 cm (length) metal tray that can float in the water bath (for instance, an aluminum tray for making ice cubes in domestic freezer or the lid of a biscuit tin).
2. Apply 10 μl of metaphase preparation to the middle of the slide using a P20 Gilson pipette.
3. Immediately float the metal tray with slide inside in a 50 °C circulating water bath (such as Grant W14, Grant Instruments, Royston, UK). Close the lid for at least 1 min (until the spreading process is complete and fixative has fully evaporated from the slide surface).
4. Evaluate the quality of metaphase spread under a phase-contrast microscope (as detailed in Sect. 3.1.1).
5. After determining optimized conditions for spreading the metaphase chromosomes, prepare a large number of slides by either placing one drop of metaphase preparation in fixative in the middle or two drops side by side on the same slide. The latter can be hybridized under two separate probes.
6. Check the spreading of metaphase chromosomes under phase-contrast microscope. The chromosomes should appear dark gray and non-reflective.
7. Label the slides using an HB pencil on the frosted ends (sample ID and position of metaphase spreads, date, etc.).

3.1.2 Pretreatment of Slide Specimen with Pepsin

This is a critical step (*see* **Note 4**).

1. Immerse the slides in a 0.01 % pepsin solution for 3–5 min at RT or 37 °C (*see* **Notes 5–6**).
2. Rinse twice in 2 × SSC (3 min each) to stop the pepsin reaction.
3. Dehydrate the slides in an ethanol series (70 %, 70 %, 90 %, 90 %, 100 %, 2 min each), and air-dry the slides by placing the slides in a rack or tilt them against a vertical surface (e.g., the Coplin jars used for dehydration).
4. Bake the slides in a 65 °C oven for at least 1 h if you want to set up the hybridization on the same day. Alternatively, the dehydrated slides can be left on the bench to age overnight to further harden the chromosomes.

3.1.3 Preparation of Paint Probes

For *single-color FISH*, on a 22 × 32 mm^2 hybridization area, the probe mixture is made by combining the following (see Table 1 for details):

- 1–3 μl of probe (i.e., labeled DOP-PCR products)
- 2 μl of blocking DNA in hybridization buffer
- 10 μl of Hyb-Z/or Hyb-P

For multicolor chromosome painting, if more than 3 μl of labeled DNA probes are required, the probes should be precipitated down in the presence of carrier (e.g., salmon sperm DNA) or

Table 1
Probe mixtures for cross-species chromosome painting in mammals

Category of Zoo-FISH		Cross order	Cross families	Cross genera	Cross species
Probe mixtures	Probes (labeled DOP-PCR products)	3 μl (200–300 ng)	2 μl (150–200 ng)	1–2 μl (70–200 ng)	0.5–1 μl 35–100 ng)
	Blocking DNA in hybridization buffer	2 μl (4 μg)	2 μl (4 μg)	2 μl (4 μg)	2 μl (4 μg)
	Hybridization buffer	10 μl Hyb-Z	10 μl Hyb-Z	10 μl Hyb-P	10 μl Hyb-P
Incubation time at 37°		63–68 h (3 days)	39–44 h (2 days)	16 h (overnight)	16 h (overnight)
Temperature of post-hybridization wash		39–42 °C	42 °C	42–45 °C	42–45 °C
Stringent washing solution		50 % formamide/ 2 × SSC (v/v)	50 % formamide/ 2 × SSC (v/v)	50 % formamide/ 1 × SSC (v/v)	50 % formamide/ 1 × SSC (v/v)

blocking DNA in ethanol and resuspended in hybridization buffer (Hyb-Z) in order to maintain the stringency of the hybridization mixture and to reduce the nonspecific binding of DNA probes (*see* **Note 7**).

Ethanol Precipitation of DNA Probes

1. Determine the volume of probes needed, add to a 1.5 ml Eppendorf tube, and precipitate the labeled DNA by adding the following:
   (a) 1/10 volume of 3 M NaAc, pH 5.2
   (b) 2.0–2.5 volume of ice-cold absolute ethanol
2. Mix thoroughly by vortexing and incubate at −20 °C for 2 h or −70 °C for 30 min.
3. Spin in a precooled microcentrifuge (e.g., Eppendorf R5417R) at 13,000× *g* at 4 °C for 25 min.
4. Discard supernatant and invert the tube on a paper towel to drain for a few seconds.
5. Add 100 μl of ice-cold 80 % ethanol and centrifuge for 5 min at 13,000× *g*.
6. Remove the supernatant using a P1000 pipette.
7. Respin at 13,000 g for 1 min. Remove remaining supernatant with P100 and P10 pipette tips but avoid touching the pellets.
8. Dry the pellets at 37 °C for 5 min or by using a vacuum spin dryer.
9. Resuspend pellets directly in hybridization buffer (usually the same volume as the starting DNA solution by vigorous vortexing or pipetting.
10. Spin the tube briefly to collect the probe mixture into the bottom of tubes and then incubate at 65–72 °C for 10 min to ensure the pelleted probes are fully resuspended (*see* **Note 8**).

3.1.4 Denaturation of Probes

1. Denature the probe mix in a 65–75 °C water bath or heating block or PCR machine for 10 min and then incubate at 37 °C for 30–60 min, until the denatured metaphase slides are ready to use (*see* **Note 9**).

3.1.5 Denaturation of Metaphase Chromosome Spreads on Microscopic Slides

1. Prepare a Coplin jar of denaturation solution that consists of 70 % formamide and 30 % 2 in 2 × SSC (*see* **Note 10**), and prewarm the solution to 67 °C. It may require up to 30 min for the denaturation solution to reach the required temperature if the denaturing solution was stored at 4 °C. Start the denaturation of slide specimens as soon as the denaturation of the probes is complete, but make sure that the temperature of the denaturation solution has reached the designated temperature.

2. Immerse ten slides (with two slides back to back) into 70 % formamide for 1.5–2 min to denature the chromosomal target DNA on the slides. This is the most critical step; start the timer as soon as the first pair of slides is immersed into the formamide solution.
3. After 1.5–2 min, transfer the denatured slides into a Coplin jar containing ice-cold 70 % ethanol, following the same order that each slide pair was immersed into the denaturation solution.
4. Dehydrate the slides through an ethanol series as described above, and then dry the slides in air (*see* **Note 11**).

3.1.6 Applying the Denatured Probes onto the Denatured Slide

1. Add 12 μl of pre-annealed paint probe mixture to the middle of each hybridization area and immediately cover with a clean 22 × 32 mm glass coverslip, avoiding the creation of air bubbles. Gently squeeze out any air bubbles by tapping the coverslips on top of the bubbles with a forceps with fine tips or a dissection needle (*see* **Note 12**). Before adding the probes, check for glass particles or dust on the surface of denatured slides and coverslip. Otherwise, you may find the coverslips fail to lay down flat.
2. Seal the edges of the coverslips with coverslip sealant.
3. Placed the sealed slides in a humidity box and incubate at 37 °C for 16–68 h, depending on the divergence time of species being compared (for general guidance see Table 1). For chromosome painting between closely related species, an overnight incubation is more than enough. However, for distantly related species such as species from different families and orders, you will need to increase the time to 2–3 days.

3.1.7 Post-hybridization Washing

1. Prepare 3 Coplin jars of 2 × SSC (50 ml each) and 2 Coplin jars of 50 % formamide/50 % 1–2 × SSC, prewarmed to 42–45 °C, and 3 Coplin jars of 4 × SSCT, prewarmed to 37 °C.
2. Remove the slides from the humidity box and remove the coverslip sealant with forceps but avoid lifting off the coverslips.
3. Remove the coverslips by soaking the slides in the first Coplin jar, containing 2 × SSC, for 5–10 min.
4. Transfer the slides into the first jar containing 50 % formamide/50 % 1–2 × SSC and incubate for 5 min.
5. Transfer the slides into the second jar containing 50 % formamide/50 % 1–2 × SSC and incubate for 5 min.
6. Transfer the slides into the second jar containing 2 × SSC and incubate for 5 min.
7. Transfer the slides into the third jar containing 2 × SSC and incubate for 5 min.

8. If probes directly labeled with fluorochromes are used, proceed with counterstaining and mounting (see below).
9. If indirectly labeled probes are used, transfer the slides into a Coplin jar containing 4 × SSCT.

3.1.8 Fluorescence Detection

Detecting Biotin-Labeled Probes with Cy3 or Cy5

1. Dilute the Cy3- or Cy5-avidin (or streptavidin, Amersham) with 4 × SSCT to a final concentration of 1 mg ml^{-1} (1:1,000 dilution) just before use. Prepare 200 μl for each slide.
2. Add 200 μl of diluted Cy3 or Cy5-avidin onto each slide and cover with 24 × 50 mm parafilm, and then incubate at 37 °C for 15–20 min.
3. Remove parafilm coverslip and wash the slides three times for 5 min each time in 4 × SSCT at 37 °C.
4. Transfer the slides into one Coplin jar of 2 × SSC and proceed with counterstaining and mounting (*see* **Note 13**).

Further Amplification of Hybridization Signals from FITC-Labeled Probes

The signals from FITC-labeled probes are generally too weak to be visualized properly for cross-species applications and thus may require further amplification of signals by two layers of antibodies (*see* **Note 14**).

1. Dilute the rabbit-anti FITC IgG antibody (1:200, molecular probes/Invitrogen) with 4 × SSCT; dilute goat anti-rabbit (Vector Labs) IgG antibody (1:200) with 4 × SSCT, 200 μl per slide.
2. Add 200 μl diluted rabbit-anti FITC IgG antibody onto each slide and cover with 24 × 50 mm parafilm, and then incubate at 37 °C for 15–20 min.
3. Remove parafilm coverslip and wash the slides three times for 5 min each time in 4 × SSCT at 37 °C.
4. Add 200 μl diluted goat anti-rabbit IgG antibody onto each slide and cover with 24 × 50 mm parafilm, and then incubate at 37 °C for 15–20 min.
5. Remove parafilm coverslip and wash the slides three times for 5 min each time in 4 × SSCT at 37 °C.
6. Transfer the slides into a Coplin jar of 2 × SSC and proceed with counterstaining and mounting.

Detecting Digoxigenin-Labeled Probes with FITC

1. Make a 1:500 dilution of mouse anti-digoxigenin monoclonal antibody (Sigma-Aldrich D8156) and a 1:200 dilution of FITC-conjugated goat anti-mouse IgG antibody (goat anti-mouse FITC conjugate (Sigma-Aldrich, F0257)) with 4 × SSCT just before use (*see* **Note 15**).

2. Apply 200 μl of diluted mouse anti-digoxigenin monoclonal antibody onto each slide and cover with 24 × 50 mm parafilm, and then incubate at 37 °C for 15–20 min.
3. Proceed to step 4 from point 3.1.7.2.

Two-Color Detection: Detection of Biotin-Labeled Probes with Avidin-Cy3 and FITC-Labeled Probes with FITC

1. Prepare detection solution:
   (a) Layer 1: Make a 1:200 dilution of rabbit anti-FITC IgG with 4 × SSCT, 200 μl for each slides.
   (b) Layer 2: Make a 1:200 dilution of FITC-conjugated goat anti-rabbit IgG and a 1:1000 dilution of avidin-Cy3 with 4 × SSCT, 200 μl for each slide.
2. Apply 200 μl Layer 1 detection solution to each slide and cover with 24 × 50 mm parafilm, and then incubate at 37 °C for 15–20 min.
3. Remove parafilm coverslip and wash the slides three times for 5 min each time in 4 × SSCT at 37 °C.
4. Apply 200 μl Layer 2 detection solution to each slide and cover with 24 × 50 mm parafilm, and then incubate at 37 °C for 15–20 min.
5. Remove parafilm coverslip and wash the slides three times for 5 min each time in 4 × SSCT at 37 °C.
6. Transfer the slides into one Coplin jar of 2 × SSC and proceed with counterstaining and slide mounting.

Two-Color Detection: Detection of Biotin-Labeled Probes with Avidin-Cy3 and Digoxigenin-Labeled Probes with FITC

1. Preparation of detecting solution:
   (a) Layer 1: make a 1:200 dilution of mouse monoclonal anti-digoxigenin antibody with 4 × SSCT, 200 μl for each slide.
   (b) Layer 2: make a 1:200 dilution of FITC-conjugated goat anti-mouse IgG and a 1:1000 dilution of avidin-Cy3 with 4 × SSCT, 200 μl for each slide.
2. Apply 200 μl Layer 1 detecting solution to each slide and cover with 24 × 50 mm parafilm, and then incubate at 37 °C for 15–20 min.
3. Remove parafilm coverslips and wash the slides three times for 5 min each time in 4 × SSCT at 37 °C.
4. Apply 200 μl Layer 2 detection solution to each slide and cover with 24 × 50 mm parafilm, and then incubate at 37 °C for 15–20 min.
5. Remove parafilm coverslip and wash the slides three times for 5 min each time in 4 × SSCT at 37°.
6. Transfer the slides into a Coplin jar of 2 × SSC and proceed with counterstaining and slide mounting.

Three-to-Seven-Color Detection of Probes Combinatorially Labeled with FITC, Biotin, and Dig-dUTPs

1. Preparation of detection solution:
   (a) Layer 1: make a 1:200 dilution of rabbit anti-FITC IgG with 4 × SSCT, 200 μl for each slide.
   (b) Layer 2: make a 1:200 dilution of FITC-conjugated goat anti-rabbit IgG and a 1:500 dilution of avidin-Cy5 with 4 × SSCT, 200 μl for each slide.
2. Apply 200 μl Layer 1 detection solution to each slide and cover with 24 × 50 mm parafilm, and then incubate at 37 °C for 15–20 min.
3. Remove parafilm coverslip and wash the slides three times for 5 min each time in 4 × SSCT at 37 °C.
4. Apply 200 μl Layer 2 detection solution to each slide and cover with 24 × 50 mm parafilm, and then incubate at 37 °C for 15–20 min.
5. Remove parafilm coverslip and wash the slides three times for 5 min each time in 4 × SSCT at 37 °C.
6. Transfer the slides into a Coplin jar of 2 × SSC and proceed with counterstaining and mounting.

3.1.9 Counterstaining with DAPI and Mounting the Slides with Antifade Solution

1. Remove slides from 2 × SSC solution, dry the underneath of the slide (i.e., the side without samples) using lint-free tissue, and allow slides to drain on paper towels for a few seconds (*see* **Note 16**).
2. Place approximately 30 μl of mounting medium (*see* **Note 17**) with DAPI along the middle of the hybridization area and cover with a 22 × 50 mm clean glass coverslip. Remove big bubbles by gentle tapping with forceps.
3. Seal the edges of coverslips with nail vanish and store the slides in 40 °C in the dark till the slides are ready for examination using a fluorescence microscope with proper filters.

3.2 Sequential G-Banding and Zoo-FISH

As an alternative to the simultaneous DAPI-banding and multicolor FISH protocol, we also include in this chapter a general Zoo-FISH protocol based on sequential G-banding and chromosome painting (*see* **Note 18**) (Fig. 1).

3.2.1 Slide Preparation

1. To obtain high quality of slide preparation, use high-quality glass microscopic slides, such as the 26 × 76 mm, double-frosted slides with a thickness of 0.8–1 mm and ground edges. The slides are then acid-washed, cleaned, and polished as below.
2. Place the slide into an acid cleaning solution (concentrated sulfuric acid saturated by salts of potassium dichromate) for 20 min (*see* **Note 19**).
3. Rinse thoroughly with a large quantity of running tap water then with distilled water; store the cleaned slides in distilled water in a 4 °C refrigerator.

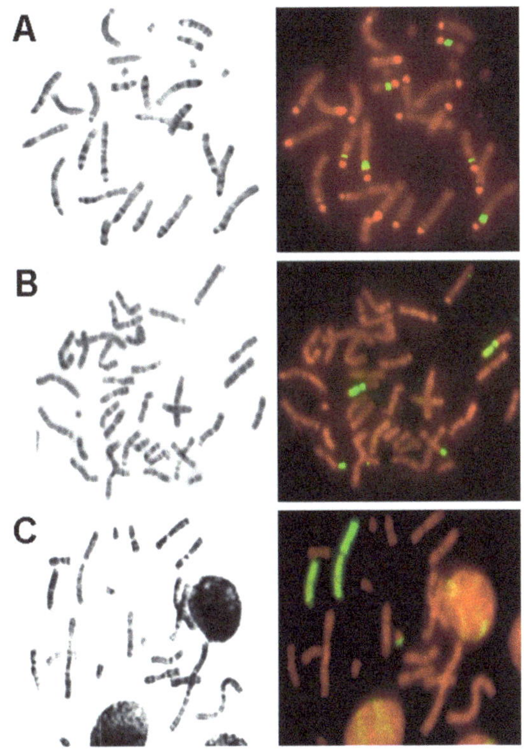

Fig. 1 Examples of chromosome painting onto G-banded metaphases. (**a**) Paint specific for Syrian hamster (*Mesocricetus auratus)* chromosome 9 hybridized onto rat-like hamster *(Tscherskia triton)* chromosomes. (**b**) Paint specific for red fox (*Vulpes vulpes*) chromosome 16 hybridized onto Corsac fox (*Vulpes corsac*) chromosomes. (**c**) Paint specific for Iberian shrew (*Sorex granarius)* X chromosome hybridized onto common shrew (*Sorex araneus)* chromosomes

4. Before spreading the metaphase chromosomes, change the fixative in the metaphase suspension by centrifuging for 5 min at 400× *g*, and replace the old fixative with a freshly made fixative (3 volumes of absolute methanol : 1 volume of glacial acetic acid). This suspension can last for 1 month without the need to change the fix again if it is stored in a −20 °C freezer immediately after use. About slide preparation see also [11] and point 3.1.

Air-Dried Slide Preparation

1. Place one drop of ice-cold suspensions (10 μl) onto a wet cold slide and leave the slide to dry. Dry slides may also be used (dip the wet slide into alcohol and then dry slides in the air).
2. If the spreading of metaphase chromosomes is insufficient, add 10 μl of fresh fixative before the fixative has completely evaporated (i.e., when grainy spots start appearing).

3. If the addition of extra fixative remains insufficient to achieve good spreading of chromosomes, after dropping suspension, hold the slide in a 70 °C water bath (at a distance of about 10–20 cm from the water) until complete evaporation of fixative.
4. If the quality of spreading remains poor, change the proportion of fixative in suspension (as described in Sect. 3.1.1). For example, for optimal spreading, try using methanol: acetic acid at a ratio of 1:2.
5. View the slide under a phase-contrast microscope to evaluate the mitotic index, metaphase quality, and chromosome morphology. For high-quality slide preparation, all chromosomes should be separated from each other and with a very thin layer of cytoplasm around chromosomes. The chromosomes should have a gray-colored appearance.

Flame-Dried Slide Preparation

1. Drop 10 μl of ice-cold metaphase suspension onto a wet cold slide. Immediately pass the slide through a flame. Do not keep the slide in the flame for more than 1 s.

3.2.2 Storage of Slide Preparations

- *Air-dried slide preparation for G-banding*: air-dried slide preparations are stored in a vacuum chamber for 3 days before being subject to G-banding treatment.
- *Flame-dried slide preparation for G-banding:* prepare flame-dried slide preparations, and store them in vacuum chamber for 4–5 days or in a slide box at room temperature for 5–7 days.

Before setting up the hybridization, metaphase spreads are treated with trypsin (see above). We routinely perform all modes of banding using slide specimens prepared from metaphase suspensions of fibroblast, peripheral blood, and bone marrow cultures. Large chromosomes can sometimes be identified based on un-enhanced DAPI-banding patterns, but for short chromosomes, especially those from bone marrow samples, G-banding patterns are required.

3.2.3 G-Banding

For G-banding we use air-dried slides (1–5 days old) and flame-dried slides (4–7 days old).

1. Immerse the slide in trypsin solution (RT) for approximately 2 min (*see* **Note 20**).
2. Transfer the slides to a 2 × SSC solution to stop the reaction of trypsin.
3. Immediately transfer the slides (without drying) into a Coplin jar containing Giemsa working solution and stain for 1–5 min.
4. Rinse slide with distilled water.
5. Dry slide by blowing off the surface water using a rubber bulb.

6. Check the quality of staining using microscope.
7. Perform microscopic examination and capture images of G-banded metaphases. After capturing images, record the coordinates of each metaphase on the microscope stage.

3.2.4 Fixation After Trypsin Treatment

1. After the microscopic examination and image recording of the G-banded slides, remove the immersion oil from the slide by washing the slides at RT in xylene (*see* **Note 21**) twice for 5 min each.
2. Dry the slides with a rubber bulb or by placing the slide vertically against a Coplin jar or on a slide tray.
3. Destain and remove any residual xylene by washing twice (RT) for 5 min each in fixative (3:1 methanol/acetic acid).
4. Dry slides on the table in vertical position in a slide tray with the frosted edge facing down.
5. Incubate the slides in 2 × SSC (RT) for 3 min, and then transfer into PBS, 50 mM $MgCl_2$ (RT) for 3 min.
6. Fix slides in 0.5 % formaldehyde in PBS, 50 mM $MgCl_2$ (RT) for 10 min (*see* **Note 22**).
7. Rinse slides in PBS and wash in 2 × SSC for 3 min.
8. Dehydrate slides through an ethanol series (70 %, 70 %, 90 %, 90 % and 100 %) for 2 min each.
9. Dry slides at RT.

3.2.5 RNase Treatment (Optional)

1. Put the slide in 0.01 % RNase A solution in 2 × SSC for 40 min at 37 °C.
2. Rinse the slide in 2 × SSC for 5 min.
3. Pass through an ethanol series (RT) for 3 min each (70 %, 80 %, 96 %).
4. Dry slides at RT.

3.2.6 FISH

Chromosome Denaturation

1. Immerse the slides in denaturing solution (70 % formamide in 2 × SSC, pH 7.0) preheated to 70 °C for 1–2 min. Ensure that the 70 % formamide has reached temperature before use (*see* **Note 23**).
2. Immediately quench the denatured slides in ice-cold 70 % ethanol for 2 min.
3. Dehydrate the slides through an ice-cold 90 % and 96 % ethanol series for 2 min each. Dry the slide at RT or using a rubber bulb.

Probe Preparation

1. For each hybridization area, combine 0.1–0.5 μg of biotin-labeled PCR product and 7–20 μg of Cot 1 DNA.
2. Add NaCl up to final concentration 0.5 M and mix by vortexing.

3. Add 2.5 volume of 96 or 100 % ethanol.
4. Leave the DNA to precipitate for at least 2 h at −20 °C.
5. Centrifuge the samples for 25 min at 12,000× *g* (Eppendorf centrifuge), remove the supernatant, and then dry at 37 °C for 30 min (do not over dry) or using a spin-vacuum drier.
6. Thoroughly resuspend the probes in 5 μl of TE containing 2 % Tween 20 by vigorous pipetting, and then spin briefly to collect the probe mixture into the bottom of the tubes. Heat denatures the probes at 96 °C for 5 min.
7. Combine 5 μl of biotin-labeled chromosome-specific paints with 15 μl of hybridization buffer, to the final concentration of formamide 40 %, dextran sulfate 10 % and 2 × SSC.

Probe Denaturation and Hybridization Incubation

1. Denature the probe mixture in either a heating bock or thermocycler (PCR machine) for 3 min at 96 °C. For in situ suppression of repetitive sequences in the probes, leave the denatured probe mixture to pre-anneal at 37–42 °C for 1 h. After 1 h, apply 15–20 μl of hybridization mixture onto the denatured slides. Cover the probe mixture with 22 × 32 mm glass coverslips and seal the edges of the coverslips with coverslip sealant. Put the slide in a humidified box and incubate at 37–42 °C for 16–18 h.

Post-hybridization Washes

1. Carefully remove the coverslip sealant using a needle or forceps; remove coverslip by jogging or by a dissection needle. From now onward, keep hybridization area wet until the detection and post-hybridization washing are completed.
2. Wash slides after hybridization for 3 min at 46 °C: in 40 % formamide, 2 × SSC three times, in 2 × SSC one time, in 0.1 × SSC three times.
3. Temperature of wash solutions can vary from 42 °C up to 50 °C (to ensure good signal/background ratio). Keep the concentration of formamide in the washing solution the same as used in the hybridization mix.

Pre-detection Blocking

1. Apply the 25 μl of blocking solution onto the slide, cover with a 22 × 32 mm coverslip and put the slide in a moisture chamber. Or, if a large number of slides are used, immerse the slides in a Coplin jar with 50 ml blocking solution. Incubate at 42 °C for 30 min.

Detection of Biotin-Labeled Probe with FITC

Biotin-labeled probes were visualized using the conventional sandwich detection system of FITC avidin and biotinylated goat anti-avidin IgG.

1. Remove the coverslip after blocking and immediately place 25 μl of avidin-FITC solution, cover with 22 × 32 mm glass coverslip

(but avoid the creation of air bubbles). Put in Petri dish as described above (see Humidified chamber). Incubate at 42 °C for 30 min.

2. Remove the coverslip, and rinse the slides in 4 × SSCT solution three times for 3 min each at 46 °C.
3. Place 25 μl of biotinylated anti-avidin solution onto the slide. Cover with coverslip. Place the slide in Petri dish and place in thermostat at 42 °C for 30 min.
4. Remove coverslips, and rinse the slide in 4 × SSCT solution three times for 5 min each at 46 °C.
5. Add 25 μl of avidin-FITC solution, cover with coverslips and incubate at 42 °C for 30 min as described above.
6. Remove the coverslips, and rinse the slide in 4 × SSCT solution three times for 5 min each at 46 °C.

Counterstaining

1. Before counterstaining, rinse the slide in 0.2 × SSC for 5 min at RT.
2. Stain the slides in DAPI solution (0.1 μg ml^{-1}) for 1–2 min at RT.
3. Rinse the slide in 0.2 × SSC for 10 s.
4. Dry the slide using a rubber bulb.
5. Add 8 μl of antifade solution or DABCO solution.
6. Carefully apply the coverslips while avoiding the creation of air bubbles.
7. Place the slides in a dark box to prevent fading. For long-term storage, keep the slides in a box in the refrigerator.

Image Capture and Processing

1. Visualize the DAPI counterstaining using DAPI filter, and then relocate the G-banded metaphases using their coordinates recorded above.
2. Capture FISH images using FITC filter for hybridization signals and DAPI filter for counterstaining.
3. Assign hybridization signals to specific chromosome(s) regions defined by DAPI-banding or by G-banding patterns previously recorded.

4 Notes

1. Note all microscopic slides were further cleaned by sonication in 2 % Decon 90 solution for 10 min, then rinsed thoroughly in a large quantity of running tap water and distilled water. Store in 96 % ethanol in an airtight container. Before use, remove

slides from 96 % ethanol and polish to dry with lint-free paper tissue).

2. High-quality metaphase spreads are critical for all cytogenetic analysis; however, it is difficult to standardize the procedures. The thickness of cytoplasm background covering the target metaphase chromosomes significantly affects the accessibility of painting probes and hybridization efficiency. The importance of optimized metaphase spreads can never be overemphasized and particularly so for cross-species chromosome painting.
3. In doing so, to prevent the formation of water condensation on the inner wall of test tubes when they are suddenly exposed to the air.
4. Although cross-species chromosome painting works with untreated samples, a brief treatment with pepsin generally improves the hybridization efficiency and signal/background ratio regardless of the thickness of the cytoplasm layer. This treatment can remove the cytoplasm background cover and debris and thus improve accessibility of probes by exposing more chromosomal target DNA for efficient in situ hybridization to occur.
5. Critical: as with GTG-banding, overtreatment will lead to significant loss of chromosomal DNA and morphology at the denaturation step. Although fixation using 0.5–1 % formaldehyde may help to preserve the chromosomal morphology, we found that the best hybridization results are obtained with unfixed slides if the optimized denaturation of target chromosomes can be achieved.
6. If nonspecific cross-hybridization is persistent due to underdenaturation and/or too much cytoplasm debris, immerse the slides in a 50 % acetic acid/50 % dH_2O solution for 3 min and then fix in 100 % methanol for 3 min before pepsin treatment. Note after the combined acetic acid/pepsin treatment, the specimen will become more sensitive to denaturation in the formamide solution. One should consider lowering the denaturation temperature at least 2 °C.
7. Most probes (apart from FITC-labeled probes) can be stored in hybridization buffer for a few years without obvious loss in the intensity of hybridization signals. To facilitate multicolor applications, we routinely ethanol precipitate down probes and resuspend them in hybridization buffer. The stringency of the probe mixture can be lowered by the addition of a few μl of TE buffer when making the probe mixture. In general, probes further cleaned by ethanol precipitation and resuspended in hybridization buffer give slightly improved background/signal

ratio compared to un-cleaned DOP-PCR products, particularly when Cot-1 DNA is absent.

8. Critical step: if probes that are not fully dissolved are used, a lot of intense, nonspecific spots may be seen among relatively dim hybridization signals, which will make the subsequent image capturing difficult.
9. Alternatively probes can denature at 94 °C for 3 min on a thermocycler.
10. This denaturation solution is not the same as the 70 % formamide in 2 × SSC which is less stringent (see Sect. 3.2)
11. The suggested denaturation condition of metaphase spreads (70 % formamide/2 × SSC at 67 °C for 1.5–2 min) is a general guide. Although this condition works with most metaphase preparations derived from fibroblast cultures and peripheral blood cultures, the optimized denaturation temperature can vary, depending on samples, from 60 to 72 °C. Whenever possible, one should try to optimize the denaturation condition experimentally for each metaphase preparation. In general, metaphase spreads derived from peripheral blood culture and normal fibroblast cell lines are more resistant than metaphases derived from EBV-transformed lymphoblastoid cell lines as well as from embryonic stem cells. For each species and each type of metaphase preparations, the optimized denaturation temperature should be determined experimentally before denaturing a large batch of slides. In addition, the number of slides co-denatured is an important factor to consider, as one slide can cause the temperature of the formamide solution to drop by up to 0.5 °C. A test denaturation step is thus recommended before subjecting all sample slides to the same denaturation condition; for instance, denature one test slide together with 5–9 dummy slides (i.e., blank slides without metaphase spreads). Control the denaturation by counterstaining the slide with DAPI and check under an epifluorescence microscope. A Coplin jar has a capacity of 10 slides (with slides placed back to back,) while a Hellendahl jar has a 16 slide capacity. We routinely denature 10 slides in a 50 ml Colpin jar or 16 slides in a 70 ml Hellendahl jar in one go without any problem. Have all slides ready in pairs before subjecting them to denaturation; start the timer as soon as the first pair of slides is placed into the formamide solution.
12. Check for glass particles or dust on the surface of denatured slides and coverslip before adding the probes. Otherwise, you may find the coverslips fail to lay down flat.
13. Although it is more time consuming, biotin-labeled probes can also be detected with either Texas red avidin or FITC avidin (Vector Labs) in combination with biotinylated goat anti-

avidin IgG antibody using the conventional three-layer "sandwich" detection system (see point 3.2). Do not attempt to use Cy3-avidin in the "sandwich" detection system because it will bring up undesirable strong background signals.

14. The FITC-labeled probes can also be detected with one layer of Alexa 488 conjugated rabbit anti-FITC (molecular probes/ Invitrogen) at a 1:200 dilution with 4 × SSCT.
15. Instead of the two-layer detection used above, digoxigenin-labeled paint probes can be detected with one layer of either FITC- or rhodamine-conjugated sheep-anti-digoxigenin Fab fragments (Roche, Basel, Switzerland) at a 1:500–1000 dilution.
16. Double check the labels on the frosted end; otherwise, you may end up with wiping off the metaphase spreads with signals.
17. For slides with Cy3, FITC, and DAPI fluorescence signals, mount the slides with Vectashield with DAPI (Vector Laboratories). For slides with Cy5 fluorescence, mount the slides in either DAPI II (Abbott Molecular/Vysis Inc.) mounting solution or SlowFade Gold with DAPI (Invitrogen/Molecular Probes) as the Vectashield mounting solution apparently offers little protection against the fading of Cy5 fluorescence.
18. For every new type of suspension, we experimentally choose the type of slide preparation, time of trypsin treatment, time of formaldehyde fixation, and the percentage of formamide that gives best results. The duration and stringency of treatment and post-hybridization washing also depend on the type of metaphase preparations as well as the type of hybridization. Hybridization between distantly related species always demands that all details given in the protocol are followed strictly.
19. Caution: this is a highly corrosive, oxidizing solution; wear protective clothing, gloves, and goggles and handle with extra care. Follow local guidance on waste disposal.
20. The optimized time needs to be determined experimentally for every suspension and for every kind of slide preparation; it could vary from 40 s to 3 min.
21. Caution: xylene is flammable and is of modest acute toxicity; handle with care in a fume hood.
22. If over-denaturation keeps occurring due to insufficient fixation, the 0.5 % formaldehyde can be replaced with 1 % or, on very rare occasions, with 3.7 % formaldehyde.
23. The duration of incubation in denaturation solution is critical and will vary according to samples and the number of slides co-denatured. Thus, for each suspension, it is recommended that

the optimal denaturation time should be experimentally titrated to within seconds.

Acknowledgments

The development of these protocols has been relying on the continuous efforts of our colleagues, graduate students, and collaborators. We are particularly grateful to Malcolm A. Ferguson-Smith and Nigel P. Carter who introduced us to the field of chromosome painting in the early 1990s. We thank Patricia CM O'Brien for a critical reading of the manuscript.

References

1. Scherthan H, Cremer T, Arnason U et al (1994) Comparative chromosome painting discloses homologous segments in distantly related mammals. Nat Genet 6:342–347
2. Pinkel D, Landegent J, Collins C et al (1988) Fluorescence *in situ* hybridization with human chromosome-specific libraries – detection of trisomy-21 and translocations of chromosome-4. Proc Natl Acad Sci USA 85:9138–9142
3. Wienberg J, Jauch A, Stanyon R et al (1990) Molecular cytotaxonomy of primates by chromosomal in situ suppression hybridization. Genomics 8:347–350
4. Jauch A, Wienberg J, Stanyon R et al (1992) Reconstruction of genomic rearrangements in great apes and gibbons by chromosome painting. Proc Natl Acad Sci USA 89:8611–8615
5. Telenius H, Carter NP, Bebb CE et al (1992) Degenerate oligonucleotide-primed PCR: general amplification of target DNA by a single degenerate primer. Genomics 13:718–725
6. Telenius H, Pelmear AH, Tunnacliffe A et al (1992) Cytogenetic analysis by chromosome painting using DOP-PCR amplified flow-sorted chromosomes. Genes Chromosomes Cancer 4:257–263
7. Rabbitts P, Impey H, Heppell-Parton A et al (1995) Chromosome specific paints from a high resolution flow karyotype of the mouse. Nat Genet 9:369–375
8. Yang F, Muller S, Just R et al (1997) Comparative chromosome painting in mammals: human and the Indian muntjac (*Muntiacus muntjak vaginalis*). Genomics 39:396–401
9. Yang F, O'Brien PCM, Milne BS et al (1999) A complete comparative chromosome map for the dog, red fox, and human and its integration with canine genetic maps. Genomics 62:189–202
10. Deng W, Tsao SW, Lucas JN et al (2003) A new method for improving metaphase chromosome spreading. Cytometry A 51:46–51
11. Henegariu O, Heerema NA, Wright LL et al (2001) Improvements in cytogenetic slide preparation: controlled chromosome spreading, chemical aging and gradual denaturing. Cytometry 43:101–109

Three-Dimensional Immunofluorescence In Situ Hybridization in Preimplantation Mouse Embryos

Tiphaine Aguirre-Lavin and Nathalie Beaujean

Abstract

It is now well established that three-dimensional organization of chromatin in the nucleus plays a vital role in regulating the genome. In this context, three-dimensional fluorescence in situ hybridization (3D FISH) has become a major technique used to visualize the location of DNA sequences in the nuclei and several variations of the technique have been published. In this article we describe a protocol which has been optimized for embryos to suit this peculiar experimental model without altering the 3D shape. Moreover, we will describe how to carry immuno-3D FISH to simultaneously check the localization of proteins and DNA sequences in embryos. The protocol is applicable to all preimplantation stages with several probes, thereby allowing multicolor FISH. With this method it is therefore possible to investigate nuclear localization of several genomic sequences together and apply automated imaging analyses.

Keywords 3D FISH, Immunostaining, Development, Imaging, Microscopy, Murine probes, Mouse embryo

1 Introduction

Fluorescence in situ hybridization (FISH) is a cytogenetic technique that allows the direct visualization and mapping of specific DNA sequences in individual cells, including genes and chromosomes. This technique has been commonly used to detect chromosomal abnormalities, numerical aberrations, and genetic disorders, especially in the case of preimplantation genetic diagnosis ([1, 2], chapter by Thomas Liehr and Anja Weise "Background"). On the other hand, it is now well established that the organization of the genome inside the interphase nucleus is a central determinant of genome functions and that knowing the sequence of a genome is insufficient to understand its physiological function ([3], chapter by Thomas Liehr and Nadezda Kosyakova "Three-Dimensional Inter phase Analysis Enabled by Suspension FISH"). Therefore, many researchers are now studying the non-random spatial distribution

Thomas Liehr (ed.), *Fluorescence In Situ Hybridization (FISH)*, Springer Protocols Handbooks,
DOI 10.1007/978-3-662-52959-1_43,

of chromosomes and genes within the nuclear space with FISH on 3D-preserved nuclei (3D FISH) approaches.

3D FISH has indeed become a major tool for analyzing three-dimensional organization. Whereas FISH can be performed on metaphase chromosomes to determine the chromosomal localization of a gene, telling within which chromosome and which specific region of a chromosome the gene is located, it can also be used in cells that are in interphase. Using interphase nuclei with preserved 3D shape, one can then determine the three-dimensional arrangements of chromosomes, providing detailed information about the spatial arrangement of chromosomes into territories or about the subcellular localization of specific genomic sequences [3].

3D FISH in combination with 3D microscopy (i.e., confocal laser scanning microscopes), image reconstruction, and voxel-based algorithms from the serial optical sections obtained is indeed an efficient tool to analyze the spatial position/organization of targeted DNA sequences in the nucleus [4, 5]. Moreover, with most confocal microscopes, it is now possible to visualize at least four different fluorochromes within one experiment and distinguish several nuclear targets. This multicolor 3D FISH procedure allows following numerous DNA sequences simultaneously and analyzing their spatial relationships on the level of individual nuclei [4]. Similarly, it is also possible to combine immunofluorescence detection of nuclear proteins and 3D FISH, thereby providing a unique opportunity to assess specific colocalization of DNA sequences and proteins (e.g., modified histones) within the nucleus [6, 7]. The major challenge in combining immunostaining and FISH (immuno-FISH) is to preserve the nuclear protein epitope detected by the antibody; we therefore recommend to perform the immunostaining first and then the FISH procedure.

According to the experimental design, different types of FISH probes can be used to target the genomic sequences of interest: (1) gene-specific probes (10–40 kb length) will bind to a particular locus of a given chromosome with high sequence homology, (2) whole-chromosome probes correspond to a collection of smaller probes binding to different sequences along a given chromosome, and (3) short monomer probes are sufficient to detect continuous repetitive sequences. Labeling of larger DNA probes is typically performed by random primed label PCR or nick translation with modified nucleotide analogues that carry either a fluorescent dye or a hapten for indirect detection (e.g., biotin, digoxigenin), while small oligonucleotides are typically 5′-end labeled. For indirect detection the signal is detected and amplified with fluorescent antibodies or immunoenzymatic reactions. Using probes labeled with different fluorochromes (multicolor FISH) will allow to distinguish each chromosome/each sequence of interest in its own unique color within the same nuclei [4, 8].

3D FISH and 3D immuno-FISH on fixed nuclei should be considered as complementary approaches to biochemical and biophysical assays for the assessment of specific DNA/protein location and their interactions. Chromatin immunoprecipitation (ChIP) and 3C-based techniques (chromosome conformation capture) indeed allow the investigation of molecular crosstalk between many thousands of different loci at a higher resolution than fluorescence microscopy [9, 10]. Unlike these techniques that are based on large mixed cell populations, 3D FISH and 3D immuno-FISH provide information at the single cell level, making it an indispensable tool, especially when studying rare material like the mammalian embryo.

However, the study of preimplantation embryos poses some limitations. Unlike most cells in culture, embryos do not attach to plates/dishes when cultured in vitro. Moreover, the nuclei of early preimplantation embryos are larger than those of differentiated cells. It is therefore very important to pay attention to probe quality, fixation, and pre- and post-hybridization steps during the 3D FISH procedure in order to preserve nuclear structure as much as possible while making DNA sufficiently accessible for probe hybridization. Here we present a protocol for multicolor 3D FISH and 3D immuno-FISH in mouse embryos, which is also applicable to other species.

2 Materials

2.1 Immunostaining Procedure

- Aspirator tube (~40 cm length) and thin glass pipettes for embryo manipulation (*see* **Note 1**).
- Dissecting microscope with appropriate working distance and heating block set at 27 °C (*see* **Note 2**).
- Embryo glass dishes (*see* **Note 3**).
- Phosphate Buffered Saline (PBS): 137 mM NaCl, 2.7 mM KCl, 8.1 mM Na_2HPO_4, 1.5 mM KH_2PO_4 (*see* **Note 4**).
- Fixative: 20 % paraformaldehyde (PFA) (*see* **Note 5**).
- Permeabilization solution: 10 % Triton X-100 (*see* **Note 6**).
- PBS-BSA 2 % solution (*see* **Note 7**).
- Specific first antibody and fluorescent secondary antibodies (*see* **Note 8**).
- Mineral oil, embryo tested: keep at room temperature (RT) and protect from direct light.
- Petri dishes, 60 mm diameter.

2.2 FISH Procedure

- Phosphate Buffered Saline (PBS): 137 mM NaCl, 2.7 mM KCl, 8.1 mM Na_2HPO_4, 1.5 mM KH_2PO_4 (*see* **Note 4**).
- Tyrode's solution stored in single-use aliquots of 1 ml at −20 °C (*see* **Note 9**).
- 20 × saline-sodium citrate (SSC) buffer containing 0.3 M sodium citrate and 3 M NaCl at a pH around 7.0. The working solutions of 0.1 × and 2 × SSC at a pH of 6.3 are prepared by diluting the 20 × stock in PBS and lowering the pH with the appropriate volume of HCl.
- DNA-specific probes (*see* **Note 10**).
- RNase A stored at 1 M in single-use aliquots of 100 μl at −20 °C.
- Heating block.
- Deionized formamide for molecular biology (purity greater than 99.5 %).
- Hybridization buffer: 50 % formamide, 10 % dextran sulfate (from 50 % w/v stock solution), 1 × Denhardt, 40 mM NaH_2PO_4 in 2 × SSC, pH 7. Store in single-use aliquots of 1 ml at −20 °C.
- Incubator set at 37 °C.
- Non-transparent humidified chambers: plastic boxes with lids wrapped with aluminum foil and humidified tissue paper inside.
- Embryo glass dishes (*see* **Note 3**).
- Glass bottom dishes, 35 mm diameter.
- Nuclear counterstain and antifading solution (*see* **Note 11**).

3 Methods

Preimplantation embryos can be collected in vivo just before 3D FISH or cultured according to standard protocols [11]. Three days are required to perform immuno-FISH, and we advise to start by the immunostaining procedure before in situ hybridization (Fig. 1). All steps are performed at RT unless otherwise specified.

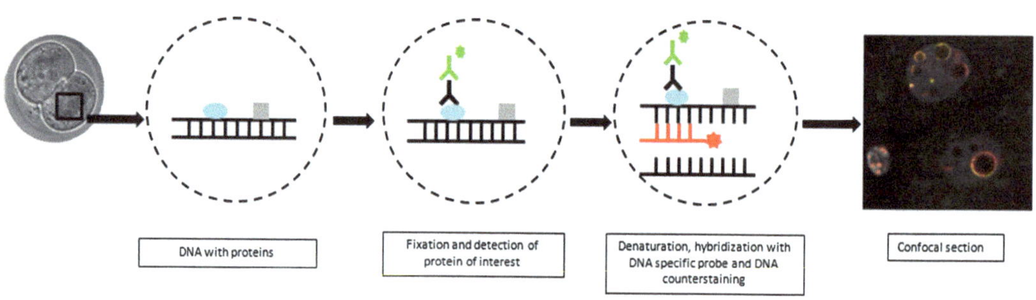

Fig. 1 Illustration of the successive steps for 3D immuno-FISH on embryos

3.1 Immunostaining Procedure

1. To fix the embryos, transfer them group by group with a thin glass manipulation pipette into 2 % PFA freshly diluted in PBS (PBS-PFA 2 %) (*see* **Note 12**).
2. Incubate in PBS-PFA 2 % for 20 min at RT (22–25 °C or 27 °C on the heating plate) (*see* **Note 13**).
3. After fixation, permeabilize the embryos for 30 min with 0.5 % Triton X-100 diluted in PBS just before use (*see* **Note 14**).
4. Saturation of non-specific sites is performed by 1 h incubation of the embryos in PBS containing 2 % BSA (PBS-BSA 2 %).
5. Dilute in a micro test tube the primary antibody directed against your target protein with PBS-BSA 2 % (*see* **Notes 15** and **16**). You will need a total of 20 μl per group and 20 μl for equilibration. In a Petri dish, prepare 20 μl drops of the primary antibody for each group and add one drop for equilibration. Cover the drops with mineral oil.
6. Embryos are transferred group by group in the primary antibody: first in the equilibration drop and then in their respective antibody incubation drops.
7. Incubate overnight at 4 °C.
8. On day 2, transfer the embryos in glass dishes with PBS and incubate three times for 10 min to wash and remove any first antibody excess (*see* **Note 17**).
9. Incubate with the fluorescence-labeled secondary antibody diluted in PBS-BSA 2 % (as above), for 1 h at RT. From this step to the end, always carefully protect dishes from light, with aluminum foil, for example.
10. Transfer the embryos in glass dishes with PBS. Wash 3 × 10 min to remove any excess of antibody (*see* **Note 18**).
11. Postfix with PBS-PFA 2 % for 15 min (*see* **Note 19**).

3.2 FISH Procedure

For 3D FISH without immunostaining, directly start from this point after in vivo collection or in vitro culture of the embryos. In the case of immuno-FISH, remember to keep your samples after the immunostaining procedure always carefully protected from light during the FISH procedure to preserve fluorescence.

The first step of the FISH procedure will be to denude the embryo by removing the zona pellucida in order to improve access of the probes within the embryos (*see* **Note 20**).

1. Rinse in PBS before removing the zona pellucida with Tyrode acid.
2. Load the manipulation pipette with fresh Tyrode's solution, and then transfer embryos into a glass dish containing 200 μl of Tyrode's solution. Immediately reload the manipulation pipette with fresh Tyrode and then transfer embryos into a

second glass dish containing another 200 μl of Tyrode's solution.

3. Incubate the embryos for no more than 90 s (*see* **Notes 21** and **22**).
4. Fix the embryos with PBS-PFA 2 % for 20 min.
5. Permeabilize the embryos for 30 min with 0.5 % Triton X-100 diluted in PBS just before use.
6. Rinse embryos in PBS and then in 2 × SSC (pH 6.3) for 5 min each time.
7. To decrease unspecific background, RNAs are digested by incubation in 200 μM of RNAse diluted in 2 × SSC (pH 6.3) for 30 min at 37 °C.
8. Rinse embryos in 2 × SSC (pH 6.3) and then in PBS for 5 min each time.
9. Equilibrate in hybridization buffer for 1–2 h (*see* **Note 22**).
10. In a 0.5 ml tube, mix 1 μl of each of the probe solutions (around 100 ng) with 12 μl of hybridization buffer and denature at 85 °C for 10 min (*see* **Note 23**). Transfer the tube into ice straight after.
11. At the same time, put your glass dish with the embryos (still in the hybridization buffer) on the heating block to denature the DNA at 85 °C for 10 min.
12. Transfer the embryos in the hybridization mix and place the dishes in the humidified chamber and then in the incubator at 37 °C for 24 h.
13. On day 3, several washing steps are required to remove the remaining unbound probes and decrease the unspecific background. First, rinse embryos in prewarmed (37 °C) 2 × SSC three times for 5 min at 37 °C. Then, in prewarmed (42 °C) 0.1 × SSC three times for 5 min at 42 °C (*see* **Note 24**).
14. Further rinse the embryos at RT in 0.1 × SSC and then in PBS for 5 min each time.
15. Rinse embryos two times for 5 min with PBS and postfix with PBS-PFA 2 % for another 5 min. Postfixation helps maintaining the fluorescent signal until observation.
16. After removal of the PFA, rinse in PBS.
17. Transfer embryos in a drop of PBS-BSA 2 % covered with oil in a micro-dish with glass bottom.
18. Replace the drop of PBS-BSA 2 % with a drop of 20 % antifading solution containing the DNA counterstain (*see* **Note 25**).
19. Incubate in this solution for 15 min at 37 °C.
20. Successively increase the percentage of antifade solution (with DNA counterstain) every 15 min (40 %, 60 %, 80 %) to have at the end a drop of 100 % antifade solution.

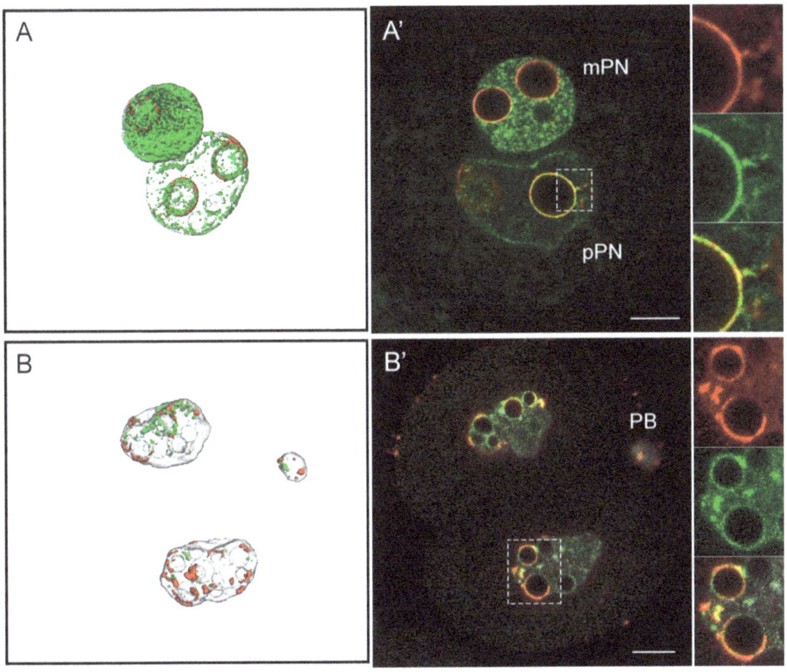

Fig. 2 Examples of 3D immuno-FISH on mouse preimplantation embryos. 3D immuno-FISH on a late 1-cell stage (**a**/**a**′) and mid 2-cell stage (**b**/**b**′) mouse embryos with the corresponding 3D reconstruction obtained by AMIRA software on the *left* and single Z-sections on the *right* (*scales bars*: 10 μm). (**a**/**a**′) shows immunodetection of replication sites in *green* and pericentromeric DNA repeats targeted by major satellite probes (FISH) in *red*, with DNA counterstaining in *gray*. While the maternal pronucleus (mPN) shows mid-replication pattern, the paternal one (pPN) is clearly showing a late replication pattern with few replication sites. On a selected region (*dashed rectangle*) that was zoomed (*last column*), we clearly see strong colocalization of the immunostaining (*green*) and FISH (*red*) signals. (**b**/**b**′) shows immuno-FISH for a specific epigenetic modification (H3S10P) in *green* and pericentromeric DNA repeats targeted by major satellite probes (FISH) in *red*, with DNA counterstaining in *gray*. On a selected region (*dashed rectangle*) that was zoomed (*last column*), we can see that H3S10P only partially overlaps with the pericentromeric probes in some condensed clusters in the nucleoplasm (that correspond to the so-called chromocenters) and around the nucleoli. *PB* polar body

21. The 3D-preserved embryos can then be observed with an appropriate microscope allowing z-scans and equipped with a high-magnification objective adapted to fluorescent observations [e.g., 63× Plan-Neofluar (*see* **Note 26**)]. Example images obtained this protocol are shown in Fig. 2.

4 Notes

1. Aspirator tubes can be handmade with HPLC clear tubing (diameter 2 mm) but are also commercially available (e.g., from Sigma-Aldrich). To this system you can add a filter tip and aspirate with mouth, or add a 1 or 2 ml syringe and control the valve with your hand. To get embryo glass manipulation

pipettes, stretch a glass capillary (or tip of a Pasteur pipette) in the flame in order to get very thin pipettes with diameters similar to the embryo ones. It is also easier to work with quite short pipettes (no more than 8 cm in total).

2. Ideally RT should be stable, between 22 and 25 °C; otherwise, we recommend using a heating block set at 27 °C.
3. Immunostaining is usually performed in commercially available glass dishes with one or several wells. Plastic 4-cell plates can also be used but are usually less convenient for small preimplantation embryos.
4. We usually prepare PBS by dissolving commercially available tablets in deionized water according to manufacturer's instructions, autoclave and store at 4 °C. Filter through a 0.22 μm sterile cellulose acetate membrane before use.
5. PFA solution EM grade with adjusted pH (~7.4 adjusted with 0.2 M NaOH) and stored at 4 °C maximum for 1 month with light protection. Before each experiment, pH should be checked with a pH indicator strip; discard if below pH 7.0 or above pH 7.4. Working solutions of 4 % are prepared by dilution with PBS just before use.
6. Prepare 10 % Triton X-100 stock solution with ultrapure water (w/v) and mix slowly to avoid bubbles. We recommend storage at 4 °C for a maximum of 1 month. The 0.5 % working solution is prepared by dilution with PBS just before use.
7. Dissolve bovine serum albumin (BSA) powder slowly in PBS (w/v) and filter through 0.22 μm sterile cellulose acetate membranes. Aliquots of 1 ml or 2 ml can be kept at −20 °C for several months and thawed just before use.
8. The dilution of the primary antibody should be determined in preliminary experiments to find the lowest concentration that still provides an immunostaining signal with a good signal-to-noise ratio. Primary antibodies recognizing specific antigens can be raised in various species (mouse, rabbit, rat, human autoantibodies). For secondary antibodies, the range of available fluorochromes (such as FITC, Rhodamine, Alexa Fluor) allows flexibility. If you want to use several antibodies simultaneously, make sure that: (1) primary antibodies were produced in different species, (2) each corresponding secondary antibody is coupled to different fluorochromes, and (3) each chosen fluorochrome can be distinguished by the microscope that will be used. Secondary antibodies are diluted according to the manufacturer's recommendations; we commonly use 1/200–1/300 dilutions with antibodies from Jackson ImmunoResearch or Molecular Probes.

9. Do not use Tyrode's solution more than 15 min after thawing as it loses its activity and becomes ineffective on embryos.
10. For example, to detect major satellites (pericentromeric heterochromatin), we used a probe prepared by PCR on genomic mouse DNA with the following primers:

    5′-CATATTCCAGGTCCTTCAGTGTGC-3′ and 5′-CACTTTAGGACGTGAAATATGGCG-3′ and Cy3 or Cy5 labeling by random priming. Similarly, for minor satellites detection (centromeric heterochromatin), we used the 2 following primers: 5′-ACTCATCTAATGTTCTACAGTG-3′ and 5′-AAAACACATTCGTTGGAAACGCG-3′.
11. An extensive assortment of nucleic acid stains is commercially available for DNA counterstaining of the embryos. Their fluorescence absorption and emission spectra span the visible-light spectrum from blue to near infrared, making them compatible with many different types of secondary antibodies. We often use propidium iodide (red fluorescence) or ethidium homodimer 2 (red fluorescence), which bind to DNA even after HCl denaturation in embryos. The antifading solution is used to preserve the fluorescent signals in the samples before and during analysis. We usually use the Vectashield Antifade Mounting Medium.
12. Transfer directly the embryos in PFA from the collection/culture medium without rinse step (embryos are known to be very sensitive to environmental changes).
13. The initial fixation step can be performed at overnight at 4 °C in PBS-PFA 4 %.
14. In some cases, especially in other species than mouse: should the immunodetection not produce a good signal, permeabilization may be either (a) extended to 1 h, (b) performed with 1 % Triton X-100 or 3) performed at higher temperature, e.g., 37 °C.
15. You may combine several primary antibodies if they are not derived from the same species (to avoid cross-reactivity upon addition of the secondary antibodies). However, the combined use of several antibodies should be very carefully tested beforehand.
16. Antibodies have to be prepared in sterile conditions, either under an appropriate hood or next to a flame. We recommend finding the appropriate dilution of the first antibody before starting immuno-FISH. If the immunostaining results are not satisfactory, showing too much background or big fluorescent dots due to antibody aggregates, we recommend centrifuging the antibody (10 min at 11,500× g with a microcentrifuge) after the dilution, before taking the supernatant to prepare the drops.

17. Tween 20 (0.05 %) may be added in these washing steps in order to reduce the background signal if necessary.
18. If the fluorescence signal is too weak, amplification of the signal may be necessary. Secondary components such as fluorescently tagged secondary antibodies can be used to provide a more pronounced signal
19. The "postfixation" is used to preserve the immunostaining during the FISH procedure.
20. Just before the depellucidation, thaw on the bench one aliquot of Tyrode's solution per group of embryos.
21. When dealing with several groups, we advise to carry out the denudation process for one group at a time as it is a very critical step in this experiment. An extended incubation with the acid will damage the embryos and their nuclei, while a short incubation time might not be enough to remove the zona pellucida completely.
22. Under the dissection microscope, you can observe the zona pellucida being degraded by the acid activity of the Tyrode's solution.
23. It is recommended to switch-on the heating block when equilibration starts in order to reach the desired temperature for the subsequent steps.
24. SSC stringency can be increased by increasing the temperature over 42 °C in order to decrease any unspecific binding of the probe in the cytoplasm and thereby background staining.
25. Use nuclear counterstain according to the choice of fluorochromes used to label the probes (*see* **Note 11**).
26. Confocal or grid microscopy is necessary for three-dimensional imaging. The optimal parameters to obtain the best 3D reconstruction is to perform Z-sectioning at no more than 0.3 μm between sections and to take two to three additional sections on top and below the limits of the nucleus.

References

1. Jehan Z, Uddin S, Al-Kuraya KS (2012) In-situ hybridization as a molecular tool in cancer diagnosis and treatment. Curr Med Chem 19:3730–3738
2. Fernández SF, Toro E, Colomar A et al (2015) A 24-chromosome FISH technique in preimplantation genetic diagnosis: validation of the method. Syst Biol Reprod Med 61:171–177
3. Bridger JM, Arican-Gotkas HD, Foster HA et al (2014) The non-random repositioning of whole chromosomes and individual gene loci in interphase nuclei and its relevance in disease, infection, aging, and cancer. Adv Exp Med Biol 773:263–279
4. Aguirre-Lavin T, Adenot P, Bonnet-Garnier A et al (2012) 3D-FISH analysis of embryonic nuclei in mouse highlights several abrupt changes of nuclear organization during preimplantation development. BMC Dev Biol 12:30
5. Andrey P, Kiêu K, Kress C et al (2010) Statistical analysis of 3D images detects regular spatial distributions of centromeres and chromocenters in

animal and plant nuclei. PLoS Comput Biol 6: e1000853

6. Ribeiro-Mason K, Boulesteix C, Fleurot R et al (2012) H3S10 phosphorylation marks constitutive heterochromatin during interphase in early mouse embryos until the 4-cell stage. J Reprod Dev 58:467–475
7. Salvaing J, Aguirre-Lavin T, Boulesteix C et al (2012) 5-Methylcytosine and 5-hydroxymethylcytosine spatiotemporal profiles in the mouse zygote. PLoS One 7:e38156
8. Bonnet-Garnier A, Feuerstein P, Chebrout M et al (2012) Genome organization and epigenetic marks in mouse germinal vesicle oocytes. Int J Dev Biol 887:877–887
9. De Wit E, de Laat W (2012) A decade of 3C technologies: insights into nuclear organization. Genes Dev 26:11–24
10. Collas P (2010) The current state of chromatin immunoprecipitation. Mol Biotechnol 45:87–100
11. Boulesteix C, Beaujean N (2015) Fluorescent immunodetection of epigenetic modifications on preimplantation mouse embryos. Methods Mol Biol 1222:113–126

Fish-FISH: Molecular Cytogenetics in Fish Species

Cassia Fernanda Yano, Luiz Antônio Carlos Bertollo, and Marcelo de Bello Cioffi

Abstract

Fishes exhibit the greatest biodiversity among the vertebrates, making them an extremely attractive group to study a number of evolutionary questions. Over the last years, the development and improvement of cytogenetic FISH analyses have substantially expanded the methods of chromosome studies and have played an important role in the precise characterization of the structure of fish genomes. Here, besides presenting the current fish-FISH protocol, which is frequently applied in many laboratories for freshwater and marine species, we also include details about the isolation and preparation of the sequences most commonly used as probes in fish-FISH experiments. Moreover, considering the quality of the chromosomal preparations in fishes, some critical steps that are crucial for the success of the experiments are also highlighted.

Keywords Fluorescence in situ hybridization (FISH), Mitotic chromosomes, Meiotic chromosomes, Whole chromosome painting (WCP), Microdissection, Zoo-FISH, Repetitive DNAs, rDNA-directed probes, U2 small nuclear ribonucleoproteins (snRNAs), Transposable elements (TEs), Fish (*Pisces*)

1 Introduction

Fishes (*Pisces*) offer broad opportunities for cytogenetic and taxonomic researches. In fact, they include 64 orders and approximately 33,800 species, of which more than 4,000 have been described in the last 10 years [1]. In addition, growing numbers of studies have shown the occurrence of supernumerary chromosomes, simple and multiple sex chromosome systems, polymorphisms and polyploidy among other chromosomal features, highlighting the genomic diversity inside this group [2].

More and more fish cytogenetic studies have been conducted since 1970, mainly using conventional methods to describe karyotypes. Indeed, the establishment of direct chromosome preparations from fish tissues [3–5], the silver-staining method to reveal nucleolar organizer regions—Ag-NORs [6], and the C-banding technique to detect the constitutive heterochromatin on chromosomes [7] promoted chromosomal investigations in fish species for

Thomas Liehr (ed.), *Fluorescence In Situ Hybridization (FISH)*, Springer Protocols Handbooks,
DOI 10.1007/978-3-662-52959-1_44,

many years. However, in a very short period, molecular cytogenetics gained strength in fish research. In recent years, research in fish chromosomes was highly improved, especially with the development of the fluorescence in situ hybridization (FISH) technique. In fact, this methodology allows integrating the molecular information of DNA sequences with their physical location on chromosomes, and, thus plays an important role in the precise characterization of the structure of fish genomes. Important data on chromosomal differentiation, evolutionary relationships and biodiversity have been supplied by FISH in fish.

Many studies have been conducted using dispersed or in tandem organized repetitive DNA sequences as probes for FISH cytogenetic mapping in distinct fish species. Additionally, comparative genomic hybridization (CGH) and application of whole chromosome painting (WCP) probes have shown to be successful methodologies for the characterization of fish genomes in Zoo-FISH experiments (Fig. 1; see chapter by Fengtang Yang et al., "Animal Probes and ZOO-FISH"). These methods have been used for different purposes, including investigations about karyotype evolution and origin and differentiation of B and sex chromosomes [8–11].

Therefore, FISH has become an indispensable tool in fish investigations and in this chapter, we present protocols for fish-FISH and go through them step by step as applied for freshwater and marine species. We also include details about the isolation and preparation of the most commonly used probes in fish-FISH experiments. Moreover, considering that the quality of the fish chromosome preparations is a crucial condition for a successful hybridization, some critical steps related to this issue are also highlighted.

2 Materials

Apart from standard cell biological and molecular cytogenetic equipment, including standard solutions (e.g., ethanol, methanol, formamide, formaldehyde, etc.), more specialized items required are listed below. Solutions and equipment needed for FISH itself are listed in the chapter by Thomas Liehr et al., "The Standard FISH Procedure" and below just some more special things one needs for fish-FISH can be found.

2.1 FISH Probe Generation and Labeling

- AmpliTaq® DNA Polymerase, Stoffel Fragment (Cat. No.: N8080038, Applied Biosystems, USA).
- Biotin 14-dATP (Cat. No.: 19524016, Invitrogen, USA).
- Biotin Nick Translation Mix (Cat. No.: 11745824910, Roche Diagnostics, Switzerland).

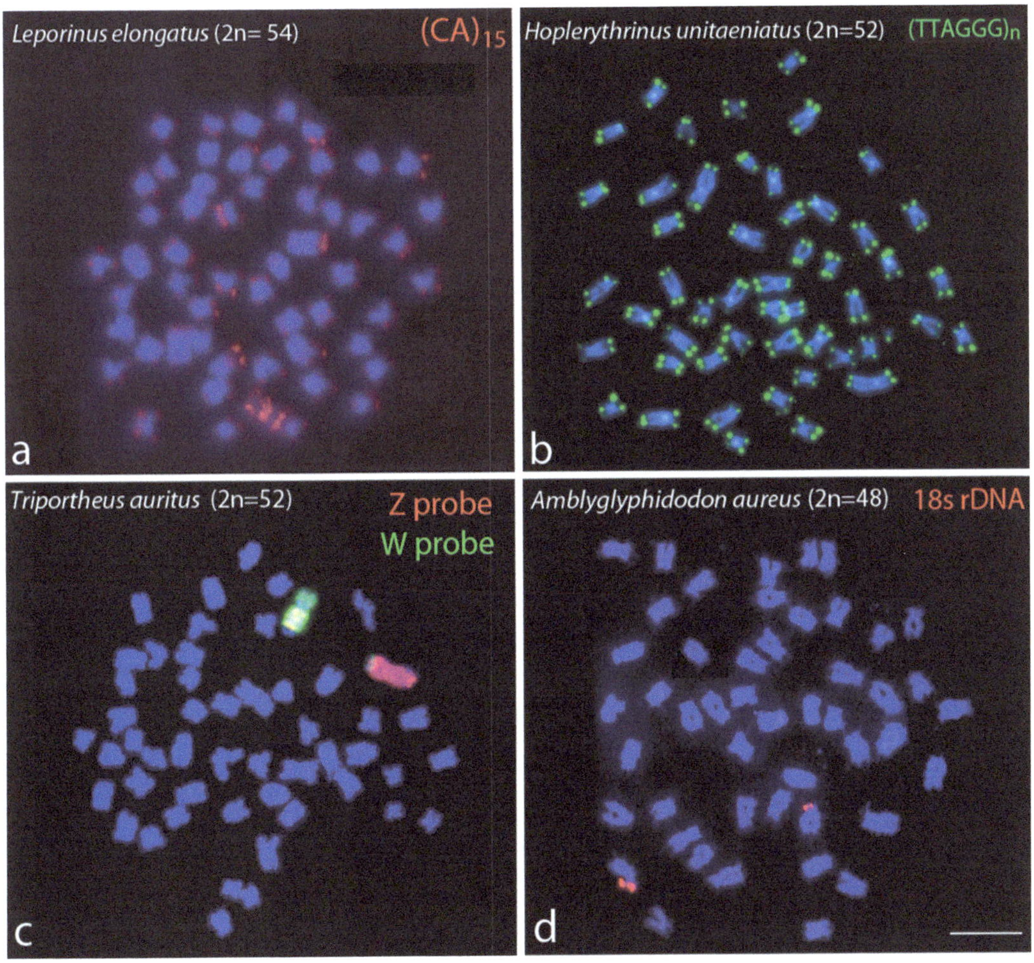

Fig. 1 Fluorescence in situ hybridization to metaphase chromosomes of distinct freshwater (**a–c**) and marine (**d**) fish species. (**a**) Simple $(CA)_{15}$ sequence repeats (*red*) in *Leporinus elongatus*; (**b**) telomeric $(TTAAGG)_n$ repeats (*green*) in *Hoplerythrinus unitaeniatus*; (**c**) whole chromosome painting (WCP) with a Z (*green*) and W (*red*) probe in *Triportheus auritus*; and (**d**) 18S rDNA sites (*red*) in *Amblyglyphidodon aureus*. *Bar* = 5 μm

- Cy5-dUTP (Cat. No.: 25005446, GE Healthcare Life Sciences, USA).
- Deionized formamide (Cat. No.: 1 09684 2500, Merck, Germany; aliquot and store at +4 °C (*see* **Note 1**).
- Deoxynucleotides 100 mM (Cat. No.: DNTP100-1KT, Sigma, USA).
- Dextran sodium sulfate (Cat. No.: D8906-10G, Sigma, USA).
- DIG Nick Translation Mix (Cat. No.: 11745816910, Roche Diagnostics, Switzerland).
- Digoxigenin-11-dUTP (Cat. No.: 11093088910, Roche Diagnostics, Switzerland).
- dNTP label mix (2 mM dATP/dCTP/dGTP, 1 mM dTTP).

- DOP primer (5′-CCG ACT CGA GNN NNN NAT GTG G-3′).
- Double-distilled water (ddH_20) = Aqua ad iniectabilia (Cat. No.: 2351544, Braun, Germany; aliquot and store at −20 °C).
- EDTA 0.5 M (e.g., Merck, Germany; store at −20 °C).
- Hybridization buffer: Dissolve 2 g of dextran sodium sulfate in 10 ml 50 % deionized formamide/2 × SSC/50 mM phosphate buffer for 3 h at 70 °C; pH adjusted to 7 with phosphate buffer; hydrochloric acid destabilizes buffer solution. Aliquot and store at −20 °C.
- Nick Translation Mix (Cat. No.: 11 745 808 910, Roche Diagnostics, Switzerland).
- Ribonucleic acid transfer—tRNA (Cat. No.: R8508, Sigma, USA).
- Salmon sperm DNA (Cat. No.: 15632-011, Invitrogen, USA).
- Sodium acetate solution (3 M, pH 5.2; e.g., Merck, Germany; store at −20 °C).
- Spectrum-Orange-dUTP (Cat. No.: 6J9415, Abbott, USA).
- Spectrum-Green-dUTP (Cat. No.: 6J9410, Abbott, USA).
- Texas Red-dUTP (Cat. No.: C-7631, Molecular Probes, USA).
- Texas Red-dUTP/Spectrum-Orange-dUTP/Spectrum-Green-dUTP/Streptavidin-Cy3/Streptavidin-Cy5 and Avidin-FITC working solutions: Reconstitute the 1 mg with 1.0 ml of double-distilled water (ddH_20); dispense into suitable aliquots and freeze.

2.2 Slide Pretreatment

- PBS 1 × (phosphate buffered saline—Cat. No.: L1825, Biochrom, Germany; store at room temperature = RT).
- Pepsin stock solution (Cat. No.: P-7012, Sigma, USA).
- Pepsin working solution (0.005 %): 99 μl H_2O, 10 μl HCl, and 2.5 μl pepsin (20 mg ml^{-1}).
- Post-fixation solution (10 ml, 1 % paraformaldehyde) (*see* **Note 2**): mix 50 ml of 2 % paraformaldehyde (e.g., Merck, Germany) with 45 ml of 1 × PBS and 5 ml 1M $MgCl_2$ (make fresh as required).
- RNase A (Cat. No.: R4642, Sigma, USA).
- RNase working solution (10 μg ml^{-1}): 1.5 μl RNase A (10 mg ml^{-1}) and 1.5 ml 2 × SSC.

2.3 FISH Procedure

- Avidin-FITC (Cat. No.: A2901, Sigma, USA).
- Anti-digoxigenin fluorescein (Cat. No.: 11207741910, Roche Diagnostics, Switzerland).
- Anti-digoxigenin rhodamine (Cat. No.: 11207750910, Roche Diagnostics, Switzerland).

- Denaturation buffer: 70 % (v/v) deionized formamide (*see* **Note 1**), 20 % (v/v) filtered double-distilled water, 10 % (v/v) 20 × SSC; make fresh as required.
- Detection solution 1: 994 μl of 3 % NFDM/4 × SSC, 2 μl of avidin-FITC working solution (1 mg ml^{-1}), and 5 μl of anti-digoxigenin rhodamine (200 μg ml^{-1}).
- Detection solution 2: 995 μl of 3 % NFDM/4 × SSC, 10 μl of Streptavidin-Cy3, and 5 μl of anti-digoxigenin fluorescein (200 μg ml^{-1}).
- Detection solution 3: 995 μl of 3 % NFDM/4 × SSC, 10 μl of Streptavidin-Cy5, and 5 μl of anti-digoxigenin fluorescein (200 μg ml^{-1}).
- Detection solution 4: 995 μl of 3 % NFDM/4 × SSC, 10 μl of Streptavidin-Cy5, and 5 μl of anti-digoxigenin rhodamine (200 μg ml^{-1}).
- Formamide solution: 2 × SSC/50 % deionized formamide, pH 7.0 (*see* **Note 1**).
- Hybridization buffer: Dissolve 2 g of dextran sodium sulfate in 10 ml of 50 % formamide/2 × SSC/50 mM phosphate buffer for 1 h at 70 °C. Aliquot and store at −20 °C.
- NFDM = Nonfat dried milk powder (Cat. No.: 9999, Cell Signaling Technology, USA).
- NFDM3 %/4 × SSC: 40 ml 20 × SSC, 160 ml ddH_20, and 5 g of NFDM. Prepare the solution in constant agitation to dissolve the NFDM.
- PBS 1 × (phosphate buffered saline—Cat. No.: L1825, Biochrom, Germany; store at RT).
- Phosphate buffer: prepare 0.5 M Na_2HPO_4 and 0.5 M NaH_2PO_4, mix these two solutions (1:1) to get pH 7.0, and then aliquot and store at −20 °C.
- SSC 20 × = saline sodium citrate (Cat. No.: 15557-036; Invitrogen, USA; store at RT); set up 1 × and 2 × SSC before use.
- Streptavidin-Cy5 (Cat. No.: 25800881, GE Healthcare Life Sciences, USA).
- Streptavidin-Cy3 (Cat. No.: S6402, Sigma, USA).
- Tween 20 = polyoxyethylene sorbitan monolaurate (Cat. No.: 10670-1000, Sigma, Germany, store at RT).
- Vectashield Mounting Medium with DAPI FR/10 ml (Cat. No.: H-1200, Vector, USA).
- Washing buffer (4 × SCCT): 4 × SSC, 0.05 % Tween 20; make fresh as required.

3 Methods

3.1 Slide Preparation: Dropping the Cell Suspension

Fish mitotic chromosomes can be directly obtained from tissues of different organs, including the anterior kidney, spleen, and gills. The general protocol for mitotic chromosome obtainment consists of an initial step of in vivo mitotic stimulation, followed by colchicine injection according to the method described in [12]. Meiotic chromosomes are obtained from gonadal tissues by sectioning the testes into small fragments and subjecting them to hypotonic and fixation treatments, according to the method described in [13]. Such procedures usually require previous approval by an Animal Experimentation Ethics Committee.

The quality of chromosomal preparations is quite important for the success of FISH experiments. If not accurately done, the cytoplasm and other cellular materials can remain on the metaphase plates impairing the proper probe hybridization and successful results. Besides quality, the amount of metaphase plates and the correct spreading of the chromosomes are also essential for good results. Thus, in order to obtain high-quality metaphase spreads, some conditions, such as humidity and temperature, should be taken into account while dropping the mitotic/meiotic cells on the slides [14].

1. Wash and thoroughly clean a glass slide and prepare a fresh fixative solution (3:1 methanol: acetic acid).
2. Place a humid filter or absorbent paper on a hot plate at 50 °C.
3. Put a dry slide on the humid paper and drop about 20 μl of the cell suspension on the slide using a micropipette.
4. Add three drops of the fresh fixative (~50 μl) on the cell suspension.
5. Dry the preparation directly in the air.

3.2 FISH Probes

There are two kinds of commercially available probes for fish-FISH:

3.2.1 Commercially Available Probes

1. Microsatellites: Several labeled oligonucleotides containing mono-, di- and trinucleotide microsatellites can be used as probes. These sequences can be ordered as directly labeled with Cy3 or Cy5 at 5′ terminal during synthesis.
2. Telomeric probes: Telomeric $(TTAGGG)_n$ repeats can be detected by FISH using a Telomere PNA FISH Kit/FITC (Cat. No.: K5325, DAKO) or Telomere PNA FISH Kit/Cy3 (Cat. No.: K5326, DAKO).

3.2.2 Homemade Probes

Several methods are available to obtain probes for FISH experiments. Usually, the probes can be isolated from the target species' genome by PCR, followed by cloning in plasmid vectors. Microdissection procedure has also been used to obtain whole or partial chromosomal probes. Below, we list the methods for obtaining the most commonly used probes in fish-FISH experiments.

3.2.3 How to Obtain Probes

Multigene families like rDNAs, U2 small nuclear ribonucleoproteins (snRNAs), or transposable elements can be obtained from whole genomic fish-DNA as outlined below. Multigene families are composed of hundreds to thousands of copies of a set of genes derived by duplication of an ancestral gene and displaying >50 % similarity [15].

rDNAs

The distribution of the 45S and 5S ribosomal genes has been widely investigated in fish species, being a useful marker for evolutionary studies [16]. These genes are highly conserved and organized in multigene families consisting of many copies, in which the 45S rDNA encodes for 18S, 5.8S, and 28S rRNAs, and the 5S rDNA encodes for 5S rRNA [15].

18S-rDNA-Directed PCR

Amplification can be obtained using 18SF (5′-CCGAGGACCT-CACTAAACCA-3′) and 18SR (5′-CC-GCTTTGGTGACTCTT-GAT-3′) as primers, according to [17].

Reagents for the reaction as shown below:

25 μl reaction master mix (per sample)	
ddH_2O	16.45 μl
AmpliTaq Buffer	2.50 μl
$MgCl_2$ (50 mM)	0.75 μl
dNTP mix (dATP, dTTP, dCTP, dGTP, 2.5 mM each)	3.1 μl
Primer 18SR (10 mM)	0.5 μl
Primer 18SF (10 mM)	0.5 μl
DNA (~50 ng μl^{-1})	1.0 μl
AmpliTaq® DNA Polymerase (5 u μl^{-1})	0.2 μl

The thermocycler needs to run the following program:

1. 95 °C—5 min
2. 95 °C—1 min
3. 60 °C—1 min
4. 72 °C—90 s
5. Go to step 2—35×
6. 72 °C—7 min
7. 4 °C—hold

5S-rDNA-Directed PCR

Amplification can be obtained using 5SF (5′-TACGCCCGA-TCTCGTCCGATC-3′) and 5SR (5′-CAGGCTGGTATGGCCG-TAAGC-3′) as primers, according to [18].

Reagents for the reaction:

50 μl reaction master mix (per sample)	
ddH_2O	36.6 μl
AmpliTaq Buffer	5.0 μl
$MgCl_2$ (50 mM)	3.0 μl
dNTP mix (dATP, dTTP, dCTP, dGTP, 10 mM each)	1.0 μl
Primer 5SR (10 mM)	1.0 μl
Primer 5SF (10 mM)	1.0 μl
DNA (~ 50 ng μl^{-1})	2.0 μl
AmpliTaq® DNA Polymerase (5u μl^{-1})	0.4 μl

The thermocycler needs to run the following program:

1. 94 °C—5 min
2. 94 °C—45 s
3. 59 °C—45 s
4. 72 °C—1 min
5. Go to step 2—35×
6. 72 °C—10 min
7. 4 °C—hold

U2 snDNA

The multigene snRNA family is involved in the mRNA maturation by splicing of the pre-mRNA and removing introns in the spliceosome complex [19]. The U2 small nuclear ribonucleoprotein (snRNA) is one of the most abundant snRNAs in higher eukaryotes [20], with the coding sequence being highly conserved [19]. Despite this, the number and organization of U2 snRNA genes can differ among fish species, showing the inherent variation of the genes that encode for this snRNA. The U2 snDNA gene can be amplified using two primers, U2F: CAAAGTGTAGTATCTGTTCTTATCAGC and U2R: CTTAGCCAAAAGGCCGAGA, designed by [19].

Reagents for the reaction are as for 5S-rDNA-directed PCR using U2F and U2R primers instead of 5SR and 5SF ones, and the thermocycler program is the same, as well.

Transposable Elements

Transposable elements (TEs) represent another important class of repetitive DNAs that is widely studied in the genome of many organisms. Although fish genomes are more compact than those of mammals, a higher diversity of TEs is found in their genomes [21]. Among them, the retrotransposable elements from the *Rex* group, isolated from the genome of the swordtail fish *Xiphophorus* [22], represent the most commonly TEs mapped on fish chromosomes, since they are ubiquitously present in the genome of many species.

The retrotransposable elements *Rex1*, *3*, and *6* can be obtained using the following primers, according to [22, 23]: Rex1F (5′-TTCTCCAGTGCCTTCAACACC-3′) and Rex1R (3′-TCCCTCAGCAGAAAGAGTCTGCTC-5′), Rex3F (5′-CGGTGAYAAAGGGCAGCCCTG-3′) and Rex3R (3′-TGGCAGACNGGGGTGGTGGT-5′), and Rex6F (5′-TAAAGCATACATGGAGCGCCAC-3′) and Rex6R (3′-GGTCCTCTACCAGAGGCCTGGG-5′).

Reagents for the reaction:

25 μl reaction master mix (per sample)	
ddH_2O	17.55 μl
AmpliTaq Buffer	2.50 μl
$MgCl_2$ (50 mM)	0.75 μl
dNTP mix (dATP, dTTP, dCTP, dGTP, 2 mM each)	2.0 μl
Primer rexR (10 mM)	0.5 μl
Primer rexF (10 mM)	0.5 μl
DNA (~ 50 ng μl^{-1})	1.0 μl
AmpliTaq® DNA Polymerase (5 u μl^{-1})	0.2 μl

Program in the Thermocycler:

1. 95 °C—5 min
2. 94 °C—40 s
3. 55 °C—40 s
4. 72 °C—2 min
5. Go to step 2—34×
6. 72 °C—5 min
7. 4 °C—hold

Probe Obtainment by Microdissection

See chapter by Fengtang Yang et al., "Generation of Paint Probes from Flow-Sorted and Microdissected Chromosomes" (*see* **Note 3**).

3.2.4 Probe Labeling

Probe labeling is an essential step for a successful FISH experiment. Different labeling methods are effective, such as nick translation and PCR protocols, which are described below. The nick translation procedure follows the manufacturer's instructions and is appropriate when the DNA sequence of interest was already isolated by PCR and amplified. In PCR labeling, the sequence is labeled while it is isolated. In addition, PCR labeling is also applied after the microdissection procedure to label sequences of whole or partial chromosomal regions (see chapter by Fengtang Yang et al., "Generation of Paint Probes from Flow-Sorted and Microdissected Chromosomes"). By these methods, probes can be directly or indirectly labeled. In the first case, nucleotides carrying a specific fluorochrome are incorporated into the probe DNA. In the second case, nucleotides carrying haptens (like biotin and digoxigenin) are incorporated, requiring a further detection step with appropriate fluorochrome-conjugated antibodies to enable the visualization of the hybridization sites.

Nick Translation

1. Add 1 μg of probe DNA to 4 μl Biotin- or DIG Nick Translation Mix and fill up to 20 μl with ddH_2O; continue with step 3.
2. OR: add 1 μg of probe DNA to 4 μl of Nick Translation Mix and 4 μl of 5× concentrated fluorophore labeling mixture* and fill up to 20 μl with ddH_2O.

   *Fluorophore-labeling mixture:
   - 5 μl 2.5 mM dATP
   - 5 μl 2.5 mM dCTP
   - 5 μl 2.5 mM dGTP
   - 3.4 μl 2.5 mM dTTP
   - 4 μl of either 1 mM of Cy5 dUTP/Spectrum-Green-dUTP/Spectrum-Orange-dUTP/Texas Red-dUTP
   - 27.6 μl ddH_2O
3. Mix carefully using the tip of a 20 μl pipette, and incubate at 15 °C for 90 min.
4. Stop the reaction by adding 2 μl of EDTA 0.5M (pH 8.0) and incubating at 65 °C for 10 min.

PCR-Labeling

Here, labeling reaction using DOP primer is described.

If the labeling PCR is also isolating the sequence of interest, instead of using DOP primer, it is to be included specific forward or reverse primers in the reaction, as well the proper program in the thermocycler.

20 μl reaction master mix (per sample)	
ddH_2O	9.8 μl
AmpliTaq Buffer	2.0 μl
$MgCl_2$ (25 mM)	2.0 μl
dNTP label mix	2.0 μl
DOP primer (20 μM)*	1.0 μl
DNA (~ 50 ng μl^{-1})	1.0 μl
AmpliTaq® DNA Polymerase (5 u μl^{-1})	0.2 μl
1 mM of Cy5 dUTP/Spectrum-Green-dUTP/Spectrum-Orange-dUTP/Texas Red-dUTP	2.0 μl

Program in the Thermocycler for DOP Primer

1. 95 °C—3 min
2. 94 °C—1 min
3. 56 °C—1 min
4. 72 °C—2 min
5. Repeat, 2–4 for 19 cycles—35×
6. 72 °C—5 min
7. 4 °C—hold

3.2.5 Precipitation of Probe DNA

The procedure is the same to precipitate the probes labeled by nick translation or PCR. First, per μg of probe it is necessary to add 2.5 vol of ethanol (100 %) and 0.1 vol of sodium acetate (3 M, pH 5.2) to the final volume of labeled probes.

1. Mix carefully and incubate at −80 °C for 30 min or at −20 °C for at least 2 h.
2. Centrifugate at 13,000 rpm for 20 min at −4 °C and remove the supernatant with a micropipette without touching the pellet.
3. Dry the pellet in a vacuum centrifuge or let it dissolve at RT until it dries.
4. Add 1 μl of ddH_20; spin it and let it shake at 37 °C for 5 min to help the dissolving process.
5. Add 20 μl of the hybridization buffer, vortex it, and let it shake at 65 °C for 5 min. In the end, a final concentration of 50 ng $μl^{-1}$ of labeled probe is obtained.

3.3 Slide Pretreatment

3.3.1 Pretreatment with Pepsin and RNase A

The pretreatment step of the slides is essential to achieve high-quality hybridization signals, especially in fish chromosomal preparations where cytoplasm or other cellular materials usually remain on the metaphase plates. A pretreatment step with RNase A is necessary to remove nuclear RNA, preventing nonspecific hybridization between probes and single-strand sequences. Similarly, additional pretreatment using pepsin is also an important step because it removes the remaining cytoplasm reducing the background in the slides (*see* **Note 4**):

1. Dehydrate the slides in ethanol series (70, 90, and 100 %) for 2 min each (RT).
2. Air-dry and incubate the slides at 60 °C for 1 h.
3. Add 100 μl of RNase (10 μg ml^{-1}) on the slides, cover with a 22 × 50-mm coverslip, and incubate at 37 °C for 1 h in a wet chamber.

4. Remove the coverslip and incubate the slides in a Coplin jar containing 1 × PBS for 5 min on a shaker at RT.
5. Add 100 μl of a pepsin solution 0.005 % on the slides, cover with a 22 × 50 mm coverslip, and keep at 37 °C for 10 min.
6. Remove the coverslip and incubate the slides in a Coplin jar containing 1 × PBS. Keep for 5 min at RT.
7. Incubate the slides for 10 min at RT in a Coplin jar containing 100 ml post-fixation solution.
8. Throw away the post-fixation solution, add 1 × PBS in the same Coplin jar, and incubate the slides for 5 min at RT.
9. Dehydrate the slides in an ethanol series (70, 90 and 100 %, 2 min each) and air-dry.

3.4 Fluorescence In Situ Hybridization (FISH)

1. Incubate the slides in a formamide solution at 75 °C, for 3 min, for DNA chromosomal denaturation.
2. Transfer the slides to a Coplin jar filled with cold 70 % ethanol (−20 °C; 2 min) to conserve target DNA as single strands. Proceed the slides into the ethanol series (90 and 100 %, RT, 2 min each) and air-dry.
3. Denature the hybridization mix (containing 20 μl of the hybridization buffer and 100 ng of the labeled probe) in a thermocycler at 85 °C for 10 min. (This step can be due while the previous ethanol series is performed).
4. Add 20 μl of the hybridization mix to the slides, cover with a 22 × 50 mm coverslip, and incubate at 37 °C for 14 h in a darkened moist chamber (wet paper with 2 × SSC solution) (*see* **Note 5**).
5. Remove the coverslip and wash the slides in 1 × SSC in a temperature ranging from 42 to 65 °C, depending on the applied probes (*see* **Note 6**).
6. Put the slides in a Coplin jar containing 100 ml of 4 × SSCT and keep them for 5 min on a shaker at RT. The detection steps [7–10] are necessary if the probes were indirectly labeled. In the case of directly labeled probes, proceed to step 11 and 12.
7. Incubate the slides in 3 % NFDM/4 × SSC for 10 min at RT.
8. Add 100 μl of the detection solution 1 (or 2–4) to each slide, cover with a 22 × 50 mm coverslip, and incubate at 37 °C for 1 h in a darkened moist chamber.
9. Remove the coverslip and wash three times for 5 min in 4 × SSCT (RT) under agitation (shaker).
10. Wash the slides briefly in 1 × PBS and dehydrate them in an ethanol series (70, 90, and 100 %), 2 min each at RT.

11. Add 17 μl of Vectashield Mounting Medium with DAPI on each slide, cover with a 22 × 50 mm coverslip, and press gently. The slides are ready to be analyzed in a fluorescence microscope. For long-term storage, slides can be kept at 4 °C to prolong the fluorescence intensity.

4 Notes

1. Remember to discard the formamide solution as hazardous waste, as it is a teratogen.
2. Remember to discard the paraformaldehyde solution as hazardous waste.
3. In general, the commercial salmon sperm can work very well as a DNA blocker; however, in some cases, it is not sufficient to block the hybridization of high-copy repeat sequences. Thus, a good alternative is to use Cot1-DNA directly isolated from the studied species prepared according to [24].
4. A successful fish-FISH experiment depends on high-quality probes and metaphase plates. If the chromosomal preparations contain excess of cytoplasm, some adjustments in the pretreatment step need to be done (Fig. 2). Thus, in order to remove or reduce background, the time of pepsin and/or RNase A pretreatment must be increased. The success of the pretreatment can be checked by microscope inspection, and an additional pretreatment step can be done if necessary. Further, background can also be the result of insufficient post-hybridization washes which is solved by raising the washing temperature or the washing solution concentration.
5. In WCP experiments, a pre-hybridization step of 15–30 min at 37 °C can be performed in order to avoid nonspecific hybridizations. In whole chromosome painting (WCP) experiments,

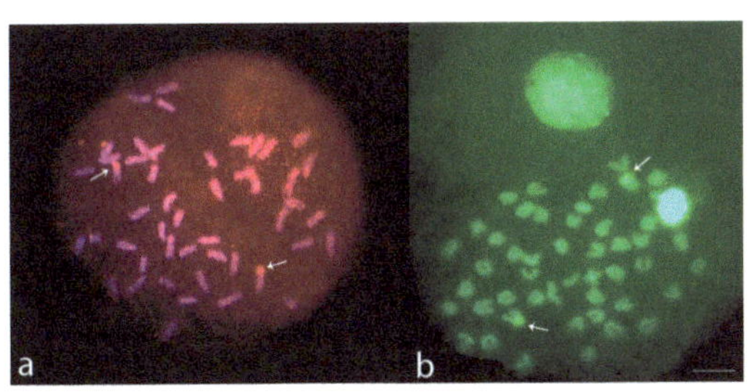

Fig. 2 Metaphase plates of the marine fish *Abudefduf bengalensis* hybridized with an 18S rDNA probe, directly labeled with (**a**) Spectrum-Orange-dUTP and (**b**) Spectrum-Green-dUTP, highlighting the large amount of cytoplasm interfering in the proper probe hybridization. *Arrows* indicate the 18S rDNA sites on the chromosomes. *Bar* = 5 μm

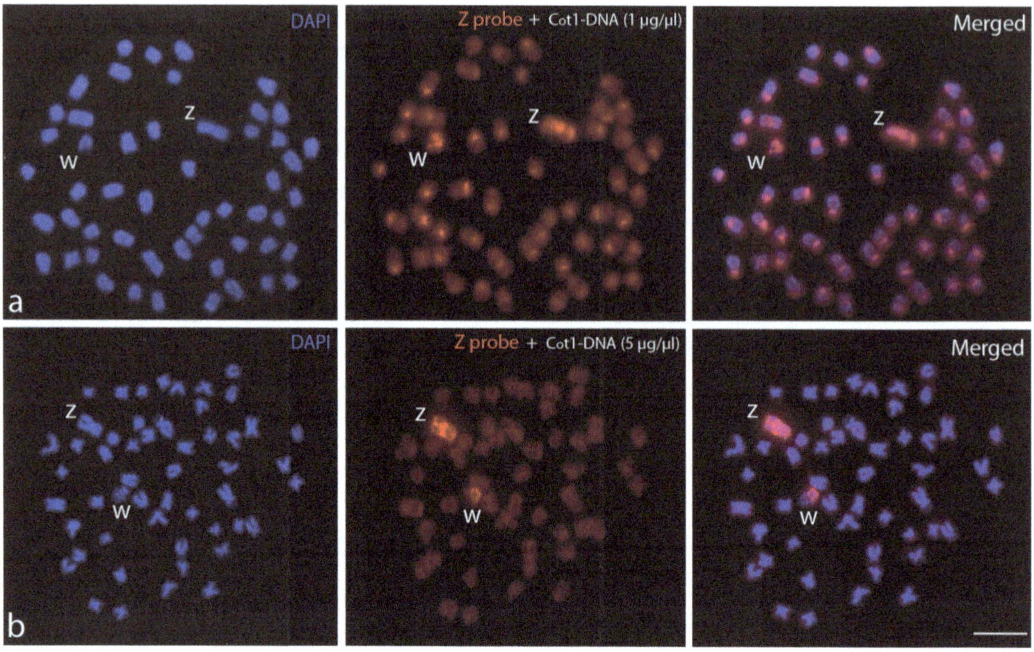

Fig. 3 Whole chromosome painting (WCP) in *Triportheus albus*, using a Z chromosome probe (*red*) highlighting the use of two different concentrations of Cot1-DNA as a competitor. Note that the probe mixed with 1 μg μl^{-1} of Cot1-DNA produced nonspecific signals (**a**), while the use of 5 μg μl^{-1} of Cot1-DNA (**b**) was sufficient to block the nonspecific signals, giving clear hybridization signals on the Z and W chromosomes. *Bar* = 5 μm

the amount of competitor (salmon sperm DNA or Cot1-DNA) is essential to obtain a successful hybridization pattern. In order to reduce nonspecific hybridizations, the concentration of Cot1-DNA can be increased (from 1 up to 5 μg μl^{-1}), as shown in Fig. 3.

6. For repetitive DNA probes, a temperature of 42 °C is sufficient to wash the slides, while for WCP probes the optimum temperature is around 62–65 °C.

Acknowledgments

We wish to thank the Brazilian funding agencies FAPESP (Fundação de Amparo à Pesquisa do Estado de São Paulo, Process 2014/23172-4) and CNPq (Conselho Nacional de Desenvolvimento Científico e Tecnológico, Process 304992/2015-1) for the financial support.

References

1. Eschmeyer WN, Fong JD (2016) Species by Family/subfamily. http://researcharchive.calacademy.org/research/ichthyology/catalog/SpeciesByFamily.asp. Accessed 01 Feb 2016
2. Oliveira C, Foresti F, Hilsdorf AWS (2009) Genetics of neotropical fish: from chromosomes to populations. Fish Physiol Biochem 35:81–100
3. Bertollo LAC, Takahashi CS, Moreira-Filho O (1978) Cytotaxonomic considerations on *Hoplias lacerdae* (Pisces, Erythrinidae). Brazil J Genet 1:103–120
4. Gold JR, Li YC, Shipley NS et al (1990) Improved methods for working with fish chromosomes with a review of metaphase chromosome banding. J Fish Biol 37:563–575
5. Foresti F, Oliveira C, Almeida Toledo LF (1993) A method for chromosome preparations from large specimens of fishes using in vitro short treatment with colchicine. Experientia 49:810–813
6. Howell WM, Black DA (1980) Controlled silver-staining of nucleolus organizer regions with a protective colloidal developer: a 1-step method. Experientia 36:1014–1015
7. Sumner AT (1972) A simple technique for demonstrating centromeric heterochromatin. Exp Cell Res 75:304–306
8. Phillips RB, Konkol NR, Reed KM et al (2001) Chromosome painting supports lack of homology among sex chromosomes in *Oncorhynchus*, *Salmo* and *Salvelinus*. Genetica 111:119–123
9. Cioffi MB, Liehr T, Trifonov V et al (2013) Independent sex chromosome evolution in lower vertebrates: a molecular cytogenetic overview in the Erythrinidae fish family. Cytogenet Genome Res 141:186–194
10. Silva DMZA, Pansonato-Alves JC, Utsunomia R et al (2014) Delimiting the origin of a B chromosome by FISH mapping, chromosome painting and DNA sequence analysis in *Astyanax paranae* (Teleostei, Characiformes). PLoS One 9:e94896
11. Symonová R, Sember A, Majtánová Z et al (2015) Characterization of fish genomes by GISH and CGH. In: Ozouf-Costaz C, Pisano E, Foresti F, Almeida Toledo LF (eds) Fish cytogenetic techniques. Ray-fin fishes and Chondrichthyans. CCR Press, Boca Raton
12. Bertollo LAC, Moreira-Filho, Cioffi MB (2015) Direct chromosome preparations from freshwater teleost fishes. In: Ozouf-Costaz C, Pisano E, Foresti F, Almeida Toledo LF (eds) Fish cytogenetic techniques. Ray-fin fishes and Chondrichthyans. CCR Press, Boca Raton
13. Kligerman AD, Bloom SE (1977) Rapid chromosome preparations from solid tissues of fishes. J Fish Res Board Can 34:266–269
14. Claussen U, Michel S, Mühlig P et al (2002) Demystifying chromosome preparation and the implications for the concept of chromosome condensation during mitosis. Cytogenet Genome Res 98:136–146
15. López-Flores I, Garrido-Ramos MA (2012) The repetitive DNA content of eukaryotic genomes. Genome Dyn 7:1–28
16. Cioffi MB, Bertollo LAC (2012) Chromosomal distribution and evolution of repetitive DNAs in fish. Genome Dyn 7:197–221
17. Cioffi MB, Martins C, Centofante L et al (2009) Chromosomal variability among allopatric populations of Erythrinidae fish *Hoplias malabaricus*: mapping of three classes of repetitive DNAs. Cytogenet Genome Res 125:132–141
18. Martins C, Galetti PM Jr (1999) Chromosomal localization of 5S rDNA genes in *Leporinus* Fish (Anostomidae, Characiformes). Chrom Res 7:363–367
19. Cross I, Rebordinos L (2005) 5S rDNA and U2 snRNA are linked in the genome of Crassostrea angulata and Crassostrea gigas oysters: does the (CT)n(GA)n micro-satellite stabilize this novel linkage of large tandem arrays? Genome 48:1116–1119
20. Pan ZQ, Prives C (1989) U2 snRNA sequences that bind U2-specific proteins are dispensable for the function of U2 snRNP in splicing. Genes Dev 3:1887–1898
21. Volff JN, Bouneau L, Ozouf-Costaz C et al (2003) Diversity of retrotransposable elements in compact pufferfish genomes. Trends Genet 19:674–678
22. Volff JN, Korting C, Sweeney K et al (1999) The non-LTR retrotransposon *Rex3* from the fish *Xiphophorus* is widespread among teleosts. Mol Biol Evol 16:1427–1438
23. Volff JN, Korting C, Froschauer A et al (2001) Non-LTR retrotransposons encoding a restriction enzyme-like endonuclease in vertebrates. J Mol Evol 52:351–360
24. Zwick MS, Hanson RE, McKnight TD et al (1997) A rapid procedure for the isolation of Cot-1 DNA from plants. Genome 40:138–142

FISH in Lampbrush Chromosomes

Anna Zlotina and Alla Krasikova

Abstract

Conventional protocols of fluorescence in situ hybridization (FISH) usually imply the use of mitotic metaphase chromosomes obtained from somatic cells. In contrast, here we provide a protocol of FISH on giant lampbrush chromosomes (LBCs) that represent meiotic diplotene half- bivalents from growing oocytes and can be found in many species including amphibians and birds. Due to the peculiarities of their organization, LBCs can serve as a relevant model for high-resolution molecular cytogenetic and cytological investigations. In this chapter we describe the basic protocol of FISH on LBCs that allows revealing both DNA and nascent RNA-transcript targets. Additionally, key variations in hybridization protocol and the examples of their application are also outlined.

Keywords FISH, High-resolution mapping, Hybridization protocol, Lampbrush chromosomes (LBCs), Nascent transcripts, RNA-FISH

1 Introduction

The majority of published guides for fluorescence in situ hybridization (FISH) contain the protocols of hybridization on mitotic metaphase chromosomes obtained from somatic tissues (chapter by Thomas Liehr and Anja Weise "Background"). At the same time, the cytogenetic analysis of meiotic chromosomes from germline cells is of particular interest (chapter by Maria Bonet Oliver "Sperms, Spermatocytes and Oocytes"). In this chapter, we present a protocol of FISH on giant lampbrush chromosomes (LBCs) typical for diplotene stage of oogenesis of some animal groups including amphibians and birds [1–3]. At this stage, the homologous chromosomes are united in bivalents at chiasmata and represent arrays of compact chromomeric granules with looped out regions that are the sites of active transcription. Being transcriptionally active and considerably decondensed, LBCs are much longer than their mitotic counterparts. Besides, LBCs of many investigated species are enriched in so-called marker structures and special loops with unusual morphology (reviewed in [1–3]).

Thomas Liehr (ed.), *Fluorescence In Situ Hybridization (FISH)*, Springer Protocols Handbooks,
DOI 10.1007/978-3-662-52959-1_45,

On the whole, due to the peculiarities of organization, LBCs represent a useful model system for comprehensive investigation of key cell nuclear processes (such as transcription and gene regulation, RNA processing and packaging, heterochromatin formation, structural genome organization, recombination, etc.) and for high-resolution physical mapping.

The great progress in LBC investigation became possible since the method of their microsurgical isolation and spreading on a microscope slide was improved for amphibian [4, 5] and adapted for avian [6–9] oocytes. For more details, see also http://projects.exeter.ac.uk/lampbrush/protocols.htm [10]. It is worth noting that for the first time in situ hybridization technique itself was applied to nuclear content preparations of *Xenopus* oocytes which resulted in detection of the highly amplified ribosomal DNA [11, 12]. This study and further initial hybridization experiments had involved ^{3}H-labeled probes with subsequent autoradiography detection until the late 1970s to early 1980s when fluorescent probes were introduced. Since then, the great variety of DNA and nascent RNA sequences have been successfully localized on LBCs spreads using FISH (e.g., [13–25]).

The DNA sequences of different nature and size can serve as molecular probes for FISH on LBCs. For example, the whole-chromosome or chromosome arm-painting probes obtained by flow sorting (chapter by Fengtang Yang et al., "Generation of Paint Probes from Flow-Sorted and Microdissected Chromosomes") or chromosome microdissection techniques (chapter by Fengtang Yang et al., "Generation of Paint Probes from Flow-Sorted and Micro dissected Chromosomes"; chapter by Nadezda Kosyakova et al., "FISH-Microdissection") are used for FISH (chapter by Thomas Liehr and Anja Weise "Background") or Zoo-FISH (chapter by Fengtang Yang et al., "Animal Probes and ZOO-FISH") to identify the chromosomes in a LBC set [18, 24–27]. For precise physical positioning of molecular markers or for comparative evolutionary studies, the cloned genomic sequences such as bacterial (BAC; chapter by Thomas Liehr "Homemade Locus-Specific FISH Probes: Bacterial Artificial Chromosomes") or P1-derived (PAC) artificial chromosomes are usually chosen for mapping on LBCs of a corresponding animal or related species [17–19, 23, 24, 28] (Fig. 1c, d). Alternatively, to localize individual genes or various kinds of repetitive sequences, the FISH probes can be prepared by standard polymerase chain reaction (PCR) with target-specific primers or short oligonucleotides that can be synthesized ([13–15, 18–21, 29]) (Fig. 1a, b).

The protocol modifications allow detecting different hybridization patterns on LBC spreads. In particular, different strategies of pre-hybridization pretreatments and denaturation are used for

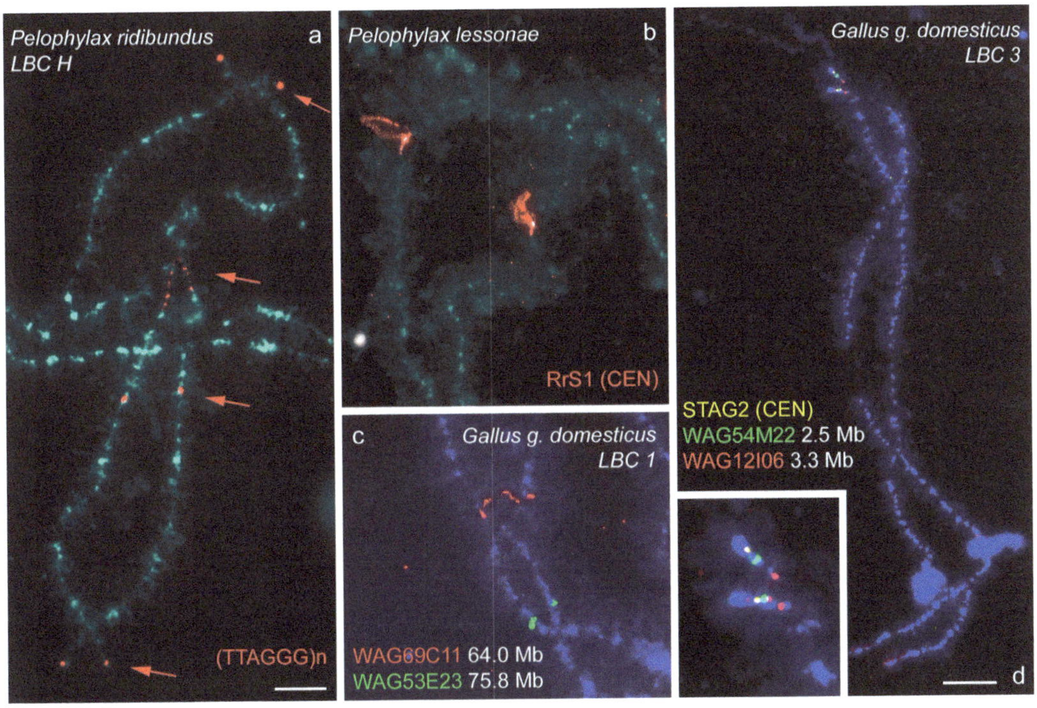

Fig. 1 Examples of FISH with different kinds of DNA probes on LBCs. (**a**) FISH mapping of $(TTAGGG)_n$ repeat to LBC H of the lake frog (*Pelophylax ridibundus*). Telomeres and interstitial (TTAGGG)-repeat sites (*red*) are pointed by *arrows*. As a probe, biotinylated oligonucleotide is used. (**b**) DNA/(DNA + RNA) FISH with RrS1 centromeric repeat to a LBC (*fragment*) of the pool frog (*Pelophylax lessonae*). The hybridization signal is revealed both in centromere chromomeres and RNP-matrix of extended lateral loops (*red*). As a probe, a target-specific PCR product from genomic DNA is used. (**c**) FISH mapping of BAC clones WAG69C11 (*red*) and WAG53E23 (*green*) to chicken (*Gallus g. domesticus*) LBC 1 (*fragment*). The hybridization signals are detected in the pairs of lateral loops. Chromosome coordinates of BACs are given in megabases (Mb) according to the chicken genome assembly Gallus_gallus-4.0. BACs are kindly provided by Richard Crooijmans and Martin Groenen (Wageningen chicken BAC library). (**d**) Immunofluorescent detection of the centromere with an antibody against STAG2 (*yellow*) followed by FISH with BACs WAG54M22 (*green*) and WAG12I06 (*red*) on chicken LBC 3. *Insert* shows enlarged chromosome region with fluorescent signals. Chromosomes are counterstained with DAPI. *Scale bar* = 10 μm

simultaneous detection of DNA and RNA targets and for selective detection of DNA component or RNA component on a slide. In Fig. 2, the different hybridization protocols and corresponding hybridization patterns on LBCs are schematically presented. As an example, the usage of the DNA/RNA hybridization protocol made it possible to characterize transcriptional activity status of some target loci, both unique sequences [28] or particular families of non-coding tandem repeats (e.g., [13, 14, 18–21, 29]).

To co-localize the probes of various size and nature (e.g., BACs and oligonucleotides), re-hybridization (reFISH) of two sets of differentially labeled probes can be applied to the same slide

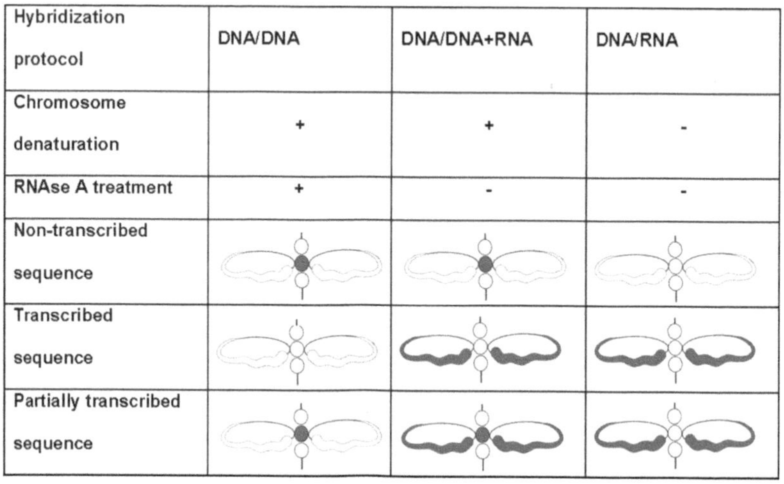

Fig. 2 Schematic presentation of different in situ hybridization protocols and corresponding hybridization patterns on LBCs. The usage of different hybridization protocols allows to investigate the transcriptional activity status of target loci on LBCs. RNAse A pretreatment step and chromosome denaturation can be applied (*plus sign*) or omitted (*minus sign*). DNA/DNA hybridization protocol allows to detect non-transcribed sequences; DNA/RNA hybridization protocol displays transcribed sequences with nascent RNA; DNA/(DNA + RNA) hybridization protocol enables to reveal partially transcribed sequences and nascent RNA. *Grey*—hybridization signal in a chromomere and/or in lateral loops. A single pair of loops is shown for simplicity

sequentially with corresponding hybridization conditions (e.g., [17, 23, 24]) (*see* **Note 1**).

The hybridization procedure can also be applied to LBCs after immunostaining with specific antibodies. That is, Immuno-FISH (chapter by Christine Ye et al., "Simultaneous Fluorescence Immunostaining and FISH") enables to co-localize particular proteins with a DNA sequence or RNA sequence of interest in LBCs or in LBC-associated structures with a high-resolution (Fig. 1d) which helps to shed light on molecular composition and functioning of distinct chromosomal regions and domains (e.g., [19, 25, 26, 30]). Moreover, by means of some variations in the protocol, it is possible to carry out FISH not only on chromosomal spreads but on intact growing oocyte nuclei (germinal vesicles) as well. Specifically, such approach was successfully applied to investigate 3D organization of RNA-processing machinery in avian growing oocyte nucleus [28] (*see* **Note 2**).

Here we provide a basic DNA/(DNA + RNA-transcript) protocol of FISH on LBC spreads. The standard procedure usually implies the usage of homemade hapten-modified DNA probes generated by PCR. At the same time, some key modifications of the protocol are also provided.

The main steps of the FISH-protocol are as follows:

- Prepare the slides with LBC spreads.
- Generate the probe DNA. In case of chromosome paints or extended cloned genomic fragments, amplify and label DNA by PCR with a universal degenerate primer. In case of short target sequences, amplify and label DNA by PCR with a pair of target-specific primers. Prepare the hybridization mix.
- Co-denature the probe and chromosomal DNA on the slide and hybridize overnight.
- Perform post-hybridization washes.
- Perform detection of indirectly labeled probes.
- Mount the slides with DAPI/antifade solution.
- Examine the slides under a microscope.

2 Materials

For more details on preparation and storage of basic FISH reagents and solutions, see also the chapter by Thomas Liehr et al., "The Standard FISH Procedure".

2.1 FISH Probe Labeling

- DNA amplification Kit for polymerase chain reaction (PCR) (Cat. No.: K0131, Sileks, Moscow, Russia)
- Biotin-dUTP (Cat. No.: N1701, Sileks, Moscow, Russia)
- Digoxigenin-dUTP (Cat. No.: 11 573 179 910, Roche Diagnostics)
- 6-MW degenerate universal primer for DOP-PCR [31]
- Target-specific primers (synthesized by a manufacturer, e.g., SYNTOL, Moscow, Russia)
- Blocking solution (e.g., Calbiochem). Final concentration: 1 % in 4 × SSC/0.1 % Tween 20; aliquot and store at −20 °C.
- Streptavidin-alkaline phosphatase (Cat. No.: 11093266910, Roche Diagnostics)/anti-digoxigenin-alkaline phosphatase (Cat. No.: 11093274910, Roche Diagnostics). Stocks: 150 U, store at +4 °C. Working dilution 1:3,000–1:5,000 in blocking solution, make fresh as required.
- NBT (nitro blue tetrazolium, Cat. No.: R0841, ThermoFisher Scientific). Stock solution: 75 mg ml^{-1} in 70 % dimethylformamide. Store in the dark.
- BCIP (5-bromo-4-chloro-3-indolyl phosphate, Cat. No.: R0821, ThermoFisher Scientific). Stock solution: 50 mg/ml in dimethylformamide. Store in the dark.

2.2 Slide Pretreatment

Optional (Depending on a Hybridization Protocol)

- Milli-Q water (ultrapure deionized; aliquot and store at −20 °C).
- 20 × SSC buffer (3 M NaCl/0.3 M sodium citrate: pH 7.0), store at room temperature (RT). Set up 2 × SSC before use.
- RNAse A (Cat. No.: EN0531, ThermoFisher Scientific). Stock solution: 10 mg ml^{-1}, store at −20 °C. Working solution: 150–200 μg ml^{-1} in 2 × SSC, make fresh as required.
- PBS 1 × (phosphate buffered saline; 1.47 mM KH2PO4, 4.29 mM Na2HPO4x7H2O, 137 mM NaCl, 2.68 mM KCl, pH 7.3); store at RT.
- Triton X-100 (e.g., Sigma-Aldrich); working solution: 0.1 % in 1 × PBS.
- Ethanol 96 % and 70 %, 50 ml, each (store at RT).

2.3 Hybridization Procedure

- Salmon sperm DNA (Cat. No.: 15632-011, Invitrogen), 50-fold excess with respect to a final concentration of the probe.
- NaCl, 5 M (Cat. No.: S5886, Sigma-Aldrich), store at RT.
- Ethanol 96 % (store at −20 °C).
- Hybridization buffer (50 % deionized formamide (Cat. No.: 80 0686, ICN; teratogen!), 2 × SSC, 10 % dextran sulfate (Cat. No.: 67578, Sigma-Aldrich)). Aliquot and store at −20 °C.
- Rubber cement: Fixogum™ (Marabu, Tamm, Germany); store at RT.

2.4 Post-hybridization Washings and Probe Detection

- Post-hybridization washing solutions (diluted from stock 20 × SSC): 0.2 × SSC, 2 × SSC; make fresh as required.
- Tween 20 (Cat. No.: P7949, Sigma-Aldrich), store at RT.
- Blocking solution (see Sect. 2.1).
- Detection solution I: Avidin-Alexa 488 (Cat. No.: A21370, Molecular Probes®) and/or mouse anti-dig-Cy3 (e.g., Jackson ImmunoResearch Laboratories) in blocking solution. Make conjugate dilutions according to manufacturer's recommendations; make fresh as required.
- Detection solution II: Goat biotinylated antiavidin (e.g., Vectorlabs) and/or anti-mouse IgG + IgM(H + L)-Cy3 (e.g., Jackson ImmunoResearch Laboratories) in blocking solution. Make conjugate dilutions according to manufacturer's recommendations; make fresh as required.
- Detection solution III: Avidin-Alexa 488 in blocking solution. Make fresh as required.

- Washing buffer (diluted from stock 20 × SSC): 4 × SSC, 0.1 % Tween 20. Make fresh as required.
- DAPI (4,6-diamidino-2-phenylindol)/antifade solution (store at −20 °C). Antifade solution: DABCO (1,4-Diazabicyclo [2.2.2]octane; Cat. No.: D2522, Sigma-Aldrich), 65 % glycerol (Cat. No.: G5516, Sigma-Aldrich), 0.01 M TE, pH 7.5.
- DAPI (Cat. No.: D9542, Sigma-Aldrich, store at −20 °C) in antifade solution, final concentration 1 μg ml^{-1}.
- Ethanol 96 % and 70 %, 50 ml, each

3 Methods

3.1 FISH Probes Preparation

Both custom (*see* **Note 3**) and homemade probes (either directly labeled by fluorochrome or hapten-modified) are successfully applied for FISH on LBCs. Here we focus on FISH with homemade hapten-modified DNA probes generated by polymerase chain reaction. As an example, the material of whole chromosomes or chromosomal arms can be amplified and labeled by PCR with a degenerate universal primer [31]. In a similar manner, we prepare the DNA probes on the basis of genomic sequences cloned in molecular vectors such as bacterial artificial chromosomes. Alternatively, the probes to individual genes or to any short chromosomal sequences can be prepared by PCR with target-specific primers designed using the appropriate software (e.g., FastPCR 6.0 software) according to consensus sequences.

1. DNA is amplified by DOP-PCR using a standard kit for DNA amplification (e.g., Sileks) and a degenerate universal primer [31]. The reaction mix comprises 1 × Taq-polymerase buffer, 2.5 mM $MgCl_2$, 0.2 mM dNTP, 0.5 pM primer, 1–2 U of Taq-polymerase, 100–200 ng of template DNA, and deionized water, with the total volume being 20 μl. Mix carefully using the tip of a 20 μl pipette.
2. Place the 0.2 ml microtube with PCR reaction mix in a thermocycler equipped with a heated lid and run the following program:
   (a) 94 °C, 5 min
   (b) 94 °C, 1 min
   (c) 30 °C, 1.5 min (ramp at 0.5 °C s^{-1} to 72 °C)
   (d) 72 °C, 3 min
   (e) Repeat b–d for 5 cycles
   (f) 94 °C, 1 min
   (g) 55 °C, 1 min
   (h) 72 °C, 1.5 min

(i) Repeat f–h for 25–30 cycles
(j) 72 °C, 10 min
(k) 4 °C, hold

3. Run 1–3 μl of the PCR product in a 1 % agarose gel (60–70 V, 40 min). The sample should appear as a smear with an average size of 0.2–0.7 kb.
4. The DOP-PCR product is usually labeled with hapten (biotin or digoxigenin) by PCR with the same degenerate primer. The reaction mix comprises 1 × Taq-polymerase buffer, 2.5 mM $MgCl_2$, dNTP mix (0.2 mM dATP, 0.2 mM dGTP, 0.2 mM dCTP, 0.16 mM dTTP), 0.04 mM bio-dUTP or dig-dUTP, 0.5 pM primer, 1–2 U of Taq-polymerase, 100–200 ng of DOP-PCR product as a template, and deionized water, with the total volume being 20 μl (*see* **Note 4**). Mix carefully using a tip of a 20 μl pipette.
5. Place the 0.2 ml microtube with PCR reaction mix in a thermocycler equipped with a heated lid and run the following program:

(a) 94 °C, 5 min
(b) 94 °C, 1 min
(c) 55 °C, 1 min
(d) 72 °C, 1.5 min
(e) Repeat b–d for 27 cycles
(f) 72 °C, 8 min
(g) 4 °C, hold

6. Run 0.5–1 μl of the PCR product in a 1 % agarose gel. The sample should appear as a smear with an average size of 0.2–0.7 kb (*see* **Note 5**).
7. Evaluate the efficiency of labeling by spotting the probe dilutions ranging from 100 pg μl^{-1} to 0.1 pg μl^{-1} onto a nylon membrane strip followed by an exposure of the membrane to UV radiation, incubation of the membrane with a blocking agent, and a detection procedure with streptavidin or anti-digoxigenin antibodies conjugated with alkaline phosphatase (AP) (for more details, see elsewhere, e.g., [32]). Add AP substrates: NBT (nitro blue tetrazolium) and BCIP (5-bromo-4-chloro-3-indolyl phosphate). The DNA probe should be regarded as a well-labeled one if at least 1–5 pg of DNA can be detected by this method.
8. Precipitate 0.5–1 μg of the labeled probe together with 25 μg of salmon sperm DNA with 0.25 volume of sodium chloride (5 M) and 2.5 volumes of cold 96 % ethanol. Precipitation can be done for either 2 h or overnight at −20 °C. Pellet the DNA by centrifugation at 12–13,000 rpm for 15 min, discard the

supernatant, and air-dry the pellet. Dissolve the pellet in 25 μl of hybridization buffer, mix well by vortexing, and spin down (*see* **Note 6**).

3.2 Slide Pretreatment

For FISH analysis, amphibian and avian LBC preparations are carried out according to standard protocols [10]. FISH can be applied for LBCs preparations either directly or after immunofluorescent staining (*see* **Note 7**).

It is worth noting, that conventional FISH protocols generally include a pepsin digestion step (see the chapter by Thomas Liehr et al., "The Standard FISH Procedure"). At the same time, even if extremely low concentrations of pepsin are used, it proves to be damaging for LBC morphology. For this reason, we do not recommend to pretreat the LBCs preparations with pepsin.

Slide pretreatment strategy varies depending on the goals (Fig. 2). To reveal both DNA targets and nascent RNA targets (DNA/DNA + RNA-transcript hybridization protocol) as well as to investigate an RNA target only (DNA/RNA hybridization protocol), no slide pretreatments should be done (*see* **Note 8**). Alternatively, to investigate a DNA target only, and hence, to get rid of potential RNA targets on the slide, DNA/DNA hybridization protocol must be applied, which implies a ribonuclease pretreatment step, more often with RNase A (*see* **Notes 9–11**).

3.3 Fluorescence In Situ Hybridization (FISH)

Depending on hybridization protocol, chromosomal DNA and DNA probes can be denatured together (see below in this section) or separately, or a denaturation step can be omitted (Fig. 2, *see* **Note 12**).

1. Drop 5 μl of a hybridization mix on a slide, cover with an 18 × 18 mm coverslip, and seal with rubber cement. The region containing LBCs sets should be preliminary marked by a diamond blade (*see* **Note 13**).
2. Denature the slides by heating at 82 °C for 5 min and leave it to hybridize at 37 °C overnight in a humid chamber (*see* **Notes 14** and **15**).
3. Take the humid chamber out of a thermostat, remove the rubber cement from the slides with forceps, and put them in a coplin jar with 2 × SSC buffer to let the coverslips swim off.
4. Wash the slides in two changes of 0.2 × SSC solution in a coplin jar at 60 °C for 5 min each, after that in two changes of 2 × SSC solution at 45 °C for 5 min each (*see* **Note 16**).
5. Add 40 μl of blocking solution on a marked region of the slide, cover with a parafilm peace, and incubate at 37 °C for 50 min (*see* **Note 17**).
6. Add 40 μl of a detection solution I on the marked region of the slide, cover with a parafilm peace, and incubate at 37 °C for 50 min.

7. Wash the slides in three changes of washing buffer (4 × SSC, 0.1 % Tween 20) at 42 °C for 5 min each.
8. Add 40 μl of a detection solution II on the marked region of the slide, cover with a parafilm peace, and incubate at 37 °C for 40 min.
9. Wash the slides in three changes of washing buffer (4 × SSC, 0.1 % Tween 20) at 42 °C for 5 min each.
10. Add 40 μl of a detection solution III on the marked region of the slide, cover with a parafilm peace, and incubate at 37 °C for 20–30 min. If the probe is modified by digoxigenin, this step is not needed.
11. Wash the slide in three changes of washing buffer (4 × SSC, 0.1 % Tween 20) at 42 °C for 5 min each.
12. Rinse the slides in deionized water, dehydrate in an ethanol series, (70 %, 96 %) and air-dry.
13. Counterstain the slides with 7 μl of DAPI/antifade solution, cover with a coverslip 24 × 24 mm. Examine the results under a fluorescence microscope with corresponding filter cubes. Despite the giant size of LBCs, FISH results (especially with locus-specific probes) should be analyzed at ×100 magnification.

4 Notes

1. After the first round of FISH with a long probe (e.g., PCR- or DOP-PCR product), wash the slides in three changes of 4 × SSC for 10 min each at 37 °C, dehydrate in an ethanol series, and air-dry. Use the slides for FISH.
2. A protocol modification for 3D FISH on intact germinal vesicles was also developed. To ensure the maintenance of the 3D architecture of the giant nucleus, it is strongly recommended to apply DNA/RNA hybridization protocol (to avoid a genomic DNA denaturation step) and to use directly labeled oligonucleotides as probes. In brief, fixed nuclei are incubated in a special hybridization buffer for 1 h at RT, after that they are transferred into the hybridization buffer containing an oligonucleotide probe followed by incubation for 1 h at 48 °C in the dark. After multiple washings in 0.4 × SSC buffer (3 M NaCl, 0.3 M sodium citrate, pH 7) at 48 °C, the nuclei are mounted in glass chambers containing PBS and DAPI (1 μg ml^{-1}) and immediately analyzed by confocal laser scanning microscope (CLSM). For details, see [28].
3. As an example, directly labeled or hapten-modified DNA oligonucleotides or locked nucleic acid (LNA) probes can be

designed by a researcher and then ordered from a manufacturer (e.g., SYNTOL, Moscow, Russia).

4. Amplification and labeling of a short genomic sequence (usually about 0.15–2 kb in size) are carried out by PCR with a pair of sequence-specific primers (forward/reverse) under standard conditions. The primer annealing temperature is chosen depending on a target sequence.
5. When a short genomic sequence is labeled by standard PCR with sequence-specific primers, a PCR product should appear as a band of a corresponding size.
6. Oligonucleotide probes are dissolved in a hybridization buffer containing 30–40 % deionized formamide.
7. In case of Immuno-FISH, the slides that have been subjected to immunofluorescent staining should be washed in 4 × SSC, 0.1 % Tween 20 at 42 °C, dehydrated in 96 % ethanol, air-dried, and then used for FISH.
8. For reliable high-quality FISH results, it is important to preserve RNA component on the slide. The instruments and the solutions for chromosome isolation should be autoclaved; all manipulations should be carried out in sterile laboratory gloves.
9. Put the slides in a coplin jar with 2 × SSC and rinse for 5 min at RT. Add 100–200 μl of RNAse A working solution on the slide, cover with a peace of parafilm, put in a humid chamber, and incubate at 37 °C for 1 h. Wash the slide in three changes of 2 × SSC for 5 min each at RT. Rinse the slides in deionized water, dehydrate in an ethanol series (70 %, 96 %), and air-dry.
10. Depending on the goals, it is also possible to pretreat LBCs slides with different ribonucleases (e.g., S7 nuclease, RNAse R, RNAse H, RNAse III). The reaction buffers, and pretreatment conditions should be usually applied according to manufacturer's recommendations.
11. Optionally, to ensure better permeabilization of compact chromomeres, chromosomes can be pretreated with 0.1 % Triton X100.
12. In case of DNA/RNA hybridization protocol where only RNA molecules represent a target, chromosomal DNA is not subjected to the denaturation procedure. A probe mix is denatured separately at 100 °C for 10 min, after that it is immediately applied to chromosomes followed by a hybridization step. If a single-strand oligonucleotide probe is used, the denaturation step is omitted for both chromosomal and probe DNA.
13. According to standard procedure of LBC preparation, there are four chromosomal sets isolated from four oocytes per slide. It is possible to hybridize a different probe to each chromosomal set using small round coverslips with a diameter of 8 mm.

14. In contrast to FISH on metaphase chromosomes, it is not necessary to do a pre-hybridization (pre-annealing) of the probes (e.g., painting probes or BACs) to block DNA repetitive elements. Hybridization of such probes on LBCs usually gives a specific and reliable signal due to DNA + RNA target detection.
15. In case of short oligonucleotide probes, a hybridization procedure is carried out at RT.
16. In case of short oligonucleotide probes, post-hybridization washings are carried out in three changes of 2 × SSC for 5 min each at RT.
17. If the probes are directly labeled, the detection step is not needed. The slides should be immediately dehydrated in an ethanol series and air-dried and mounted in an antifade solution with DAPI.

Acknowledgments

The authors would like to thank Dmitry Dedukh (St. Petersburg State University) for the images presented in Fig. 1a, b. The research was supported by Russian Science Foundation (grant #14-14-00131). The work was partially performed using experimental equipment of the Environmental Safety Observatory and the Research Resource Centers "Chromas" and "Molecular and Cell Technologies" of St. Petersburg State University.

References

1. Callan HG (1986) Lampbrush chromosomes. Mol Biol Biochem Biophys 36:1–254
2. Morgan GT (2002) Lampbrush chromosomes and associated bodies: new insights into principles of nuclear structure and function. Chromosome Res 10:177–200
3. Gaginskaya E, Kulikova T, Krasikova A (2009) Avian lampbrush chromosomes: a powerful tool for exploration of genome expression. Cytogenet Genome Res 124:251–267
4. Gall JG (1954) Lampbrush chromosomes from oocyte nuclei of the newt. J Morphol 94:283–352
5. Callan HG, Lloyd L (1960) Lampbrush chromosomes of crested newts *Triturus cristatus* (Laurenti). Philos Trans R Soc Lond Ser B 243:135–219
6. Koecke HU, Müller M (1965) Change of form and number of chromosomes in chicken and duck. Naturwissenschaften 52:483
7. Kropotova EV, Gaginskaya ER (1984) Lampbrush chromosomes from the Japanese quail oocytes: light and electron microscopic study. Tsitologiia 26:1008–1015
8. Hutchison N (1987) Lampbrush chromosomes of the chicken. J Cell Biol 105:1493–1500
9. Solovei I, Gaginskaya E, Allen T et al (1992) A novel structure associated with a lampbrush chromosome in the chicken, Gallus domesticus. J Cell Sci 101:759–772
10. Lampbrush chromosome protocols. http://projects.exeter.ac.uk/lampbrush/protocols.htm. Accessed 07.02.2016
11. Gall G, Pardue ML (1969) Formation and detection of RNA-DNA hybrid molecules in cytogenetical preparations. Proc Natl Acad Sci USA 63:378–381
12. Gall JG (2016) The origin of in situ hybridization – a personal history. Methods 98:4–9
13. Solovei I, Gaginskaya ER, Macgregor HC (1994) The arrangement and transcription of telomere DNA sequences at the ends of lampbrush chromosomes of birds. Chromosome Res 2:460–470

14. Solovei IV, Joffe BI, Gaginskaya ER et al (1996) Transcription on lampbrush chromosomes of a centromerically localized highly repeated DNA in pigeon (Columba) relates to sequence arrangement. Chromosome Res 4:588–603
15. Saifitdinova A, Derjusheva S, Krasikova A et al (2003) Lampbrush chromosomes of the chaffinch (*Fringilla coelebs L.*). Chromosome Res 11:93–113
16. Edwards NS, Murray AW (2005) Identification of xenopus CENP-A and an associated centromeric DNA repeat. Mol Biol Cell 16:1800–1810
17. Galkina S, Deryusheva S, Fillon V et al (2006) FISH on avian lampbrush chromosomes produces higher resolution gene mapping. Genetica 128:241–251
18. Krasikova A, Deryusheva S, Galkina S et al (2006) On the positions of centromeres in chicken lampbrush chromosomes. Chromosome Res 14:777–789
19. Krasikova A, Fukagawa T, Zlotina A (2012) High-resolution mapping and transcriptional activity analysis of chicken centromere sequences on giant lampbrush chromosomes. Chromosome Res 20:995–1008
20. Deryusheva S, Krasikova A, Kulikova T et al (2007) Tandem 41-bprepeats in chicken and Japanese quail genomes: FISH mapping and transcription on lampbrush chromosomes. Chromosoma 116:519–530
21. Dedukh D, Mazepa G, Shabanov D et al (2013) Cytological maps of lampbrush chromosomes of European water frogs (*Pelophylax esculentus complex*) from the Eastern Ukraine. BMC Genet 14:26
22. Dedukh D, Litvinchuk S, Rosanov J et al (2015) Optional endoreplication and selective elimination of parental genomes during oogenesis in diploid and triploid hybrid European water frogs. PLoS One 10:e0123304
23. Zlotina A, Galkina S, Krasikova A et al (2010) Precise centromere positioning on chicken chromosome 3. Cytogenet Genome Res 129:310–313
24. Zlotina A, Galkina S, Krasikova A et al (2012) Centromere positions in chicken and Japanese quail chromosomes: *de novo* centromere formation versus pericentric inversions. Chromosome Res 20:1017–1032
25. Zlotina A, Kulikova T, Kosyakova N et al (2016) Microdissection of lampbrush chromosomes as an approach for generation of locus-specific FISH-probes and samples for high-throughput sequencing. BMC Genomics 17:126
26. Derjusheva S, Kurganova A, Krasikova A et al (2003) Precise identification of chicken chromosomes in the lampbrush form using chromosome painting probes. Chromosome Res 11:749–757
27. Griffin DK, Robertson LB, Tempest HG et al (2008) Whole genome comparative studies between chicken and turkey and their implications for avian genome evolution. BMC Genomics 9:168–184
28. Krasikova A, Khodyuchenko T, Maslova A (2012) Three-dimensional organisation of RNA-processing machinery in avian growing oocyte nucleus. Chromosome Res 20:979–994
29. Krasikova A, Vasilevskaya E, Gaginskaya E (2010) Chicken lampbrush chromosomes: transcription of tandemly repetitive DNA sequences. Genetika 46:1329–1334
30. Kulikova T, Chervyakova D, Zlotina A, et al (2015) Giant poly (A)-rich RNP aggregates form at terminal regions of avian lampbrush chromosomes. Chromosoma 2016, DOI: 10.1007/s00412-015-0563-4
31. Telenius H, Carter NP, Bebb CE et al (1992) Degenerate oligonucleotide-primed PCR: general amplification of target DNA by a single degenerate primer. Genomics 13: 718–725
32. Boyle A, Perry-O'Keefe H (2003) Labeling and colorimetric detection of nonisotopic probes. In: Ausubel FM (ed) Current protocols in molecular biology. Wiley, Hoboken, pp 3.18.1–3.18.9

General Protocol of FISH for Insects

Ana Paula Alves-Silva, Luísa Antônia Campos Barros, and Silvia das Graças Pompolo

Abstract

The class Insecta comprises a widely distributed and diverse group of organisms. This diversity also extends to cytogenetic data, such as chromosome number, sex determination systems and peculiar kinds of chromosomes. Information on fluorescence in situ hybridization applied in insects is available, and the approach has been increasingly used over recent years. However, corresponding data is still scarce for some groups. In this chapter, we provide a detailed FISH protocol with varied use in insects, including reagent preparation instructions, adaptations and discussion. We also provide information about homemade C_0t1 DNA.

Keywords C_0t1 DNA, Molecular cytogenetics, FISH protocol for insects, Class Insecta

1 Introduction

The class Insecta harbours more than half of the described eukaryote species on Earth. The estimates of the number of insect species already described range from 720,000 [1] to around one million species [2]; however, real numbers of these species can range from around 5 to 6 million [3]. Insects are distributed over 29 orders and can be considered cosmopolitan, being found from freshwater streams up to very dry deserts.

The taxonomic diversity observed can be extended to the cytogenetic data of the group. The diploid number varies between $n = 1$ in *Myrmecia croslandi* (Hymenoptera: Formicidae) [4] and $n = \sim224–226$ in *Polyommatus* (*Plebicula*) *atlanticus* (Lepidoptera: Lycaenidae) [5]. Moreover, all types of sex determination systems of the animal kingdom are represented in insects. For example, the sex determination "ratio of X chromosomes to sets of autosomes", where females are 2X:2A and males are X:2A (A is a haploid group of autosomes), is what determines the sex in some orthopterans and dipterans [e.g. crickets, grasshopper and some

Thomas Liehr (ed.), *Fluorescence In Situ Hybridization (FISH)*, Springer Protocols Handbooks,
DOI 10.1007/978-3-662-52959-1_46,

population of *Drosophila* (chapter by Amanda Larracuente "FISH in *Drosophila*")]. The XY and ZW sexual system can be found in the genus *Musca* and the lepidopterans, respectively (e.g. butterflies and moths). The order Hymenoptera organizes its sex determination by haplodiploidy, where females are diploid originated from fertilized eggs and males are haploid, originated from unfertilized eggs [6]. Beyond these, some insects still present polytene, lampbrush and holocentric chromosome.

In 1977, the first non-radioactive in situ hybridization was performed using labelled probes by indirect immunofluorescence. This protocol was applied on polytene chromosomes, in *Drosophila melanogaster* using the 5S rRNA probe [7]. Before this improvement in the protocol, radioactively labelled RNA-DNA hybrids were exclusively used and visualized using the autoradiographic method. That method was slow and it was necessary to wait several days to visualize the results [8].

Currently, the fluorescence in situ hybridization (FISH) is a tool used in many groups of insects, from studies focussing on insects of medical and agricultural interest (e.g. methods for introduction of exogenous genes) to descriptive and evolutionary research, using specific probes (e.g. 18S and repetitive DNA [9, 10]). Despite its increasing application during the last 30 years, in some groups such as the ants, which already include 13,986 described species [11], only 23 species have been studied by FISH, all of them using ribosomal DNA and/or telomeric probes [12–20].

In this chapter, we provide a detailed FISH protocol with varied use in insects, including reagent preparation instructions, adaptations and discussion. We also provide information about homemade C_0t1 DNA.

2 Materials

2.1 Homemade C_0t DNA

- Autoclaved Milli-Q water
- S1 nuclease and 10 × S1 nuclease buffer (Ref. M5761, Promega)
- 2-Propanol ACS reagent, ≥99.5 % (e.g. Sigma; stored at room temperature = RT)
- Sodium acetate solution (3 M, ph 5.2; e.g. Merck; stored at RT)
- TE buffer solution (e.g. Promega; stored at RT)
- PureLink® Genomic DNA Mini Kits (Ref. K1820-02, Invitrogen; stored at 15 to 30 °C)

2.2 FISH Probe Labelling

- DIG-Nick Translation Mix (Ref. 11 745 816 910, Roche Diagnostics; stored at −15 to −25 °C).

- EDTA 0.5 M (e.g. Merck; stored at RT).
- Ethanol 100 % (e.g. Merck; stored at RT).
- Hybridization buffer: dissolve 2 g dextran sulphate in 10 ml 50 % deionized formamide/2 × SSC/50 mM phosphate buffer for 3 h at 70 °C. pH adjusted to 7 with phosphate buffer or hydrochloric acid destabilizes buffer solution. Aliquot and store at −20 °C.
- Autoclaved Milli-Q water.
- Sodium acetate solution (3 M, ph 5.2; e.g. Merck; stored at −20 °C).
- Dextran sulphate (Ref. 67578, Sigma; stored at RT).
- Formamide (Ref. F7503, Sigma).

2.3 RNAse and Pepsin Pretreatment

- RNAse stock (Ref. 10 109 142 001, Roche Diagnostics GmbH; stored at 2–8 °C).
- RNAse solution: add 0.05 vol. RNAse at 10 mg ml^{-1} in 0.95 vol. of 2 × SSC and mix well; make fresh as required.
- 20 × SSC = saline sodium citrate (Ref. 15557-036, Gibco BRL; store at RT) or homemade: dilute 175.6 g of NaCl (Ref. S3014, Sigma) and 88.2 g of sodium citrate (Ref. 6132-04-3, Sigma) in 1 l of Milli-Q water. If necessary, use HCl to reach pH 7.0. Set up 0.4 ×, 1 × and 2 × before use.
- PBS 10 × = phosphate-buffered saline (Ref. L1825, Biochrom; stored at RT) or homemade: dilute 75.8 g NaCl (Ref. S3014, Sigma), 9.93 g Na_2HPO_4 (Ref. S7907, Sigma) and 4.14 g NaH_2PO_4 (Ref. S8282, Sigma) in 1 l of Milli-Q water.
- Ethanol 70 %, 95 % and 100 % each (e.g. Merck; stored at RT).
- Pepsin stock (Ref. P7012, Sigma).
- Pepsin solution: mix 10 μl of 1 M HCl and 2.5 μl vol. 20 ng ml^{-1} pepsin in 990 μl of Milli-Q water; make fresh as required.
- Postfix solution: mix 5 ml of 2 % paraformaldehyde (e.g. Merck), 4.5 ml of 1 × PBS and 0.5 ml 1 M $MgCl_2$; store at 4 °C.

2.4 Denaturation

- Denaturation buffer: 0.7 vol. of formamide, 0.2 vol. of Milli-Q water and 0.1 vol. 20 × SSC; store at 4 °C.
- Formamide (Ref. F7503, Sigma).
- Ethanol 70 % (at −20 °C), 95 % and 100 % (e.g. Merck; stored at RT).

2.5 Washing Slide

- Washing buffer: 0.2 vol. of 20 × SSC, 0.8 vol. Milli-Q water and 0.05 vol. of Tween 20.

- Tween 20 = polyoxyethylene-sorbitan monolaurate (Ref. 10670-1000, Sigma; stored at RT).
- Ethanol 70 %, 95 % and 100 % each (e.g. Merck; stored at RT).
- Marvel solution: dilute 0.1 g of milk powder Marvel in 2 ml of washing buffer (can use milk powder Molico®).
- Anti-digoxigenin-rhodamine (Ref. 11 207 750 910, Roche Diagnostics GmbH; stored at 2–8 °C).
- Antibody solution: 10 μl of anti-digoxigenin-rhodamine (200 μl ml^{-1}) and 990 μl of Marvel solution, make fresh as required.
- Fluoroshield™ with DAPI (Ref. F6057, Sigma; stored at 2–8 °C) or homemade: dissolve 1.5 μl of 1 M DAPI stock solution (4,6-diamidino-2-phenylindole · 2HCl: Ref. 124653, Merck) in 1 ml antifade VECTASHIELD (Ref. H1000, Vector Laboratories/Biozol; stored at 4 °C).

3 Methods

3.1 Homemade C_0t DNA

The C_0t-1 DNA is an unlabelled repetitive DNA fraction that is used to block nonspecific hybridization in FISH assays. Its use allows better and clearer results in some probes, such as genomic bacterial artificial chromosome (BAC) clone or whole chromosome painting.

1. Extract ~500 μg of the genomic DNA from the species of interest using PureLink® Genomic DNA Mini Kits, and dilute to 100–1,000 ng μl^{-1} DNA solution in Milli-Q water.
2. For denaturation and fragmentation of the DNA, use a heat block preheated at 120 °C for 2 min and 30 sec, using 0.5 μl tubes with safe-lock (*see* **Note 1**).
3. Reassociate the DNA at 60 °C for 15–150 min (*see* **Note 2**). Then place the tube with DNA on ice for 2 min.
4. Transfer the DNA tube to 42 °C and add preheated 10 × S1 nuclease buffer and S1 nuclease, then incubate for 1 h (*see* **Note 3**).
5. Precipitate DNA by adding 0.1 vol. of 3 M sodium acetate and 1 vol. of 2-propanol. Centrifuge at 14,000 rpm for 20 min at 4 °C.
6. Discard the supernatant, and dry the DNA pellet at room temperature (RT).
7. Add 100 μl DNA in 70 % ethanol, shake for 30 s and centrifuge again at 14,000 rpm for 10 min at 4 °C.
8. Discard the supernatant carefully, and dry the pellet.
9. Dissolve the DNA pellet in 50–100 μl of TE buffer solution.

10. Measure the DNA concentration and visualize by gel electrophoresis (*see* **Note 4**).

3.2 Self-Labelled and Denaturation Probes for FISH

The probe DNA can be obtained by flow sorting, microdissection or PCR (chapter by Fengtang Yang et al. "Generation of Paint Probes from Flow-Sorted and Microdissected Chromosomes"; chapter by Nadezda Kosyakova et al. "FISH-Microdissection"; chapter by Thomas Liehr "Homemade Locus-Specific FISH Probes: Bacterial Artificial Chromosomes") and labelled directly (e.g. SpectrumOrange, SpectrumGreen, TexasRed) or indirectly (e.g. biotin, digoxigenin). In our lab, we use indirect labelling, but there are available Nick translation kits for direct and indirect labelling.

3.2.1 Indirect Labelling Probes

1. Put in a microtube 3.5 μl of the probe DNA (recommended concentration 50 ng μl^{-1}) in 12.5 μl of Milli-Q water, and add 4 μl of the DIG-Nick Translation Mix. Mix and spin briefly.
2. Incubate for 90 min at 16 °C.
3. To stop the reaction, add 1 μl of 0.5 M EDTA and put at 65 °C for 10 min.
4. Store at −20 °C.

3.2.2 Denature Probe Solution for FISH

If it is not necessary to use C_0t DNA:

1. For each slide use 2 μl of the labelled probe diluted in 18 μl of hybridization buffer.
2. Denature for 5 min at 85 °C, and then keep the solution in ice or at 4 °C until use.

If it is necessary to use C_0t DNA:

1. Precipitate the labelled probe and C_0t DNA together using 1 vol. of 2-propanol and 0.1 vol. of sodium acetate (3 M, pH 5.2). For better results the use of proportions higher than 1:20 probe to C_0t1 DNA fraction is recommended.
2. Centrifuge at 14,000–15,000 rpm for 20 min at 4 °C, discard the supernatant and dry the DNA pellet at RT.
3. Dilute the pellet in 20 μl of hybridization buffer.
4. Denature the probe solution at 85 °C for 5 min. Then, the pre-hybridization step is done at 37 °C for 30 min.
5. Keep the solution in ice or at 4 °C until it is used.

3.2.3 RNAse and Pepsin Pretreatment

Most of the protocols in insects use RNAse and pepsin treatment, followed by postfixation, to reduce the background. However, in FISH made in ants and wasps, we see no significant change in the background amount and less degraded chromosomes when we skipped the RNAse and pepsin treatment, thus starting the procedure at step 4.

1. Put 100 μl RNAse solution on slide and cover with coverslip, and then incubate in humid chamber at 37 °C for 1 h.
2. Remove the coverslip and wash in 2 × SSC for 5 min and air-dry.
3. Add 50 μl of 0.005 % pepsin and cover with coverslip for 10 min at RT.
4. Wash in 1 × PBS for 5 min at RT.
5. Incubate the slide in 100 μl of postfix solution under coverslip for 10 min at RT.
6. Remove the coverslip and wash for 5 min in 1 × PBS.
7. Dehydrate slides in ethanol series (70 %, 95 % and 100 %, 3 min each one) and air-dry.

3.3 Denaturation

1. Put 100 μl of denaturation buffer on slide and cover with coverslip, and then denature on preheated hotplate at 73 °C for 3 min.
2. Remove the coverslip and place the slide in 70 % ethanol at −20 °C for 3 min; after, pass the slide through ethanol series, 95 % and 100 %, for 3 min each one at RT (*see* **Note 5**). Air-dry.
3. Add 20 μl of probe solution, already denatured, on the slide and cover with coverslip. Incubate the slide overnight at 37 °C in a humid chamber (*see* **Note 6**).

3.4 Detect Probe and Washing Slides

When labelled indirectly, the use of antibody conjugated with fluorochrome is necessary (e.g. anti-digoxigenin or anti-biotin attached with Cy3, Cy5 or rhodamine). If you use direct labelling, there is no need to make detection steps of the probes, and skip steps 3 to 6 of this Section.

1. Take slide from humid chamber at 37 °C and remove coverslip, and then wash the slide in 100 ml of 0.4 × SSC at 65–68 °C for 3–4 min in a Coplin jar placed into a water bath.
2. Transfer the slides into 100 ml of 4 × SSC/0.2 % Tween for 5 min at RT in a shaker.
3. Add 100 μl Marvel solution under coverslip at 37 °C for 10–15 min in a humid chamber.
4. Remove coverslip and wash the slide in 4 × SSC/Tween for 2 min at RT.
5. The detection is made through 100 μl of antibody solution for each slide covered with coverslip, and incubate for 20–35 min at 37 °C in a humid chamber.

6. Remove the coverslip and put the slide in 4 × SSC/0.2 % Tween for 5 min on the shaker; repeat this step three times using a new solution in each one.
7. Rinse in Milli-Q water briefly, and pass through ethanol series (70 %, 95 % and 100 %) for 3 min each one at RT. Air-dry.
8. Apply 20 μl of Fluoroshield™ with DAPI and cover with coverslip.
9. Evaluate through microscope.

4 Notes

1. The fragmentation of DNA can be made by autoclaving; autoclave for 5 min at 121 °C (1 atm). To finish this step, it is important to take a sample and check in gel if the fragmentation worked well.
2. The reassociation time can be calculated following the formula $t = \frac{C_0 t X \times 4.98}{C_0}$ where t is the time of incubation, X is the fraction of C_0t ($C_0t1 = 1$; $C_0t2 = 2$, $C_0t3 = 3$, etc.) and C_0 is the concentration of initial DNA in μg μl^{-1} [21]. The C_0t1 fraction is the most commonly used in FISH.
3. The amount of the S1 nuclease and 10 × S1 nuclease buffer is associated with the quantity of DNA and with the final total volume, respectively. For each 1 μg of DNA, 1U of S1 nuclease is used, and the buffer represents 10 % of the total volume (e.g. for 500 μg of DNA + 5.6 μl S1 nuclease (89 U μl^{-1}) + 50 μl of 10 × S1 nuclease buffer + 444.4 μl of water, the final total volume is 500 μl).
4. As C_0t DNA are fragments of repetitive DNA, normally, with sizes between 50 and 500 bp, you should see in gel a long band with more DNA concentration around 200 bp size (use a 100 bp DNA ladder, e.g. New England Biolabs, to help identify the size).
5. It is important that 70 % ethanol is very cold in this step. Then, make sure that it has been placed under refrigeration at −20 °C some hours before starting the protocol.
6. The time of incubation can vary; while some probes show better results when incubated for 8 h (overnight), others will need up to 48 h. However, longer times can cause an increase of the background. To reduce background, if the probe was labelled directly, incubate the slide upside down only in this step; if the probe was labelled indirectly, incubate the slide upside down in this step and in the antibody solution step.

References

1. May RM (2000) The dimensions of life on earth. In: Raven PH (ed) Nature and human society: the quest for a sustainable world. National Academy Press, Washington, DC
2. Brusca RC, Brusca GJ (2003) Invertebrates, 2nd edn. Sinauer Associates, Sunderland, MA
3. Raven HP, Yeates DK (2007) Australian biodiversity: threats for the present, opportunities for the future. Aust J Entomol 46:177–187
4. Crosland MWJ, Crozier RH (1986) *Myrmecia pilosula*, an ant with only one pair of chromosomes. Science 231:1278
5. Lukhtanov VA (2015) The blue butterfly *Polyommatus* (*Plebicula*) *atlanticus* (Lepidoptera, Lycaenidae) holds the record of the highest number of chromosomes in the non-polyploid eukaryotic organisms. Comp Cytogenet 9: 683–690
6. Sánchez L (2008) Sex-determining mechanisms in insects. Int J Dev Biol 52:837–856
7. Rudkin GT, Stollar BD (1977) High resolution detection of DNA-RNA hybrids in situ by indirect immunofluorescence. Nature 265: 472–473
8. Pardue ML, Gall JG (1969) Molecular hybridization of radioactive DNA to the DNA of cytological preparations. Proc Natl Acad Sci USA 64:600–604
9. Grozeva S, Kuznetsova VG, Anokhin BA (2011) Karyotypes, male meiosis and comparative FISH mapping of 18S ribosomal DNA and telomeric $(TTAGG)_n$ repeat in eight species of true bugs (Hemiptera, Heteroptera). Comp Cytogenet 5:355–374
10. Palacios-Gimenez OM, Cabral-de-Mello DC (2015) Repetitive DNA chromosomal organization in the cricket *Cycloptiloides americanus*: a case of the unusual X_1X_20 sex chromosome system in Orthoptera. Mol Genet Genomics 290:623–631
11. Antbase (2016) http://www.antbase.org, accessed 18 May 2016
12. Hirai H, Yamamoto MT, Ogura K et al (1994) Multiplication of 28S rDNA and NOR activity in chromosome evolution among ants of the *Myrmecia pilosula* species complex. Chromosoma 103:171–178
13. Meyne J, Hirai H, Imai HT (1995) FISH analysis of the telomere sequences of bulldog ants (*Myrmecia*: Formicidae). Chromosoma 104: 14–18
14. Hirai H, Yamamoto MT, Taylor RW et al (1996) Genomic dispersion of 28S rDNA during karyotypic evolution in the ant genus *Myrmecia* (Formicidae). Chromosoma 105: 190–196
15. Lorite P, Aránega AE, Luque F et al (1997) Analysis of the nucleolar organizing regions in the ant *Tapinoma nigerrimum* (Hymenoptera, Formicidae). Heredity 78:578–582
16. Mariano CSF, Pompolo SG, Barros LAC et al (2008) A biogeographical study of the threatened ant *Dinoponera lucida* Emery (Hymenoptera: Formicidae: Ponerinae) using a cytogenetic approach. Insect Conserv Diver 1:161–168
17. Santos IS, Costa MA, Mariano CSF et al (2010) A cytogenetic approach to the study of neotropical *Odontomachus* and *Anochetus* ants (Hymenoptera: Formicidae). Ann Entomol Soc Am 103:424–429
18. Aguiar HJAC, Barros LAC, Mariano CSF (2011) 45S rDNA localization for the giant ant *Dinoponera gigantea* (Perty, 1833) with evolutionary inferences for *Dinoponera* genus (Formicidae: Ponerinae). Sociobiology 57: 607–620
19. Barros LAC, Aguiar HJAC, Andrade-Souza V et al (2012) Occurrence of pre-nucleolar bodies and 45S rDNA location on the chromosomes of the ant *Mycocepurus goeldii* (Forel) (Formicidae, Myrmicinae, Attini). Hereditas 149:50–54
20. Barros LAC, Aguiar HJAC, Teixeira GA et al (2015) Cytogenetic data on the threatened leafcutter ant *Atta robusta* Borgmeier, 1939 (Formicidae: Myrmicinae: Attini). C R Biol 338:660–665
21. Timoshevskiy VA, Sharma A, Sharakhov et al (2012) Fluorescent in situ hybridization on mitotic chromosomes of mosquitoes. J Vis Exp 67:e4215. doi:10.3791/4215

FISH in *Drosophila*

Amanda M. Larracuente

Abstract

Cytogenetic studies of genome structure and evolution across taxa often depend on mapping specific tandemly repeated sequences to chromosome arms. Large blocks of tandem repeats are detected by hybridizing fluorescent probes to mitotic chromosomes. This protocol offers a simple approach to mapping medium-to-high copy number tandem repeats in *Drosophila* genomes using mitotic chromosomes found in larval neuroblasts.

Keywords *Drosophila*, Mitotic chromosomes, Tandem repeats, Chromosomes from brain tissue, Polytene chromosomes

1 Introduction

Tandemly repeated sequences can make up a large fraction of eukaryotic genomes. These sequences change rapidly over short periods of evolutionary time—closely related species often differ in their distribution and abundance of tandemly repeated sequences [1]. Mapping specific repetitive sequences to chromosomes is important for cytogenetic studies of genome structure and evolution (chapter by Thomas Liehr et al. "cenM-FISH Approaches"; chapter by Thomas Liehr et al. "Heterochromatin Directed M-FISH (HCM-FISH)"; chapter by Fengtang Yang et al. "Animal Probes and ZOO-FISH"; chapter by Cassia Yano et al. "Fish-FISH: Molecular Cytogenetics in Fish Species"; chapter by Anna Zlotina and Alla Krasikova "FISH in Lampbrush-Chromosomes"; chapter by Ana Paula Alves-Silva et al. "General Protocol of FISH for Insects"; chapter by Ekaterina Badaeva et al. "In Situ Hybridization to Plant Chromosomes"; chapter by Harry Scherthan "Yeast Chro mosome Dynamics Revealed by FISH"; chapter by Benedetta Bottari et al. "FISHing for Food Microorganisms"). In *Drosophila*, fluorescence in situ hybridization to squashed brain tissue offers an effective approach to studying the localization of specific repeat families. Larval neuroblasts are the preferred tissue for mapping

Thomas Liehr (ed.), *Fluorescence In Situ Hybridization (FISH)*, Springer Protocols Handbooks,
DOI 10.1007/978-3-662-52959-1_47,

heterochromatic sequences along mitotic chromosomes [2]. While some large blocks of tandem repeats are easy to probe using fluorescently labeled oligonucleotides, others are more difficult due to the composition of the locus. One important factor to consider when choosing a probe type is the sequence diversity among repeats in a tandem array. For complex repeats with a higher-order structure and/or sequence diversity, longer probes generated by PCR followed by nick translation or random priming tend to work the best (for probe instructions, see Thomas Liehr et al. "The Standard FISH Procedure"). The amount of sequence diversity in a tandem array of repeats can be determined using molecular methods, such as modifying hybridization conditions in a southern blot or using computational methods with genomic data [3].

The following protocol is modified from Larracuente and Ferree [4] and Pimpinelli et al. [2, 5] to offer a simple and effective method for detecting medium-to-high copy-number tandem repeats in *Drosophila* genomes. This includes relatively small blocks of repeats (<50 kb) with sequence diversity among repeats in the array. This protocol also works well on other squashed tissues in several insect species (e.g., *Drosophila*, *Nasonia*, and *Photinus* testes).

Outline

- Dissect and fix tissue on a microscope slide
- Hybridize and wash slides
- Block and detect probe
- Wash and perform any amplification steps
- Mount and examine slide

2 Materials

Probe Types: Nick-translated PCR products, oligonucleotide probes, otherwise directly fluorescently labeled probes. Includes signal amplification steps for biotin probes. See the chapter by Thomas Liehr et al. "The Standard FISH Procedure", for probe recommendations. Here we will focus on biotinylated probes, as these are inexpensive to make for nearly any sequence.

2.1 Tissue Types

- Although specifically developed for squashed *Drosophila* brains, this protocol works well for several other squashed insect tissues (e.g., *Drosophila*, *Nasonia*, or *Nasonia* testes, *Drosophila* salivary glands, and polytene chromosomes).

2.2 Sequence Type

- Tandem repeats, bacterial artificial chromosomes (BACs), or sequences corresponding to pooled long PCR products can be used.

2.3 Specific Chemicals

- Biotinylated Anti-Avidin (Cat. No.: BA-0300, Vector Laboratories)
- Rhodamine-Avidin D (Cat. No.: A-2002,Vector Laboratories)
- Vectashield Antifade Mounting Medium with DAPI (Cat. No.: H-1200, Vector Laboratories)

2.4 Solutions to Be Prepared

- 10 × PBS (1 l): 80 g NaCl, 2.0 g KCl, 14.4 g Na_2HPO_4, 2.4 g KH_2PO_4 pH to 7.4 with H_2O
- SBT (10 ml): 2 ml 20 × SSC, 0.01 g bovine serum albumin (BSA), 10 μl Tween, and 7.9 ml H_2O
- 20 × SSC: 175.3 g NaCl, 88.2 g Na citrate in 800 ml H_2O. pH to 7, H_2O, to 1 l
- Hybridization buffer (20 μl): 10 μl formamide, 4 μl 50 % dextran sulfate, 2 μl 20 × SSC, 4 μl H_2O
- Blocking solution: 0.3 g BSA, 10 μl Tween, 2 ml 20 × SSC, 8 ml H_2O
- 4 × SSC with 0.1 % Tween (4 × SSCT): 200 ml 20 × SSC, 1 ml Tween, 799 ml H_2O
- 0.1 × SSC: 5 ml 20 × SSC in 995 ml H_2O
- 10 mM phosphate buffer (100 ml): 0.026 g KH_2PO_4, 0.217 g Na_2HPO_4, 0.871 g NaCl in 80 ml H_20. pH to 7.4. H_2O to 100 ml
- Hypotonic solution: 0.5 % sodium citrate
- 45 % glacial acetic acid
- Fixative: 1.8 % paraformaldehyde in 45 % acetic acid, prepared fresh

3 Methods

For a schematic outline of the method, see Fig. 1.

3.1 Dissect and Fix Tissue

1. Place four actively crawling 3rd instar *Drosophila* larvae in a drop of 1 × PBS. With sharp forceps, grab hold of the mouthparts and ~2/3 of the way down the body. Gently tug with the forceps holding the mouthparts to expose the brain.
2. Separate the brain from the surrounding tissue and place in a drop of clean 1 × PBS. Continue dissecting larvae to collect four brains.
3. Place the brains in a hypotonic solution of 0.5 % sodium citrate for 5–10 min. Incubating for longer than 10 min can cause complete separation of sister chromatids.
4. Move brains to a drop of 45 % glacial acetic acid on a coverslip for 1 min (*see* **Note 1**).

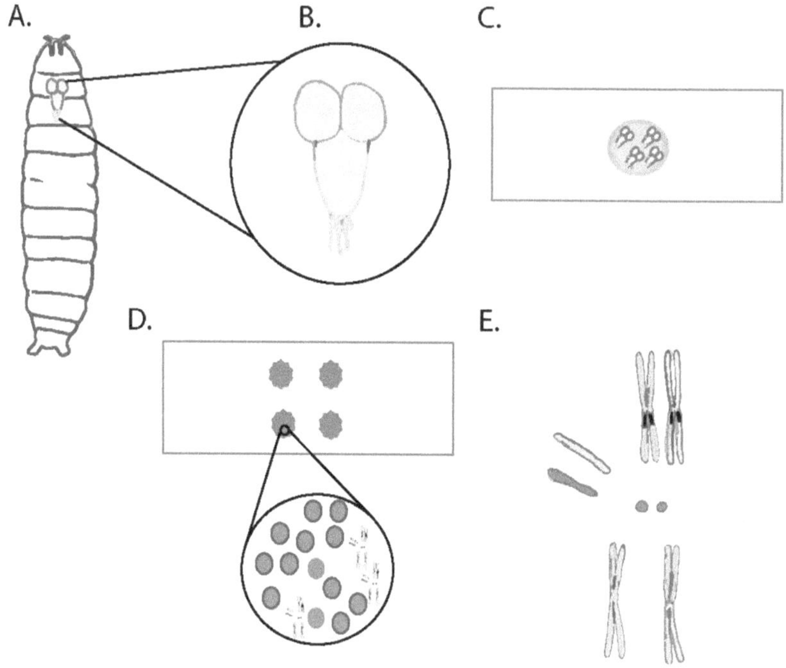

Fig. 1 FISH in *Drosophila* larval neuroblasts: (**a**) Choose 3rd instar larvae that are actively crawling up the side of the vials. The brain is located just below the mouthparts. (**b**) The *Drosophila* brain has a distinct structure that is easy to distinguish from the surrounding tissue. (**c**) Fix four brains on a poly-L-lysine-coated slide prior to squashing. (**d**) Mitotic chromosomes are found among neuroblast nuclei from squashed brains. (**e**) *Drosophila* mitotic figures with probe hybridized to pericentric heterochromatin

5. Place a drop of fixative on a clean poly-L-lysine-coated slide.
6. Turn the slide over and gently pick up the coverslip containing brains with the slide and let sit for 1 min (*see* **Note 2**).
7. Place slide in a folded piece of filter paper and gently blot off excess fixative without letting the coverslip shift positions. Press down firmly with a thumb for 20–30 s (*see* **Note 3**).
8. Gently lower slide into liquid nitrogen for 1 min. Flick off coverslip with a razor blade (*see* **Note 4**).
9. Place slide in 100 % ethanol for 20 min and air-dry (if using fresh slides may proceed to step 3 described under Sect. 3.2).

3.2 Hybridize and Wash

1. Dehydrate slides in 70 % EtOH for 3 min, 90 % EtOH for 3 min, followed by 100 % EtOH for 3 min.
2. Air-dry slides for at least 30 min.
3. For each slide, mix 80–100 ng of each probe with 20 μl of hybridization buffer. Add all probes (and all probe types) at the same time (*see* **Note 5**).
4. Add 20 μl of hybridization buffer containing probes to each slide, cover with a coverslip, and transfer slides to 95 °C heat block for 6 min.

5. Allow slide to cool slightly, wrap in parafilm, and place in humid chamber (*see* **Note 6**).
6. Incubate overnight at 30 °C (*see* **Note 7**).
7. Gently lift coverslip off slide.
8. Wash slides 3 × in 4 × SSCT at 42 °C for 5 min each wash.
9. Wash slides 3 × in 0.1 × SSC at 60 °C for 5 min each wash. When only using directly fluorescently labeled probes, stop here and skip to Sect. 3.5. Otherwise, continue for biotin (or DIG) probes.

3.3 Block and Detect

1. Blot excess liquid off sample, apply 100 μl blocking solution, cover with coverslip, wrap slide in parafilm, and incubate for 30 min in a humid chamber at 37 °C.
2. Blot excess liquid off sample, apply 80 μl of avidin-rhodamine diluted 1:100 in SBT solution, cover with coverslip, wrap in parafilm, and incubate in a dark humid chamber at 37 °C for 30 min.
3. Wash slides 3 × in 4 × SSCT at 42 °C for 5 min each wash.
4. Wash slides 3 × in 0.1 × SSC at 60 °C for 5 min each wash. If using a high copy-number probe, skip to Sect. 3.5. Otherwise, continue with protocol for amplification of signal (*see* **Note 8**).

3.4 Signal Amplification

1. Blot excess liquid off sample, apply 100 μl of biotinylated anti-avidin (3 μg ml^{-1} in 10 mM phosphate buffer), and incubate in dark at 37 °C for 30 min.
2. Wash slides 3 × in 4 × SSCT for 5 min each wash.
3. Wash slides 3 × in 0.1 × SSC for 5 min each wash.
4. Block by adding 10 % normal goat serum to sample, cover with coverslip, wrap slide in parafilm, and incubate in dark humid chamber at 37 °C for 30 min.
5. Blot excess liquid off sample, apply 80 μl of avidin-rhodamine diluted 1:100 in SBT solution, cover with coverslip, wrap in parafilm, and incubate in a dark humid chamber at 37 °C for 30 min.
6. Wash slides 3 × in 4 × SSCT for 5 min each wash.
7. Wash slides 3 × in 0.1 × SSC for 5 min each wash.

3.5 Rinse and Mount

1. Rinse slide in double-distilled H_2O.
2. Air-dry and mount slide in Vectashield with DAPI.
3. Secure coverslip to slide with clear nail polish and let dry for 30 min.
4. Image slides on fluorescence microscope.

4 Notes

1. If there are problems with tissue sticking to coverslips after step 4, then coverslips can be siliconized with Sigmacote.
2. Invert slide and lower to the point where it just barely touches the drop of 45 % acetic acid and lifts coverslip toward the slide. Quickly invert the slide with the coverslip attached.
3. Be sure to apply downward pressure to prevent the coverslip from shifting and smearing tissue. Before proceeding to the next step, it may be useful to turn the slide face down and mark the slide on opposite ends of the coverslip with a glass etcher. This is helpful for identifying correct placement of the coverslip for steps 2–5.
4. Take care not to scrape the tissue with the razor blade.
5. If after the FISH procedure there are problems with background, precipitate probes as follows: mix biotinylated (or DIG-labeled) probe with 3 μg of salmon sperm DNA, 1/10 volume 3 M sodium acetate (pH 4.5), and 2 volumes cold ethanol. Chill at −80 °C for 15 min. Centrifuge at 13,000× *g* for 15 min, remove ethanol, and air-dry pellet. Resuspend pellet in 20 μl hybridization buffer (per slide) and vortex.
6. A humid chamber can easily be constructed from a Tupperware container or empty tip box with dampened tissue paper at the bottom.
7. 30 °C works well for AT-rich probes, for less AT-rich probes, try 37 °C. Some experimentation with hybridization temperatures may be required.
8. For most heterochromatic probes corresponding to large blocks of tandem repeats, it will be fine to stop here. Some probes corresponding to lower copy-number tandem repeats benefit from a signal amplification step.

References

1. Yunis JJ, Yasmineh WG (1971) Heterochromatin, satellite DNA, and cell function. Structural DNA of eucaryotes may support and protect genes and aid in speciation. Science 174:1200–1209
2. Pimpinelli S, Bonaccorsi S, Fanti L et al (2000) Preparation and analysis of mitotic chromosomes. Cold Spring Harbor Laboratory Press, Cold Spring Harbor, NY, pp 3–24
3. Larracuente AM (2014) The organization and evolution of the Responder satellite in species of the *Drosophila melanogaster* group: dynamic evolution of a target of meiotic drive. BMC Evol Biol 14:233
4. Larracuente AM, Ferree PM (2015) Simple method for fluorescence DNA in situ hybridization to squashed chromosomes. JoVE 95: e52288
5. Pimpinelli S, Bonaccorsi S, Fanti L, et al (2010) Fluorescent in situ hybridization (FISH) of mitotic chromosomes from *Drosophila* larval brain. Cold Spring Harbor protocols 2010(3): pdb prot5391

FISH on Insect Cells Transfected with Heterologous DNA

Thomas Liehr

Abstract

Insect cells transfected with heterologous DNA can be used for the assessment of functional properties of genes that have been isolated or subjected to in vitro mutagenesis. An example for such a cell system is embryonic cells derived from *Drosophila melanogaster*. These can be transfected with special plasmids and used for heterologous protein expression. FISH is an elegant technique to perform the control of transfection efficiency on a single-cell level. The background and approach are described here in detail.

Keywords Schneider cells, Transfection, Insect cells, *Drosophila melanogaster*, Heterologous DNA, Plasmid

1 Introduction

Insect cells transfected with heterologous DNA, such as plasmids or BACs (bacterial artificial chromosomes), can be used to assess the functional properties of genes that have been isolated or subjected to in vitro mutagenesis [1, 2]. Schneider cells provide an example of a cell system that is suited to such an approach [3]. These are derived from embryonic cells of *Drosophila melanogaster* (chapter by Amanda Larracuente "FISH in Drosophila") and can be transfected with special plasmids [4] and used for heterologous expression. Several generally reliable in vitro methods can be performed to monitor the transfection rate and efficiency in such cells (i.e., PCR and Southern, Northern, or Western blot approaches). However, all of these protocols are applied to monitor transfection within a cell population and not in the single cell. FISH is an approach that is suited to evaluating single cells and characterizing mosaics (chapter by Ivan Iourov et al. "Interphase FISH for Detec tion of Chromosomal Mosaicism"). Also, in contrast to approaches testing heterologous expression at the protein level (e.g., by green fluorescent protein [5]), the FISH approach also allows the number of integrated heterologous DNA copies to be determined. Here, a

Thomas Liehr (ed.), *Fluorescence In Situ Hybridization (FISH)*, Springer Protocols Handbooks,
DOI 10.1007/978-3-662-52959-1_48, © Springer-Verlag Berlin Heidelberg 2017

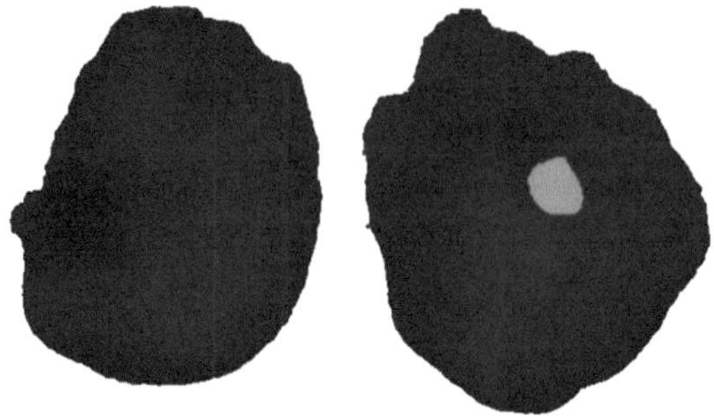

Fig. 1 FISH of Schneider cells that are transfected (*right nucleus*) or not (*left nucleus*) with plasmid pRmHa-3. The plasmid pRmHa-3 was labeled with biotin and detected with avidin-FITC. The nuclei are counterstained with DAPI (4,6-diamidino-2-phenylindole · 2HCl). Images were captured on a Zeiss Axioplan microscope using the ISIS digital FISH imaging system (MetaSystems, Altlussheim, Germany) using a XC77 CCD camera with on-chip integration (Sony). The transfected Schneider cells were kindly provided by Dr. B. Rautenstrauss, Erlangen, Germany

fast and reliable method for assessing the transfection rate of Schneider cells at both the DNA and the single-cell levels is outlined. It is based on the FISH technique and was first described by Rautenstrauss et al. [6].

The protocol was successfully used to transfect Schneider cells with pRmHa-3. Different mutated versions of myelin protein zero (*MPZ*) were introduced beforehand into this plasmid. FISH tests proved that transfection was successful (Fig. 1 [6]). Cell adhesion tests resulted in a decreased adhesion capability in comparison with wild-type *MPZ* [7].

2 Materials

Apart from standard cell biological and molecular cytogenetic equipment, including standard solutions (e.g., ethanol, methanol, formamide, formaldehyde, etc.), the more specialized equipment needed for FISH itself is listed in chapter by Thomas Liehr et al. "The Standard FISH Procedure".

3 Methods

3.1 Preparation of the Schneider Cells for FISH

1. Cultivate Schneider cells transfected for transient expression (e.g., with a derivate of the plasmid pRmHa-3 [4]) and wild-type Schneider cells (without any plasmid, used as negative controls) according to [3].

2. Collect cells of one tissue culture flask in a 15 ml tube and sediment by centrifugation (1,000 rpm, 5 min).
3. Carefully discard the supernatant, resuspend the pellet in 5 ml 1 % formaldehyde solution (in 1 × PBS), and incubate at room temperature (RT) for 2–3 h.
4. Sediment the cells by centrifugation (1,000 rpm, 5 min) and discard the supernatant.
5. Wash the resulting pellet twice in 5 ml 1 × PBS and resuspend in 0.5 ml 1 × PBS.
6. Transfer approx. 70 μl of fixed Schneider cells onto clean and dry slides, and fix by air-drying overnight. Afterward, the slides can be stored at −20 °C for several weeks before applied in FISH (*see* **Note 1**).

3.2 Labeling of the Probes

1. The DNA probe that was used to transfect the Schneider cells is now used as the specific FISH probe. Thus, label this DNA with, e.g., biotin, by nick translation using the kit from Roche (Basel, Switzerland).
2. Dilute 1 μg of the probe in 16 μl of double-distilled water and add 4 μl of the nick translation solution. Mix carefully with the tip from a 20 μl Eppendorf pipette and incubate the 1.5 μl microtube at 15 °C for 90 min.

3.3 Fluorescence In Situ Hybridization (FISH)

As described in chapter by Thomas Liehr et al. “The Standard FISH Procedure” (*See* **Note 2**).

4 Notes

1. Remember to prepare the cells to be tested for transfection and the wild-type cells as negative controls. The approach described here always needs a negative control.
2. Postwash the slides with formamide solution and detect non-fluorescent haptens by fluorophore-labeled hapten-directed antibody.

References

1. Rubin GM, Spradling AC (1983) Vectors for P element-mediated gene transfer in Drosophila. Nucleic Acids Res 11:6341–6351
2. Snow PM, Bieber AJ, Goodman CS (1989) Fasciclin III: a novel homophilic adhesion molecule in Drosophila. Cell 59:313–323
3. Schneider I (1972) New cell line from embryonic cells of Drosophila melanogaster. J Embryol Exp Morph 27:353–356
4. Bunch TA, Grinblat Y, Goldstein LS (1988) Characterization and use of the Drosophila metallothionein promoter in cultured Drosophila melanogaster cells. Nucleic Acids Res 16:1043–1061
5. Keith MB, Farrell PJ, Iatrou K et al (1999) Screening of transformed insect cell lines for recombinant protein production. Biotechnol Prog 15:1046–1052

6. Rautenstrauss B, Fuchs C, Ekici A et al (1998) Assay of transfection rate in insect cells on a single cell level. Genet Anal Biomol Eng 14:103–104
7. Ekici AB, Fuchs C, Nelis E et al (1998) An adhesion test system based on Schneider cells to determine genotype-phenotype correlations for mutated P0 proteins. Genet Anal 14:117–119

In Situ Hybridization to Plant Chromosomes

Ekaterina D. Badaeva, Alevtina S. Ruban, Lala Aliyeva-Schnorr, Celia Municio, Susann Hesse, and Andreas Houben

Abstract

In situ hybridization is a fundamental method in modern plant molecular cytogenetics research. It cannot be overstressed that good in situ hybridization results begin with good cytological preparations. Well-spread chromosome preparations with low amounts of cytoplasm give the best hybridization signals. Here, we present different protocols for the preparation of mitotic and meiotic plant chromosomes as well as tissue sections suitable for the detection of low-/single-copy sequences in a range of plant species. In addition a method suitable for the detection of single-copy probes is provided.

Keywords Chromosome preparation, Meiosis, Mitosis, Tissue sections, Single-copy FISH, Plant, Plant chromosomes

1 Introduction

In situ hybridization (ISH) is a fundamental method in modern plant molecular cytogenetics research. All in situ hybridization techniques follow a basic principle that is the detection of specific sequences on cytological targets like metaphase chromosomes, interphase nuclei, or extended chromatin fibers by (fluorescence) labeled short DNA probes [1, 2]. ISH of nucleic acid sequences on morphologically preserved chromosomes in plants was first achieved in the early 1970s to detect ribosomal genes in *Zea mays* and *Phaseolus*, as well as highly repetitious tandem repeats in *Vicia faba* and *Scilla siberica* [3–5]. At this time, for the detection of hybridized sequences, radioisotopes were the only labels available. Aside from safety problems regarding the use of radioisotopes, ISH suffered for limited spatial resolution, poor stability, and time-consuming autoradiography. At the beginning of the 1980s, non-radioactive ISH on wheat chromosomes using biotin-labeled probes was introduced [6]. Subsequent improvements on molecular techniques and availability of better nonradioactive labels, e.g.,

Thomas Liehr (ed.), *Fluorescence In Situ Hybridization (FISH)*, Springer Protocols Handbooks,
DOI 10.1007/978-3-662-52959-1_49,

fluorophores, allowed a huge progress on development and applications of ISH. In situ hybridization using fluorescence-based detection (FISH) was first applied to plants by Schwarzacher et al. [7] and Yamamoto et al. [8] in 1989. Since then FISH became a powerful tool for karyotyping, phylogenetic analysis, and physical mapping [9]. The most important advantage of fluorescence-based ISH over classical cytogenetics techniques is the possibility of simultaneous detection of several sequences [1]. By the choice of probes, FISH can be categorized into three groups of applications, i.e., whole chromosome/genome detection (GISH, chromosome-specific painting), detection of repetitive sequences, and locus-specific detection (single-copy FISH) ([9–12], chapter by Thomas Liehr "Classification of FISH Probes"). Whereas FISH is widely applied to locate repeated sequences ([13], chapter by Cassia Yano et al. "Fish-FISH: Molecular Cytogenetics in Fish Species"; chapter by Ana Paula Alves-Silva et al. "General Protocol of FISH for Insects"), more recently, FISH technology has been improved in plants for the detection of small and single-copy fragments [14–16]. An improvement of spatial FISH resolution can be achieved by the use of super-stretched chromosomes or DNA fibers (chapter by Sandra Louzada et al. "Fluorescence In Situ Hybridization onto DNA Fibers Generated Using Molecular Combing") rather than mitotic metaphase chromosomes. While FISH on super-stretched chromosomes provides a resolution of up to 70 kb, fiber-FISH allows for fine-mapping of up to a few kilobases [9, 17, 18]. In this chapter we present protocols for the preparation of mitotic and meiotic chromosomes as well as tissue sections suitable for the detection of low-/single-copy sequences in a range of plant species.

2 Materials

2.1 For Chromosome Preparation, Tissue Sections, and FISH

- Heating block (MEDAX GmbH 12603).
- Fine-point forceps (10–12 cm long).
- Scalpels.
- Dissecting needle (#HSO 104-03, Hammacher Instrumente).
- Straight teasing needles.
- Sharpened wooden stick.
- Microscopic slides and coverslips (18 × 18 mm, 20 × 20 mm, 24 × 24 mm, 24 × 32 mm) (*see* **Note 1**).
- Poly-L-lysine-coated slides (e.g., #P0425, Sigma).
- Razor blades (two-edge type).
- Coplin glass jars with lids (e.g., #900470 or 900630, Wheaton).
- Hellendahl plastic jars (e.g., #45000355, Auxilab).

- Embryo dishes for enzyme treatment (RA Lamb Embryo Dishes #90, Thermo Scientific).
- Diamond pen.
- Slide holders.
- 10 ml glass or plastic beakers.
- Parafilm.
- Aluminum foil.
- Rubber cement: Fixogum™ (Marabu, Tamm, Germany; store at room temperature = RT).
- Silicone casting molds (e.g., Plano-EM).
- Dry ice or liquid nitrogen.
- 20 × SSC (saline sodium citrate buffer, 3 M NaCl, 300 mM trisodium citrate): dissolve 175.3 g of NaCl and 88.2 g of Na-citric ($Na_3C_6H_5O_7*2H_2O$ or 107.4 g $Na_3C_6H_5O_7*5.5H_2O$) in ~900 ml of dH_2O. Adjust pH to 7.0 with hydrochloric acid, add H_2O to total volume of 1 l. Autoclave and store at 4 °C.
- 10 × PBS (phosphate-buffered saline): dissolve 80 g NaCl, 2 g KCl, 14.4 g Na_2HPO_4, and 2.4 g KH_2PO_4 in 800 ml dH_2O, and adjust the pH to 7.4 if necessary. Add dH_2O until a total volume of 1 l. Autoclave and store at 4 °C.
- 0.01 M citrate buffer: dissolve 2.94 g Na-citric ($Na_3C_6H_5O_7*2H_2O$) and 2.1 g citric acid ($C_6H_8O_7*H_2O$) in 1 l H_2O; adjust pH to 4.5–4.8.
- 0.01 M sodium citrate buffer: dissolve 2.94 g $Na_3C_6H_5O_7*2H_2O$ in 1 l of dH_2O; adjust pH to 6.0 with HCl.
- Enzyme mix: 0.7 % Cellulase R10 (#C8001, Duchefa), 0.7 % Cellulase (#219466, Calbiochem), 1 % Pectolyase (#P3026, Sigma), and 1 % Cytohelicase (#C8274, Sigma) diluted in 0.01 M citrate buffer. Store the enzyme mix at −20 °C and reuse up to 5×.
- Fixative 3:1 of ethanol and glacial acetic acid. Make fresh on the day of use (*see* **Note 2**).
- Paraformaldehyde (PFA) fixative solution (50 ml 3 % paraformaldehyde (precaution: toxic) needs to be discarded after use as hazardous waste): add 5 μl 1 M NaOH to 40 ml dH_2O and mix with 1.5 g paraformaldehyde (e.g., PanReac AppliChem). Heat at 60 °C until PFA is dissolved. Add 5 ml 10 × PBS and adjust pH to 7.0 with 1 M HCl. Add 250 μl 10 % Triton X-100 and adjust volume to 50 ml (make fresh as required).
- Carnoy's fixative: 6:3:1 ethanol:chloroform:acetic acid (#9065, Roth; #3313, Roth; #100063, Merck Millipore). Store at 4 °C.
- Formaldehyde 4 % (e.g., Roth; store at RT).

- Polyester wax: 90 g poly(ethylene glycol) distearate (average M_n ~930, #305413, Sigma; store at RT), heat to 60 °C and add 10 g 1-hexadecanol (cetyl alcohol, #A3595, PanReac AppliChem; store at RT). Store at 37 °C if need to be liquid.
- Potassium permanganate (0.1 mol l^{-1}, #KK69.1, Roth; toxic to aquatic life with long-lasting effects; avoid release to the environment—store in the tightly closed dark glass bottle).
- 1 % acetocarmine: 1 % carmine in 45 % acetic acid (e.g., #CB91876528, ProChem; store in dark bottle at +4°C).
- Acetic acid (glacial) 100 % (#100063 Merck Millipore) and diluted to 45 % and 60 %.
- Triton X-100 10 % solution: dissolve 1 ml of Triton X-100 (e.g., PanReac AppliChem; may be harmful by inhalation, ingestion, or skin absorption; wear appropriate protection. Store at RT) in 9 ml ddH_2O by stirring. Store at 4 °C.
- Glycerol (e.g., Merck; store at RT).
- Double-distilled water (#235 1544, Braun; aliquot and store at −20 °C).
- Colchicine (#C9754, Sigma; fatal if swallowed or inhaled, may cause genetic defects; wear appropriate protection. Store at RT).
- 8-Hydroxychinoline (#820261, Merck Millipore; harmful if swallowed. Store at RT).
- 9-Aminoacridine (#A7295, Sigma; may cause skin, eye, and respiratory irritation; wear appropriate protection. Store at 2–8 °C).
- 0.5 M EDTA (e.g., Merck; store at −20 °C).
- 100 % ethanol (e.g., Merck; store at RT).
- 10 × nick translation buffer (0.5 M Tris-HCl pH 7.5, 50 mM $MgCl_2$, 0.05 % bovine serum albumin).
- dNTPs (#ID201900, QIAGEN).
- Labeled dUTPs (1 mM) (Texas Red-12-dUTP, #C3176, Invitrogen; Alexa 488, #C11397, Invitrogen; aminoallyl-dUTP-Atto 550, #NU-803-550-S, Jena Bioscience).
- Mercaptoethanol, 0.1 M (e.g., Merck; store at RT).
- 10 × PCR buffer (Fermentas).
- DNA polymerase I (#EP0042, Fermentas).
- DNase I (0.1 U) (#EN0521, Fermentas).
- Herring sperm (#D1811, Promega).
- 3 M sodium acetate solution (pH 5.2; e.g., Merck; store at −20 °C).
- 1 × TE buffer (10 mM Tris-HCl; 1 mM EDTA).

- Nuclease-free H_2O.
- Deionized formamide (#1 09684 2500, Merck; aliquot and store at −20 °C; please remember to discard the formamide solution as hazardous waste).
- 4 × hybridization buffer (20 × SSC; 1 M Tris-HCl pH 8; 1.5 M EDTA; 10 μg μl^{-1} fish sperm (#11467140001, Roche)).
- DAPI (4′,6-diamidino-2-phenylindole dihydrochloride), 1 μg ml^{-1} in antifade mounting medium (VECTASHIELD, Vector Lab H-1200).

3 Methods

3.1 Chromosome Preparations

It cannot be overstressed that good in situ hybridization results begin with good cytological preparations. Well-spread chromosome preparations with low amounts of cytoplasm give the best hybridization signals. Most FISH analyses have been done on mitotic root tip preparations and pollen mother cells at meiosis. In addition, mitotic chromosomes can be prepared from young shoots, buds, and young spikes.

3.1.1 Mitotic Chromosomes: Squashing of Meristems After Acetocarmine Staining

1. Germinate seeds on moist filter paper in a Petri dish. The seeds can be preliminary treated in potassium permanganate solution for 15 min to avoid fungal infection (*see* **Note 3**).
2. To accumulate metaphases, treat excised roots (2–3 cm long) (*see* **Note 4**) with one of the metaphase-arresting agents. Optimal conditions may differ for different species (*see* **Note 5**).
3. Fix the roots in ethanol:acetic acid (3:1) fixative for 4 days at RT. Material can be kept in fixative solution at −20 °C up to 1 year. If longer storage of the material is required, transfer the roots into 70 % ethanol at −20 °C.
4. The method of squashing should be chosen based on the thickness and composition of cell walls in the material examined, which depends on the type of material (herbs, trees, tissue culture) and the source of meristem (i.e., roots from seedlings, roots from adult plants, shoots or buds, spikes).

   Squashing of acetocarmine-stained tissues appropriate for root meristem of cereals and other herbs and young spikes cannot be applied for material stored in ethanol.
   Stain the root tips in 1 % acetocarmine for 15 min up to 1 h at RT.
5. Place the root on slide (Fig. 1a), cover it with filter paper, and cut the root cap with a razor blade (Fig. 1b). Gently squeeze a small piece of the meristem using a dissecting needle (Fig. 1c).

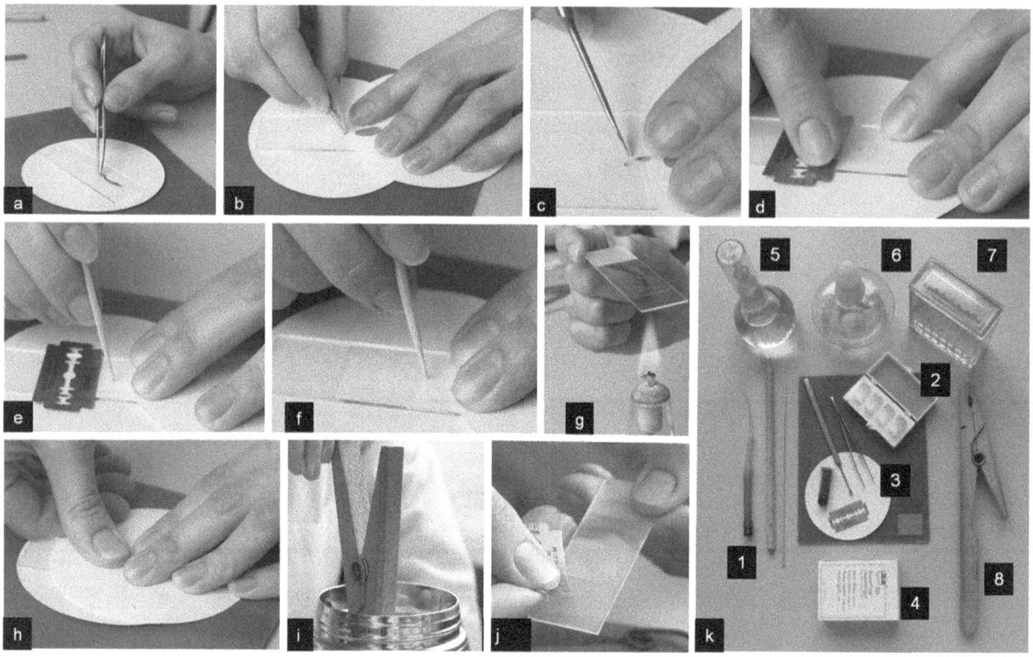

Fig. 1 Step-by-step preparation of mitotic chromosomes by squashing after acetocarmine staining: (**a–c**) squeezing of meristematic tissue from the root; (**d–f**) spreading of cells under coverslip using a sharpened wooden stick; (**g**) heating of the slide over the flame; (**h**) squashing of the cells after heating; (**i**, **j**) freezing and removing of coverslip; (**k**) equipment required for squashed chromosome preparations: *1* plastic pipette with a flexible tip and sharpened wooden sticks of two sizes; *2* thick coverslips 18 × 18; *3* tools for handling plant material—forceps and a dissecting needle, tube with acetocarmine, razor blade, and filter paper; *4* superfrost slides; *5* flask with 45 % acetic acid; *6* laboratory spirit lamp; *7* Coplin jar for eight slides; *8* the wooden peg

6. Apply a small drop of 45 % acetic acid, place a razor blade on one side of acid drop, and put on a coverslip (18 × 18 mm) (Fig. 1d). Hold a coverslip from one side with filter paper, and spread a tissue by gently tapping the coverslip with a sharpened wooden stick (Fig. 1e). Important: avoid any lateral or circular movements!
7. Remove razor blade, tap the coverslip few more times if needed (Fig. 1f), and briefly warm the slide on a flame (until the water condensate disappears from the surface of the slide, Fig. 1g).
8. Place the slide between two layers of filter paper, and squash the cells using the thumb avoiding any lateral movement (Fig. 1h).
9. Place the slide on dry ice or immerse in liquid nitrogen for several seconds (until "boiling" stops) using a wooden peg (Fig. 1i). Remove the coverslip using a razor blade (Fig. 1j) and transfer the slide in 96 % ethanol.

10. Examine the preparation under a phase-contrast microscope, and keep only those slides that carry a reasonable number of good metaphase spreads.
11. Indicate the position of preparation on the slide using a diamond pen. The borders of a preparation can easier be seen under reflected light on dark background. Slides can be kept in 96 % ethanol at −20 °C up to 1 year.

3.1.2 Mitotic Chromosomes: Squashing of Meristems After Enzyme Treatment Appropriate for Shoot and Bud Meristems of Wooden Trees

1. Place material fixed as described in Sect. 3.1.1 in an embryo dish, and wash in water 3 × 10 min to remove fixative and then in citrate buffer for 10 min on ice.
2. Macerate the meristem in enzyme mixture at 37 °C until the material is soft. Duration of treatment depends on the type of meristem and should be determined experimentally for each particular object.
3. Wash out enzyme solution first with citrate buffer (2 × 10 min) and then with distilled water (2 × 10 min) on ice.
4. Excise the meristem from shoots or buds using fine needles (the root meristem breaks off from the root after enzyme treatment). Transfer the meristem carefully into one drop of 45 % acetic acid on a glass microscope slide. Slides with sintered glass labels are best because they identify the surface that carries the biological material. Apply a coverslip. Carefully disperse the material between glass slide and coverslip by tapping the coverslip gently with a wooden stick.
5. Proceed as described above for squashing of meristems after acetocarmine staining (Sect. 3.1.1, step 8).

3.1.3 Mitotic Chromosomes: Air-Dry Dropping Chromosome Preparation Method

1. Germinate 10–20 seeds (method is optimized for barley) on two layers of moist filter paper in a Petri dish under dark conditions for 2 days at 22–24 °C. Cut off vigorous roots with the length of 1–2 cm from the seeds by using a razor blade.
2. Prepare ice-cold water by placing a 500 ml glass bottle containing cold tap water into crushed ice water. Aerate the ice-cold water and immerse root tips for 20 h to increase the frequency of metaphase cells.
3. Transfer roots from water to 50 ml of ethanol:acetic acid (3:1) fixative to fix them at RT for 2 days (*see* **Note 6**). Store roots in a freshly prepared ethanol:acetic acid (3:1) fixative at 4 °C until use up to a year.
4. Wash the 10–20 roots with 30 ml ice-cold tap water for 5 min twice using a 50 ml glass beaker. Transfer roots one by one into 30 ml citrate buffer using forceps, and wash by shaking the glass beaker for 5 min twice. Place roots on filter paper to remove the

liquid completely, and cut off undesired non-meristematic tissue using a razor blade.

5. Incubate up to 20 root tips in 1 ml enzyme mixture at 37 °C for about 50 min to soften the plant tissue (*see* **Note** 7) in an embryo dish.
6. Remove the enzyme by pipetting, and wash the root tips on ice with 5 ml citrate buffer twice to replace the residual enzyme.
7. Wash root tips with 1 ml 96 % ethanol twice carefully in the same embryo dish. Replace ethanol with freshly prepared fixative (75 % acetic acid/25 % ethanol). Use 10–15 μl fixative per root tip.
8. Transfer root tips together with the fixative into a 2 ml tube and disintegrate root meristems with a dissecting needle or forceps. Tap the tube 20 times to resuspend cells to obtain a cell suspension. Store the cell suspension at −20 °C up to 2 months.
9. Place 2–3 layers of water-soaked paper tissue on a hot plate at 50 °C (see Fig. 2). Immerse microscopic slides in ice-cold tap water in the fridge for 30 min and place slides on top of the moist paper tissue.
10. Pipette 7–10 μl of cell suspension and drop it from a distance of 20 cm onto the cooled slide placed on the hot plate (*see* **Note 8** and [19]). Pipette 10 μl of acetic acid-ethanol (3:1) mixture on the same place as cell suspension on the slide, and keep the slide on the hot plate for additional 2 min. Place the slide on the hot plate without the wet tissue and let it dry for 1 min.
11. Check slides using a phase-contrast microscope to control the quality of the chromosome spread. Use slides either the same day or store by immersing in 96 % ethanol in a Coplin jar at −20 °C.

3.1.4 Meiotic Chromosomes of Cereals: Squashing Method

1. Collect spikes on appropriate stage of development (*see* **Note 9**), and fix them entirely at RT in ethanol:acetic acid (3:1) fixative (in the case of long storage, transfer spikes into 70 % ethanol at −20 °C after fixation) at least for 1 day. If plants are growing in the pots and may be easily transferred, it is convenient to identify and isolate directly anthers containing pollen mother cells at the appropriate stage of meiosis (*see* **Notes 9** and **10**) and fix them.
2. Stain the anthers in 1 % acetocarmine for 15 min up to 1 h at RT.
3. Gently squeeze pollen mother cells into 10 μl of 45 % acetic acid and remove empty anther. Put on coverslip (it is not necessary to use a razor blade in this case), and proceed as described above for squashing of meristems after acetocarmine

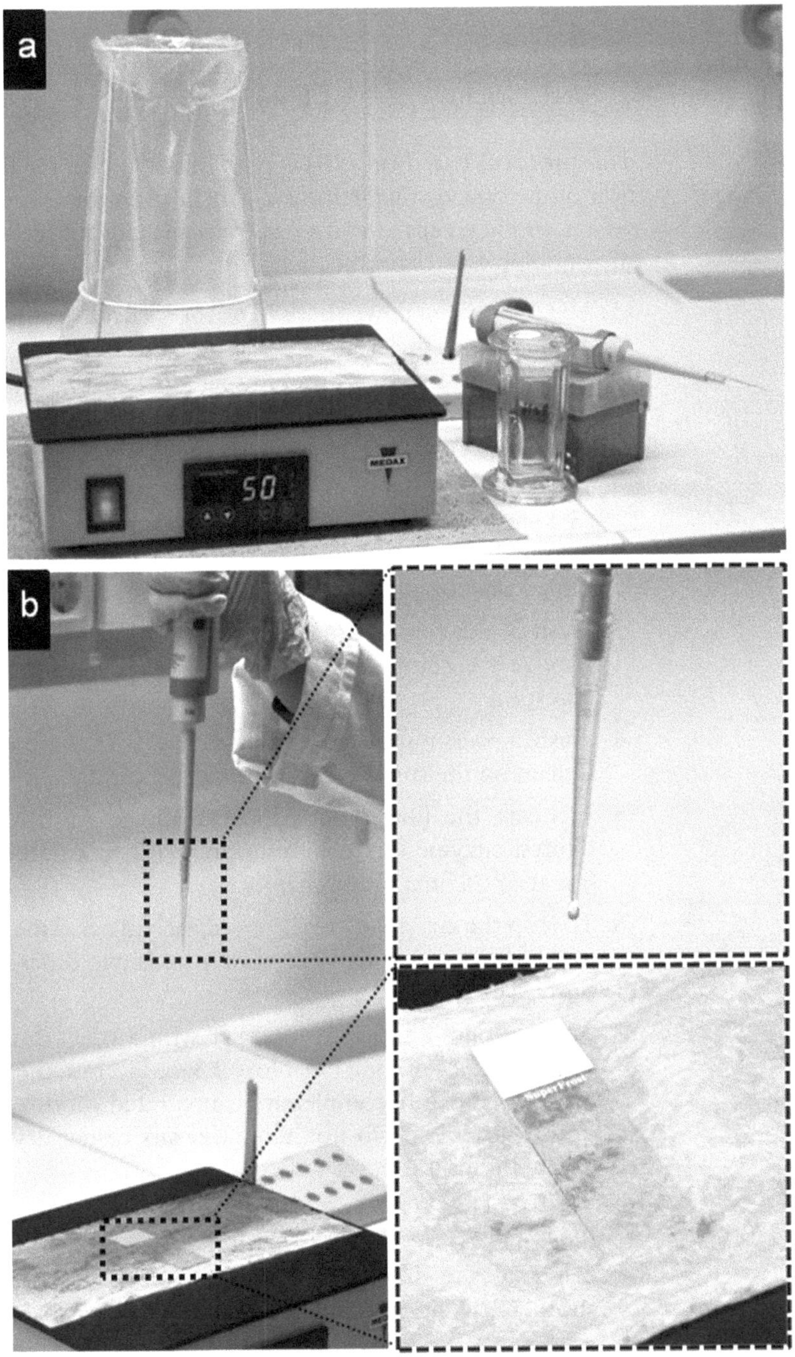

Fig. 2 Air-dry dropping chromosome preparation method. (**a**) Hot plate at 50 °C with a water-soaked paper tissue. (**b**) Pipetting of 7–10 μl of cell suspension on the cold slide. The cold slide is placed on top of the moist paper tissue

staining (Sect. 3.1.1, step 7 and further). If squeezing is not possible, leave the entire anther and proceed exactly as described in Sect. 3.1.1 from step 6.

3.1.5 Meiotic Chromosomes: Spreading Method

This protocol based on [20] is particularly useful for the preparation of meiotic *Arabidopsis thaliana* chromosomes. In addition to pollen mother cells undergoing meiosis, mitotic cells from the surrounding tissues can be prepared at the same time.

1. Fix *A. thaliana* flower buds in Carnoy's fixative at RT (*see* **Note 11**). Change the fixative every day during the next 2–3 days until the fixative becomes transparent. Let flower buds in fixative at 4 °C for at least 1 week before using them. For long time storage, keep at −20 °C.
2. Place the material in an embryo dish with ethanol:acetic acid (3:1) fixative and individualize the flower buds with the help of forceps and needle. Discard old flowers, leaves, and other parts. Only keep complete flower buds (Fig. 3a, b).
3. Wash 3 × 5 min in ethanol:acetic acid (3:1) fixative at RT to remove the chloroform; it could interfere with the enzymatic digestion.
4. Wash 3 × 5 min in citrate buffer at RT. If some of the buds remain on the surface, try to sink them carefully with a needle.
5. Incubate the flower buds in 300–500 μl (until covered) of diluted enzyme mixture (dilution is 1:2 in citrate buffer) for 2 h at 37 °C in a moist chamber.
6. To stop the enzymatic reaction, replace the enzyme mixture by ice-cold citrate buffer. Wash flower buds two times with citrate butter (*see* **Note 12**).
7. Transfer one flower bud together with a small drop of citrate buffer to a clean slide (*see* **Note 13**). Macerate the bud with a teaser needle until completely disintegrated whitish cell suspension (Fig. 3e–g). Do not allow the mix to get dry; add more citrate buffer if necessary.
8. Add 10 μl of ice-cold 60 % acetic acid to the mix, and place the slide on a hot plate at 43 °C for 1 min. Mark with a diamond marker where the mix is. Remove from the hot plate and add another 10 μl of ice-cold 60 % acetic acid.
9. Refix the preparation with 100 μl ice-cold 3:1 fixative by surrounding the mix and with another 100 μl of fixative by dropping them on top of the mix (*see* **Note 14**). Drain off the excess of fixative and air-dry the slides.
10. The slides can be stored in a dry container at 4 °C, used directly for FISH, or stained with DAPI to check the meiosis.

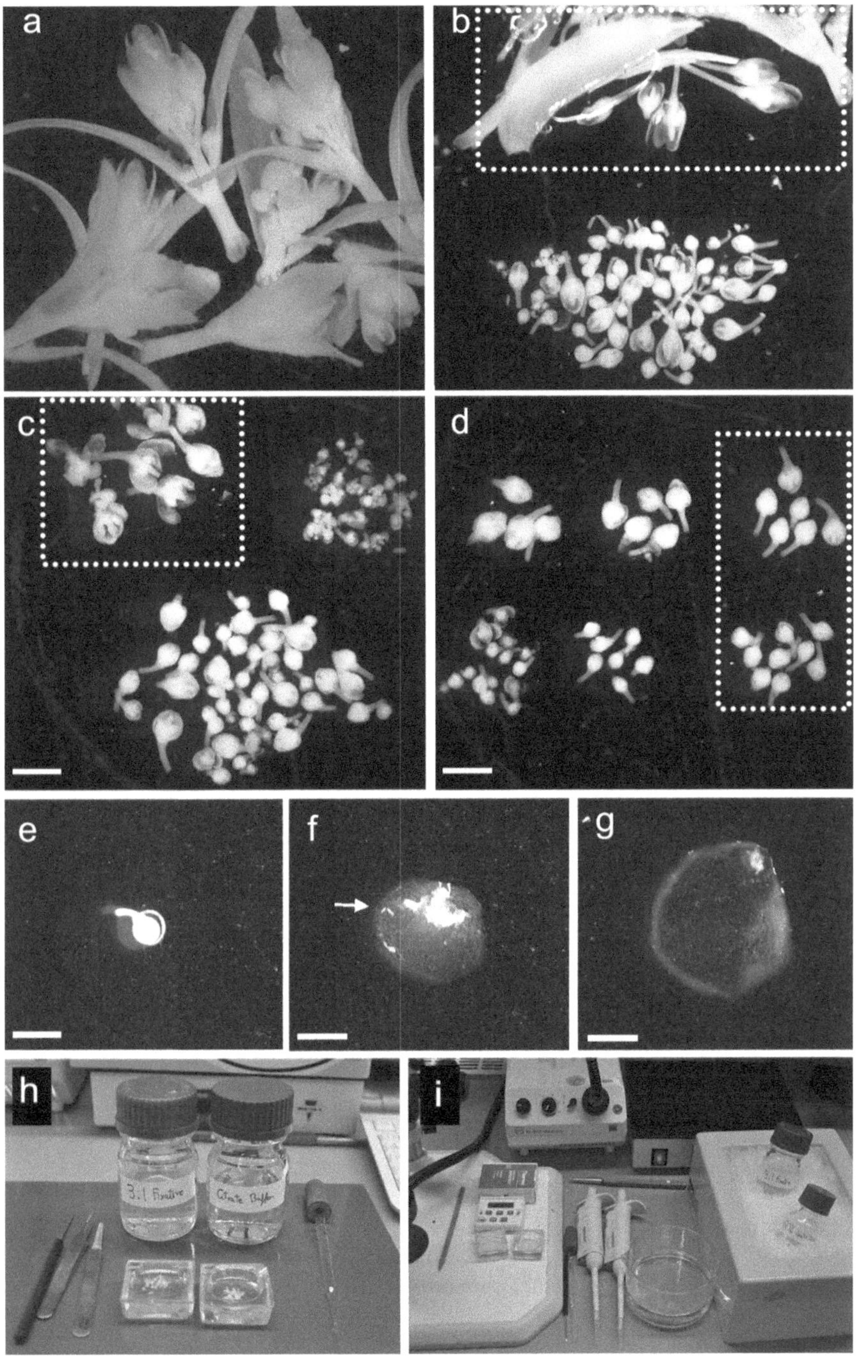

Fig. 3 Meiotic chromosome spreading method. (**a**) Fixed whole inflorescences of *A. thaliana*. (**b**) Individualized flower buds. Discard old flowers and leaves (in *white rectangle*). (**c**) After digestion remove flowers at pollen stage (in *white rectangle*). (**d**) Flower buds are ordered by size. Correct sizes are between 0.3 and 0.4 mm (in *white rectangle*). (**e**) Flower bud within a small volume of citrate buffer. (**f**) Macerate the flower buds completely. Disrupt well the anthers (in *white arrows*). (**g**) Completely disintegrated flower bud (*whitish cell suspension*). (**h**) Material needed for steps 2–4. (**i**) Material needed for steps 7–9. *Scale bar* 1 mm

3.1.6 Preparation of Tissue Sections

Tissue sections are commonly used to analyze undisturbed tissues and cell layers. Many different embedding media exist and are commercially available for different types of analysis. The polyester wax of Steedman [21] has an advantage of low melting temperature, which allows avoiding different heat-induced artifacts. Here, we describe tissue section protocol suitable for FISH and immunostaining and for combination of both methods. The modified protocol based on [22] can be used for the preparation of tissue sections from various plant materials, such as roots, embryos, and young seeds.

1. Fix material immediately after collection in ice-cold 3 % paraformaldehyde for a few hours, depending on the type and size of material (around 2 h for roots and 5–7 h for young cereal kernels). It is necessary to remove all hard covering tissues and undesirable parts of the sample to ensure better penetration of fixative solution and embedding medium. Application of a mild vacuum will help to increase the penetration. Volume of fixative solution should exceed the material volume at least 10×.
2. Wash material twice for 15 min in 1 × PBS under agitation.
3. Dehydrate in ethanol series: 30, 50, 70, 90, and 100 % for 30 min in each at RT and in 100 % for 30 min at 37 °C. For short storage of material, *see* **Note 15**.
4. Infiltrate material with polyester wax at 37 °C. Infiltration is performed in the polyester wax/ethanol series with increasing wax concentration (1/2, 1/1, 2/1 v/v) and in pure wax for 12–24 h in each solution.
5. Embed material in small casting molds and let it polymerize at RT overnight. After hardening, blocks should be taken out and stored at 4 °C in the closed plastic bags. It is possible to store blocks for approximately 1 year.
6. Cut the blocks into 2–10 μm sections using a microtome with appropriate knife. Cutting may be performed at temperature condition between 10 and 22 °C. If material and wax differ in hardness, *see* **Note 16**.
7. Transfer sections onto poly-L-lysine-coated slides using fine tweezers and brush. Add a small drop of water (around 1 μl) over each slice and let them dry overnight at RT.
8. Remove the rest of wax by washing slides twice for 10 min in 100 % ethanol and twice for 10 min in 90 % ethanol. In case of need to check quality of sections under microscope, *see* **Note 17**.
9. Wash slides in distilled water twice for 5 min.
10. Transfer up to eight slides into plastic Hellendahl jar with 90 ml of sodium citrate buffer, and microwave at 800 W for 60 s (*see* **Note 18**).

11. Transfer the slides immediately after microwave treatment into 1 × PBS (in the case of subsequent immunostaining) or into 2 × SSC (in the case of subsequent FISH) for 5 min.
12. Start immediately with standard FISH or immunostaining procedure.

3.2 Probe Labeling for Single-Copy FISH

The DNA template for single-copy probes is amplified by the polymerase chain reaction (PCR) to provide sufficient quantity for the direct labeling with fluorochromes. Labeled nucleotides (dUTPs) are incorporated by nick translation. Though a wide range of fluorochromes is available, red fluorescence dye (e.g., Texas Red-12-dUTP) established itself as the best choice for the detection of the unique sequences.

1. Place 3 μg of high-quality template DNA into a 0.5 ml microtube, and add calculated volume of sterile, nuclease-free water (*see* **Note 19**).
2. Add first 4 μl of nick translation buffer and then 4 μl of non-labeled dNTPs and 4 μl of 0.1 M mercaptoethanol; mix by vortexing.
3. Add 0.8 μl of labeled dUTP (e.g., Texas Red-12-dUTP, #C3176, Invitrogen), and mix gently using pipette tip.
4. Add 4 μl of DNA polymerase I and mix gently.
5. Add calculated amount (*see* **Note 20**) of DNase I and mix gently.
6. Incubate for 120 min at 15 °C.
7. Check the size of the products of nick translation by electrophoresis with 1 % agarose gel using 3 μl of the labeled probe. Run the gel for 30 min at 80 V. The optimal size of labeled nick translation fragments is approx. 50–200 bp.
8. Precipitate labeled probe by adding 163 μl of 1 × TE buffer, 3 μl of herring sperm DNA, 20 μl of NaOAc, and 500 μl of a cold 96–100 % ethanol; mix well by vortexing.
9. Store the microtube at −20 °C overnight.
10. Centrifuge at 4 °C at 30,000 rpm for 30 min, remove the supernatant, and wash the pellet by carefully adding 0.5 ml of ice-cold 70 % ethanol, and centrifuge again for 10 min.
11. Discard the supernatant and dry the pellet (*see* **Note 21**).
12. Dissolve the labeled DNA in 12 μl water. The labeled probe can be stored at −20 °C up to 2 years.

3.3 Fluorescence In Situ Hybridization (FISH)

1. Place slides in a Coplin jar containing 50 ml of 2 × SSC for 5 min. Using forceps, transfer slides to a Coplin jar containing 50 ml of 45 % acetic acid for 3–10 min.

2. Transfer slides to a Coplin jar containing 50 ml of 2 × SSC for 10 min. Transfer slides to a Coplin jar containing 50 ml of 4 % formaldehyde, and immerse slides for 10 min to fix chromosomes.
3. Remove formaldehyde by rinsing the slides 3 times for 4 min each, in a Coplin jar containing 50 ml 2 × SSC. Dehydrate slides in a Coplin jar for 2 min in series of 70 %, 90 %, and 100 % ethanol, respectively, and dry slides in a vertical position.
4. For each slide, prepare a hybridization solution of 20 μl in total using 10 μl of deionized formamide, 5 μl of 4 × hybridization buffer, 3 μl of the probe, and 2 μl of DNase-free water.
5. Add 20 μl of hybridization solution per slide, and cover with a 24 × 32 mm coverslip, and arrest the coverslip with rubber cement. Denature slides with probes simultaneously at 80 °C for 2 min on a hot plate.
6. Transfer slides to a moist chamber and incubate slides at 37 °C overnight avoiding light. Remove coverslips by rinsing the slides in a Coplin jar with 2 × SSC. Place slides in a Coplin jar containing 55–60 °C 2 × SSC and incubate for 20 min.
7. Place the slides to 2 × SSC in a Coplin jar for 2 min at RT. Dehydrate slides in a Coplin jar for 2 min in series of 70 %, 90 %, and 100 % ethanol, respectively.
8. Air-dry the slides, and counterstain with DAPI solution (anti-fade mounting medium added), and cover with a coverslip, avoiding intense light.
9. Analyze the slides using an epifluorescence microscope. The selection of filter depends on the fluorochrome used for probe labeling. If necessary, store slides at 4 °C under dark conditions up to a year.

4 Notes

1. Coverslips are provided by manufacturer in different thickness. The thickest coverslips (i.e., No 2 according to Assistant catalogue) are preferable for squashing technique, while the standard size is suitable for hybridization.
2. To prepare fixative solution (i.e., 3:1 of ethanol and glacial acetic acid), it is important to use glacial acetic acid provided in glass bottles. The acid kept in plastic bottles can absorb minor amount of plastic, resulting in hampering of chromosome squashing or spreading, thus decreasing the quality of chromosomal preparation.
3. In some gene banks, the seeds can be heavily contaminated with fungal infection, which can inhibit or even block seed

germination. In such cases brief treatment of dry seeds with potassium permanganate solution (concentration 0.1 mol l^{-1}) can significantly reduce fungal background. Immerse the seeds before germination in potassium permanganate solution for 15–20 min, and then wash them in a glass under tap water until color disappears, and germinate seeds in Petri dishes, as usual.

4. The roots can also be taken from adult plants. In this case the plant should be thoroughly watered a day before taking the roots, and only young roots with clear meristematic parts should be used. Mitotic cells can also be obtained from other plant tissues containing meristem, i.e., from shoots, young spikes, or buds. Pretreatment of these tissues is the same as for root meristem.

5. Plant species significantly differ in chromosome number ($2n = 4$ to ~1,500) and size (the average chromosome size, defined by dividing genome size by haploid chromosome number, varied from 3.4×10^5 bp in *Cardamine amara* to 7.33×10^9 bp in *Trillium rhombifolium*). In addition, plant species differ in the duration of cell cycle. Therefore, no uniform method for accumulation of metaphase cells exists in plants, and in each case pretreatment should be adapted for particular species. For example, in conifers (chromosome size is larger than 10^9 bp), the roots are treated in 0.2 % colchicine solution for 14–16 h (or 1 % colchicine for 4 h) followed by ice water for 30 min. In herbs belonging to *Alliaceae* and *Trilliaceae* families with similar chromosome size, the roots are treated in 0.2 % colchicine solution for 3–4.5 h. In cereals, legumes, and other species with large chromosomes (average size $5–9.5 \times 10^8$ bp), the best method of metaphase arrest is pretreatment of roots, shoots, or young spikes (flowering buds) in ice water for 16–26 h or in 0.05 % colchicine solution for 2–3 h. Germinating buds of foliage trees with small chromosome size ($3–7 \times 10^7$ bp) are treated in 0.002 M aqueous solution of 8-hydroxychinoline at 10–16 °C for 3–4 h or in ice water for 1.5 h. The roots of species with small chromosomes may be treated in ice-cold water supplemented with 1 μg ml^{-1} 9-aminoacridine for 12–20 h to elongate chromosomes [23].

6. Mild vacuum treatment can help fixative to penetrate the root tissue better. Vacuum treatment is carried out in a vacuum concentrator at RT, for 2–5 min.

7. Incubation time in the enzyme mixture depends, i.e., on the length of the fixation time. The longer the roots were fixed, the longer it takes to digest the roots. Therefore, it is important to control the "softness" of the roots using forceps after 50 min of incubation and increase the time if needed.

8. Drop the cell suspension on the moist slide. After dropping, the bottom of the slide can be wiped with a tissue paper to fasten the drying process on the hot plate.
9. In cereals like wheat and rye, meiosis proceeds in spikes which are located approximately 1–2 cm below the last leaf prior flag leaf. However, it is highly dependent on growing conditions. When checking spikes for anthers in an appropriate stage of meiosis, it is important to remember that wheat and rye start to flower from the middle of the spike.
10. It may be convenient to fix/store whole flower buds at various stages of meiosis and then remove and identify anthers containing the required meiotic stages prior to the next stage. A single young anther, or often all the anthers of a single flower bud, normally contains pollen mother cells at the same (synchronous) stage of meiosis. The size and color of the bud, though often difficult to determine accurately, may be the only clues to the progress of meiosis. Some plants, such as many of the grasses and cereals, have flowering spikes with a series of florets that exhibit a temporal sequence of maturity, and each floret therefore differs slightly in meiotic stage. This can be very useful in finding the best chromosomes for analysis. In addition, grasses often contain three meiotically synchronized anthers, so the examination of one another indicated the appropriateness of the other two for the in situ procedure. Variants of this approach may be useful in preparing meiotic chromosomes from a variety of different species, but only experience within a particular species will yield reliable results.
11. For fixation it is not required to fix individual flower buds, it is better to fix the whole inflorescence and to select afterward the buds with the correct size (Fig. 3a).
12. Digested flower buds can be kept in citrate buffer at 4 °C for 1 day before continuing the slide preparation.
13. It can be difficult to find the right flower bud size with the desired stages; the correct size should be between 0.3 and 0.4 mm [24]. Once the buds are digested, it can help to remove flowers with yellow anthers (at pollen stage) and to order the remaining buds from biggest to smallest (Fig. 3c, d). Start the slide preparation with the biggest buds and when finished, check it using a phase-contrast microscope. If pollen is found, move to a smaller bud size until the meiocytes at the desired stage are found.
14. For refixing the preparation, it is convenient to do it over a bowl with water; this way the excess of fixative will fall in the bowl, and the water will dilute the acetic smell (Fig. 3i).

15. Material could be stored in 70 % ethanol overnight at 4 °C. In this case perform a short wash (10 min) in 70 % ethanol at RT at the next day, and continue with 90 % ethanol.
16. Blocks with soft material should be kept at RT overnight before cutting. For hard material it is useful to additionally chill the blocks on the surface of melting water ice. Do not use a freezer to avoid cracking of blocks.
17. Briefly dry slides after the first washing in 90 % ethanol. Check the quality of sections under microscope and proceed with washing immediately. Avoid overdrying of slides.
18. This step is highly recommended to reduce background and increase penetration of antibodies in the case of immunostaining/FISH combination. Microwave treatment may be replaced by boiling of slides in sodium citrate buffer in glass jar for 10–20 min. Properly prepared sections won't be lost from slides during these treatments.
19. Total labeling reaction per probe contains 40 μl. The amount of used water depends on (I) the applied DNA volume (X μl) and (II) DNase I volume (Y μl). Thus, water amount is calculated as 23.2 μl—X μl—Y μl.
20. Applied DNase I volume (Y μl) depends on the length of the labeled probe. The longer is the fragment, the higher is the applied amount of the DNase I.
21. It is possible to accelerate the drying of the pellet by the additional centrifugation step of 1 min at max. speed and pipette supernatant gently out of the microtube avoiding the contact between the pellet and the pipette tip. Probe pellet is visible at one corner of the microtube.

Acknowledgments

The authors thank the members of the laboratories for a helpful discussion. C.M. is the recipient of a Marie Curie Initial Training Network fellowship (FP7-PEOPLE-2013-ITN, CHIP-ET).

References

1. Lavania UC (1998) Fluorescence in situ hybridization in genome, chromosome and gene identification in plants. Curr Sci India 74:126–133
2. Speicher MR, Carter NP (2005) The new cytogenetics: blurring the boundaries with molecular biology. Nat Rev Genet 6:782–792
3. Brady T, Clutter ME (1972) Cytolocalization of ribosomal cistrons in plant polytene chromosomes. J Cell Biol 53:827–832
4. Timmis JN, Deumling B, Ingle J (1975) Localisation of satellite DNA sequences in nuclei and chromosomes of two plants. Nature 257:152–155

5. Wimber D, Duffey P, Steffensen DM et al (1974) Localization of the 5S RNA genes in Zea mays by RNA-DNA hybridization in situ. Chromosoma 47:353–360
6. Rayburn AL, Gill BS (1985) Use of biotin-labeled probes to map specific DNA-sequences on wheat chromosomes. J Heredity 76:78–81
7. Schwarzacher T, Leitch AR, Bennett MD et al (1989) In situ localization of parental genomes in a wide hybrid. Ann Bot 64:315–324
8. Yamamoto M, Mukai Y (1989) Application of fluorescence in situ hybridisation to molecular cytogenentics of wheat. Wheat Inf Serv 69:30
9. Jiang JM, Gill BS (2006) Current status and the future of fluorescence in situ hybridization (FISH) in plant genome research. Genome 49:1057–1068
10. Jiang JM, Gill BS (1994) Nonisotopic in situ hybridization and plant genome mapping – the first 10 years. Genome 37:717–725
11. Silva GG, Souza MM (2013) Genomic in situ hybridization in plants. Genet Mol Res 12:2953–2965
12. Lysak MA, Fransz PF, Ali HB et al (2001) Chromosome painting in Arabidopsis thaliana. Plant J 28:689–697
13. Gill B, Friebe B (1998) Plant cytogenetics at the dawn of the 21st century. Curr Opin Plant Biol 1:109–115
14. Kato A, Albert PS, Vega JM et al (2006) Sensitive fluorescence in situ hybridization signal detection in maize using directly labeled probes produced by high concentration DNA polymerase nick translation. Biotech Histochem 81:71–78
15. Danilova TV, Friebe B, Gill BS (2014) Development of a wheat single gene FISH map for analyzing homoeologous relationship and chromosomal rearrangements within the Triticeae. Theor Appl Genet 127:715–730
16. Aliyeva-Schnorr L, Beier S, Karafiátová M et al (2015) Cytogenetic mapping with centromeric BAC contigs shows that this recombination-poor region comprises more than half of barley chromosome 3H. Plant J 84:385–394
17. Valárik M, Bartos J, Kovárová P et al (2004) High-resolution FISH on super-stretched flow-sorted plant chromosomes. Plant J 37:940–950
18. Fransz PF, Alonso-Blanco C, Liharska TB et al (1996) High-resolution physical mapping in Arabidopsis thaliana and tomato by fluorescence in situ hybridization to extended DNA fibres. Plant J 9:421–430
19. Aliyeva-Schnorr L, Ma L, Houben A (2015) A fast air-dry dropping chromosome preparation method suitable for FISH in plants. J Vis Exp 106:e53470
20. Sanchez Moran E, Armstrong SJ, Santos JL (2001) Chiasma formation in Arabidopsis thaliana accession Wassileskija and in two meiotic mutants. Chromosome Res 9:121–128
21. Steedman HF (1957) Polyester wax; a new ribboning embedding medium for histology. Nature 179:1345
22. Braszewska-Zalewska AJ, Wolny EA, Smialek L et al (2013) Tissue-specific epigenetic modifications in root apical meristem cells of Hordeum vulgare. PLoS One 8:e69204
23. Muravenko OV, Amosova AV, Samatadze TE et al (2003) 9-Aminoacridine: an efficient reagent to improve human and plant chromosome banding patterns and to standardize chromosome image analysis. Cytometry A 51:52–57
24. Armstrong SJ, Jones GH (2003) Meiotic cytology and chromosome behaviour in wild-type Arabidopsis thaliana. J Exp Bot 54:1–10

Yeast Chromosome Dynamics Revealed by Immuno FISH

Harry Scherthan

Abstract

Fluorescence in situ hybridization (FISH) provides an effective means to delineate chromosomes and their subregions during all stages of the cell cycle. This makes FISH particularly useful for studying chromosome behavior in species with minute genomes and/or weak chromosome condensation at metaphase, which is the case for model organisms such as the budding yeast *Saccharomyces cerevisiae* or *Schizosaccharomyces pombe*. Since its introduction in 1992, yeast FISH with composite whole chromosome or locus-specific probes in combination with immunofluorescence staining has become an indispensable tool in the analysis of chromosome behavior in metaphase and interphase cells, and especially of meiotic chromosome pairing of wild-type and mutant yeast strains.

Keywords Chromosome dynamics, Chromosome painting, DNA labeling, FISH, Immunofluorescent staining, *Saccharomyces cerevisiae*, *Schizosaccharomyces pombe*

1 Introduction

As compared to multicellular eukaryotes, the baker's yeast *Saccharomyces cerevisiae* ($n = 16$) challenges classical cytology, since metaphase occurs without nuclear envelope breakdown. Highly condensed chromosomes are absent, and yeast nuclei lack a lamina, which makes them vulnerable to distortions during isolation procedures. Despite its powerful genetics, *S. cerevisiae* cytology is further hampered by the tininess of the object: diploid nuclei are only 2–4 μm in diameter and harbor 32 chromosomes with a total DNA content of a mere 24 Mbp of predominantly unique DNA sequences [1, 2]. For these reasons, light and electron microscopic studies of yeast metaphase chromosomes at best revealed tiny chromatin lumps that are hardly reminiscent of the metaphase chromosomes of more complex eukaryotes [3, 4]. However, a reasonable chromosome (bivalent) structure can be obtained by silver or fluorochrome staining of chromosome spreads of meiotic cells [5–7], in which homologues are connected by a joint protein zipper, the synaptonemal complex (SC). Staining and tracking of particular

Thomas Liehr (ed.), *Fluorescence In Situ Hybridization (FISH)*, Springer Protocols Handbooks,
DOI 10.1007/978-3-662-52959-1_50,

chromosomes in cytological preparations from time-course experiments became possible by the introduction of FISH in budding yeast cytology [8]. The combination of the above methods has since facilitated cytological studies in this model organism [9] as well as in fission yeast [10].

Since FISH provides a powerful means to delineate individual chromosomes and their subregions in spreads and nuclei (chapter by Thomas Liehr and Anja Weise "Background"), it has become a major tool for studying chromosome behavior in wild-type and mutant yeast strains, e.g., [11–17]. Furthermore, the delineation of specific chromosome regions by binding of fluorescent protein-tagged inducer or repressor molecules to chromosomally integrated LacO or TetO arrays has emerged as a valuable tool to study chromosomes and subregions, particularly in live cells [18–20]. FISH, in comparison, is less laborious with respect to strain construction, since any chromosomal site can be delineated by the appropriate choice of probes facilitated by the genome sequence. Moreover, FISH allows to rapidly adjust the size of the chromosome regions to be labeled according to demand, so that single-copy regions, chromosome arms, or even entire chromosomes can be marked (painted) by, in the latter two cases, large (composite) DNA probes (Fig. 1).

Furthermore, genomic in situ hybridization (GISH) using the DNA of two different yeast species can be applied to delineate single chromosomes or entire chromosome (sub)sets. GISH utilizes the preferential hybridization of homologous genomic DNA under suppression conditions to the chromosomes of one yeast species in a different species background or a chromosome set of one species in a species hybrid [21]. In the following a protocol for FISH is outlined (Fig. 2), which has proven robust in the analysis of yeast chromosome behavior.

2 Materials

Apart from standard cell biological and molecular cytogenetic equipment, including standard solutions (e.g., ethanol, methanol, formamide, formaldehyde, xylene), a few more specialized items are required here. For FISH basic material requirements, see also chapter by Thomas Liehr et al. "The Standard FISH Procedure".

2.1 Cell Growth and Preparation

- Rich medium (YPD), liquid: 1 % yeast extract, 2 % peptone, 2 % glucose in distilled water, autoclaved.
- Presporulation medium (YPA), liquid: 1 % yeast extract, 2 % peptone, 1 % potassium acetate in distilled water, autoclaved.
- Sporulation medium (SPM), liquid: 2 % potassium acetate in distilled water, autoclaved.

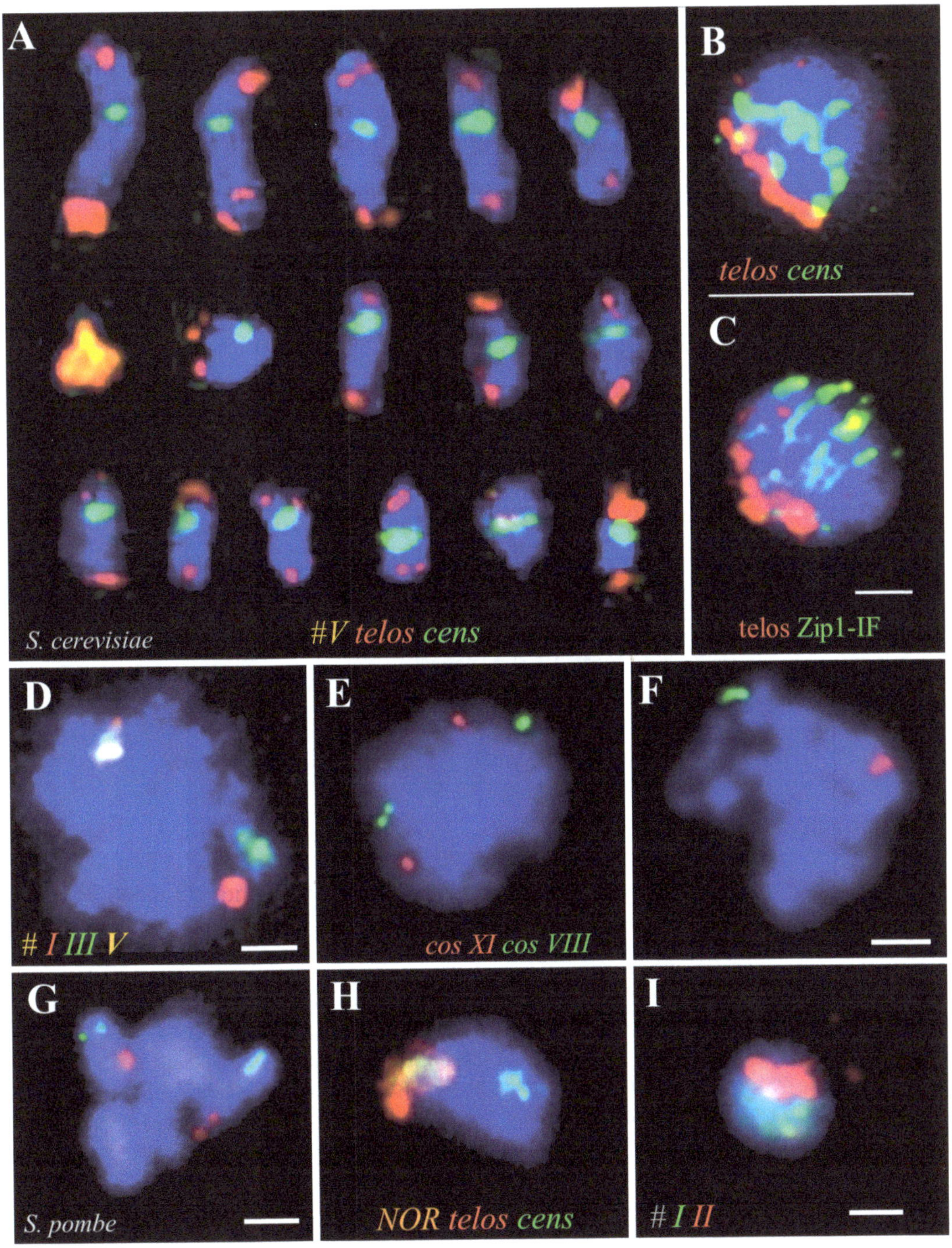

Fig. 1 (**a**) Pachytene karyotype (DNA *blue*, DAPI) of *S. cerevisiae* ($2n = 32$) showing FISH staining of centromeres (lambda phage clones, *green*), telomeres (XY repeat probes, *red*), and painting of the chromosome V bivalent by a composite DNA probe (*orange*). This magnified pachytene karyotype was assembled from extensively spread meiotic nuclei. (**b**) Meiotic bouquet nucleus of *S. cerevisiae* displaying telomeres (*red*, XY repeat probe) aligned at the nuclear periphery at the *lower left* and centromeres (*green*, lambda probes) across a mildly spread nucleus. (**c**) Immuno-FISH experiment revealing the progress of synapsis (IF with anti-ZIP1

- Zymolyase: Enzyme for dissolving the yeast cell wall. Prepare a 10 mg ml^{-1} stock solution of Zymolyase 100 T (Seikagaku Co., Tokyo, Japan) in distilled water (*see* **Note 1**).
- Solution I: 0.8 M sorbitol with 10 mM dithiothreitol (prepared from a frozen 1 M stock).
- Digestion solution: Add 7 μl of Zymolyase stock solution (see above) per 500 μl solution I.
- Stop solution: 0.1 M 2-(*N*-morpholino)ethanesulfonic acid (MES), 1 mM EDTA, 0.5 mM $MgCl_2$, 1 M sorbitol; adjust pH to 6.4 using dilute 0.1 M NaOH.
- 2 % sodium *N*-lauroylsarcosine.
- 37 % acid-free formaldehyde (Merck No. 1.03999.1000).
- Fixative I: 4 % formaldehyde solution made from 37 % acid-free formaldehyde stock solution (Merck) by diluting with distilled water. *Caution: Formaldehyde is a hazardous chemical; handle always with care in a fume hood and wear protective clothes!*
- Detergent: Prepare a 1 % solution of "Lipsol liquid concentrate" (www.camlab.co.uk or www.labdepotinc.com) in distilled water. The working solution can be stored for several months in the refrigerator (*see* **Note 2**).
- Fixative II: 4 % paraformaldehyde supplemented with 3.6 % sucrose (*see* **Note 3**).
- PBS: 130 mM NaCl, 7 mM Na_2HPO_4, 3 mM NaH_2PO_4, in distilled water (pH 7.5).
- Antifading agent such as VECTASHIELD (Vector Laboratories Inc., Burlingame, CA) is a useful tool to prevent fading of the fluorescent signals during microscopy. Antifade agents of different brands should be tested for performance on yeast FISH preparations.

Fig. 1 (Continued) antibodies, *green*) in a meiotic bouquet nucleus with clustered telomeres (*red*) at the lower left (for details see [25]). (**d**) Multicolor FISH delineates bivalents of paired chromosomes *I* (*red*), *III* (*green*), and *V* (*orange*) in a spread pachytene nucleus (DNA, *blue*). (**e**) Spread meiocyte nucleus (DNA, *blue*) hybridized with cosmids for chromosomes IIIV and XI; four signals indicate the absence of homologous pairing, while a split signal (*green*, lower left) is caused by replicated DNA in prophase I nuclei. (**f**) Same experiment but pachytene nucleus with two large signals indicating homologous pairing. (**g–i**) FISH to spreads of diploid *Schizosaccharomyces pombe* cells (wild-type fission yeast is haploid, $n = 3$). (**g**) Mitotic metaphase figure (DNA, *blue*) of a diploid *S. pombe* strain with telomere regions of chromosomes II differentially marked by FISH with two telomeric cosmid probes (*red*, *green*). (**h**) Horsetail nucleus obtained at meiotic prophase. Telomeres (*red*) and adjacent rDNA (*orange*, by red/green labeling) mark the leading edge of this bouquet nucleus, while centromeres (*green*) locate in the trailing end of the nucleus. (**i**) Painting of chromosomes 1 (*green*) and 2 (*red*) in a mildly spread diploid nucleus. Homologue association leads to one signal that delineates a joint territory of each chromosome pair (for details on *S. pombe* FISH, see [30]). *Bars* in this figure, 2 μm. Images d to i represent digitized conventional fluorescence microscopy images.

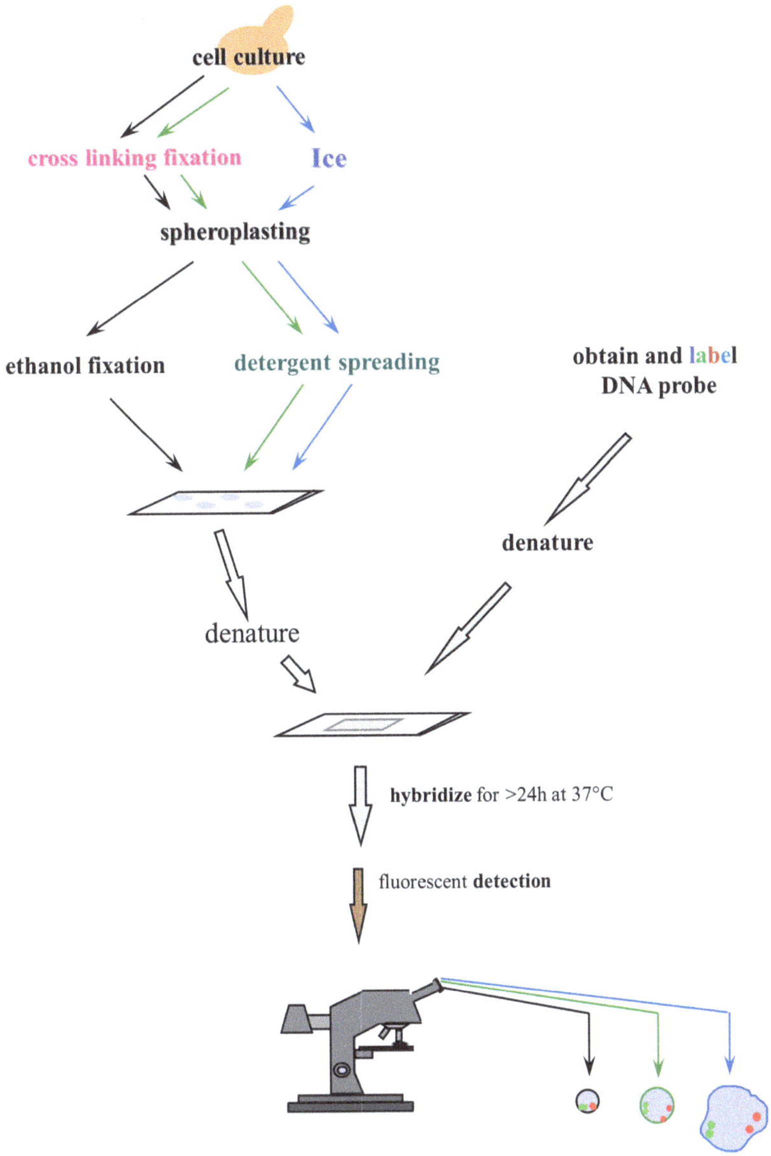

Fig. 2 Flowchart of the preparation procedure for undisrupted (*black-arrowed route*) and spread nuclei (*green-* and *blue-arrowed routes*) for yeast FISH. Depending on the mode of fixation prior to detergent spreading, either moderately spread (*green arrows*) or extensively spread nuclei (*blue arrows*) may be obtained. Nuclear spreads offer a higher spatial resolution and thereby facilitate visual analysis of FISH signals. It should be noted that particular three-dimensional chromosome arrangements are disrupted by extensive nuclear spreading (see Fig. 1)

2.2 Combined FISH and Immunostaining

- Antibodies against various components of the yeast synaptonemal complex (SC), kinetochores, and several chromosomal proteins have been raised in various labs, and several are commercially available. For instance, immunolabeling of microtubules with the YOL1/34 monoclonal rat anti-yeast tubulin antibody [22, 23] produces good results. It can be obtained, e.g., from Serotec, Kidlington, UK.

- Secondary antibodies: obtained from commercial suppliers like Jackson Labs, Sigma-Aldrich, or others. Always perform tests without primary antibodies to ensure specificity.
- DAPI (4′6-diamidino-2-phenylindole) is used as a DNA-specific counterstain. It may be purchased ready-made in antifade solution (see Sect. 2.3).
- Rubber cement for sealing cover slips, e.g., Fixogum (Marabuwerke GmbH, Tamm, Germany).

2.3 FISH

- Probes for FISH can be obtained or generated in various ways: (1) Clones for the desired chromosomal loci can be selected from the *Saccharomyces* Genome Database [1] (www.yeastgenome.org) as cosmid or λ-phage clones and be purchased from the American Type Culture Collection (Rockville, MD) (*see* **Note 4**). The DNA of these clones can be combined to produce composite chromosome painting probes (*see* **Note 5**) [8]. (2) PCR fragments may be used as probes. Fragments of approx. 5–10 kb size are generated by long-range PCR [suitable kits are available from various companies: e.g., Expand™ Long Template PCR System (www.sigmaaldrich.com); TaKaRa Ex Taq (TaKaRa Shuzo Co., Ltd, Otsu, Japan)] with appropriate primers chosen from the *Saccharomyces* Genome Database (www.yeastgenome.org; *see* **Note 4**).
- Labeled nucleotides, e.g., Cy3-dUTP, Cy5-dUTP (www.gelifesciences.com), fluorescein-dUTP, tetramethylrhodamine-dUTP, digoxigenin-dUTP, or biotin-dUTP (www.sigmaaldrich.com). Choose one specific nucleotide for each DNA probe. These can be combined in a single hybridization solution and differentially visualized.
- RNAase, DNAase-free (e.g., Sigma-Aldrich).
- BT buffer: 0.15 M $NaHCO_3$, pH 8.3, 0.1 % Tween 20.
- Hybridization solution: 50 % formamide, 2 × SSC, 10 % dextran sulfate, 1 μg μl^{-1} salmon sperm carrier DNA.
- Blocking buffer: 3 % BSA in BT buffer.
- Detection buffer: 0.05 % BSA, 0.1 % Tween 20 in BT buffer.
- Detection reagents: Avidin-FITC or avidin-Cy3 conjugate (e.g., ExtrAvidin® FITC conjugate, ExtrAvidin® Cy3 conjugate; Sigma-Aldrich) for bion labelled probes, biotin-conjugated anti-avidin antibody (Vector Labs), anti-digoxigenin-fluorophore-conjugated antibodies for digoxigenin-labelled probes (fluorescein, rhodamine, AMCA can be obtained from, e.g., Sigmaaldrich.com or www.jenabioscience.com).

3 Methods

3.1 Cell Growth and Sporulation

1. For the study of mitotic cells, inoculate 5 ml YPD with a small colony from a plate and grow to a concentration of ~2 × 10^7 cells per ml in a shaker at 30 °C (usually overnight).
2. To obtain meiotic cells, inoculate 50 ml YPA with a small colony of a diploid strain from a plate and grow to a concentration of ~2 × 10^7 cells per ml in a shaker at 30 °C (*see* **Note 6**). Centrifuge cell suspension for 4 min at 2000 rpm (700× *g*). Resuspend cells at a density of about 4 × 10^7 cells ml^{-1} in SPM and incubate until 2–4 % four nucleated cells appear in DAPI-stained preparations. At this point the majority of cells will be at pachytene which is the most favorable stage for cytological examination. Sporulation time varies considerably between strains. Budding yeast strain SK1 [24], which is widely used for meiotic studies, shows a maximum of pachytene cells ~4–5 h after transfer to sporulation medium.

3.2 Cell Preparation

Here we describe preparation procedures to study yeast chromosomes cytologically. Ethanol fixation (see Sect. 3.2.1) is useful in combination with DAPI staining and inspection of GFP-labeled chromosomes and other cellular components. Formaldehyde fixation (see Sect. 3.2.3) provides good preservation of cellular morphology and astral microtubules. It can be used in combination with GFP-tagged protein detection or immunostaining [23, 25]. Nuclear spreading (see Sect. 3.2.2) offers enhanced cytological resolution, but it will disrupt cells. It is suited for the visualization of SCs by Ag staining or immunolabeling of SC components and may also be used in combination with FISH (Fig. 1). Semi-spreading (see Sect. 3.2.3) is a good compromise for obtaining a good spatial resolution of nuclear contents and a reasonable preservation of cell morphology for FISH. For a flowchart of the different steps, see Fig. 2.

3.2.1 Ethanol Fixation

1. Remove 1 ml of cells from the culture.
2. Centrifuge the cell suspension for 10 s at 700× *g*.
3. Discard the supernatant and resuspend pellet in PBS.
4. Repeat step 2.
5. Resuspend pellet in 1 ml 70 % ethanol.
6. Repeat step 2.
7. Resuspend in 50 μl 70 % ethanol.
8. Place 20 μl ethanol-fixed cells on a clean glass slide and streak out with the edge of a cover slip, without touching the surface of the slide.
9. Air-dry; add 18 μl antifade solution containing DAPI.

10. Cover with a cover slip.
11. Inspect under a phase-contrast microscope.

3.2.2 Spreading

The spreading protocol described here is a modification by Loidl et al. [7] of the method of Dresser and Giroux [6].

1. Take 5 ml of a cell suspension obtained according to the procedure in Sect. 3.1.
2. Spin the cell suspension for 20 s at 700 g.
3. Resuspend the cell pellet in digestion solution (see Sect. 2.1).
4. Allow spheroplasting to proceed for 20 min at 37 °C.
5. Put the cell suspension on ice.
6. Check degree of cell wall degradation by placing 5 μl of the digested suspension on a glass slide.
7. Mix with an equal volume of 2 % sodium *N*-lauroylsarcosine.
8. Immediately place a cover slip on the cell suspension and instantly observe the sample under a phase-contrast microscope at low-power magnification.
9. After a few seconds, the cells should be seen bursting (initially bright cells become dark and then fragment). Approx. 80 % of cells should rupture instantly, which is a good indicator for a cell preparation suitable for FISH.
10. Stop the digest by adding an equal volume of ice-cold stop solution; spin the cells down and resuspend them in 250 μl stop solution (*see* **Note 7**).
11. Place 20 μl of the cell suspension on a slide; add sequentially: 40 μl fixative I, 80 μl detergent, and 80 μl of fixative II (*see* **Notes 8–9**).
12. Disperse the mixture with a glass rod over the slide without touching its surface. Put slides in a chemical hood and let dry for >2 h—overnight (*see* **Note 10**).
13. Continue with one of the procedures described in Sect. 3.3.2. Slides may be stored frozen at −20 °C for months until further use.

3.2.3 Semi-spreading for Generation of Structurally Preserved Nuclei

This procedure is used when preservation of nuclear structure has to be combined with good access of the DNA probes (e.g., for combinatorial FISH and immunostaining [26]).

1. Transfer a sample from the sporulating culture (see Sect. 3.1) at the desired time point immediately to 1/10 vol. of ice-cold 37 % acid-free formaldehyde (e.g., 1 ml culture to 0.1 ml 37 % formaldehyde pre-cooled on ice in an Eppendorf tube).
2. Incubate for 30 min on wet ice.

3. Sediment cells at 700 g for 2 min and wash them once in an excess of deionized H_2O.
4. Sediment cells at 700 g for 2 min and resuspend them in 1/10 of the Sect. 3.2.2 sample volume in Solution I.
5. Spheroplast cells with Zymolyase in digestion solution by following procedure described in Sect. 3.2.2, steps 2–4.
6. Drop 20 μl of the spheroplast suspension onto a slide and add sequentially 80 μl detergent and 80 μl of fixative. (Handle with care, the fixative contains formaldehyde; use protective equipment!)
7. Spread out the mixture with a glass rod and leave slides for drying in a chemical hood for >2 h—overnight (*see* **Note 10**).
8. Continue with the procedures described in Sect. 3.3. Alternatively, slides may be stored frozen at −20 °C until use.

3.3 FISH

3.3.1 DNA Labeling and Probe Preparation

Nick translation is the preferable method to label DNA probes for FISH. DNA probes of choice can be directly fluorochrome or hapten labeled by the incorporation of fluorochrome-, biotin-, or digoxigenin-labeled deoxynucleotides (*see* **Note 11**) using commercial nick translation kits. Nick translation has the advantage that the fragment size of the labeled probe molecules is ~50–300 bp, which is imperative for access of the probe to the target molecules (*see* **Note 12**). Target regions as small as 2.4 kb can be successfully detected by yeast FISH (HS, unpublished observations).

1. Label DNA by nick translation according to the instructions of the supplier. (This typically requires 90 min of incubation at 16 °C.)
2. Stop the reaction by adding 0.5 M EDTA to result in a final conc. of >30 mM and incubate for 10 min at 65 °C.
3. Store the labeled DNA at −20 °C until use.
4. Ethanol-precipitate labeled probe DNA with 1/10 volume 3 M NH_4 acetate and 2.5 volumes of ethanol for >30 min at −20 °C.
5. Spin in microcentrifuge at >12,000 RPM (>11,500× *g*) for 30 min.
6. Discard supernatant and briefly dry the DNA pellet by incubating the open tube for a few minutes at 65 °C.
7. Resuspend the DNA pellet in hybridization solution to result in a final concentration of 30 ng μl^{-1}.
8. Denature the DNA probe in hybridization solution at 95 °C for 5 min.
9. Place the tube for 5 min on ice or in a fridge. Keep chilled until further processing.

3.3.2 Hybridization

1. Place a slide produced by the procedure described in Sect. 3.2.2 or 3.2.3 in a Coplin jar filled with deionized water until the sucrose layer has dissolved.
2. Replace the deionized water once.
3. Drain and briefly air-dry the slide by standing it upright on a paper towel.
4. Apply 50 μl RNAase (100 μg ml^{-1} in 2 × SSC) to each slide, cover with a cover slip, and incubate for 60 min (up to 120 min) at 37 °C in a moist chamber.
5. To denature chromosomal DNA, place 100 μl 70 % formamide, 30 % 2 × SSC (pH 7.0) on top of the slides and cover with a 24 × 60 mm cover slip. Place slide on the surface of a hot plate, thermocycler, or drying block for 5–10 min at 83 °C.
6. Add a few microliters of the formamide/2 × SSC solution (see Sect. 3.3.2 step 5) to the rim of the cover slip every 3 min to replace evaporating liquid.
7. Rinse off the cover slip with ice-cold deionized water and let the slide air-dry, standing in an upright position on a paper towel.
8. (Optional) Place a drop of antifade containing DAPI on the slide and locate a region with well-spread nuclei using the fluorescence microscope.
9. Remove the cover slip from the slide (see step 8) by placing a drop of BT buffer to the rim of the cover slip and let it sit for ~2 min, then lift off the slide carefully with tweezers.
10. Wash briefly in deionized water and shake off excess liquid from slide; air-dry.
11. Apply 5 μl of denatured probe mixed onto each slide (this depends on the size of the cover slip/region to be hybridized). Combine the probes in equal amounts, if two or more probes have to be hybridized simultaneously on the same slide.
12. Place a cover slip (18 × 18 mm) over the region of interest containing the sample and seal with rubber cement.
13. Let the rubber cement dry until it appears clear.
14. Incubate at 37 °C (e.g., in a moisturizing cell culture incubator or moist box) for 24–48 h to allow for hybridization to proceed (*see* **Note 13**).
15. Peel off the rubber cement and gently rinse off the cover slip by placing the slides in a Coplin jar with 0.05 × SSC at 37 °C.
16. Wash slides 3 × 5 min in 0.05 × SSC at 37 °C (*see* **Note 14**).
17. Wash once with BT buffer.
18. After hybridization with directly fluorophore-labeled DNA, apply 18 μl of antifade solution with DAPI and seal under a cover slip.

19. When digoxigenin- or biotin-labeled DNA probe molecules have to be detected after hybridization, refer to Sect. 3.3.3.

3.3.3 Signal Detection of Hapten-Labeled Probes

1. Put a large drop (100 μl) of blocking buffer to the slides and incubate slide for >10 min at 37 °C in a moist chamber; for biotinylated probes continue with step 2, for digoxigenated ones with step 3.
2. To detect biotinylated probes, float off cover slip in detection buffer and incubate slides with 100 μl FITC-conjugated avidin (green, diluted 1:400 in BT) under a cover slip or a slide-sized piece of autoclave bag for 45 min at 37 °C. Continue with steps 4 or 8.
3. To detect digoxigenin-labeled probes, apply rhodamine-conjugated anti-digoxigenin antibody (red; e.g., Sigmaaldrich.com) diluted 1:200 in BT buffer and incubate for 45 min at 37 °C. Anti-digoxigenin and avidin-FITC may be mixed if two probes have to be differentially detected on the same slide.
4. Wash 3 × 3 min in BT buffer at 37 °C.
5. If a biotin-labeled probe displays only weak fluorescent signals on the specimen, amplify its fluorescence as follows: rinse away cover slip with BT buffer (see Sect. 3.3.2 steps 8–10).
6. Wash slides twice for 5 min in BT buffer and drain excess liquid.
7. Incubate preparation with a biotin-conjugated anti-avidin antibody (diluted 1:250 in BT) for 35 min at 37 °C.
8. Rinse away cover slip with BT buffer, wash slides 3 × 3 min in BT, and drain excess liquid.
9. Incubate preparation with FITC-conjugated avidin (diluted 1:450) as in step 2.
10. Rinse away cover slip with BT buffer, wash slides 3 × 3 min in BT buffer, and drain excess liquid.
11. Mount preparation in antifade medium supplemented with DAPI and inspect under the microscope (Fig. 1).

3.4 Combined Immunostaining and FISH

1. Wash slides, which have been obtained by spreading or semi-spreading (see Sect. 3.2.3), twice for 5 min in PBS, 0.05 % Tween 20 at RT.
2. Drain excess liquid.
3. To the slide add 50 μl of primary antibody (diluted in 1 × PBS, 0.05 % Tween 20, 0.5 % casein at the appropriate dilution, usually 1:50 to 1:250; this has to be tested empirically).
4. Cover with a cover slip or an appropriately sized piece of autoclave bag and incubate at 4 °C (e.g., in a fridge) overnight.
5. Float cover slip off with 1 × PBS, 0.05 % Tween 20.

6. Wash slides twice for 5 min in PBS, 0.05 % Tween 20, and drain excess liquid.
7. Dilute fluorochrome-conjugated secondary antibody 1 × PBS, 0.05 % Tween 20, 0.5 % casein, according to the instructions of the supplier. Usually 1:250–400 works fine for FITC-labeled antibodies, and Cy3-labeled antibodies are diluted 1:500–1500.
8. Add 100 μl of antibody solution under a cover slip or slide-sized piece of autoclave bag, and incubate for 90 min at room temperature.
9. Rinse away cover slip with 1 × PBS, 0.05 % Tween 20.
10. Wash slides twice for 5 min in 1 × PBS, 0.05 % Tween 20, and drain excess liquid.
11. At this point, the preparation can be mounted with 18μl antifade with DAPI under a cover slip. Images are recorded with a fluorescence microscope.
12. After recording of images and cell coordinates, rinse off the cover slip and antifade solution with 1 × PBS, 0.05 % Tween 20 (*see* **Note 15**).
13. Fix cells for 1 min in 1 % formaldehyde in 1 × PBS.
14. Rinse the cells twice in PBS, 0.05 % glycine to quench unsaturated aldehyde groups and apply the standard FISH procedure (see Sects. 3.3.2 and 3.3.3).
15. If the IF was destroyed by the denaturation steps or has faded, relocate the cells from step 6 and record the corresponding FISH images (*see* **Note 14**). Merge the two images electronically [27].

3.5 Microscopic Evaluation of Fluorescent Signals

An epifluorescence microscope equipped with appropriate filter sets for the excitation and emission of fluorescence spectra characteristic for the fluorochromes used is necessary to visualize signals. The optimal combinations of excitation filter, beam splitter, and emission filter have to be chosen according to the experimental requirements. For high specificity, narrow-bandwidth filters should be selected. This increases the specificity but sometimes at the cost of dimmer signals. Narrow-bandwidth filters, however, minimize the bleed-through of the fluorescence of other fluorochromes in multicolor (m)FISH experiments. Images are best recorded with a cooled CCD camera with high sensitivity to a wide spectrum of wavelengths, including far-red, e.g., as emitted by Cy5.

4 Notes

1. The Zymolyase stock can be stored at −20 °C for several months and repeatedly refrozen. The powder does not dissolve completely; therefore, there is usually a pellet after thawing, which should be stirred up before use.
2. Lipsol is a laboratory cleaning agent, a mixture of nonionic and anionic detergents plus a chelating agent and builders (information by the manufacturer). Several standard laboratory detergents (Nonidet, Triton X-100, sodium dodecyl sulfate, *N*-lauroyl sarcosine) failed to produce comparable results [28].
3. Add 4 g paraformaldehyde to 90 ml distilled water and heat the suspension on a magnetic stirrer to 80 °C (*caution*: heating creates hazardous formaldehyde vapors! Work in a fume hood with protective equipment!). After 20–30 min, the solution should become clear. If it stays opaque, add 5 M NaOH in small drops until it becomes clear. After cooling, add 3.4 g sucrose to the solution. If NaOH has been added, the solution has to be titrated back to pH 8.5 using HCl. Thereafter, fill up volume to 100 ml. If the fixative is not completely clear, it may be filtered. It can be stored for several months in the refrigerator.
4. FISH probes should be carefully selected, and their sequence be checked against the database (e.g., http://genome-www.stanford.edu) to avoid that they contain repetitive genomic elements (such as the Ty1 transposon), which would result in unspecific speckled background staining (unpublished observations).
5. Attempts to isolate uncontaminated chromosome-specific DNA from pulsed-field gels have failed in several labs. Hence, it is recommended to use contiguous sets of long-range PCR products and/or cosmid or P1 clones to paint extended chromosome regions. Pooled-probe DNAs can be labeled by nick translation.
6. Presporulation growth in YPA [29] improves the synchrony of sporulation.
7. This suspension can be kept on ice for up to 1 day, during which it should be used for the preparation of slides.
8. Fixative is added to the slide before and after the detergent. A small amount of fixative present during detergent spreading prevents the disruption of spheroplasts but does not interfere too much with spreading. The relative amounts and order of application of nuclear suspension, detergent, and fixative should be optimized by testing, since the optimal spreading depends on the density of nuclei in the suspension, the degree of spheroplasting, and the age of solutions. The process of

spreading can be watched under the phase-contrast microscope at low-power magnification without a cover slip. Spheroplasts should swell slowly and gradually turn from white to black and then to gray. They should not disintegrate instantly!

9. The presence of sucrose in the fixative has the advantage that the mixture is hygroscopic and does not dry out completely. Therefore, this kind of preparations can be used for immunostaining even after prolonged storage in the refrigerator or in the freezer.
10. During the spreading or semi-spreading procedures (see Sects. 3.2.2 and 3.2.3), chromatin tightly adheres to the surface of the slides owing to intimate charge interactions of DNA and glass. Coating of slides (with, e.g., poly-L-lysine, aminosilane) is not necessary.
11. Directly fluorochrome-conjugated deoxynucleotides (e.g., Cy3-dUTP, Cy5-dUTP, fluorescein-dUTP/fluorescein-dATP) in conjunction with nick translation generate direct fluorescing DNA probes, which render FISH signals of sufficient intensity for most applications. However, when several probes are to be combined for multicolor FISH applications, or if fluorescence signal intensity is too weak (in the case of single-copy probes of a few kilobases), biotin or digoxigenin labeling of probe DNAs together with indirect immunodetection is recommended, since this renders brighter signals [8, 25, 28].
12. Labeling reactions should be optimized to result in labeled products of 100–500 bp length, which should be monitored by agarose gel electrophoresis. If the DNA fragments are still too long after the first 90 min of nick translation, they can be chopped down, e.g., by adding 1 μl of a dilute (2 U/μl) DNAase I solution followed by 10 min of incubation. The effectiveness of this treatment should be first checked with test DNA by agarose gel electrophoresis. Commercial nick-translation kits (e.g., BioNick kit of Life technologies or DIG-Nick Translation Mix of Roche Applied Science) have performed well in our hands and are recommended for the beginner.
13. In some circumstances FISH may fail, which is often due to closed-chromatin state due to formaldehyde fixation. In such cases a second denaturation step, e.g., for 5 min at 75 °C with the probe sealed under a cover slip, is sufficient to bring about a successful hybridization.
14. It has been found that 0.05 × SSC efficiently replaces the 70 % formamide, 2 × SSC solution usually used in standard procedures for post-hybridization washes, as it delivers the same stringency [8, 26]. This replacement also reduces the potential

exposure to the fumes of the hazardous and teratogenic chemical formamide.

15. When the signal from immunofluorescent (IF) staining has vanished after the FISH procedure, recoding of images and position of IF cells with the help of the coordinate system of the microscope stage is recommended prior to FISH. After FISH, these cells are relocated and the FISH signals photographed. Finally, IF and FISH images are merged [27]. If the immunostaining signal is maintained after FISH [26] (which is often the case for SC proteins, but has to be tested empirically), slides are immediately subjected to FISH [25].

Acknowledgments

I thank J. Loidl, E. Trelles-Sticken, and A. Lorenz for stimulating discussions. The work in the lab of HS was partly supported by the DFG.

References

1. Cherry JM, Adler C, Ball C et al (1998) SGD: saccharomyces genome database. Nucleic Acids Res 26:73–79
2. Goffeau AEA et al (1997) The yeast genome directory. Nature 387:S1–S105
3. Kater JM (1927) Cytology of *Saccharomyces cerevisiae* with special reference to nuclear division. Biol Bulletin 52:436–449
4. Wintersberger U, Binder M, Fischer P (1975) Cytogenetic demonstration of mitotic chromosomes in the yeast *Saccharomyces cerevisiae.* Mol Gen Genet 142:13–17
5. Kuroiwa T, Kojima H, Miyakawa I et al (1984) Meiotic karyotype of the yeast *Saccharomyces cerevisiae.* Exp Cell Res 153:259–265
6. Dresser M, Giroux C (1988) Meiotic chromosome behavior in spread preparations of yeast. J Cell Biol 106:567–573
7. Loidl J, Nairz K, Klein F (1991) Meiotic chromosome synapsis in a haploid yeast. Chromosoma 100:221–228
8. Scherthan H, Loidl J, Schuster T et al (1992) Meiotic chromosome condensation and pairing in *Saccharomyces cerevisiae* studied by chromosome painting. Chromosoma 101:590–595
9. Loidl J (2003) Chromosomes of the budding yeast *Saccharomyces cerevisiae.* Int Rev Cytol 222:141–196
10. Loidl J, Lorenz A (2009) Analysis of *Schizosaccharomyces pombe* meiosis by nuclear spreading. Methods Mol Biol 558:15–36
11. Guacci V, Hogan E, Koshland D (1994) Chromosome condensation and sister chromatid pairing in budding yeast. J Cell Biol 125: 517–530
12. Loidl J, Klein F, Scherthan H (1994) Homologous pairing is reduced but not abolished in asynaptic mutants of yeast. J Cell Biol 125: 191–200
13. Weiner BM, Kleckner N (1994) Chromosome pairing via multiple interstitial interactions before and during meiosis in yeast. Cell 77:977–991
14. Rockmill B, Sym M, Scherthan H et al (1995) Roles for two RecA homologs in promoting meiotic chromosome synapsis. Genes Dev 9: 2684–2695
15. Gotta M, Laroche T, Gasser SM (1999) Analysis of nuclear organization in *Saccharomyces cerevisiae.* Methods Enzymol 304:663–672
16. Trelles-Sticken E, Adelfalk C, Loidl J et al (2005) Meiotic telomere clustering requires actin for its formation and cohesin for its resolution. J Cell Biol 170:213–223
17. Admire A, Shanks L, Danzl N et al (2006) Cycles of chromosome instability are associated with a fragile site and are increased by defects in DNA replication and checkpoint controls in yeast. Genes Dev 20:159–173
18. Straight AF, Belmont AS, Robinett CC, Murray AW (1996) GFP tagging of budding yeast chromosomes reveals that protein-protein

interactions can mediate sister chromatid cohesion. Curr Biol 6:1599–1608
19. Michaelis C, Ciosk R, Nasmyth K (1997) Cohesins: chromosomal proteins that prevent premature separation of sister chromatids. Cell 91:35–45
20. Aragon-Alcaide L, Strunnikov AV (2000) Functional dissection of in vivo interchromosome association in *Saccharomyces cerevisiae.* Nat Cell Biol 2:812–818
21. Lorenz A, Fuchs J, Trelles-Sticken E et al (2002) Spatial organisation and behavior of the parental chromosome sets in the nuclei of *Saccharomyces cerevisiae* x *S. paradoxus* hybrids. J Cell Sci 115:3829–3835
22. Kilmartin JV, Wright B, Milstein C (1982) Rat monoclonal antitubulin antibodies derived by using a new nonsecreting rat cell line. J Cell Biol 93:576–582
23. Pringle JR, Adams AE, Drubin DG et al (1991) Immunofluorescence methods for yeast. Methods Enzymol 194:565–602
24. Kane SM, Roth R (1974) Carbohydrate metabolism during ascospore development in yeast. J Bacteriol 118:8–14
25. Trelles-Sticken E, Loidl J, Scherthan H (1999) Bouquet formation in budding yeast: initiation of recombination is not required for meiotic telomere clustering. J Cell Sci 112: 651–658
26. Trelles-Sticken E, Dresser ME, Scherthan H (2000) Meiotic telomere protein Ndj1p is required for meiosis-specific telomere distribution, bouquet formation and efficient homologue pairing. J Cell Biol 151: 95–106
27. Loidl J, Jin Q-W, Jantsch M (1998) Meiotic pairing and segregation of translocation quadrivalents in yeast. Chromosoma 107: 247–254
28. Scherthan H, Loidl J (2010) FISH as a tool to investigate chromosome behavior in budding yeast. Methods Mol Biol 659:363–377
29. Roth R, Halvorson HO (1969) Sporulation of yeast harvested during logarithmic growth. J Bacteriol 98:831–832
30. Scherthan H, Bahler J, Kohli J (1994) Dynamics of chromosome organization and pairing during meiotic prophase in fission yeast. J Cell Biol 127:273–285

FISHing for Food Microorganisms

Benedetta Bottari, Andrea Mancini, Danilo Ercolini, Monica Gatti, and Erasmo Neviani

Abstract

FISH has the potential to make the visualization of microorganisms in food matrices possible and to allow for the enumeration, location, and distribution of positive, spoilage, and pathogenic microorganisms via nondestructive methods. Innovative techniques and methodical improvements have boosted the potential of FISH to study food microorganisms. The better understanding of the functioning of microbial communities is a challenging and crucial issue in the field of food microbiology, as it constitutes a prerequisite to the optimization of positive and technological microbial population functioning, as well as for the better control of pathogen contamination of food. As it enables the detection of most bacteria, even in samples where the proportion of cultivable bacteria among the total microbial population is relatively low, FISH has been applied for the specific detection of food spoilers as well as an early enumeration and identification of specific contamination sources in factory processes, including food pathogens and food-borne parasites. In this chapter, we present an updated overview on FISH protocols for the microbiological analysis of food.

Keywords Fluorescence in situ hybridization, Food microorganisms, Nondestructive methods, Innovative techniques, Methodical improvements

1 Introduction

Foods are teeming with microorganisms that may be innocuous, pathogenic threats (chapter by Alexander Swidsinski and Vera Loening-Baucke "Evaluation of Polymicrobial Involvement Using Fluorescence In Situ Hybridization (FISH) in Clinical Practice), spoilage agents, or beneficial microorganisms that drive fermentations or act as biocontrol agents. Regardless of their complexity, microbial populations in foods are the object of one of the widest attention from the scientific community because of the great influence they can have on quality and safety issues that are pivotal in food science [1]. The better understanding of the functioning of microbial communities is a challenging and crucial issue in the field of food microbiology, as it constitutes a prerequisite to the optimization of positive and technological microbial population

Thomas Liehr (ed.), *Fluorescence In Situ Hybridization (FISH)*, Springer Protocols Handbooks, DOI 10.1007/978-3-662-52959-1_51, © Springer-Verlag Berlin Heidelberg 2017

functioning, as well as for the better control of pathogen contamination of food [2]. The last 30 years have been characterized by a significant change in the approaches used for the microbiological examination of food [3], particularly due to the well-known bias of cultivation [4–9]. Traditional methods of studying microbial populations, like plating on selective media, commonly detect the most frequently occurring organisms, which grow to detectable levels by forming colonies on the selective media. However, environmental modifications during cultivation on synthetic media could affect the structure of the microbial community and thus limiting our view of the full ecosystem considered [10]. Cells that are stressed and injured are often not able to grow on synthetic media that contain agents, such as antibiotics, to make them selective toward a specific microorganism, and this can lead to false-negative results. Finally, populations that are numerically less important are not detected by means of traditional methods, because they are masked on the plates [4]. These straightforward methods provide a very simplistic, often biased, view of the physiological state of microbial populations in which several subpopulations characterized by various levels of "viability" and metabolic activity may coexist [11]. In terms of viability in fact (both qualitative and quantitative), it has been long recognized that microbial cells may exist in different states (i.e., cryptobiotic, dormant, moribund, latent) in which they do not form colonies on a solid media but potentially still having other metabolic activities able to have an important role in food quality and safety [12–16]. Besides, understanding the ecology of complex microbial communities, such as those present in foods, also requires studies on the activities and distributions of microbes that should be performed in situ or in minimally disturbed samples [17, 18].

The emergence of molecular techniques has opened new opportunities to characterize the numerous intermediate states of microbial cells and to study the metabolic activities of microorganisms in food [6, 9]. Recently, Cocolin and Ercolini [1] described how the culture-independent analysis has revolutionized the study of the food microbiota, leading to what could be defined as a "cultural" evolution. During the years, several target-based molecular methods that avoid cultivation have been developed, allowing the detection of taxonomically defined group of microbes, specific species, or specific strains. By the core methodology on which they are based, these methods can be used to track functionally important members of a given ecosystem [5]. Molecular methods encompass fluorescence in situ hybridization (FISH), flow cytometry, "omics," and polymerase chain reaction (PCR)-based technologies [6]. FISH is based on the identification of cells containing specific nucleic acid sequences. Oligomer probes conjugated to fluorescent molecules hybridize to their target DNA or RNA, and fluorescence is detected by microscopic observation [10]. Initially applied in the medical, ecological, and developmental biology domains, its popularity depends on the possibility to profile microbial communities in a quantitative manner, overcoming culturing and PCR-based methods, with the possibility of a direct observation of targeted cells within their native environment [5, 19]. In particular, FISH has found a large number of applications, including the investigation of microbial symbiosis, the analysis of microbial diversity in environmental samples, the evaluation of the presence of bacteria in wastewater treatment plants [20], the identification of bacteria relevant in diagnostic medicine, and the detection of pathogens within human and animal tissues [21]. Generally based on the use of a labeled oligonucleotide probe directed on a specific region of rRNA (16S or 23S), this method allows to range the phylogenetic specificity from the domain to the subspecies level [22–24] with the impossibility to differentiate between strains of prokaryotic species, due to the relatively slow rate of mutation in rRNA [25]. By the time some modifications and adaptations from the initial approach have been also considered, especially to overcome some limits such as species-strain specificity, hybridization performance of nucleotide rRNA-directed probes and fluorescence intensity dependence by the targeted region(s) or rRNA(s) copy number availability [26–28]. Particularly, to increase the potentiality of this technique [9, 24], peptide nucleic acid (PNA) probes, substitution of fluorophore with enzyme (catalyzed reporter deposition-FISH, CARD-FISH)[24, 29, 30], multiple-probe utilization [31, 32], and association with flow cytometry have been extensively adapted, allowing better and faster results [6, 28, 33–36]. Along with these improvements, also specific associated limits as extended procedure time and probe/equipment cost have been reported [9].

FISH represents a promising alternative method in food microbiology although, due to sometimes low repeatability and artifacts and interferences that can occur with the food matrix, it is still difficult to set up and is not yet routinely used to analyze and monitor food products. The method has been used to study microbial communities (both bacteria and yeasts) of various unfermented and fermented foods, such as dairy products [32, 37–43], olives [44], wine [45–49], and beer [50]. As it enables the detection of most bacteria, even in samples where the proportion of cultivable bacteria among the total microbial population is relatively low, FISH has been applied also for the specific detection of food spoilers [40, 42, 51–53] as well as an early enumeration and identification of specific contamination sources in factory processes [54]. Alongside with technological and ecological application purposes of this methodology, FISH is also highly useful for the in situ detection of food pathogens, especially *Campylobacter*, *Salmonella*, *Listeria*, *Escherichia coli*, *Shigella*, *Vibrio*, and *Yersinia* [9, 33, 42, 55–62], and food-borne parasites like *Giardia* and *Cryptosporidium* [63]. However, despite the huge background of knowledge about it and despite innovative techniques and methodical improvements have boosted the potential of this technique to study food microorganisms, FISH has not yet been intensively used in food microbiology. One reason for that is the food matrix itself, in which complexity is high and prone to the loss of information (e.g., data regarding the distribution of bacterial cells in the food product) due to the needed pretreatments. Examples of that are the removal of components with natural fluorescence activity (as chlorophyll, hemoglobin, or other pigments), homogenization [42, 53, 64], or use of unspecific protease treatment and mechanical disintegration that may affect the final microbe detection [9, 65]. Studies aimed at improving in particular the bacteria localization/sample integrity have been proposed by Ercolini et al. [66], who developed a FISH method to identify microbes resident in different locations within a cheese matrix, and by Bisha and Brehm-Stecher who used transparent adhesive tapes (tape-FISH) for the detection of food surface microbes [33, 67]. After all, the use of FISH has the potential to make the visualization of microorganisms in food matrices possible and to allow for the enumeration, location, and distribution of positive, spoilage, and pathogenic microbes via nondestructive methods.

FISH with rRNA target probes for the in situ analysis of single microbial cells was developed as a culture-independent "non-PCR-based" molecular technique for the simultaneous visualization, identification, enumeration, and localization of individual microorganisms from all fields of microbiology [20, 21, 23]. Generally a typical FISH procedure includes four main steps: (1) fixation and permeabilization of the sample, (2) hybridization, (3) washing steps that remove unbound probe, and (4) detection of labeled cells by microscopy or flow cytometry [20]—for possibilities of evaluation of ISH experiments in microbiology by electron microscope, see chapter by Hannes Schmidt, Thilo Eickhorst "Gold-FISH: In situ Hybridization of Microbial Cells for Combined Fluorescence and Scanning Electron Microscopy".

The fixation and permeabilization of the bacterial cells precede the hybridization phase, allowing the penetration of the fluorescent probes into the cell as well as protecting the RNA molecules from degradation by endogenous RNase [21]. The fixing agent can be used directly in order to cover the sample when it has settled on a membrane filter [68], or it can be mixed with the sample before incubation, sedimentation by centrifugation, and resuspension and then spotting on the glass slides followed by air dehydration [23]. An enzymatic treatment is sometimes necessary, e.g., in gram-positive bacteria [29, 69, 70]. Furthermore, in order to avoid cell loss or the insufficient adhesion of specimens to glass slides, the glass surfaces are often treated with coating agents such as gelatin [23], poly-L-lysine [71], or silanating agents [72]. The hybridization takes place in a dark humid chamber, usually at temperatures of between 37 and 50 °C, for times ranging from 30 min to several hours. Slides are subsequently briefly rinsed with distilled water in order to remove unbound probe, mounted in anti-fading agents to prevent fluorescence "bleaching" [21], visualized, and documented.

In food microbiology, additional steps for the sample preparation and homogenization, pre-enrichment procedures, or bacterial separation might be required. Low numbers of the target organism can require the concentration of the bacteria, and huge amounts of protein, fat, or components with natural fluorescence have to be removed since they might disturb the hybridization of the probes or cause strong background in the consecutive microscopic examination [9].

Overall, FISH with rRNA-targeted probes and quantitative microscopy is today a standard tool for revealing the identity, abundance, and spatial localization of microbial cells in complex samples. Innovative techniques and methodical improvements have boosted the potential of FISH to study food microorganisms. With additional characteristics such as low costs per sample, the feasibility of high-throughput analyses, and, ideally, a high degree of simplicity concerning the performance of test, FISH could be transferred to routine use in food microbiology [9].

2 Materials

A wide variety of probes are currently being used to examine natural bacterial communities, such as those in food matrices. DNA probes require stringent hybridization conditions, specific to each individual probe, necessitating the optimization of the hybridization protocol on a case-by-case basis. In recent years, free in silico-modeling software tools have considerably simplified this optimization process [73–75]. The chemicals and solutions reported here refer to the universal probe Eub338 [23], used here as an example. Apart from the standard equipment required for FISH, the following list presents the more specialized reagents that are needed (listed in alphabetical order):

2.1 Chemicals

The probes that can be used for FISH in different bacteria are listed in Table 1.

- EDTA (ethylenediaminetetraacetic acid)
- Ethanol
- Formamide (methanamide)
- Lysozyme
- NaCl (sodium chloride)
- Na citrate (sodium citrate)

Table 1
Some of the oligonucleotide probes used for FISH analysis in food samples

Probe	Specificity	Sequence (5′–3′) of probe	Target site (rRNA position)	%FA[a]	References
Bet42a	*Betaproteobacteria*	GCCTTCCCACTTCGTTT	23S (1027–1043)[b]	35	[76]
Bif164	*Bifidobacterium*	CATCCGGCATTACCACCC	16S (164–181)[b]	0	[77]
BRE1239	*Brevibacterium*	TCTCTCTGTACCAGCCAT	16S (1239–1257)[b]	30	[37]
CAMP653	*Campylobacter*	CTGCCTCTCCCTYACTCT	16S (653–670)	35	[56]
Eco1482	*Escherichia coli*	TACGACTTCACCCCAGTC	16S (1482–1499)[b]	30	[27]
ENT	*Enterobacteriaceae*	TGCTCTCGCGAGGTCGCTTCTCTT	16S (1251–1274)[b]	20	[78]
ESA452	*Enterococcus italicus*	CATTCTCTTCTCATCCTT	16S (453–470)	10	[53]
EUB	Most bacteria	GCTGCCTCCCGTAGGAGT	16S (338–355)[b]	0–60	[23]
EUK	Eukarya	ACCAGACTTGCCCTCC	18S (502–517)[b]	20	[23]
Gam42a	*Gammaproteobacteria*	GCCTTCCCACATCGTTT	23S (1027–1043)[b]	35	[76]
HGC69a	*Actinobacteria*	TATAGTTACCACCGCCGT	23S (1901–1918)[b]	25	[79]
Lab158	*Lactobacillus/Enterococci*	GGTATTAGCAYCTGTTTCCA	16S (158–177)[b]	0	[80]
Lac663 (PNA)	*Lactobacillus* spp.	ACATGGAGTTCCACT	16S	0	[42]
LactV5	*Lactococcus lactis*	GCTCCCTACATCTAGCAC	16S (821–839)[b]	25	[81]
Lbh-1	*Lactobacillus helveticus*	ACTTACGTACATCCACAG	16S	0	[Tailliez Patrick, Oriane Matte-Tailliez (1997) Unpublished data]
LbpV3	*Lactobacillus plantarum*	CCGTCAATACCTGAACAG	16S (468–486)[b]	25	[81]
Lbrev	*Lactobacillus brevis*	CATTCAACGGAAGCTCGTTC	16S (64–83)[b]	0	[46]

(continued)

Table 1
(continued)

Probe	Specificity	Sequence (5′–3′) of probe	Target site (rRNA position)	%FA[a]	References
LDE23	*Lactobacillus delbrueckii subsp. bulgaricus*	GCGTGTTCCRTCCTTAAGC	23S	20	[82]
LGC354ab	Firmicutes	YGGAAGATTCCCTACTGC	16S (354–371)[b]	35	[83]
Lis-637	*Listeria*	CACTCCAGTCTTCCAGTTTCC	16S	35	[65]
Lis-1255	*Listeria, Brochothrix*	ACCTCGCGGCTTCGCGAC	16S (1255–1272)[b]	35	[69]
Lis-1400	*Listeria* spp.	CGCACATTTCCATTCGTGCGATTCC	23S (1491–1515)[c]	50	[58]
Liv-16S-5 (PNA)	*Listeria ivanovii*	ACGCATGTCATCACT	16S (142–156)[d]	30	[60]
Lis-16S-1 (PNA)	*Listeria* spp.	ACTGTTGTTAGAGAAG	16S (440–455)[d]	30	[60]
Lm-16S-2 (PNA)	*Listeria monocytogenes*	TAGTACAAAGGGTCG	16S (1247–1261)[d]	30	[60]
Lpara	*Lactobacillus casei—L. paracasei*	GTTCCATGTTGAATCTCGG	16S (94–113)[b]	0	[46]
LU2	*Leuconostoc*	GATCCATCTCTAGGTGACGCCG	16S (221–242)[b]	0	[84]
OENOS 5/2	*Oenococcus oeni*	TACTTTGGGCCCTGACA	5S[f]	0	[85]
OENOS 5/3	*Oenococcus oeni*	ACCTTGCAACAGGCGTT	5S[f]	0	[85]
Pae	*Pseudomonas spec.*	TCTGGAAAGTTCTCAGCA	16S (997–1014)[b]	0	[86]
Pap446	*Propionibacterium acidipropionici*	ACACCCCAAAACGATGCCTTCGCC	16S	0	[40]

(continued)

Table 1 (continued)

Probe	Specificity	Sequence (5′–3′) of probe	Target site (rRNA position)	%FA[a]	References
Pfr435	*Propionibacterium freudenreichii*	CTTGCGCTTCGTCATGGATGAAAG	16S	0	[40]
Pj446	*Propionibacterium jensenii*	CACCCCGATAGGCACTTCGTC	16S	0	[40]
PS	*Pseudomonas* sp.	GATCCGGACTACGATCGGTTT	16S (1284–1304)[b]	0	[54]
Sal3	*Salmonella* spp.	AATCACTTCACCTACGTG	23S (1713–1730)	0	[61]
SalPNA1873	*Listeria monocytogenes*—*Salmonella enterica*	AGGAGCTTCGCTTGC	23S (1873–1887)[e]	30	[87]
St4	*Streptococcus thermophilus*	TTATCCCCCGCTACAAGG	16S	0	[Tailliez Patrick, Oriane Matte-Tailliez (1997) Unpublished data]
Sth	*Streptococcus thermophilus*	CATGCCTTCGCTTACGCT	16S (69–87)[b]	25	[88]
STH23	*Streptococcus thermophilus*	CATGCCTTCGCTTACGCT	23S	20	[82]

[a]Formamide concentration (FA) in the in situ hybridization buffer (vol/vol)
[b]*Escherichia coli* rRNA numbering [89]
[c]*L. monocytogenes* (X68420)
[d]Gene bank (FJ434468)
[e]*S. enterica* subsp. *enterica* serovar *Typhimurium* LT2 (ATCC 43971) 23S rRNA gene sequence (accession number U77920)
[f]*Bacillus subtilis* rRNA sequence comparison [85, 90]

- Paraformaldehyde
- PBS (phosphate-buffered saline: 8 g NaCl, 0.2 g KCl, 1.44 g Na_2HPO_4, 0.24 g KH_2PO_4 in 800 ml of distilled H_2O)
- Proteinase K
- SDS (sodium dodecyl sulfate)
- TE buffer (10 mM Tris, bring to pH 7.5 with HCl, 1 mM EDTA)
- Tris-HCl (Tris hydrochloride)

2.2 Solutions to Be Prepared

- Hybridization buffer:
  - 20 mM Tris-HCl, pH 7.2
  - 0.01 % SDS
  - 40 mM NaCl
  - 5 mM EDTA
- Lysozyme solution: dissolve 1 ml of lysozyme 69,490 U in 1 ml of 5 mM l^{-1} EDTA, 100 mM l^{-1} Tris-HCl, pH 7.5
- Proteinase-K solution: dissolve 10 mg in 1 ml of redistilled sterile water
- Washing buffer:
  - 0.9 M NaCl
  - % SDS
  - 20 mM Tris-HCl, pH 7.2
  - Formamide at the appropriate concentration

3 Methods

Food samples, whether liquid or solid, may require different upstream sample processing. For milk and other dairy products, the pretreatment may involve the use of a homogenizer, simple or multiple centrifugation steps, and addition of an appropriate buffer to obtain a bacterial pellet with only a few interfering substances [42, 53, 64]. An additional treatment with unspecific proteases (e.g., Proteinase K) can be done for milk clearing and to reduce the background and autofluorescence of the food matrices [32, 54]. Solid food samples, when analyzed as suspensions, have to be reduced to small pieces, followed by further steps of mechanical disintegration, e.g., by stomaching [91]. Finally, a filtration of the sample through a membrane transmissible for microbes or a short centrifugation can be used to remove larger particles of the food matrix, which might disturb efficient microscopic evaluation [53].

However, these treatments may have an impact on the detection of microorganisms by FISH [65].

Hybridization and washing require precise temperature control in order to prevent nonspecific hybridization at low temperatures and the loss of correctly hybridized probes at high temperatures. The optimal hybridization and washing temperature must be chosen based on the melting point of the selected probe and the accessibility of the target site. In fact, temperature affects not only the dissociation of the probe but also the conformation of the targeted rRNA or DNA and thus the accessibility of the targeted rRNA to oligonucleotide probes [27, 92]. In general, the higher the temperature, the more stringent the hybridization and washing temperatures must be. The temperatures and the hybridization and washing times presented here are examples and must be optimized case by case.

3.1 Preparation of Liquid Samples or Suspensions of Solid Samples for FISH Analysis

In order to evaluate the microbial populations in foods, we must distinguish between at least two kinds of food: liquid and solid. Moreover, we have to consider fermented foods with or without a microbial starter, foods where beneficial microorganisms drive fermentations or act as biocontrol agents, and unfermented foods where microorganism may be innocuous, pathogenic threats or spoilage agents. FISH on liquid food samples or on suspensions of food solid samples is quite simple and can be used, for example, to follow population dynamics in complex ecosystems. After efficient sample preparation, an in situ analysis of complex sample materials can be performed on morphologically intact cells. This is an appropriate approach when studying dominant microbial populations, such as those found in fermented foods, where their dominance over specific species of other microbes that are naturally present can be studied in order to improve knowledge of the fermentation processes [4, 38, 40, 53, 91]. In particular, FISH can quite easily and effectively be used to analyze liquid starter cultures. Starter cultures are of great industrial significance, since they play a crucial role in the manufacturing of fermented foods and in the development of their flavors and textures. The FISH technique can be a useful approach for studying a specific starter culture and subsequently evaluating its effectiveness and/or performance during the production process. It can be even more effective in studying highly variable natural starter cultures, such as those traditionally used to produce PDO (protected designation of origin) Italian cheeses such as Parmigiano-Reggiano and Grana Padano [39; Tailliez Patrick, Oriane Matte-Tailliez (1997) Unpublished data]. For example, with the simultaneous use of two species-specific probes [93] labeled with different dyes, it has been possible to highlight in natural whey starter the abundance of *Lactobacillus helveticus* with respect to *Streptococcus thermophilus* (Fig. 1a, b).

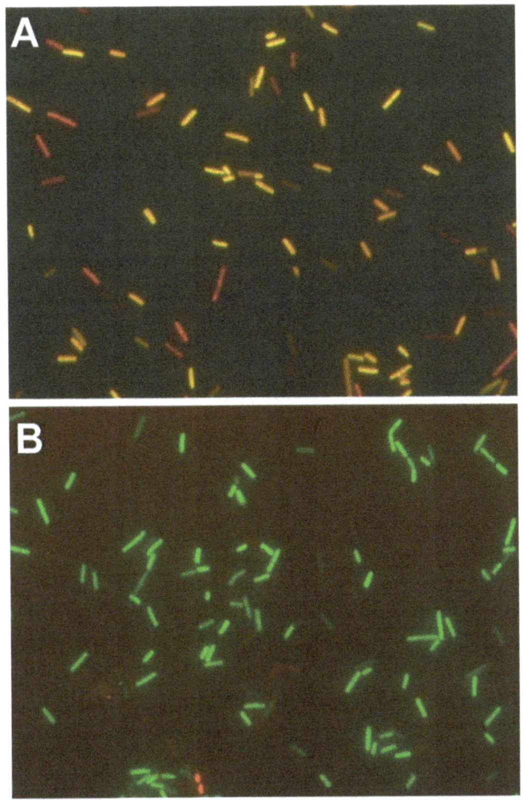

Fig. 1 FISH of a natural whey starter. (**a**) Simultaneous use of the probes Eub338 (*red label*) and Lbh1 (*green label*). *Lactobacillus helveticus* cells appear *orange* (hybridization by both probes). Other bacteria appear *red* (hybridized only by Eub338 probe). (**b**) Simultaneous use of the probes Lbh1 (*green label*) and St4 (*red label*). *Lactobacillus helveticus* cells appear *green*, while *Streptococcus thermophilus* cells are *red*

3.1.1 Fresh Liquid Samples

This could, for example, be whey starter or milk, and it should be prepared as follows:

1. Centrifuge an aliquot (i.e., 0.3–0.5 ml) of the sample (7,000× *g* for 5 min).
2. Discard the supernatant, resuspend, and wash the cell pellets in an equal volume of TE buffer. Repeat step 1.
3. Discard the supernatant and resuspend the pellet in an equal volume of 1 × PBS.
4. Add paraformaldehyde (4 %, −20 °C) 1:3 and fix for 1 h at +4 °C; repeat step 1.
5. Wash the pellet with 1 ml of 1 × PBS. Repeat step 1.
6. Discard supernatant and resuspend the pellet in 50 % (v/v) ethanol/PBS.
7. Store at −20 °C until further FISH analysis.

3.1.2 Suspension of Solid Samples

Cheese or meat, for example, should be prepared as follows:

1. Disperse 5–10 g of each sample in 1:10 Na citrate solution, pH 7.2, by stomaching for 2 min.
2. Centrifuge 1 ml of stomached mixture at 7000× *g* for 5 min.
3. Discard the supernatant and resuspend the pellet in an equal of TE buffer. Repeat step 2.
4. Wash the pellet with 1 ml of 1 × PBS. Repeat step 2.
5. Add paraformaldehyde (4 %, −20 °C) 1:3 and fix for 12 h at +4 °C.
6. Centrifuge at 10,000× *g* for 5 min.
7. Discard the supernatant and resuspend the pellet in 50 % (v/v) ethanol/PBS.
8. Store at −20 °C until further FISH analysis.

3.2 Preparation of Solid Samples for FISH Analysis

The use of a sample-embedding procedure that withstands the hybridization reactions enables the microorganisms to be detected in situ within the food matrix (Fig. 2). The section-FISH approach permits the location of specific groups of bacteria within the food matrix and the investigation of relationships between specific groups of bacteria [56, 66].

Solid samples can be prepared in different ways depending on the nature of the matrix. Basically, the solid samples must be sliced into very thin sections (5–30 μm thick) in order to be analyzed by FISH. The sections should:

- Endure the hybridization conditions (solutions, temperature, detergents, etc.)
- Ensure the integrity of the matrix
- Allow probe entry and a good hybridization yield
- Not alter the microbial colony distribution within the sample

Therefore, different sectioning procedures can be taken into account. For resistant matrices, a simple cryo-sectioning procedure can be sufficient. Alternatively, embedding agents such as paraffin can be used. The latter do not always work for delicate but microbiologically complex samples such as solid foods. However, FISH can be applied to tissues or foods embedded in a cold polymerizing

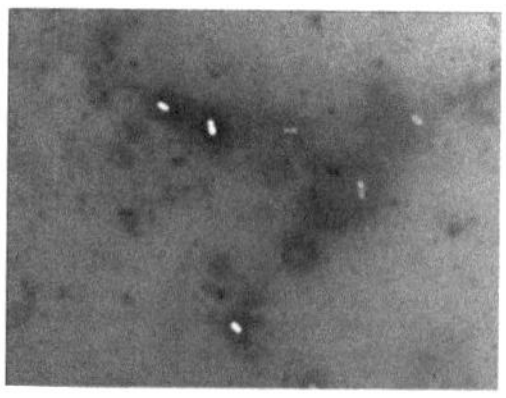

Fig. 2 FISH of a cheese section performed with the eubacterial probe Eub338

resin. The sections have a long shelf life, and it is often possible to avoid enzymatic pretreatment of the samples. Some protocols have been developed for tissues [66] and cheese [81, 94] using an embedding procedure with a cold polymerizing glycol methylacrylate (GMA) resin (Technovit 8100, Kuzler, Wehereim, Germany) according to the manufacturer's instructions.

3.3 FISH on Liquid Samples or Suspensions of Solid Samples

1. Spot about 20 μl of fixed cell suspension onto slides coated with poly-L-lysine.
2. Dry in an oven at 46 °C for 10 min.
3. Dehydrate in an ethanol series by covering the spots with about 50 μl of 50, 80, and 100 % ethanol solutions for 3 min each and then air-dry.
4. Enzymatic treatment: specimens can be treated by covering the spots either with 10 μl of proteinase K (10 mg ml^{-1}) for 10 min at 37 °C or with 30 μl of lysozyme (1 mg ml^{-1}) for 5 min at room temperature (*see* **Note 1**).
5. Stop the reaction by washing with ice-cold PBS before drying.
6. Add 10 μl of the hybridization buffer containing 10 ng of the selected rRNA probe onto the dry specimen (*see* **Notes 2–3**).
7. Incubate the slides in a dark, humid chamber at 45 °C overnight (*see* **Note 4**).
8. Remove unbound oligonucleotides by incubating the slides in prewarmed washing buffer at 45 °C for 15 min (*see* **Note 5**).
9. Rinse the slides by pipetting about 500–1,000 μl of sterile water onto the surface.
10. Air-dry.
11. Embed the samples in mounting oil.
12. Evaluate the slides with an epifluorescence microscope equipped with a 100× objective and appropriate filter sets (*see* **Notes 6, 7**).

4 Notes

1. Technical problems can arise when optimizing the cell permeabilization conditions, which are affected by different cell growth phases [18, 95] and by the simultaneous presence of gram-positive and gram-negative bacteria [40, 44] in the sample studied. For this reason, different permeabilization treatments should be screened in order to figure out which is the most effective in each case. For example, an extended lysozyme treatment can result in the hybridization of all of the cells, but, as a drawback, the cells can often display a diffuse appearance, suggesting the loss of cell structure and the leakage of cell content including rRNA.

2. The accuracy and reliability of FISH are highly dependent on the specificity of the oligonucleotide probe, which is strictly correlated with the stringency of the protocols applied. Formamide and sodium chloride are used to adjust the stringency of the hybridization buffer and the wash solution, respectively. It is necessary to choose appropriate concentrations of these chemicals in order to achieve the proper annealing of the oligonucleotide probes to the target site. In particular, formamide decreases the melting temperature by weakening the hydrogen bonds, thus enabling lower temperatures to be used with higher stringency. The more concentrated the formamide, the more stringent and the more specific the hybridization. However, further addition leads to a drastic drop in bound probe and signal intensity [74, 96]. Hybridization washes can even be conducted under more stringent conditions in order to remove excess probe and to disrupt all duplexes other than those between very closely related sequences. The stringency of the washing buffer can also be regulated by varying the concentration of salt instead of using formamide, thus reducing the amount of toxic waste [97]. Increasing the concentration of NaCl enhances the stability of mismatched heteroduplexes, and so lowering the salt concentration encourages the dissociation (denaturation) of mismatched heteroduplexes and gives higher washing stringency. The lower the salt concentration and the higher the wash temperature, the more stringent the wash. In general, greater specificity is obtained when hybridization is performed at a high stringency and washing at similar or lower stringency, rather than hybridizing at low stringency and washing at high stringency.
3. When more than one probe is used for the same specimen, they must be mixed at suitable ratios in the hybridization buffer up to 10 ng [81].
4. Depending on the cell wall characteristics, the penetration of the probes into bacteria can sometimes be variable and insufficient. This problem can be overcome by using peptide nucleic acids (PNAs) [45], which hybridize to target nucleic acid molecules more rapidly and with higher affinity and specificity compared to DNA probes [98]. In fact, PNA molecules are uncharged DNA analogs that bind to nucleic acids much more strongly than oligonucleotides, because there is no electrostatic repulsion between the PNA probe and the negatively charged sugar phosphate backbone of the target molecule [99]. The use of PNA probes for FISH analysis of food products has increased significantly in recent years [62, 100–102].
5. Washing steps have to be performed carefully in order to avoid the loss of cells or sample. For this reason, the first aliquots of the washing buffer or water must be poured softly at the edges of the spots or the food sections, since they are completely covered.

6. To overcome the problem of the autofluorescence of the food matrix and of the microorganisms themselves, the use of narrowband filter sets, monochromators, and signal amplification systems is recommended [29, 103]. With a food matrix causing a huge background signals and autofluorescence, as well as to counteract a weak fluorescence signal, the use of catalyzed reporter deposition-FISH (CARD-FISH) may be preferred, considering though that permeabilization protocols must be optimized for the microorganisms of interest, as probes labeled with this horseradish peroxidase penetrate fixed cells very poorly [76].
7. FISH results are easily interpreted when the number of microbial cells present in these food samples is about 10^8–10^9 cells ml^{-1}. Otherwise, FISH has a significant limit of detection when coupled with fluorescence microscopy (10^6 cells ml^{-1}), which can hamper the monitoring of bacterial cells that are present in very low amounts (e.g., pathogens). For example, a few hybridized cells per field have been detected in an experimental cheese sample artificially inoculated with 10^6 cells ml^{-1} of *E. coli* and analyzed by FISH with an *E. coli*-specific probe (Fig. 3). FISH-based testing kits containing probes that are specific for pathogens [55] are commercially available, and

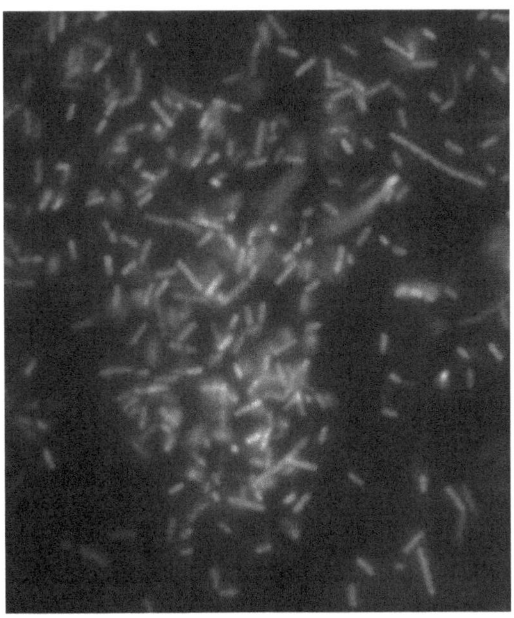

Fig. 3 FISH analysis of experimental cheese inoculated with 10^6 cells g^{-1} of *Escherichia coli*. The micrographs show the microscopic field after FISH analysis of the cheese suspension with the species-specific probe Eco1482 (see Table 1). *Brightest cells*: hybridized *E. coli* cells. *Darkest cells*: unhybridized lactic acid bacteria

several FISH methods have been optimized for pathogen detection [9], but they all require pre-enrichment of the sample in nutrient substrate. Thus, these FISH-based kits and approaches should not be considered "culture-independent" procedures. To improve the detection limit, other modifications to the general FISH protocol have been proposed such as FISH on filter or Flow-FISH. FISH-on-filter technique is based on filters with a small pore size where large volumes of liquid samples with only a few particles and microbes can be analyzed, thus permitting an increased sensitivity due either to concentration or cultivation of cells [104]. Flow-FISH is the combination between FISH and flow cytometry and has the advantage of avoiding microscopy limits [54], but it has not yet yielded many applications in routine testing of food products, as food samples are still too challenging for this analysis and because of higher costs [9].

References

1. Cocolin L, Ercolini D (2015) Zooming into food-associated microbial consortia: a "cultural" evolution. Curr Opin Food Sci 2:43–50
2. Bridier A, Hammes F, Canette A et al (2015) Fluorescence-based tools for single-cell approaches in food microbiology. Int J Food Microbiol 213:2–16
3. Cocolin L, Alessandria V, Dolci P et al (2013) Culture independent methods to assess the diversity and dynamics of microbiota during food fermentation. Int J Food Microbiol 167:29–43
4. Cocolin L, Ercolini D (2008) Molecular techniques in the microbial ecology of fermented foods. Springer, New York
5. Bokulich NA, Mills DA (2012) Next-generation approaches to the microbial ecology of food fermentations. BMB Rep 45:377–389
6. Sohier D, Pavan S, Riou A et al (2014) Evolution of microbiological analytical methods for dairy industry needs. Front Microbiol 5:16
7. Davis C (2014) Enumeration of probiotic strains: review of culture-dependent and alternative techniques to quantify viable bacteria. J Microbiol Methods 103:9–17
8. Kuchta T, Knutsson R, Fiore A et al (2014) A decade with nucleic acid-based microbiological methods in safety control of foods. Lett Appl Microbiol 59:263–271
9. Rohde A, Hammerl JA, Appel B et al (2015) FISHing for bacteria in food-a promising tool for the reliable detection of pathogenic bacteria? Food Microbiol 46:395–407
10. Bottari B, Ercolini D, Gatti M, Neviani E (2006) Application of FISH technology for microbiological analysis: current state and prospects. Appl Microbiol Biotechnol 73:485–494
11. Davey HM (2011) Life, death, and in-between: meanings and methods in microbiology. Appl Environ Microbiol 77:5571–5576
12. PostGate JR (1967) Viability measurements and the survival of microbes under minimum stress. Adv Microb Physiol 1:1–23
13. Kaprelyants AS, Kell DB (1993) Dormancy in stationary-phase cultures of micrococcus luteus: flow cytometric analysis of starvation and resuscitation. Appl Environ Microbiol 59:3187–3196
14. Wayne LG (1994) Dormancy of *Mycobacterium tuberculosis* and latency of disease. Eur J Clin Microbiol Infect Dis 13:908–914
15. Kaprelyants AS, Mukamolova GV, Davey HM et al (1996) Quantitative analysis of the physiological heterogeneity within starved cultures of *Micrococcus luteus* by flow cytometry and cell sorting. Appl Environ Microbiol 62:1311–1316
16. Mukamolova GV, Murzin AG, Salina EG et al (2006) Muralytic activity of *Micrococcus luteus Rpf* and its relationship to physiological activity in promoting bacterial growth and resuscitation. Mol Microbiol 59:84–98
17. Amann R, Kühl M (1998) In situ methods for assessment of microorganisms and their activities. Curr Opin Microbiol 1:352–358

18. Amann R, Fuchs BM (2008) Single-cell identification in microbial communities by improved fluorescence in situ hybridization techniques. Nat Rev Microbiol 6:339–348
19. Amann RI, Ludwig W, Schleifer KH (1995) Phylogenetic identification and in situ detection of individual microbial cells without cultivation. Microbiol Rev 59:143–169
20. Amann R, Fuchs BM, Behrens S (2001) The identification of microorganisms by fluorescence in situ hybridisation. Curr Opin Biotechnol 12:231–236
21. Moter A, Göbel UB (2000) Fluorescence in situ hybridization (FISH) for direct visualization of microorganisms. J Microbiol Methods 41:85–112
22. DeLong EF, Wickham GS, Pace NR (1989) Phylogenetic stains: ribosomal RNA-based probes for the identification of single cells. Science 243:1360–1363
23. Amann RI, Krumholz L, Stahl DA (1990) Fluorescent-oligonucleotide probing of whole cells for determinative, phylogenetic, and environmental studies in microbiology. J Bacteriol 172:762–770
24. Wagner M, Haider S (2012) New trends in fluorescence in situ hybridization for identification and functional analyses of microbes. Curr Opin Biotechnol 23:96–102
25. Ouverney CC, Fuhrman JA (1997) Increase in fluorescence intensity of 16S rRNA in situ hybridization in natural samples treated with chloramphenicol. Appl Environ Microbiol 63:2735–2740
26. Frischer ME, Floriani PJ, Nierzwicki-Bauer SA (1996) Differential sensitivity of 16S rRNA targeted oligonucleotide probes used for fluorescence in situ hybridization is a result of ribosomal higher order structure. Can J Microbiol 42:1061–1071
27. Fuchs BM, Wallner G, Beisker W et al (1998) Flow cytometric analysis of the in situ accessibility of Escherichia coli 16S rRNA for fluorescently labeled oligonucleotide probes. Appl Environ Microbiol 64:4973–4982
28. Brehm-Stecher BF (2008) Methods for whole cell detection of microorganisms. In: Camesano TA, Mello CM (eds) Microbial surfaces. American Chemical Society, Washington, DC, pp 29–51
29. Schönhuber W, Zarda B, Eix S et al (1999) In situ identification of cyanobacteria with horseradish peroxidase-labeled, rRNA-targeted oligonucleotide probes. Appl Environ Microbiol 65:1259–1267
30. Cerqueira L, Azevedo NF, Almeida C et al (2008) DNA mimics for the rapid identification of microorganisms by fluorescence in situ hybridization (FISH). Int J Mol Sci 9:1944–1960
31. Takada T, Matsumoto K, Nomoto K (2004) Development of multi-color FISH method for analysis of seven *Bifidobacterium* species in human faeces. J Microbiol Methods 58:413–421
32. Yamaguchi N, Kitaguchi A, Nasu M (2012) Selective enumeration of viable *Enterobacteriaceae* and *Pseudomonas spp.* in milk within 7 h by multicolor fluorescence in situ hybridization following microcolony formation. J Biosci Bioeng 113:746–750
33. Bisha B, Brehm-Stecher BF (2009) Simple adhesive-tape-based sampling of tomato surfaces combined with rapid fluorescence in situ hybridization for *Salmonella* detection. Appl Environ Microbiol 75:1450–1455
34. Azevedo NF, Jardim T, Almeida C et al (2011) Application of flow cytometry for the identification of Staphylococcus epidermidis by peptide nucleic acid fluorescence in situ hybridization (PNA FISH) in blood samples. Antonie Van Leeuwenhoek 100:463–470
35. Manti A, Boi P, Amalfitano S et al (2011) Experimental improvements in combining CARD-FISH and flow cytometry for bacterial cell quantification. J Microbiol Methods 87:309–315
36. Neumeyer A, Hübschmann T, Müller S et al (2013) Monitoring of population dynamics of Corynebacterium glutamicum by multiparameter flow cytometry. Microb Biotechnol 6:157–167
37. Kolloffel B, Meile L, Teuber M (1999) Analysis of brevibacteria on the surface of Gruyere cheese detected by in situ hybridization and by colony hybridization. Lett Appl Microbiol 29:317–322
38. Mounier J, Monnet C, Jacques N et al (2009) Assessment of the microbial diversity at the surface of Livarot cheese using culture-dependent and independent approaches. Int J Food Microbiol 133:31–37
39. Bottari B, Santarelli M, Neviani E et al (2010) Natural whey starter for Parmigiano Reggiano: culture-independent approach. J Appl Microbiol 108:1676–1684
40. Babot JD, Hidalgo M, Argañaraz-Martínez E et al (2011) Fluorescence in situ hybridization for detection of classical propionibacteria with specific 16S rRNA-targeted probes and its application to enumeration in Gruyère cheese. Int J Food Microbiol 145:221–228
41. Branco P, Monteiro M, Moura P et al (2012) Survival rate of wine-related yeasts during

alcoholic fermentation assessed by direct live/dead staining combined with fluorescence in situ hybridization. Int J Food Microbiol 158:49–57

42. Machado A, Almeida C, Carvalho A et al (2013) Fluorescence in situ hybridization method using a peptide nucleic acid probe for identification of Lactobacillus spp. in milk samples. Int J Food Microbiol 162:64–70
43. Mikš-Krajnik M, Babuchowski A (2014) 16S rRNA-targeted oligonucleotide probes for direct detection of *Propionibacterium freudenreichii* in presence of *Lactococcus lactis* with multicolour fluorescence in situ hybridization. Lett Appl Microbiol 59:320–327
44. Ercolini D, Villani F, Aponte M et al (2006) Fluorescence in situ hybridisation detection of Lactobacillus plantarum group on olives to be used in natural fermentations. Int J Food Microbiol 112:291–296
45. Stender H, Kurtzman C, Hyldig-Nielsen JJ et al (2001) Identification of *Dekkera bruxellensis* (Brettanomyces) from wine by fluorescence in situ hybridization using peptide nucleic acid probes. Appl Environ Microbiol 67:938–941
46. Blasco L, Ferrer S, Pardo I (2003) Development of specific fluorescent oligonucleotide probes for in situ identification of wine lactic acid bacteria. FEMS Microbiol Lett 225:115–123
47. Xufre A, Albergaria H, Inácio J et al (2006) Application of fluorescence in situ hybridisation (FISH) to the analysis of yeast population dynamics in winery and laboratory grape must fermentations. Int J Food Microbiol 108:376–384
48. Andorra I, Monteiro M, Esteve-Zarzoso B et al (2011) Analysis and direct quantification of Saccharomyces cerevisiae and Hanseniaspora guilliermondii populations during alcoholic fermentation by fluorescence in situ hybridization, flow cytometry and quantitative PCR. Food Microbiol 28: 1483–1491
49. Wang C, Esteve-Zarzoso B, Mas A (2014) Monitoring of Saccharomyces cerevisiae, *Hanseniaspora uvarum*, and *Starmerella bacillaris* (synonym *Candida zemplinina*) populations during alcoholic fermentation by fluorescence in situ hybridization. Int J Food Microbiol 191:1–9
50. Yuuki T, Kagami N (2001) Novel quantitative method for detection of pectinatus using rRNA targeted fluorescent probes. Am Soc Brew Chem 53:117–121
51. Thelen K, Beimfohr C, Bohak I, Back W (2002) Specific rapid detection test for beer-spoilage bacteria using fluorescently labelled gene probes. Brauwelt Int 20:155–159
52. Thelen K, Beimfohr C, Snaidr J (2003) Specific rapid detection of Alicyclobacillus by fluorescently- labeled gene probes in fruit juices. Fruit Proc 6:416–418
53. Fornasari ME, Rossetti L, Remagni C et al (2008) Quantification of Enterococcus italicus in traditional Italian cheeses by fluorescence whole-cell hybridization. Syst Appl Microbiol 31:223–230
54. Vieira-Pinto M, Oliveira M, Bernardo F et al (2007) Rapid detection of Salmonella spp. in pork samples using fluorescent in situ hybridization: a comparison with VIDAS®-SLM system and ISO 6579 cultural method. Arq Bras Med Veterinária E Zootec 59:1388–1393
55. Gunasekera T (2003) Potential for broad applications of flow cytometry and fluorescence techniques in microbiological and somatic cell analyses of milk. Int J Food Microbiol 85:269–279
56. Stephan R, Schumacher S, Zychowska MA (2003) The VIT technology for rapid detection of Listeria monocytogenes and other Listeria spp. Int J Food Microbiol 89:287–290
57. Schmid MW, Lehner A, Stephan R et al (2005) Development and application of oligonucleotide probes for in situ detection of thermotolerant Campylobacter in chicken fecal and liver samples. Int J Food Microbiol 105:245–255
58. Fuchizawa I, Shimizu S, Kawai Y et al (2008) Specific detection and quantitative enumeration of *Listeria* spp. using fluorescent in situ hybridization in combination with filter cultivation (FISHFC). J Appl Microbiol 105:502–509
59. Fuchizawa I, Shimizu S, Ootsubo M et al (2009) Specific and rapid quantification of viable *Listeria monocytogenes* using fluorescence in situ hybridization in combination with filter cultivation. Microbes Environ 24:273–275
60. Almeida C, Azevedo NF, Fernandes RM et al (2010) Fluorescence in situ hybridization method using a peptide nucleic acid probe for identification of *Salmonella* spp. in a broad spectrum of samples. Appl Environ Microbiol 76:4476–4485
61. Oliveira M, Vieira-Pinto M, Martins da Costa P et al (2012) Occurrence of *Salmonella* spp. in samples from pigs slaughtered for consumption: a comparison between ISO

6579:2002 and 23S rRNA Fluorescent In Situ Hybridization method. Food Res Int 45:984–988

62. Zhang X, Wu S, Li K et al (2012) Peptide nucleic acid fluorescence in situ hybridization for identification of *Listeria genus*, *Listeria monocytogenes* and *Listeria ivanovii*. Int J Food Microbiol 157:309–313
63. Graczyk TK, Conn DB, Lucy F et al (2004) Human waterborne parasites in zebra mussels (*Dreissena polymorpha*) from the Shannon River drainage area, Ireland. Parasitol Res 93:385–391
64. Laflamme C, Gendron L, Turgeon N et al (2009) Rapid detection of germinating *Bacillus cereus* cells using fluorescent in situ hybridization. J Rapid Methods Autom Microbiol 17:80–102
65. Schmid M, Walcher M, Bubert A et al (2003) Nucleic acid-based, cultivation-independent detection of *Listeria spp.* and genotypes of *L. monocytogenes*. FEMS Immunol Med Microbiol 35:215–225
66. Ercolini D, Hill PJ, Dodd CER (2003) Development of a fluorescence in situ hybridization method for cheese using a 16S rRNA probe. J Microbiol Methods 52:267–271
67. Bisha B, Brehm-Stecher BF (2010) Combination of adhesive-tape-based sampling and fluorescence in situ hybridization for rapid detection of *Salmonella* on fresh produce. J Vis Exp 44:2308
68. Glöckner FO, Fuchs BM, Amann R (1999) Bacterioplankton compositions of lakes and oceans: a first comparison based on fluorescence in situ hybridization. Appl Environ Microbiol 65:3721–3726
69. Wagner M, Schmid M, Juretschko S et al (1998) In situ detection of a virulence factor mRNA and 16S rRNA in *Listeria monocytogenes*. FEMS Microbiol Lett 160:159–168
70. Krimmer V, Merkert H, von Eiff C et al (1999) Detection of *Staphylococcus aureus* and *Staphylococcus epidermidis* in clinical samples by 16S rRNA-directed in situ hybridization. J Clin Microbiol 37:2667–2673
71. Lee N, Nielsen PH, Andreasen KH et al (1999) Combination of fluorescent in situ hybridization and microautoradiography – a new tool for structure-function analyses in microbial ecology. Appl Environ Microbiol 65:1289–1297
72. Moter A, Leist G, Rudolph R et al (1998) Fluorescence in situ hybridization shows spatial distribution of as yet uncultured treponemes in biopsies from digital dermatitis lesions. Microbiology 144:2459–2467
73. Ludwig W (2004) ARB: a software environment for sequence data. Nucleic Acids Res 32:1363–1371
74. Yilmaz LS, Parnerkar S, Noguera DR (2011) mathFISH, a web tool that uses thermodynamics-based mathematical models for in silico evaluation of oligonucleotide probes for fluorescence in situ hybridization. Appl Environ Microbiol 77:1118–1122
75. Greuter D, Loy A, Horn M, Rattei T (2016) probeBase-an online resource for rRNA-targeted oligonucleotide probes and primers: new features 2016. Nucleic Acids Res 44: D586–D589
76. Manz W, Amann R, Ludwig W et al (1992) Phylogenetic oligodeoxynucleotide probes for the major subclasses of proteobacteria: problems and solutions. Syst Appl Microbiol 15:593–600
77. Langendijk PS, Schut F, Jansen GJ et al (1995) Quantitative fluorescence in situ hybridization of *Bifidobacterium* spp. with genus-specific 16S rRNA-targeted probes and its application in fecal samples. Appl Environ Microbiol 61:3069–3075
78. Ootsubo M, Shimizu T, Tanaka R et al (2002) Oligonucleotide probe for detecting *Enterobacteriaceae* by in situ hybridization. J Appl Microbiol 93:60–68
79. Roller C, Wagner M, Amann R et al (1994) In situ probing of gram-positive bacteria with high DNA G + C content using 23S rRNA-targeted oligonucleotides. Microbiology 140:2849–2858
80. Alm EW, Oerther DB, Larsen N et al (1996) The oligonucleotide probe database. Appl Environ Microbiol 62:3557–3559
81. Ercolini D, Hill PJ, Dodd CER (2003) Bacterial community structure and location in Stilton cheese. Appl Environ Microbiol 69:3540–3548
82. García-Hernández J, Moreno Y, Amorocho CM et al (2012) A combination of direct viable count and fluorescence in situ hybridization for specific enumeration of viable *Lactobacillus delbrueckii subsp.bulgaricus* and *Streptococcus thermophilus*. Lett Appl Microbiol 54:247–254
83. Meier H, Amann R, Ludwig W et al (1999) Specific oligonucleotide probes for in situ detection of a major group of gram-positive bacteria with low DNA G + C content. Syst Appl Microbiol 22:186–196
84. Nissen H, Holck A, Dainty RH (1994) Identification of *Carnobacterium* spp. and *Leuconostoc* spp. in meat by genus-specific 16S rRNA probes. Lett Appl Microbiol 19:165–168

85. Hirschhäuser S, Fröhlich J, Gneipel A et al (2005) Fast protocols for the 5S rDNA and ITS-2 based identification of *Oenococcus oeni*. FEMS Microbiol Lett 244:165–171
86. Amann R, Snaidr J, Wagner M et al (1996) In situ visualization of high genetic diversity in a natural microbial community. J Bacteriol 178:3496–3500
87. Almeida C, Azevedo NF, Santos S et al (2011) Discriminating multi-species populations in biofilms with peptide nucleic acid fluorescence in situ hybridization (PNA FISH). PLoS One 6, e14786
88. Beimfohr C, Krause A, Amann R et al (1993) In situ identification of Lactococci, Enterococci and Streptococci. Syst Appl Microbiol 16:450–456
89. Brosius J, Dull TJ, Sleeter DD, Noller HF (1981) Gene organization and primary structure of a ribosomal RNA operon from *Escherichia coli*. J Mol Biol 148:107–127
90. Cannone JJ, Subramanian S, Schnare MN et al (2002) The comparative RNA web (CRW) site: an online database of comparative sequence and structure information for ribosomal, intron, and other RNAs. BMC Bioinformatics 3:2
91. Santarelli M, Bottari B, Malacarne M et al (2013) Variability of lactic acid production, chemical and microbiological characteristics in 24-hour Parmigiano Reggiano cheese. Dairy Sci Technol 93:605–621
92. Tang YZ, Gin KYH, Lim TH (2005) High-temperature fluorescent in situ hybridization for detecting Escherichia coli in seawater samples, using rRNA-targeted oligonucleotide probes and flow cytometry. Appl Environ Microbiol 71:8157–8164
93. Tortorello M, Reineke K (2000) Direct enumeration of Escherichia coli and enteric bacteria in water, beverages and sprouts by 16S rRNA in situ hybridization. Food Microbiol 17:305–313
94. Bottari B, Agrimonti C, Gatti M et al (2013) Development of a multiplex real time PCR to detect thermophilic lactic acid bacteria in natural whey starters. Int J Food Microbiol 160:290–297
95. Jain KK (2004) Current status of fluorescent in-situ hybridisation. Med Device Technol 15:14–17
96. Bouvier T, Del Giorgio PA (2003) Factors influencing the detection of bacterial cells using fluorescence in situ hybridization (FISH): a quantitative review of published reports. FEMS Microbiol Ecol 44:3–15
97. de Vries MC, Vaughan EE, Kleerebezem M, de Vos WM (2004) Optimising single cell activity assessment of *Lactobacillus plantarum* by fluorescent in situ hybridisation as affected by growth. J Microbiol Methods 59:109–115
98. Ray A, Nordén B (2000) Peptide nucleic acid (PNA): its medical and biotechnical applications and promise for the future. FASEB J 14:1041–1060
99. Almeida C, Azevedo NF, Iversen C et al (2009) Development and application of a novel peptide nucleic acid probe for the specific detection of *Cronobacter genomospecies* (*Enterobacter sakazakii*) in powdered infant formula. Appl Environ Microbiol 75:2925–2930
100. Almeida C, Cerqueira L, Azevedo NF, Vieira MJ (2013) Detection of Salmonella enterica serovar Enteritidis using real time PCR, immunocapture assay, PNA FISH and standard culture methods in different types of food samples. Int J Food Microbiol 161:16–22
101. Almeida C, Sousa JM, Rocha R et al (2013) Detection of *Escherichia coli* O157 by peptide nucleic acid fluorescence in situ hybridization (PNA-FISH) and comparison to a standard culture method. Appl Environ Microbiol 79:6293–6300
102. Sørensen AH, Torsvik VL, Torsvik T et al (1997) Whole-cell hybridization of Methanosarcina cells with two new oligonucleotide probes. Appl Environ Microbiol 63:3043–3050
103. Kubota K (2013) CARD-FISH for environmental microorganisms: technical advancement and future applications. Microbes Environ 28:3–12
104. Bond PL, Banfield JF (2001) Design and performance of rRNA targeted oligonucleotide probes for in situ detection and phylogenetic identification of microorganisms inhabiting acid mine drainage environments. Microb Ecol 41:149–161

Evaluation of Polymicrobial Involvement Using Fluorescence In Situ Hybridization (FISH) in Clinical Practice

Alexander Swidsinski and Vera Loening-Baucke

Abstract

The involvement of microorganisms in infection is often deduced from their presence during disease and absence in healthy humans or animals, which is an oversimplification. The proof of direct involvement is decisive. Fluorescence in situ hybridization (FISH) combines the specific identification of microorganisms and the morphological aspect of the host tissues and is as a consequence especially helpful for these purposes.

The present manuscript describes FISH methods which we use in ambulatory patients for Polymicrobial Infections and Bacterial Biofilms of the Charité Hospital to visualize pathogens (pathogenic consortia) in clinical samples.

Keywords Polymicrobial infections, Pathogenic consortia, FISH

1 Introduction

The contemporary understanding of infections is based on identifying microorganisms in diseased persons that are absent in healthy persons. However, the presence of a bacterium (or bacteria) in health does not mean that it is healthy or at least harmless (chapter by Benedetta Bottari et al. "FISHing for Food Microorganisms").

Neisseria meningitidis is part of the normal nonpathogenic flora in the nasopharynx of up to 5–15 % of adults. Its causative involvement in meningitis is however beyond doubt, since it is the only bacterium found in the inflamed cerebrospinal fluid.

Detecting bacteria at the site of an infection is more appropriate for identifying the infectious agent than its absence within normal colonization. Difficulties arise when multiple organisms are present at the infection site. In this case, the criminological experiment is decisive. A transfection of the suspected bacteria to healthy animals helps to uncover potential pathogens.

Thomas Liehr (ed.), *Fluorescence In Situ Hybridization (FISH)*, Springer Protocols Handbooks,
DOI 10.1007/978-3-662-52959-1_52, © Springer-Verlag Berlin Heidelberg 2017

However, when none of the involved microorganism causes infection, does this exclude the harmful potential of a group? No.

A well-known example is the induction of Vincent's angina by Rosebury, who transferred plaque-infected material holding different components [1]. While single microorganisms were innocuous and incapable to initiate infection, it was possible to cause disease with the combination of different species. The required consortium was called the "Pathogenic Quartet" and included the following species that were isolated from a patient diagnosed with Vincent's angina: a spirochete, a fusiform *Bacillus*, a *Vibrio*, and an anaerobic *Streptococcus*. Rosebury's conclusion was that each of these species is a member of healthy indigenous flora, but they may cooperate and form an unmanageable complex structure.

In nature, microorganisms build diverse consortia in which single participants complement each other and display specific properties, which cannot be discovered in one of the participants or in other associations. Can some of these consortia be pathogenic? Yes.

We should await the presence of such consortia on surfaces which contact the outer world such as the skin, mouth, intestine, vagina, etc.

Can the role of these consortia be proved in transfection experiments? Presently, no.

Rosebury transfected not really a consortia but a mix of isolated cultured single strains. This should be only in exceptional cases successful. The problem is that until now, we are unable to cultivate polymicrobials. When more than three bacterial strains are incubated in the same culture, their growth is getting unpredictable, and one of the strains suppresses and overgrows the others. Polymicrobial culture is a challenge for future research.

In the absence of polymicrobial cultures, a link between the consortium of distinct species and their involvement in disease can be established directly by visualizing pathogenic consortia within biofilms and microbial infiltrates in host tissues via fluorescence in situ hybridization (FISH).

We have successfully used this approach in case of colonic cancer [2], inflammatory bowel disease [3], gallstones [4], tonsillitis [5], appendicitis [6], bacterial vaginosis [7], candidiasis [8], and urethritis [9].

FISH combines the specific identification of microorganisms and the morphological aspect and is especially helpful for identification of polymicrobial consortia involved in local infection. Each single bacterium possesses 10^3–10^5 ribosomes of which each ribosome owns the same copy of ribosomal RNA. Some of the regions of the rRNA are strain-specific; others are universal for species, families, or even kingdoms. Oligonucleotides synthesized complimentary to rRNA sequences and labeled with fluorescent dye are called FISH probes. When added to samples containing bacteria,

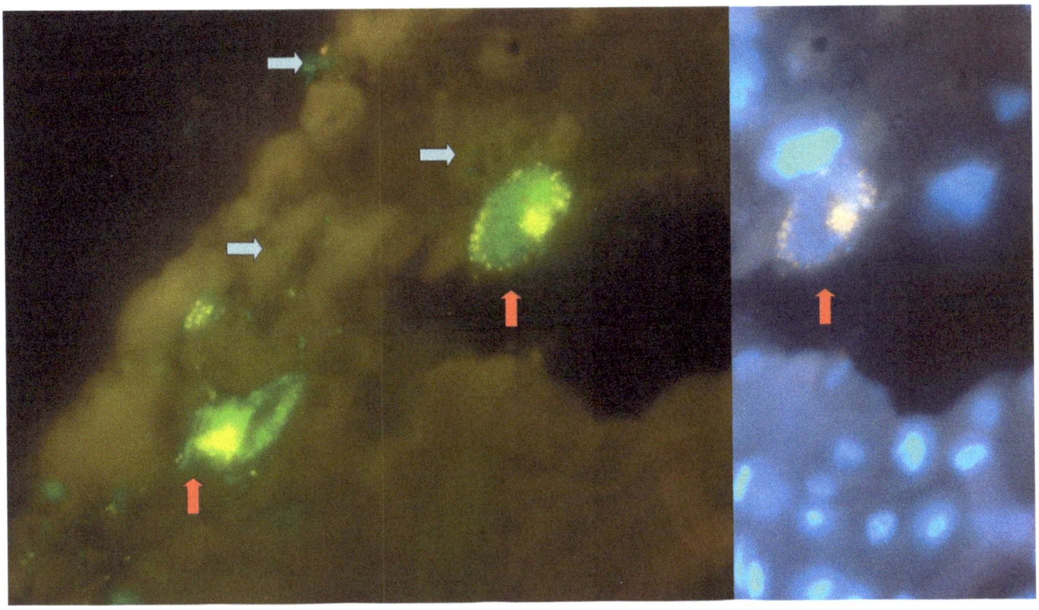

Fig. 1 Isolated islands of bacteria attached to desquamated epithelial cells ×1,000: mouth and surgically removed material; universal bacterial probes (Eub338 FITC, green fluorescence), and *Burkholderia* (Burkho-Cy3, yellow fluorescence) on the *left*. Unspecific DAPI stain of the DNA is overlaid with *Burkholderia* fluorescence on the *right*

FISH probes hybridize with the rRNA of the bacterial ribosomes. No additional enhancement of fluorescence is necessary and bacteria can be visualized directly with a fluorescence microscope due to the large number of ribosomes in each bacterium.

Multicolor FISH enables the identification of potentially all bacterial groups in spatial relation to each other and in relation to histological structures of the host. Any biological material can be studied for in situ presence of bacteria and bacterial biofilms, including smears from tonsils or vagina, desquamated epithelial cells in the urine, tissue biopsies, surgically removed tissues, saliva, perspiration, exudation, sperm samples, and medical devices removed from the body (Figs. 1, 2, 3).

FISH protocols described here are standard protocols, which are used for ambulatory patients in the Laboratory for Molecular Genetics, Polymicrobial Infections, and Bacterial Biofilms at the Charité Hospital in Berlin, Germany.

2 Materials

Apart from standard cell biological and molecular cytogenetic equipment, including standard solutions (e.g., ethanol, methanol, formamide, formaldehyde, xylene, etc.), no more specialized items

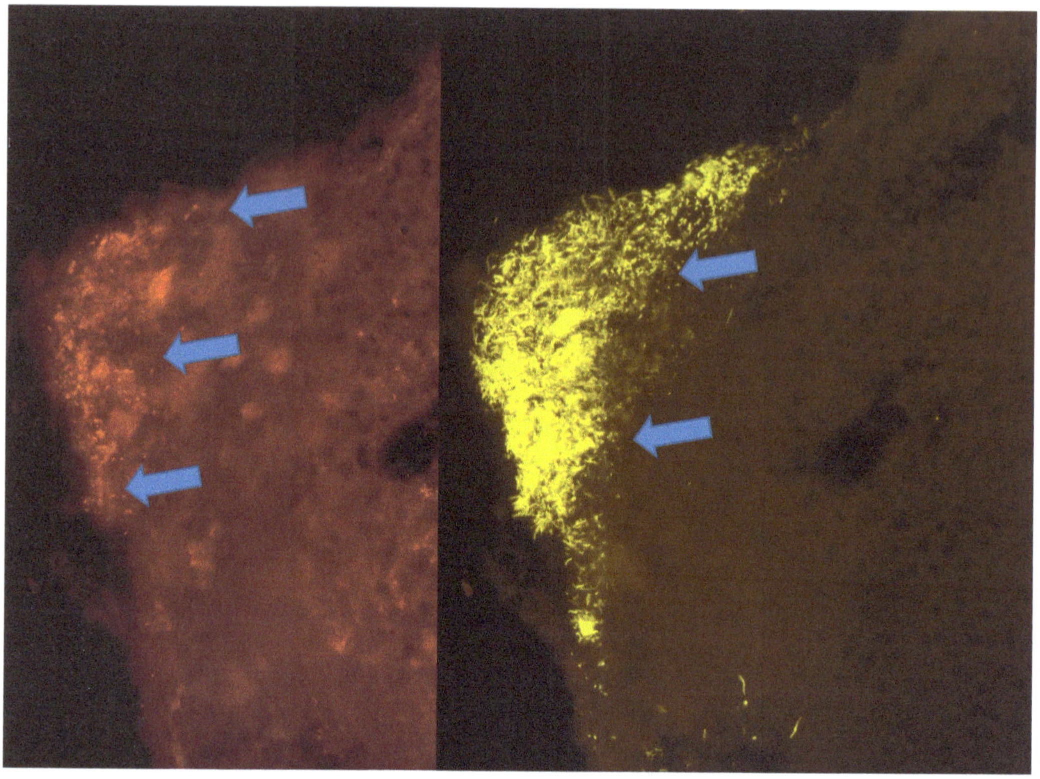

Fig. 2 Multicolor FISH of superficial tonsils infiltrates ×400. Gamma proteobacteria as a part of superficial infiltrate (Gam42-Cy5, red fluorescence). The main group involved in infiltration is a *Fusobacterium nucleatum* (Fnuc Cy3, yellow fluorescence)

are required. The equipment needed for FISH is listed in the chapter by Thomas Liehr et al. "The Standard FISH Procedure".

However, skilled laboratory staff with experience in FISH microscopy is necessary for performing this protocol. Although the techniques applied here do basically not differ from those used by pathologists, they cannot be delegated to the staff of the routine pathology department, because the preparation of tissue sections is performed with no regard to possible microbiological cross contaminations, and bacteria are massively present in the environment. This contamination is easily avoided when materials in which single steps are performed are renewed after each sample, and instruments are kept clean and are often changed.

In addition, a routine pathologic laboratory uses automatic equipment, vessels, and containers in which large parts of sterile and highly contaminated samples are processed simultaneously, leading to enrichment of bacteria in solutions, massive microbial cross contamination, and diagnostic biases.

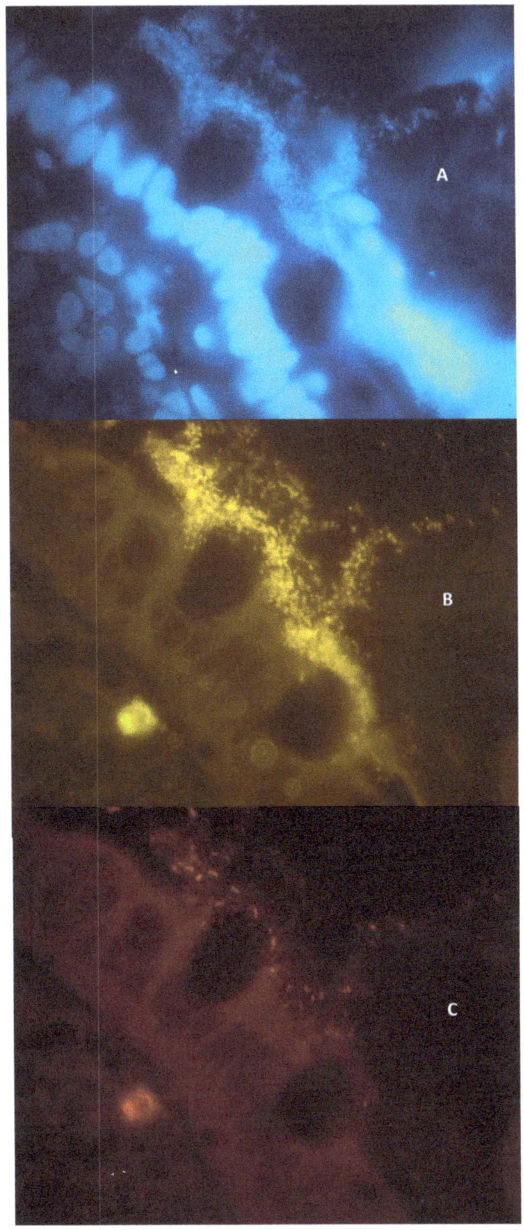

Fig. 3 Prolific bacterial biofilm covers the colonic mucosa in a patient with Crohn's disease ×1000 multicolor FISH: (**a**) DAPI stain of DNA structures; (**b**) Bac303 Cy3; Bacteroides; orange fluorescence; (**c**) EREC Cy5; *Eubacterium rectale*; and red fluorescence

3 Methods

Bacteria-specific FISH probes are inexpensive and can be purchased from many oligonucleotide manufacturers (such as MWG Biotech, Ebersberg, Germany). A probe purchased for 50 € is sufficient for

3.1 Ribosomal RNA-Based FISH for Evaluation of Polymicrobial Consortia in Clinical Settings

at least 5,000 hybridizations. Over 200 FISH probes targeting the bacterial rRNA at a domain, group, and species level are described in the literature and can be freely accessed online over resources like www.microbial-ecology.net/probebase and www.arb-silva.de/fish-probes. New probes can be developed in case of specific clinical questions.

3.1.1 Bacteria-Specific FISH Probes

Probes routinely used for evaluation of intestinal samples in clinical settings are Bac 303, EREC, Fprau, Bif 153, and EBAC representing *Bacteroides/Prevotella* and *Enterobacteriaceae.*

Probes routinely used for evaluation of urogenital samples in clinical settings are GardV, Lab, Ato, Cor, EBAC, and EUB338 probes representing *Gardnerella*, *Lactobacillus*, *Atopobium*, *Coriobacterium*, *Enterobacteriaceae*, and *Eubacteria*, respectively (*see* **Notes 1** and **2**).

The choice of FISH probes must be adjusted to the specific requirements of the biotope and the aims of the research. Probes which deliver unsure results in the microbiome should be avoided (*see* **Notes 1–3**).

FISH is an excellent tool for the assessment of spatial structure. However, results must be interpreted carefully. In case FISH probes seem to detect bacterial groups that were never described in the specific biotope or anatomical location even in case of an apparent high specificity of the probes, the presence of these groups must be confirmed using alternative methods such as culturing, polymerase chain reaction with subsequent cloning, and sequencing.

When using multicolor FISH, a large variation in practicable fluorochromes exists, but only four of them can be error-free discriminated by the human eye regardless of all possible nuances of color shades. The four colors used on a regular basis are orange, dark red, green, and blue represented by carbocyanine (Cy) 3, Cy5, fluorescein isothiocyanate (FITC), and 4′,6-diamidino-2-phenylindole (DAPI) as counterstain, respectively. The advantages of these appropriate fluorochromes are the slow bleaching, demonstrating little autofluorescent background and allowing high-quality micrographs. Cy3 is most resilient to bleaching, followed by Cy5 and FITC. Alexa fluorochromes corresponding to Cy3, Cy5, FITC, and DAPI are likewise practicable but about ten times as expensive as Cy fluorochromes (chapter by Michael Sommerauer et al. “Optical Filters and Light Sources for FISH”).

In our experience with other than the abovementioned fluorochromes, there was a massive increase of signals which could not be definitively assigned to bacteria.

3.2 Optimally Fixated Material

Paraffin embedding and preparation of histological sections is the only time-consuming and elaborate step leading to additional costs for personal. Optimal materials to perform FISH are biopsies; eluates from swabs; urine sediments; surgically removed tissues; sections of nylon-membrane strips placed into the oral cavity or prepuce/vagina or attached to skin overnight; and sections of adhesive tape attached to the anal region and removed after 60 s. Smears on glass slides are less appropriate since the arrangement of cells on each slide is artificial and unique. Hybridizations of the same sample performed under the exact same conditions using different bacterial FISH probes on different glass slides are not quantitatively comparable.

3.2.1 General Rules to Avoid Biases

In aqueous solutions, the DNA is unstable, leading to a decreased intensity of specific hybridization signals and increased background fluorescence. To achieve optimal relation between high-specific bacterial and low-background fluorescence, we recommend the following:

- Do not use formalin/paraformaldehyde fixation, but rather use water-free modified Carnoy's solution.
- Do not use any technique of freezing while preserving samples. Freezing and thawing are deleterious for many microorganisms and especially for biofilm structures.
- Shorten the exposure to water containing solutions to the absolute minimum; drop all rehydration steps which are often used in histology.
- Hold hybridization time as short as necessary (30–90 min depending on material). With longer exposure, bacteria may detach from their original place and spread over the surface of histological cuts, leading to contaminations of sterile locations.
- Use polypropylene tubes (e.g., 2 ml Eppendorf tube, 15–50 ml Falcon tube) because polypropylene is resistant to Carnoy.

3.2.2 Collecting, Fixating, Transporting, and Embedding of Samples

Different fixatives were tested, and the best results were achieved with modified nonaqueous Carnoy's solution composed of 6/6/1 vol. ethanol/glacial acetic acid/chloroform.

For Tissue Biopsies and Surgically Removed Material

1. Samples of human tissue can be directly placed in a modified Carnoy fixative and stored or shipped at room temperature (RT) to the laboratory. All mechanic pressure (flattening the sample or the use of a shaker) should be avoided because it may injure the biofilm structure. The amount of modified Carnoy's solution should exceed the volume of tissue sample by a factor of 20 (Table 1; *see* **Note 4**).
2. The minimal incubation time for biopsies in Carnoy fixative is 12 h but preferably at least 24 h. In case of larger tissue pieces, prolong the incubation time for 4 h for each additional millimeter of the tissue size. If the added Carnoy fixative is less than ten times the tissue/sample volume, the storage should not

Table 1
Amount of Carnoy depending on the sample size

Sample size (mm^3)	Volume of Carnoy (ml)
1	0.5
5	5
6–20	15

exceed 2 weeks. In case of excess of Carnoy (more than 20-fold), excellent results could still be obtained after storage for 6 months and probably longer.

3. A change of Carnoy by longer storage time is preferable.
4. After decanting the modified Carnoy solution, the same volume of ethanol (absolute) is added, and the sample is incubated at 4 °C for a time period based on the size of the sample. Biopsy samples up to 3 mm are incubated for 15 min and big tissue samples (4–20 mm) for 2 h (*see* **Notes 5–6**).
5. After cold incubation, the ethanol is poured, and the tissue is put very carefully—without squeezing—into a new tube with ca. 5 ml of xylene (p.a.) and incubated overnight at RT.
6. When the xylene is decanted the following day, the sample is transferred carefully in an embedding mold with pre-warmed, melted paraffin. When the samples are larger than 5 mm, they are first put on a paper towel for 10 min to achieve the evaporation of the xylene before putting the sample into the paraffin. Small samples (<3 mm) are incubated for 1 h at 75 °C, samples up to 10 mm for 2 h at 75 °C, and large samples (10–20 mm) overnight at 65 °C.
7. After this first incubation, the paraffin is disposed and fresh paraffin is added and incubated for 2 h. This step is not necessary for small samples. After this, the mold is taken out of the incubator and placed on ice. The sample is positioned in the middle of the mold by using a toothpick. While the paraffin is coagulating, a pre-identified embedding cassette is put on top of the mold. Some warm paraffin is poured on this cassette to connect with the mold. These molds are first stored at 4 °C during 15 min followed by storage at −20 °C for at least 15 min. After complete coagulation of the paraffin, 4 μm cut sections can be made and put on a glass slide (Superfrost Plus slide) used for histological research. These slides are incubated for 1 h at 50 °C to assure the connection between the cut section and the slide.
8. The deparaffinization of the slides is obtained by putting them four times for 2–3 min each time at RT first in successive xylene

(p.a.) baths and next in successive ethanol (absolute) baths. The slides are incubated for 25 min at 50 °C. The sections are encircled with a thin line with the pap pen and dried.

9. A lysozyme step (Carl Roth, Germany) is only necessary for some strains of mainly gram-positive bacteria (depending on the probe). The entire section is covered by lysozyme solution (1 mg ml^{-1}) and incubated at 37 °C for 15–90 min in a humid pre-warmed chamber (depending on the sample). The optimal time has to be evaluated for each kind of sample. The lysozyme can introduce biases in polymicrobial communities by destroying, for example, *Proteobacteria* completely. It is important to obtain the best results in regard to the specific target species and other bacterial groups.
10. After incubation, the slides are flushed with distilled water and dried for 5 min at 50 °C.

For Fluid Secretions, Lavage, and Urine

1. Fluid secretions and lavages are done by MDs. The collection and fixation of urine samples is performed by the patient. Women are asked not to wash the urogenital region in the evening before sample collection and to use the first part of the morning urine, which increases the number of desquamated epithelial cells in urine sediments. Men need to pull the foreskin back over the glans penis before urine collection.
2. Two milliliters of the liquid sample are mixed with 8 ml Carnoy fixative in 15 ml Falcon tubes. These samples are not stable due to the high water amount. Therefore, the time for delivery to the laboratory should not exceed 4 weeks. Shorter periods are preferable. At arrival in the laboratory, fixated fluids and secretions should be centrifuged and the fixative solution decanted. Then Carnoy fixative is added to the sediment in a proportion of at least 1:20. Such prepared sediments can be stored until 6 months at RT.
3. Then there is a 1 cm circle drawn with a pap pen on the Superfrost Plus glass slide. Then 5 μl of the stirred sample is transferred to the slide with a plastic pipette within the area of hybridization and dried at 50 °C for 30 min. 5 μl of the final aliquot are convenient to use for single hybridizations. In case of urine sediments fixated as described above, this volume represents 30 μl of the initial urine volume.

3.3 The Hybridization

The hybridization of slides from tissues and fluids follows the same protocol.

1. A hybridization solution (Table 2) is prepared, in which the amount of formamide varies depending on the FISH probe.

Table 2
Composition of hybridization solutions

% of formamide	Formamide (μl)	H_2O (μl)	NaCl 5M (μl)	Tris-HCl 1M (pH 7.4) (μl)	SDS 10 % (μl)
0	0	1,600	360	40	10
1	20	1,580	360	40	10
5	100	1,500	360	40	10
10	200	1,400	360	40	10
15	300	1,300	360	40	10
20	400	1,200	360	40	10
25	500	1,100	360	40	10
30	600	1,000	360	40	10
35	700	900	360	40	10
40	800	800	360	40	10
45	900	700	360	40	10
50	1000	600	360	40	10
55	1100	500	360	40	10
60	1200	400	360	40	10
65	1300	300	360	40	10

2. This solution is pre-warmed at 46 °C (or the probe-specific hybridization temperature; *see* **Note 7**), and 0.5 μl of probe (50 ng μl^{-1}) is added to 50 μl of hybridization buffer (per sample section). This amount of hybridization buffer is enough to cover the section completely.
3. The section is incubated for 45 min to 3 h at 46 °C in a humid pre-warmed chamber and in the dark. The incubation time should be optimized for different bacterial groups.
4. Wash buffer (Table 3) is prepared and warmed at 48 °C during this incubation period. The wash buffer composition depends on the formamide concentration used for hybridization.
5. After incubation, the slides are flushed with distilled water, and two slides are put back-to-back in a 50 ml Falcon tube completely filled with the pre-warmed wash buffer.
6. The tubes are incubated in a 48 °C water bath for 5 min; then they are flushed with distilled water.
7. The slides are dried in an upright position in an oven at 50 °C for 5 min in the dark.
8. Sections are placed in a cardboard slide folder, covered with 50 μl DAPI solution (1 μg ml^{-1}), and incubated for 5–10 min at RT in the dark.

Table 3
Composition of wash buffer

% of formamide used for hybridization	Final NaCl concentration (mM)	NaCl 5M (μl)	H_2O (ml)	Tris-HCl 1M (pH 7.4) (μl)	SDS 10 % (μl)	EDTA 0.5M (μl)
0	900	9,000	40.0	1,000	30	0
1	900	9,000	40.0	1,000	30	0
5	636	6,300	42.7	1,000	30	0
10	450	4,500	44.5	1,000	30	0
15	318	3,180	45.8	1,000	30	0
20	225	2,150	46.4	1,000	30	500
25	159	1,490	47.0	1,000	30	500
30	112	1,020	47.5	1,000	30	500
35	80	700	47.8	1,000	30	500
40	56	460	48.0	1,000	30	500
45	40	300	48.2	1,000	30	500
50	28	180	48.3	1,000	30	500
55	20	100	48.4	1,000	30	500

Materials necessary:
- Lysozyme (Carl Roth, Germany)

9. Then they are flushed with distilled water and dried for 5 min in the dark in an oven at 50 °C in an upright position.
10. The slides can be kept dry for about 6 weeks in cardboard folders at RT in the dark.

3.4 Evaluation

This is the most expensive part. The microscope can however be shared at the beginning with other research groups. In our laboratory, we use a Nikon E600 fluorescence microscope (Nikon; Tokyo, Japan; 40,000 €). We also use a Digital Microscope Camera ProgRes® CFcool (7,000 €) and accompanying software (Jenoptik, Jena, Germany). The color camera is necessary for documentation and performance of multicolor FISH pictures to demonstrate the spatial relationship between single microbial groups and to exclude cross hybridizations of unrelated FISH probes.

True color micrographs are preferred because they approach reality the most. However, evaluation of fluorescence signals based on micrographs only should be discouraged. In contrast to material composed of bacteria only, human samples contain complex DNA-bearing structures, which may non-specifically bind the oligonucleotides of FISH probes and make it difficult to distinguish them from the bacteria-specific signals. While the human eye can easily differentiate between real signal and biases, multiple irrelevant signals may appear genuine in micrographs, especially when contrast and intensity are manipulated by software. The nuances are imperceptible on micrographs taken with black and white fluorescence camera. The often used subsequent coloring of the signals is deceptive.

3.4.1 Application of FISH: The Enumeration of Bacteria

Only hybridization signals which are clear and morphologically distinguishable as bacterial cells with at least a triple color identification with universal and group-specific FISH probes and DAPI stain, in the absence of cross hybridization with taxonomically unrelated probes, can be enumerated.

We enumerate bacterial concentrations of homogeneous populations visually in one of the square fields of the ocular raster corresponding to 10×10 μm of the section surface at $\times 1{,}000$ magnification or 10^9 bacteria ml^{-1} (a 10 μl sample with a concentration of 10^7 cells ml^{-1} has on average 40 cells per microscopic field at a $\times 1{,}000$ magnification). In case of uneven distribution of bacteria over the microscopic field, the amount of positive signals is counted in ten fields of the ocular raster along the gradient of distribution and divided by ten.

For microbial populations taking smaller surfaces than 10×10 μm, the above equation is adopted to a closer 1×1 μm raster.

In case of urine sediments, bacteria and epithelial cells must be referred to the urine volume and each other. Concentrations of epithelial cells within the 5×5 mm area of hybridization (corresponding to the initial sample volume) are calculated and converted to numbers of epithelial cells per milliliter of urine. Since adherence is not even, it is recommended to determine the maximal and mean numbers of adherent bacteria per epithelial cell. The overall concentration of adherent bacteria in the urine results from multiplication of the mean number of bacteria per epithelial cell and the concentration of epithelial cells per ml of urine.

For possibilities of evaluation of ISH experiments in microbiology by electron microscope, see chapter by Hannes Schmidt, Thilo Eickhorst "Gold-FISH: In Situ Hybridization of Microbial Cells for Combined Fluorescence and Scanning Electron Microscopy".

4 Notes

1. Contrary to the expectations, none of the published FISH probes that were tested in our lab proved to be absolutely specific. Depending on the microbial community investigated (e.g., human or animal intestines, pancreatic duct, gallstones, biliary stents), all FISH probes demonstrated some cross-hybridization when conditions of optimal stringency were applied.

2. Some FISH probes that delivered highly specific results in human samples were cross hybridizing with unrelated probes in murine material. This indicates that the global diversity of bacteria is much higher than we presently accept.
3. When FISH probes for unrelated bacterial groups identify bacteria of similar morphology and equal numbers and at similar locations, the specificity of signals should be evaluated by performing multicolor FISH with probes stained with different fluorochromes. When both micrographs are overlaid, the signals detected by both should not be the same [10].
4. The use of modified Carnoy in a higher ratio does not reveal disadvantages, while a smaller volume increases the proportion of water in the solution, resulting in a decreased quality of hybridization.
5. After paraffin embedding, the fluorescence intensity declines over time (10 % during each year of storage). The reduction of 30 % after 3 years can be critical for less numerous and metabolically active microbial groups. Therefore, it is optimal to perform comparative studies within the first 2 years.
6. The fixation step is followed by embedding in paraffin which is time consuming. Microbiological cross contamination between samples should be avoided by preparing each sample on its own and by averting the use of an automated paraffin station.
7. Any hybridization oven can be used. When lacking such an oven, a microbial incubator that is able to maintain the temperature between 46 and 50 °C is an option.

References

1. Foley G, Rosebury T (1942) Comparative infectivity for guinea pigs of fusospirochetal exudates from different diseases. J Dent Res 21:375–378
2. Swidsinski A, Khilkin M, Kerjaschki D et al (1998) Association between intraepithelial Escherichia coli and colorectal cancer. Gastroenterology 115:281–286
3. Swidsinski A, Weber J, Loening-Baucke V et al (2005) Spatial organization and composition of the mucosal flora in patients with inflammatory bowel disease. J Clin Microbiol 43:3380–3389
4. Swidsinski A, Loening-Baucke V, Herber A (2009) Mucosal flora in Crohn's disease and ulcerative colitis – an overview. J Physiol Pharmacol 6:61–71
5. Swidsinski A, Schlien P, Pernthaler A et al (2005) Bacterial biofilm within diseased pancreatic and biliary tracts. Gut 54:388–395
6. Swidsinski A, Göktas O, Bessler C et al (2007) Spatial organisation of microbiota in quiescent adenoiditis and tonsillitis. J Clin Pathol 60:253–260
7. Swidsinski A, Dörffel Y, Loening-Baucke V et al (2011) Acute appendicitis is characterised by local invasion with Fusobacterium nucleatum/necrophorum. Gut 60:34–40
8. Swidsinski A, Loening-Baucke V, Mendling W et al (2014) Infection through structured polymicrobial Gardnerella biofilms (StPM–GB). Histol Histopathol 29:567–587
9. Alexander Swidsinski A, Vera Loening-Baucke V, Mendling W et al (2012) Positive effects of local therapy with a vaginal lactic acid gel on dysuria and E. coli bacteriuria question our current views on recurrent cystitis. Arch Gynecol Obstet 285:1619–1625
10. Swidsinski A (2006) Standards for bacterial identification by fluorescence In situ hybridization within eukaryotic tissue using ribosomal rRNA-based probes. Inflamm Bowel Dis 12:824–826

Gold-FISH: In Situ Hybridization of Microbial Cells for Combined Fluorescence and Scanning Electron Microscopy

Hannes Schmidt and Thilo Eickhorst

Abstract

Understanding interactions of microorganisms with their habitat has recently become an important topic in (environmental) microbiology. In this context phylogenetic identification via in situ hybridization and specific detection and visualization of single microbial cells on resolutions beyond light microscopy is a promising approach. Here we describe a protocol which is based on rRNA-targeted in situ hybridization and catalyzed reporter deposition (CARD) enabling the detection of fluorescent signals and a specific nanogold deposition on the same target cells. The gold-stained microbial cells are analyzed via electron microscopy (EM) using backscattered electron detectors (BSD) or energy-dispersive X-ray spectroscopy (SEM-EDX). This Gold-FISH protocol has been successfully applied on pure and mixed bacterial cultures and environmental samples. The combined labeling allows reliable quantification of Gold-FISH-stained cells via fluorescence microscopy and analysis of microbe-surface interactions with techniques such as elemental mapping via EDX or NanoSIMS on submicron scales. It is furthermore a sufficient proof of specific detection of single cells via nanogold and electron microscopy and can be combined with DNA-specific counterstains.

Keywords Fluorescence in situ hybridization (FISH), Catalyzed reporter deposition (CARD), Biotin tyramide, Fluorophore, Nanogold, Fluorescence microscopy, Scanning electron microscopy

1 Introduction

In recent years, several attempts utilizing metal-based labeling strategies of microorganisms in environmental microbiology have been reported [1, 2]. These techniques allow the detection and identification of archaea and bacteria with microscopic techniques beyond fluorescent microscopy at high resolution such as scanning electron microscopy (SEM) or transmission electron microscopy (TEM). For example, bacteria have been labeled with nanogold particles to facilitate their retrieval during NanoSIMS analysis and to obtain information on single cells within their environment in situ on a relevant micro-scale [3]. A comparison of in situ hybridization (ISH) strategies for this purpose has recently been reviewed by Eickhorst and Schmidt [4].

Thomas Liehr (ed.), *Fluorescence In Situ Hybridization (FISH)*, Springer Protocols Handbooks, DOI 10.1007/978-3-662-52959-1_53, © Springer-Verlag Berlin Heidelberg 2017

One of the most critical issues besides the specificity of detection via probe-based whole-cell ISH is the sufficient labeling of target sites with either fluorescent molecules or metal labels such as nanogold. The success of ISH approaches using mono-labeled oligonucleotide probes depends on the accessibility of target regions on rRNA molecules and is often limited by low contents of rRNA present in the target cells. Therefore, combinations of ISH with catalyzed reporter deposition (CARD) have shown to be promising procedures to increase detectability of single microbial cells in pure culture but also in various environmental samples [5, 6]. The main principle of this CARD-ISH technique is based on tyramide signal amplification (TSA) catalyzed by the enzyme horseradish peroxidase (HRP).

The protocol presented in this chapter combines both fluorescence and metal-based whole-cell in situ hybridization based on CARD-ISH and is an updated version of the technique referred to as Gold-FISH [1]. The procedure consists of an in situ hybridization step using HRP-conjugated probes, followed by amplification of biotinylated tyramides that serve as binding sites for streptavidin molecules linked with a fluorophore and a nanogold particle (Fig. 1). Samples may then be observed by fluorescence microscopy or directly treated via autometallographic enhancement of gold particles [7] for the detection by scanning electron microscopy (SEM). SEM combined with energy-dispersive X-ray spectroscopy (SEM-EDX) allows not only the localization of the gold-stained target cells on an ultrastructural level but also to perform elemental mapping of the substrate inhabited by these microorganisms.

2 Materials

2.1 Fixation and Pretreatment of Microbial Cells

- Fixation solution: Dilute formaldehyde (37 %, e.g., Sigma-Aldrich) to a final concentration of 4 % (v/v) in 1 × PBS buffer. Prepare fresh.
- Agarose, low melting point (gel strength around 1,000 g cm^{-2}, e.g., Invitrogen Life Technologies Cat. No. 16520): Prepare a 1 % (w/v) agarose in 1 × PBS stock solution which is autoclaved and stored at 4 °C for several months. Mix 200 μl agarose stock with 800 μl double distilled water and heat to 60 °C. Store up to 2 weeks at 4 °C.
- Polycarbonate filters (e.g., Whatman Nuclepore, 0.2 μm pore size).
- Lysozyme solution (10 mg ml^{-1}): Dissolve lysozyme (e.g., Sigma-Aldrich) in 0.05 M EDTA (pH 8.0) and 0.1 M Tris-HCl (pH 8.0) and add double distilled water. Prepare fresh.

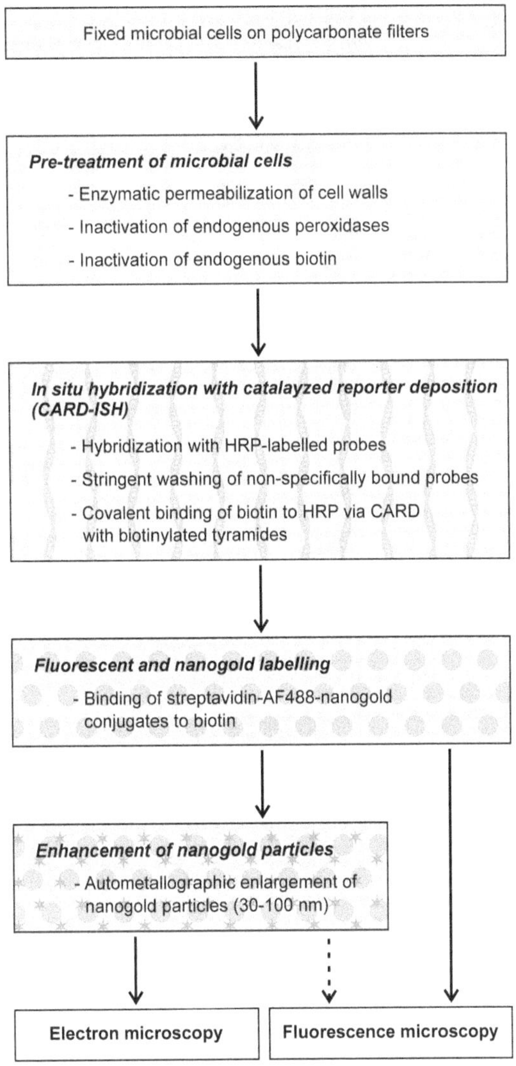

Fig. 1 Flow chart of the Gold-FISH protocol

- Achromopeptidase solution (60 U ml^{-1}): Dissolve achromopeptidase (e.g., Sigma-Aldrich) in 0.01 M NaCl and 0.01 M Tris-HCl (pH 8.0) and add double distilled water. Prepare fresh.
- Methanol (e.g., Merck): Store at room temperature (RT) in safety container.
- Hydrogen peroxide solution (30 %, e.g., Sigma-Aldrich): Store at 4 °C.
- Streptavidin/Biotin blocking kit (Vector Laboratories, Cat. No. SP-2002): Store at 4 °C.

Table 1
Volumes of formamide and double distilled water for 10 ml of hybridization buffer according to [5]

% formamide (v/v) in hybridization buffer	ml formamide	ml water
20	2.0	5.0
25	2.5	4.5
30	3.0	4.0
35	3.5	3.5
40	4.0	3.0
45	4.5	2.5
50	5.0	2.0
55	5.5	1.5
60	6.0	1.0
65	6.5	0.5

2.2 *In Situ* Hybridization Followed by Fluorescent and Nanogold Labeling

- Hybridization buffer (10 ml): Dissolve 1 g dextran sulfate (Sigma-Aldrich, Cat. No. D8906) in 1,800 μl 5 M NaCl (0.9 M), 200 μl 1 M Tris-HCl (0.02 M, pH 8.0), X μl double distilled water (see Table 1). Add X μl formamide (e.g., Fluka, Cat. No. 47671) according to the required formamide concentration for each oligonucleotide probe. Add 1,000 μl blocking reagent (Roche, Cat. No. 11096176001; stock concentration 10 % (w/v), autoclaved and stored at 4 °C for several weeks), 10 μl SDS (e.g., Sigma-Aldrich, 10 % (w/v), sterile filtered and stored at RT for several weeks). Aliquot and store at −20 °C or store up to 2 weeks at 4 °C.
- Probe working solution (50 ng μl^{-1}): Resolve HRP-labeled oligonucleotide probe (e.g., Biomers) in 1 × TE buffer (0.01 M Tris, 1 mM EDTA, sterile filtered). Store at 4 °C.
- Wash buffer (50 ml): Mix X μl 5 M NaCl (depends on the formamide concentration in the hybridization buffer; see Table 2), 1 ml 1 M Tris-HCl, 500 μl 0.5 M EDTA, 50 μl SDS (10 % (w/v), see above) and add double distilled water. Prepare fresh.
- TXP: Dissolve 0.05 % (v/v) Triton-X100 (BIO-RAD, Cat. No. 161-0407) in 1 × PBS. Prepare fresh.
- Amplification buffer (5 ml): Dissolve 0.5 g dextran sulfate (e.g., Sigma-Aldrich, Cat. No. D8906) in 2,000 μl 5 M NaCl, 50 μl blocking reagent (Roche, see above), and 2950 μl 1 × PBS

Table 2
Volumes of 5 M NaCl in 50 ml of washing buffer corresponding to the formamide concentration in the hybridization buffer. The Na^{+} concentration is calculated for stringent washing at 48 °C after hybridization at 46 °C

% formamide (v/v) in hybridization buffer	μl of 5 M NaCl
20	2,150
25	1,490
30	1,020
35	700
40	460
45	300
50	180
55	100
60	40
65	0

buffer. Aliquot and store at −20 °C or store up to 2 weeks at 4 °C.

- Hydrogen peroxide (0.15 %): Dilute 5 μl hydrogen peroxide (30 %, see above) in 1,000 μl 1 × PBS. Store at 4 °C.
- Biotinylated tyramide solution: Dissolve 1 mg 100 μl^{-1} succinimidyl D-biotin ester (e.g., Invitrogen, Cat. No. B-1513) in *N*,*N* dimethylformamide (DMF, e.g., Sigma-Aldrich, Cat. No. D4551). Incubate 1 ml of biotin-DMF stock with 330 μl tyramine HCl [10 μl triethylamine (e.g., Sigma-Aldrich) with 1 ml *N*,*N* dimethylformamide, 10 mg tyramine HCI (e.g., Sigma-Aldrich)] for 6 h in the dark at RT. Dilute with ethanol to obtain final concentration of 1 mg biotin ml^{-1} tyramide solution and add 2 % 3-iodophenylboronic acid [(w/v), e.g., Sigma-Aldrich]. Aliquot and store at −20 °C. For details on tyramide synthesis, see Hopman et al. [8].
- PGT buffer: 1 × PBS containing 0.1 % Gelatin (e.g., Fluka, Cat. No. 48720) and 0.1 % Tween-20 (e.g., Sigma-Aldrich, Cat. No. P9416).
- Alexa Fluor FluoroNanogold-Streptavidin (NanoProbes, Cat. No. 7216). Store at 4 °C, light sensitive.
- BSA: Bovine serum albumin fraction V without biotin (e.g., Roth, Cat. No. 0163).

2.3 Autometallography

- NaCl-Tween (50 mL): Mix 5 ml 5 M NaCl, 250 μl Tween-20 (Sigma-Aldrich, Cat. No. P9416), and 45 mL double distilled water. Prepare fresh.
- GoldEnhance (EM) (NanoProbes, Cat. No. 2113).
- NTS: Add 1 % (w/v) $Na_2S_2O_3$ (e.g., Sigma-Aldrich) to 50 ml double distilled water. Prepare fresh.

2.4 Microscopic Observation

- Mounting Medium with DAPI (e.g., Vector Laboraties, Cat. No. H-1200)
- Glass slides (e.g., Roth, Cat. No. H868.1) and cover slips (e.g., Roth, Cat. No. LH25.1)
- Immersion oil (e.g., Carl Zeiss Immersol 518 F)
- Adhesive carbon tabs (12 mm, e.g., Agar Scientific)
- SEM specimen stubs (12.5 mm, e.g., Agar Scientific)
- Carbon fiber for vacuum coating (e.g., Agar Scientific)

3 Methods

3.1 Fixation and Pretreatment of Microbial Cells for Gold-FISH Detection

This protocol describes the preparation of microbial cell suspensions for the application of Gold-FISH. For certain types of microorganisms and biofilms, as well as environmental samples, additional pretreatments may be required (*see* **Note 1**).

3.1.1 Fixation of Microbial Cells (In 2 ml Safe-Lock Tubes)

1. Add up to 2 ml fixation solution (4 % (v/v) formaldehyde) to microbial cells pelleted via centrifugation beforehand.
2. Resuspend by vortexing and incubate for at least 3 h at 4 °C.
3. Centrifuge for 5 min (10,000× *g*, 4 °C).
4. Discard supernatant and add 1 × PBS (up to 2 ml).
5. Resuspend pellet and repeat washing steps 3 and 4.
6. Add PBS/ethanol (1:1, v/v) and resuspend cells.
7. Store fixed cells at −20 °C or proceed.

3.1.2 Sample Preparation on Polycarbonate Filters

1. Preheat 0.2 % (w/v) agarose in a water bath (60 °C) and let cool down to approx. 40 °C.
2. Add X μL of fixed cells (depending on density) to 10 ml double distilled water.
3. Label polycarbonate filter at the edge with a pencil.
4. Filtrate samples onto individual polycarbonate filters using a vacuum filtration device (at 800 mbar).
5. Air-dry filters, dip upside in a drop (e.g., 100 μl) 0.2 % low melting point agarose and place in a petri dish with the bottom

side. Air-dry filters (alternatively in the hybridization oven at 46 °C).

6. Wash filter in absolute ethanol, air-dry, and store at −20 °C or proceed.

3.1.3 Permeabilization of Microbial Cell Walls

1. Add a drop of lysozyme solution (e.g., 100 μl) on each filter placed in a petri dish.
2. Enclose petri dish with parafilm and incubate for 60 min at 37 °C.
3. Wash filters in excess double distilled water.

For archaea additional permeabilization is recommended [9]. Therefore follow steps 4 and 5; otherwise proceed with step 7.

4. Add drop of achromopeptidase solution (e.g., 100 μl) on each filter placed in a petri dish.
5. Enclose petri dish with parafilm and incubate for 30 min at 37 °C.
6. Wash filters in excess double distilled water.
7. Immerse filters in absolute ethanol (2 min) and air-dry.
8. Store at −20 °C or proceed.

3.1.4 Inactivation of Endogenous Peroxides

1. Mix 10 ml methanol and 50 μl H_2O_2 [30 %] and transfer to a glass petri dish.
2. Add filters and incubate for 30 min at RT.
3. Wash in excess double distilled water for 5 min (twice).
4. Immerse filters in absolute ethanol (2 min) and air-dry.
5. Label and cut filters into sections using a scalpel; sections should fit into a 0.5 ml safe-lock tube.
6. Store at −20 °C or proceed.

3.1.5 Inactivation of Endogenous Biotin

1. Incubate filter section in drop of streptavidin solution (15 min, RT).
2. Wash filter sections in excess 1 × PBS (10 min).
3. Dip filters on blotting paper to remove excess liquid (with bottom side only).
4. Incubate filter section in drop of biotin solution (15 min, RT).
5. Wash filter sections in excess 1 × PBS (10 min).
6. Immerse filters in absolute ethanol (2 min), air-dry, and store at −20 °C or proceed.

3.2 In Situ Hybridization Followed by Fluorescent and Nanogold Labeling

In situ hybridization is performed with oligonucleotide probes targeting 16S rRNA in microbial cells on selected phylogenetic levels. Each probe has optimal hybridization conditions (formamide concentration in the hybridization buffer and hybridization temperature) ensuring highly stringent conditions (*see* **Note 2**).

3.2.1 In Situ Hybridization

1. Place up to 5 filter sections into 0.5 ml safe-lock tube.
2. Add 450 μl hybridization buffer and 1.5 μl probe working solution (50 ng μl^{-1}).
3. Incubate at 46 °C for at least 120 min (up to 16 h).
4. Incubate filter sections in preheated wash buffer (5 min, 48 °C).
5. Immerse filter sections in excess double distilled water (2 min, RT).
6. Immerse filter sections in excess TXP (5 min, RT).

3.2.2 Catalyzed Reporter Deposition

Catalyzed reporter deposition is used for amplification of biotinylated tyramides providing sufficient target sites for subsequent binding of streptavidin conjugates (see Sect. 3.2.3; *see* **Note 3**).

1. Dip filters on blotting paper to remove excess liquid (with bottom side only).
2. Place up to 5 filter sections into 0.5 ml safe-lock tube.
3. Add 450 μl amplification buffer, 5 μl H_2O_2 (0.15 %), 1.5 μl biotinylated tyramide solution.
4. Incubate filter sections at 46 °C for 20 min.
5. Immerse filter sections in excess PGT (10 min, RT).
6. Immerse filter sections in excess double distilled water (10 min, RT).

3.2.3 Streptavidin Binding of Fluorescent and Nanogold Labeling

Streptavidin conjugated with a fluorescent dye and a nanogold molecule is bound to biotin deposited at the hybridization sites in microbial cells (see Sect. 3.2.2; *see* **Note 3**).

1. Add 400 μl 1 × PBS containing 1 % BSA (w/v) to 0.5 ml tube.
2. Dip filters on blotting paper to remove excess liquid (with bottom side only).
3. Add 1 μl of Streptavidin-AF_{488}-Nanogold solution.
4. Incubate for at least 3 h at RT.
5. Immerse filter sections in PGT at RT (twice, 10 min each).
6. Immerse filter sections in double distilled water (10 min, RT).
7. Immerse filter sections in absolute ethanol (2 min), air-dry on microscopic slide.

For fluorescence microscopy proceed with Sect. 3.4 (*see* **Note 4**). For immediate gold enhancement omit step 7 and proceed with Sect. 3.3.

3.3 Enhancement of Nanogold Particles

Autometallographic signal amplification is used to enhance clusters of nanogold particles which are necessary for effective detection of hybridized target microorganisms via electron microscopy.

1. Immerse filter sections in NaCl-Tween (thrice, 5 min each, RT).
2. Immerse filter sections in double distilled water (thrice, 5 min each, RT).
3. Mix one drop of enhancer solution with three drops of activator solution each in glass petri dish (RT), wait 5 min (*see* **Note 5**).
4. Add one drop of initiator solution and buffer solution, respectively.
5. Immediately add filter sections and incubate for 10 min at RT (*see* **Note 6**).
6. Wash filter sections in 50 ml double distilled water while rotating (5 min each, RT; *see* **Note 7**).
7. Wash filter sections in NTS while rotating (5 min at RT).
8. Immerse filter sections in double distilled water (twice, 5 min, RT; *see* **Note 8**).
9. Air-dry filter sections in petri dish.

3.4 Microscopic Observation

3.4.1 Epifluorescence Microscopy

1. Place air-dry filter sections on microscopic slide.
2. Dip each filter section in a drop of mounting medium (with DAPI; *see* **Note 9**).
3. Add cover slip and press gently.
4. Observe fluorescent signals using blue excitation (490 nm) and green emission (510–540 nm).
5. Remove mounting medium from filter sections by washing in absolute ethanol (thrice, 3 min each).

For autometallographic enhancement of nanogold particles, proceed with Sect. 3.3.

3.4.2 Scanning Electron Microscopy

1. Place air-dry filter section on adhesive carbon tab mounted on a SEM specimen stub. Avoid warping of filter section.
2. Coat filter sections with carbon (vacuum evaporation; *see* **Note 10**).
3. Observe specific nanogold deposits using backscattered electron detector (BSD; *see* **Notes 11–12**) or energy-dispersive X-ray spectroscopy (EDX; *see* **Note 13**) along with typical SEM imaging of secondary electrons (*see* **Note 14**; Fig. 2).

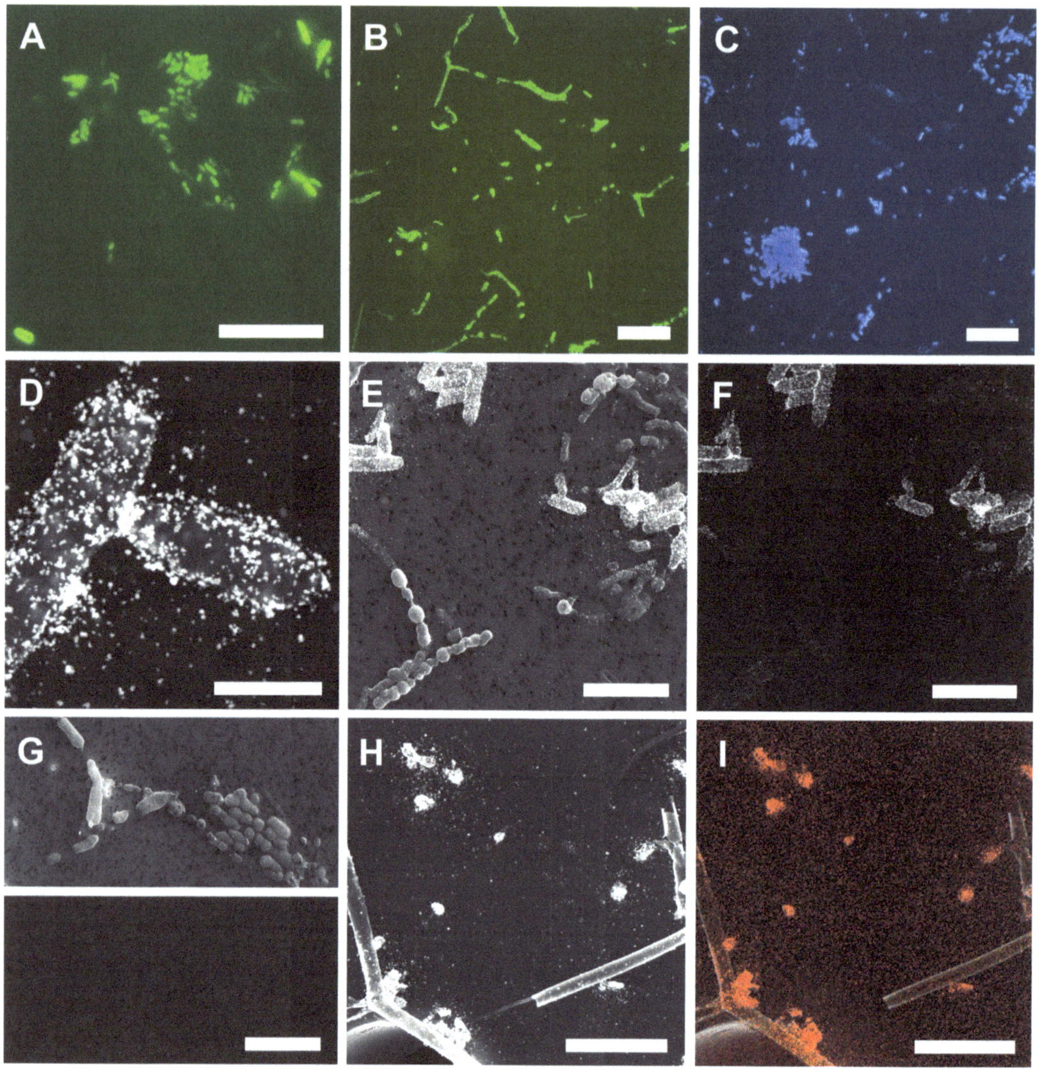

Fig. 2 Representative micrographs of Gold-FISH detected microbial cells visualized via fluorescence microscopy (A-C) and scanning electron microscopy (SEM; D-I). (A) Pure culture of *Bacillus subtilis*, probe EUB338-HRP, label AF488-nanogold; (B) Mixed culture of *E. coli* and *Roseobacter* strain AK199, probe ROS537-HRP (1), label AF488-nanogold; (C) Counterstain with DAPI corresponding to B; (D) Pure culture of *E. coli*, probe GAM42a-HRP, label AF488-nanogold followed by gold enhancement, backscattered electrons (BSE); (E) Mixed culture of *E. coli* and *Roseobacter* strain AK199, probe GAM42a-HRP, label AF488-nanogold followed by gold enhancement, SEM; (F) Corresponding BSE image of E; (G) Mixed culture of *E. coli* and *Roseobacter* strain AK199, probe NONEUB-HRP, label AF488-nanogold followed by gold enhancement, top: SEM image, bottom: corresponding BSE image; (H) *Bacteroidetes* associated with marine diatoms, probe CF319a (4), label AF488-nanogold followed by gold enhancement, SEM; (I) Elemental mapping of gold (Au-LA) corresponding to H, EDX elemental mapping merged with SEM (courtesy of Christin Bennke, MPI for Marine Microbiology Bremen). Scale bars A-C: 20 μm, D: 1 μm, E-I: 5 μm

4 Notes

1. Different cell wall structures of archaea (lacking peptidoglycan) and bacteria (thick vs. thin layers of peptidoglycan) may require the application of different fixation (formaldehyde, ethanol) and permeabilization (lysozyme, achromopeptidase, proteinase k) reagents prior to ISH. For cells clustering in biofilms, a homogenization via ultrasonication before filtration on polycarbonate membranes is recommended. For applications in sediments and soils, sonication is essential to obtain a homogenous distribution of microorganisms in the suspension before filtration.
2. As with any FISH approach, targeting microbial 16S rRNA control hybridizations should be performed using nonsense probes (e.g., NON338 [10]) to evaluate the stringency of hybridization conditions.
3. The occurrence and degree of nonspecific signals should be tested in control hybridizations without specific reagents (e.g., tyramide or streptavidin conjugate).
4. When samples should be evaluated via fluorescence microscopy in a quantitative manner, we recommend performing fluorescence microscopy prior to autometallography.
5. Unspecific accumulation of nanogold was observed on polycarbonate filters being processed in plastic petri dishes while conducting the Gold-FISH protocol probably as a result of surface charge. This effect can be avoided by using glassware throughout the entire procedure.
6. With increasing development time nanogold particles grow in size. However, gold enhancement for more than 30 min resulted in the formation of large colloids of gold clusters randomly distributed across the specimen. We therefore recommend keeping the development time to 10 min, which produces sufficient particle sizes of gold (30–100 nm).
7. Rotation during washing steps 6 and 7 (see Sect. 3.3) has been observed to reduce nonspecific background of gold particles formed during autometallography.
8. Efficient rinsing and gentle whipping of filter sections (bottom side only) is recommended when filters show higher background gold deposition near edges.
9. The simultaneous labeling with a fluorophore and nanogold allows for an analysis of the resulting FISH signals via fluorescent microscopy along with a general counterstain such as DAPI. This should be performed to verify the specificity of ISH, CARD, and streptavidin-biotin binding.

10. Carbon coating of samples may be avoided for environmental scanning electron microscopes (ESEM) preventing negative-charge accumulation. This will allow for fluorescent microscopic observations even after electron microscopic analysis.
11. When detecting the backscattered electrons representing material of higher atomic weight such as gold, clusters can be easily identified following the morphology of detected cells. Noise related to unspecific background deposition during autometallographic gold enhancement can be eliminated by using digital image analysis [4].
12. In complex environmental samples, the autometallographic enhancement of nanogold particles frequently resulted in the formation of relatively large and unspecific cocci-shaped elemental gold structures. This was observed especially in a clayey soil sample and could be the result of negatively charged clay minerals. However, the fluorescent signals resulting from the application of Gold-FISH in respective samples were clearly detectable and proven to be specific via DNA-specific DAPI staining.
13. For elemental mapping the sensitivity of the EDX detector has to be high (increased count rates at high resolution) in order to detect accumulations of enhanced nanogold particles.
14. In contrast to the fluorescent signal, the detected nanogold particles are scattered inside and around the detected microbial cell. Therefore SEM observation of nanogold-labeled cells along with secondary electron imaging is recommended or in combination with elemental mapping.

References

1. Schmidt H, Eickhorst T, Mußmann M (2012) Gold-FISH: a new approach for the in situ detection of single microbial cells combining fluorescence and scanning electron microscopy. Syst Appl Microbiol 35:518–525
2. Kenzaka T, Ishidoshiro A, Yamaguchi N et al (2005) rRNA sequence-based scanning electron microscopic detection of bacteria. Appl Environ Microbiol 71:5523–5531
3. Kubota K, Morono Y, Ito M et al (2014) Gold-ISH: a nano-size gold particle-based phylogenetic identification compatible with NanoSIMS. Syst Appl Microbiol 37:261–266
4. Eickhorst T, Schmidt H (2015) Gold-based in situ hybridization for phylogenetic single-cell detection of prokaryotes in environmental samples. In: Yates M, Nakatsu C, Miller R, Pillai S (eds) Manual of environmental microbiology, 4th edn. ASM Press, Washington, DC
5. Pernthaler A, Pernthaler J, Amann R (2002) Fluorescence in situ hybridization and catalyzed reporter deposition for the identification of marine bacteria. Appl Environ Microbiol 68:3094–3101
6. Eickhorst T, Tippkötter R (2008) Improved detection of soil microorganisms using fluorescence in situ hybridization (FISH) and catalyzed reporter deposition (CARD-FISH). Soil Biol Biochem 40:1883–1891
7. Hainfeld JF, Powell RD (2002) Silver- and gold-based autometallography of Nanogold. In: Hacker GW, Gu J (eds) Gold and silver staining: techniques in molecular morphology. CRC Press, Boca Raton, FL
8. Hopman AH, Ramaekers FC, Speel EJ (1998) Rapid synthesis of biotin-, digoxigenin-, trinitrophenyl-, and fluorochrome-labeled tyramides and their application for In situ

hybridization using CARD amplification. J Histochem Cytochem 46:771–777

9. Ishii K, Mussmann M, MacGregor BJ et al (2004) An improved fluorescence in situ hybridization protocol for the identification of bacteria and archaea in marine sediments. FEMS Microbiol Ecol 50:203–213

10. Wallner G, Amann R, Beisker W (1993) Optimizing fluorescent in situ hybridization with rRNA-targeted oligonucleotide probes for flow cytometric identification of microorganisms. Cytometry 14:136–143

Part VII

Comparative Genomic Hybridization (CGH) and Array CGH

Comparative Genomic Hybridization (CGH) and Microdissection-Based CGH (Micro-CGH)

Thomas Liehr, Anita Glaser, and Nadezda Kosyakova

Abstract

Comparative genomic hybridization (CGH) can be used for comprehensive low-resolution analysis of chromosomal imbalances of an entire human genome. Genomic DNAs from a tested specimen (test DNA) and a normal one (reference DNA) are differentially labeled and simultaneously hybridized to normal metaphase spreads. The ratio of fluorescence intensities along each normal chromosome is analyzed and allows the detection of regions that are over- or underrepresented in the tested specimen. Even though higher-resolution array CGH is nowadays widely used instead of CGH, this molecular cytogenetic approach should not be forgotten, as it can be performed also by labs only equipped with a 3-filter fluorescence microscope. In this chapter we describe the technique of CGH itself and a variant of it. The so-called microdissection-based comparative genomic hybridization (micro-CGH) technique allows the molecular cytogenetic investigation of harvested and cytogenetically fixed interphase nuclei. Micro-CGH has been already used in leukemia cytogenetics as well as clinical genetics.

Keywords Comparative genomic hybridization (CGH), Microdissection-based CGH (micro-CGH), Glass needle-based microdissection, Amplification, Loss, Tumor, Leukemia, Clinical genetics

1 Introduction

Comparative genomic hybridization (CGH), first described in 1992 [1], has since been shown to be a highly efficient tool for realizing genome-wide screening of chromosomal copy number changes within a single experiment. The standard CGH protocol relies on the availability of DNA from macroscopic samples, in most cases tumors that do not contain high proportions of non-tumor-derived (i.e., normal) cells [2]. In exceptional studies/cases, CGH has also been applied to clinical samples [3, 4]. One modification is degenerate oligonucleotide-primed polymerase chain reaction (DOP-PCR, chapter by Nadezda Kosyakova et al. "FISH-Microdissection")-based CGH, which makes it possible to survey the entire genome starting from just a few nanograms of genomic DNA [5]; as previously mentioned, up to now CGH has mainly been applied

Thomas Liehr (ed.), *Fluorescence In Situ Hybridization (FISH)*, Springer Protocols Handbooks,
DOI 10.1007/978-3-662-52959-1_54,

in solid tumor cytogenetics [2, 6, 7]. CGH studies of leukemia and lymphoma cases are much less frequently reported [8–11]; as in most cases, the bone marrow aspirate is too limited to perform DNA extraction in addition to well-established cytogenetic analysis. More recently developed array-based CGH approaches yield resolutions that are generally higher and lead to more detailed results for the gains and losses in the analyzed corresponding genomes (chapter by Eftychia Dimitriadou and Joris Vermeesch "Array CGH"; [12–16]); however, the sheer amount of obtained data [17] as well as the necessity to have an "array platform" may also favor classical CGH for specific questions. CGH can be performed even if only 3-filter-based fluorescence microscope is available.

Here we present a technique that is a combination of microdissection and DOP-PCR CGH (micro-CGH = microdissection-based comparative genomic hybridization) and which enables CGH results to be obtained from a minimum of 15 interphase nuclei from harvested and fixed cell suspensions which can be derived from malignant [8, 9, 18–20] or clinical cases (Fig. 1; [21]; chapter by Anja Weise and Thomas Liehr "Pre- and Postnatal Diagnostics and Research on Peripheral Blood, Bone Marrow, Chorion, Amniocytes, and Fibroblasts").

2 Materials

Apart from the standard cell biological and molecular cytogenetic equipment, also the specialized items listed in chapter by Nadezda Kosyakova et al. "FISH-Microdissection" are needed. The equipment recommended for FISH itself is listed in chapter by Thomas Liehr et al. "The Standard FISH Procedure".

3 Methods

3.1 Microdissection, Amplification, and Labeling of the Probes

As described in chapter by Nadezda Kosyakova et al. "FISH-Microdissection".

3.2 Preparation of the Probe Solution

As described in chapter by Thomas Liehr et al. "The Standard FISH Procedure" (*see* **Note 1**).

3.3 Slide Pretreatment

As described in chapter by Thomas Liehr et al. "The Standard FISH Procedure" (*see* **Note 2**).

3.4 Fluorescence In Situ Hybridization (FISH)

As described in chapter by Thomas Liehr et al. "The Standard FISH Procedure".

a

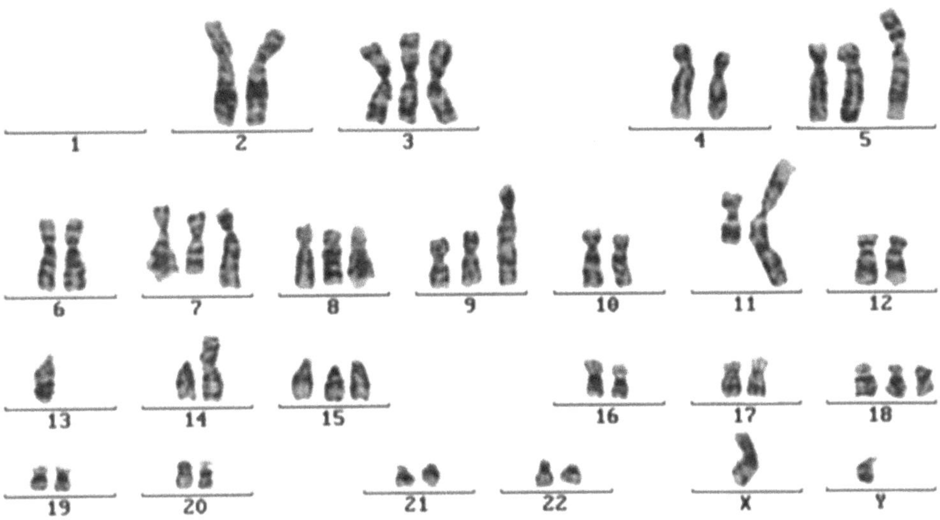

b

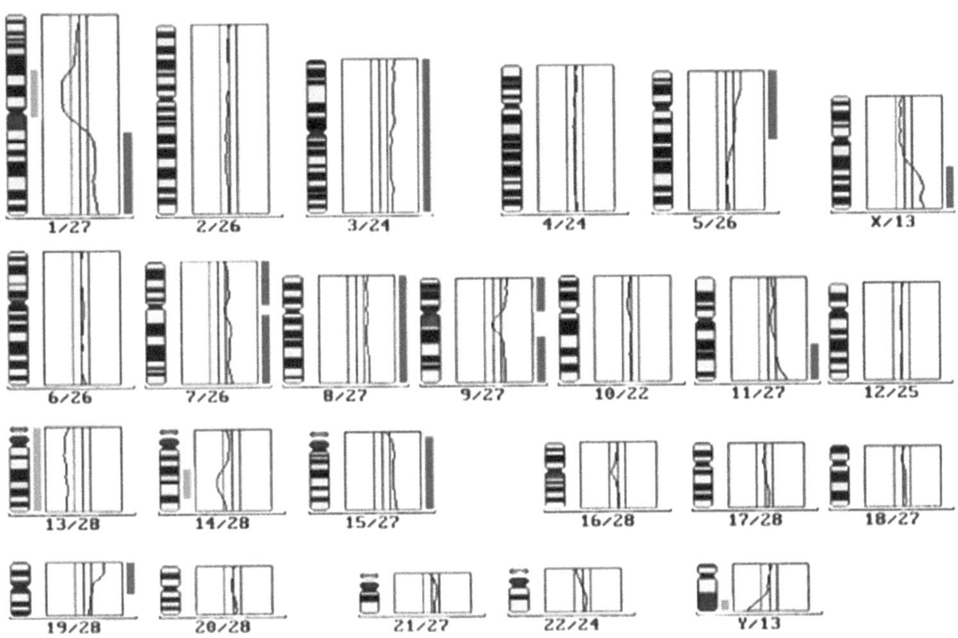

Fig. 1 (**a**) GTG banding result obtained in a case diagnosed as being plasma cell leukemia. (**b**) Micro-CGH results for this case. When these were combined with the M-FISH and multicolor banding (MCB) results (not shown here but published in Heller et al. [19]), the karyotype was found to be describable as 51,XY,-1,-1,+3, +der(5)t(5;11;1)(5pter->5q13-q14::11q24->11q25::1q12->1qter),+7 or + deryt(7;1)(7qter->7p15::1p31.1-> 1pter),+8,+der(9)t(1;9)(1qter->1q12::9q12->9pter),der(11)t(1;11;1)(1pter->1p31.1::11p15.5->11q25::1q12-> 1qter),-13,der(14)t(X;14)(Xqter->Xq21.3::14pter->14qter),+15,+18,der(19)t(9;19)(9qter->9q12::19q11->19pter), +i(19)(q10)

3.5 Evaluation

At least 15–20 well-spread metaphases should be acquired and karyotyped per case. Special software (e.g., CGH software, Metasystems, Altlussheim, Germany) is needed to analyze the CGH results. Refer to the corresponding manufacturer's instructions (*see* **Note 3**).

4 Notes

1. Labeling products derived from test DNA and reference DNA are precipitated together with higher amounts of Cot1 DNA, i.e., 5–10 μl. Prehybridization step at 37 °C should be done for at least 30 min.
2. Slides used in CGH need to carry well-spread normal metaphases to enable evaluation. So-called steal chromosomes may be prepared by doing an additional prefixation step using formaldehyde solution (same as applied later as postfixation step in chapter by Thomas Liehr et al. "The Standard FISH Procedure") previous to pepsin treatment as detailed in chapter by Thomas Liehr et al. "The Standard FISH Procedure" [22].
3. If the ratio of tumor/aberrant cells to normal cells is lower than 1:1, CGH analysis does not provide reliable results. In our experience, cases with 40 % aberrant cells (according to GTG banding results) yield analyzable ratios. Another known issue with chromosome-based CGH is that there are interpretation problems for the pericentromeric, heterochromatic and near-heterochromatin regions, as well as for chromosome 19 and 1pter. An internal control can be realized if the control and tumor samples are derived from different genders. Thus, the X chromosome can always serve as a control for the success of CGH itself, as twice the copy number of the X chromosome should always be observable in females compared to males.

References

1. Kallioniemi A, Kallioniemi OP, Sudar D et al (1992) Comparative genomic hybridization for molecular cytogenetic analysis of solid tumours. Science 258:818–821
2. Gebhart E, Liehr T (2000) Patterns of genomic imbalances in human solid tumors (review). Int J Oncol 16:383–399
3. Daniely M, Aviram-Goldring A, Barkai G et al (1998) Detection of chromosomal aberration in fetuses arising from recurrent spontaneous abortion by comparative genomic hybridization. Hum Reprod 13:805–809
4. Sanlaville D, Baumann C, Lapierre JM et al (1999) De novo inverted duplication 9p21pter involving telomeric repeated sequences. Am J Med Genet 83:125–131
5. Speicher MR, du Manoir S, Schröck E et al (1993) Molecular cytogenetic analysis of formalin-fixed, paraffin-embedded solid tumors by comparative genomic hybridization after universal DNA-amplification. Hum Mol Genet 2:1907–1914
6. Knuutila S, Björkqvist AM, Autio K et al (1998) DNA copy number amplifications in human neoplasms: review of comparative

genomic hybridization studies. Am J Pathol 152:1107–1123

7. Gebhart E (2004) Comparative genomic hybridization (CGH): ten years of substantial progress in human solid tumor molecular cytogenetics. Cytogenet Genome Res 104:352–358
8. Heller A, Chudoba I, Bleck C et al (2000) Microdissection based comparative genomic hybridization analysis (micro-CGH) of secondary acute myelogenous leukemias. Int J Oncol 16:461–468
9. Karst C, Heller A, Claussen U et al (2005) Detection of cryptic chromosomal aberrations in the in vitro non-proliferating cells of acute myeloid leukemia. Int J Oncol 27:355–359
10. Kearney L, Horsley SW (2005) Molecular cytogenetics in haematological malignancy: current technology and future prospects. Chromosoma 114:286–294
11. Gebhart E (2005) Genomic imbalances in human leukemia and lymphoma detected by comparative genomic hybridization. Int J Oncol 27:593–606
12. Albertson DG, Pinkel D (2003) Genomic microarrays in human genetic disease and cancer. Hum Mol Genet 12 (Spec No 2): R145–R152
13. Paulsson K, Heidenblad M, Mörse H et al (2006) Identification of cryptic aberrations and characterization of translocation breakpoints using array CGH in high hyperdiploid childhood acute lymphoblastic leukemia. Leukemia 20:2002–2007
14. van Beers EH, Nederlof PM (2006) Array-CGH and breast cancer. Breast Cancer Res 8:210
15. Klijn C, Holstege H, de Ridder J et al (2008) Identification of cancer genes using a statistical framework for multiexperiment analysis of nondiscretized array CGH data. Nucleic Acids Res 36:e13
16. van der Veken LT, Buijs A (2011) Array CGH in human leukemia: from somatics to genetics. Cytogenet Genome Res 135:260–270
17. Mocellin S, Rossi CR (2007) Principles of gene microarray data analysis. Adv Exp Med Biol 593:19–30
18. Chudoba I, Henn TK, Senger G et al (1997) Comparative genomic hybridization using DOP-PCR amplified DNA from small number of nuclei. Cs Pediatr 52:519–521
19. Heller A, Fricke HJ, Starke H et al (2004) Characterization of a highly aberrant plasma cell leukemia karyotype: a case report. Oncol Rep 11:89–92
20. Starke H, Raida M, Trifonov V et al (2001) Molecular cytogenetic characterization of an acquired minute supernumerary marker chromosome as the sole abnormality in a case clinically diagnosed as atypical Philadelphia-negative chronic myelogenous leukaemia. Br J Haematol 113:435–438
21. Heller A, Seidel J, Hübler A et al (2000) Molecular cytogenetic characterisation of partial trisomy 9q in a case with pyloric stenosis and a review. J Med Genet 37:529–532
22. Liehr T, Heller A, Starke H et al (2002) High quality CGH slides irrespective of used lymphocyte suspension. TTO 1(138), T02436

Array CGH

Eftychia Dimitriadou and Joris R. Vermeesch

Abstract

Array comparative genome hybridization, array CGH or aCGH, enables dense interrogation of specific loci or the entire human genome, and even allele-specific characterization of single nucleotide polymorphisms (SNPs); they are commonly referred to as oligoarrays and SNP arrays, respectively, or chromosomal microarrays in general. Here, we present an overview of the state of the field, potential applications, detailed protocols on their use, and a troubleshooting guide.

Keywords Comparative genome hybridization, SNP arrays, aCGH, Chromosomal microarrays, Copy number variation, Chromosomal imbalances

1 Introduction

Chromosome studies were initially performed using simple staining techniques that only allowed the detection of entire groups of chromosomes. The degree of precision increased in the 1970s with the introduction of chromosome-banding techniques. These techniques enabled the detection of individual chromosomes and segments (bands) within chromosomes (chapter by Thomas Liehr and Anja Weise "Background"). Although chromosomal karyotyping allows the genome-wide detection of large chromosomal abnormalities and translocations, it has a number of inherent limitations:

1. It is time consuming: It takes 4–10 days to culture the cells, visualize the chromosomes, and perform the analysis.
2. It is laborious and demanding: It requires skilled technicians to perform a Giemsa-banded karyotype analysis, which increases employment costs and can lead to organizational difficulties in small laboratories.

Thomas Liehr (ed.), *Fluorescence In Situ Hybridization (FISH)*, Springer Protocols Handbooks,
DOI 10.1007/978-3-662-52959-1_55,

3. The resolution is limited to 5–10 Mb depending on
   (a) the location in the genome
   (b) the quality of the chromosome preparation, and
   (c) the skills and experience of the cytogeneticist.
4. It requires short-term cell culture, a step that can sometimes lead to preferential growth of specific cell lines (see chapter by Thomas Liehr and Anja Weise "Background").

With the introduction of fluorescence in situ hybridization (FISH), the detection of submicroscopic chromosomal imbalances became possible (chapter by Thomas Liehr and Anja Weise "Back ground"). In FISH, labeled DNA probes are hybridized to nuclei or metaphase chromosomes in order to detect the presence, number, and location of small (submicroscopic) regions of chromosomes. Unfortunately, FISH can only detect individual DNA targets rather than the entire genome. To overcome this problem, multicolor FISH-based karyotyping [SKY, MFISH, and COBRA FISH (see chapter by Liehr and Kosyakova "Multiplex FISH and Spectral Karyotyping")] was developed, which enables the simultaneous detection of all chromosomes. Another technology allowing the genome-wide detection of copy number aberrations was introduced in 1992 and was termed "comparative genomic hybridization" (CGH). In CGH, test and reference genomic DNAs are differentially labeled with fluorochromes and then co-hybridized onto normal metaphase chromosomes. Following hybridization, the chromosomes are scanned to measure the fluorescence intensities along the lengths of the normal chromosomes in order to detect intensity ratio differences which subsequently pinpoint genomic imbalances. Overall, the resolutions at which copy number changes can be detected using these techniques are only slightly higher than those provided by conventional karyotyping (>3 Mb), and experiments are labor-intensive and time-consuming (see chapter by Thomas Liehr et al. "Comparative Genomic Hybrid ization (CGH) and Microdissection-Based CGH (Micro-CGH)").

By replacing metaphase chromosomes with mapped DNA sequences or oligonucleotides arrayed onto glass slides as individual hybridization targets, the resolution can be increased tremendously. Following the hybridization of differentially labeled test and reference genomic DNA to the target sequences on the microarray, the slide is scanned to measure the fluorescence intensities at each target on the array. The normalized fluorescence ratio for the test and reference DNA is then plotted against the position of the sequence along the chromosome. Gains or losses across the genome are identified by changes from a 1:1 ratio (log2 value of zero), and the detection resolution then depends on only the size and the number of targets on an array and the positions of these targets (their distribution) on the genome. A schematic overview of

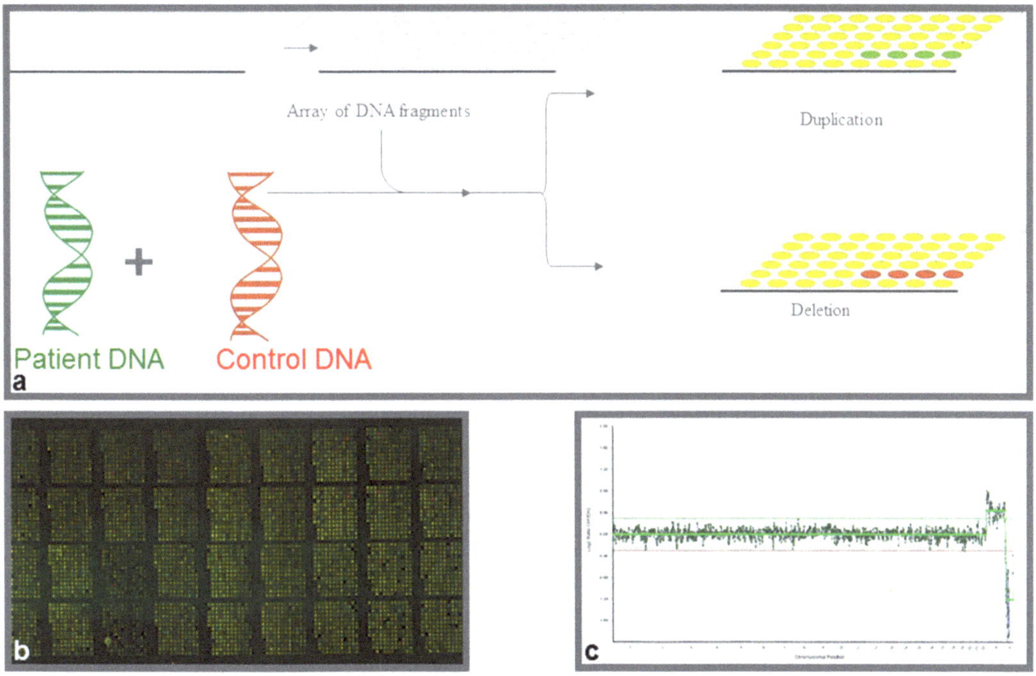

Fig. 1 (**a**) Schematic overview of the array comparative genomic hybridization technique. Test and reference DNA are labeled, each with different fluorochromes. Both DNA samples are mixed and hybridized together on a slide containing arrays of spotted DNA fragments. Excess probe is washed away, and images are taken of the bound fluorescent DNA fragments. The intensity ratios are calculated and normalized. If both the test and the control DNA have equal copy numbers for the DNA targets represented on the array, the signal intensities are equal. If a deletion or duplication is present in the test DNA, the fluorescent intensity ratios are unequal. (**b**) Screenshot from the scanned image. (**c**) Array CGH ratio profile using hybridized DNA from a female versus a male. Clones are ordered from the short arm of telomere one to the long arm of Y-chromosome. The *Y* axis shows the log-transformed intensity ratios at each locus

the technique is provided in Fig. 1. This methodology was first described in 1997 and is termed matrix or array CGH [1, 2]. Array CGH has initially been employed to analyze copy number changes in tumors in order to identify genes involved in the pathogenesis of cancers [3, 4]. Later, however, this methodology has been optimized and applied to detect unbalanced constitutional rearrangements in humans [5, 6]. With improved protocols, it rapidly became clear that not only larger insert BAC clones but also smaller-sized cDNA fragments [7], PCR products [8], and oligonucleotides were appropriate targets for array CGH [9].

Array-based approaches offer a high throughput and relatively fast way of genome-wide analysis at high resolution. The use of array CGH has significantly increased the clinical sensitivity in postnatal settings [10–12] and made it possible to uncover abnormalities leading to developmental disorders, mental retardation, intellectual disability, multiple congenital anomalies, and autism spectrum disorders [6, 13]. This led to an international consensus that genome-wide arrays are the appropriate approach for the diagnostic work-up of such patients [14]. The successful implementation of array CGH postnatally soon paved the way for the use of this technology in prenatal settings as well [15, 16] and has in the meanwhile become the method of choice for all fetuses presenting abnormal ultrasound findings [17–19]. In the absence of ultrasound abnormalities, chromosomal microarrays were reported to increase the detection rate above karyotyping by 1–2 % [17, 20]. Clinically significant findings unrelated to the initial referral reason are detected in 0.5–3.6 % of cases [19].

In the last few years, the attention of the scientific community focuses on the single-cell level [21]. Array CGH has been introduced as an alternative to targeted copy number detection for segmental and whole-chromosome aneuploidy in embryos [22–24] and has become an almost inseparable part of everyday practice in laboratories offering preimplantation genetic diagnosis (PGD) and/or screening (PGS).

While it is mainly applied in the detection of clinically relevant chromosomal imbalances, array CGH has wider applications. It has been, for instance, used to detect benign copy number variations (CNVs) among different human populations [25–27]; to detect evolutionary copy number changes among genomes of different species [28, 29]; to study copy number variation among different cell types and single cells from different populations [30]; to answer other biological questions, such as genome-wide replication timing [31, 32]; and to elucidate mechanisms leading to chromosomal instability in human embryogenesis [33, 34].

Next to array CGH, comprehensive screening techniques with a higher resolution are currently being explored. In addition to comparative hybridization using two differentially labeled DNA samples, single-sample hybridization can also be compared with various reference arrays. In such an approach, an individual is genotyped by interrogating hundreds of thousands up to millions of single nucleotide variants (SNVs), which are polymorphic in the human population. This is the basis of single nucleotide polymorphism (SNP) arrays [35] and can be used for both aneuploidy detection and genotyping, thus enabling detection of regions or chromosomes with uniparental disomy (UPD) [36]. Two main genotyping chemistries exist, which can be classified according to the design of the probes on the array: (1) allele-specific probes encompassing the SNP are deposited on the array which is hybridized with a labeled human DNA sample, enabling genotyping of genome-wide SNPs in this sample on the basis of fluorescent intensities observed above background on the probes specific for each variant allele of the SNP, or (2) the probes on the array are designed to hybridize one base upstream of the SNP locus in the DNA sample, which is followed by fluorescent one-base extension to enable SNP genotyping. Following imaging, genotyping algorithms that interpret the fluorescent signals of each probe provide the overview of all homozygous and heterozygous SNPs interrogated by the array.

In the context of PGD, SNP arrays have been used to test for monogenic diseases provided that genetic information from the parents is available and to assess copy number of thousands of SNP loci across the genome enabling aneuploidy detection for all 24 chromosomes [37–39]. Aiming to a generic approach for PGD and PGS, two genome-wide haplotyping-based methods using SNP arrays have been developed and clinically implemented for the past few years, karyomapping and siCHILD [40, 41]. These approaches offer a generic and more accurate alternative for the detection of monogenic disorders and eventual selection of those embryos that do not carry the high-risk allele for embryo transfer. To further elucidate the copy number profiles in cases of chromosomal disorders or imbalances, haplarithmisis has been developed, allowing the determination of copy number and segregational origin of the haplotypes across the genome, rendering siCHILD applicable for the diagnosis of chromosomal disorders as well [41].

In this chapter, we provide detailed protocols for DNA labeling, array hybridization, and analysis of oligoarrays and suggest protocols for SNP arrays, starting from DNA extracted from thousands or millions of cells; for an alternative protocol using microdissection-derived material for array CGH, see chapter by Maria Isabel Melaragno and Mariana Moysés-Oliveira "Breakpoint Mapping of Balanced Chromosomal Rearrangements Using Array CGH of Microdissection-Derived FISH-Probes". Protocols regarding genome-wide array-based approaches of single or small amounts of cells have been published elsewhere [42]. We also discuss problems that are occasionally encountered when performing array-based analyses. Because different platforms are now available commercially, and custom-made arrays can be ordered from several companies, we do not include protocols on the production of the arrays but instead refer the reader to other publications for more information on such protocols [43–46]. Protocols for these commercial arrays are generally derivatives of this standard protocol, but they may have array- and provider-specific modifications. No protocols are described for SNP arrays, as the approach may vary depending on the chemistry of the platform used. The reader is encouraged to follow the platform-specific protocol and is thus referred to the webpage of the respective commercial provider (Affymetrix: http://www.affymetrix.com, Agilent: http://www.genomics.agilent.com, Illumina: http://www.illumina.com).

1.1 Conclusions

Array CGH is a molecular cytogenetic tool that is largely based on the principles of the early FISH techniques. With the protocols described above in hand, it should be relatively easy for the reader to apply the technology. This technology is evolving rapidly. Whereas array production, analysis, and data interpretation were largely developed and implemented in individual laboratories, all of these aspects are now rapidly automated and commercialized. Novel and increasingly high-resolution genome-wide screening tools are being developed. The spectrum of applications for them is also expanding. Besides the detection of CNVs in healthy and diseased populations, these tools are being used to detect copy number changes in single cells [47, 48] and to detect epigenetic changes in chip-on-chip experiments that enable the visualization of DNA methylation changes and/or chromatin modifications [49].

Each application requires its own techniques, data analysis tools, and interpretation. It seems likely that with the increasing number of applications and the rapid evolution of technology, novel data analysis tools and new criteria will be developed. The protocols provided here are the original and basic protocols.

1.2 Future Directions in CNV Detection

Given that most developmental disorders are not caused by CNVs, but by single nucleotide or small insertions or deletions (indels) [50], the rapid reduction of sequencing costs over the last decade has rendered it an attractive alternative for genome-wide analyses. Whole-exome sequencing (WES) studies of children and adults with developmental disorders have shown a diagnostic yield of about 25 % [51, 52]. The use of WES in the prenatal setting, although not yet clinically implemented, is being explored as well [53, 54]. Low-coverage whole-genome sequencing is price competitive with arrays and, as a consequence, is being implemented in diagnostic laboratories for preimplantation [55] and prenatal aneuploidy detection [56, 57]; the cost for deeper sequencing for structural variation and SNV detection remains though prohibitively high.

1.3 Outline of the Array CGH Procedure and Timeline

Differentially labeled DNA derived from two samples is hybridized onto arrayed targets spotted onto slides. The difference in copy number between the two DNA samples is determined by measuring the intensity ratios of the hybridized fluorophores on each target.

Timeline

Labeling of DNA samples: 2 h to overnight

Purifying DNA and measuring the concentration: ±1 h 30 min

Preparation of hybridization solutions and blocking reagents: ±1 h 30 min

Prehybridization: ±1 h

Hybridization: 16–72 h

Washing slides: ±1 h

Scanning slides: depends on the scanning equipment

Analysis: depends on the analysis software (see Tables 1 and 2)

Table 1
Examples of commercially available software for the analysis of array CGH data

Software	Company	URL	Comments
BlueFuse	BlueGnome/ Illumina	http://www.illumina.com/	Tiff images are imported
NEXUS CGH	Biodiscovery	http://www.biodiscovery.com/	Supports several array formats
CytoGenomics	Agilent	http://www.agilent.com/	Analysis of CGH and CGH + SNP microarray
CytoSure	Oxford Gene Technology	http://www.ogt.co.uk/	Includes data annotation with links to external databases
GenomeStudio	Illumina	http://www.illumina.com/	Ability to compare data from different applications

Table 2
Examples of open-access software for the analysis of array CGH data

Software	References	URL	Comments
ViVar	[58]	https://www.cmgg.be/vivar	
CGHloop	n.a.	http://tomcat.esat.kuleuven.be/loop/	Contains automatic 2D Lowess Three-way hybridization
CGH Explorer	[59]	http://heim.ifi.uio.no/bioinf/Projects/ CGHExplorer/	Java based
CAPweb	[60]	http://bioinfo-out.curie.fr/CAPweb/	Imports gpr file
CGHnormaliter	[61]	http://www.ibi.vu.nl/programs/ cghnormaliterwww/	Normalization based on LOWESS
aCGHtool	[62]	http://www.mhh.de/acghtool.html	Uses R and Bioconductor

2 Materials

Apart from the standard equipment and chemicals, the following more specialized items are needed (listed in alphabetical order).

2.1 Equipment and Instruments

- Arrays
- Benchtop centrifuge with slide holder
- Computer with analysis software (see Tables 1 and 2)
- Dark box to keep the arrays in (typically the boxes used to hybridize FISH slides)
- Hybridization oven with rotor

- NanoDrop ND-1000 spectrophotometer (NanoDrop Technologies, Wilmington, DE, USA)
- Scanner

2.2 Chemicals, Solutions, and Consumables

- Array wash solutions:
  - Wash 1 (25 ml 20 × SSPE, 0.25 ml 20 % *N*-lauroylsarcosine, 975 ml MilliQ water) or commercially available (e.g., wash 1, Agilent)
  - Wash 2 (5 ml 20 × SSPE, 20 % *N*-lauroylsarcosine, 995 ml MilliQ water) or commercially available (e.g., wash 2, Agilent)
- Blocking reagent
- Cot-1 DNA (human)
- DNA labeling kit (e.g., BioPrime Array CGH Genomic Labeling System, Invitrogen):
  - 2.5× random primers
  - 10 × dCTP mix
  - Exo-Klenow Fragment (40 U μl^{-1}) Cy5 and Cy3-dCTP
- DNA purification columns
  - Collection tubes
  - Buffers
- Ethanol (absolute)
- Formamide
- Hybridization solution
- Salmon sperm DNA
- Sodium acetate (3 M, pH 5.2)
- SSC 20 × (3 M NaCl/0.3 M sodium citrate, pH 7.0)
- TE buffer 1 × (10 mM Tris HCl, pH = 8.0/1 mM EDTA)

3 Methods

3.1 Labeling of DNA Samples

1. Measure the concentration of DNA using NanoDrop.
2. Calculate the volume of DNA solution needed (=200–500 ng).
3. Calculate the volume of water (=9 μl - volume DNA).
4. Prepare two PCR vials for each DNA sample.
5. Add 5 μl random primers and 5 μl reaction buffer into each vial.
6. Vortex and quickly spin down.
7. Denature the samples for 20 min in a pre-warmed thermocycler at 98 °C.

8. Place the samples directly on ice or keep in the thermocycler at 4 °C for 5 min.
9. Work further on ice: Add 5 μl dCTP labeling mix and 0.5 μl Cy5 or Cy3 to each vial.
10. Vortex and spin down shortly (keep the vials in dark).
11. Add 0.5 μl Exo-Klenow to each vial (*see* **Note 1**).
12. Mix gently and shortly spin down.
13. Incubate in a pre-warmed thermocycler at 37 °C for at least 2 h (*see* **Note 2**).
14. Inactivate the enzyme by incubating the samples at 65 °C for 10 min in a thermocycler.

3.2 Purification of the Labeled DNA

Purification of the labeled DNA can be performed using different commercially available purification systems. Here, we describe two column-based approaches, Amicon columns (Amicon Ultra-0.5 ml Centrifugal Filters, Millipore) and OGT columns (part of Cyto-Sure™ Genomic DNA Labeling Kit, Oxford Gene Technologies). For some types of arrays (e.g., 24Sure BAC arrays, BlueGnome/Illumina), no purification step is required.

Variant A (Amicon Columns)

1. Spin down the PCR tubes containing the labeled DNA.
2. Mix Cy3 (red)- and Cy5 (blue)-labeled DNA (±25 μl per sample) and transfer in an Eppendorf tube. The mixture will take a purple color.
3. Add 450 μl TE buffer in each tube.
4. Vortex and spin down.
5. Put the Amicon column in a collection tube and transfer the sample in the column.
6. Centrifuge for 10 min at 13,000 rpm.
7. Throw away the elute and put the column back to the collection tube.
8. Add 450 μl TE buffer in the column.
9. Centrifuge for 10 min at 13,000 rpm.
10. Put the column upside down in a new collection tube.
11. Centrifuge for 2 min at 3,300 rpm in order to elute the labeled DNA.
12. Throw away the purification columns and keep the elute including the labeled DNA (approximately 20 μl).

Variant B (OGT Columns)

1. Vortex the purification columns.

2. Relax the lid of the column, break the plastic cover on the bottom side of the column, and put in the collection tube.
3. Centrifuge for 1 min at 2,000× *g*.
4. Transfer the column to an Eppendorf tube, and throw away the collection tube containing the elute.
5. Spin down the PCR tubes containing the labeled DNA.
6. Mix Cy3 (red)- and Cy5 (blue)-labeled DNA (±25 μl per sample). The mixture will take a purple color.
7. Carefully transfer the mixture to the center of the purification column without touching the surface of the gel.
8. Close the lid and centrifuge for 1 min at 2,000× *g*. Throw away the purification columns, and keep the elute including the labeled DNA (approximately 45 μl).

The concentration of the labeled DNA as well as the fluorescent dye incorporation can be measured by NanoDrop.

The purification of labeled DNA is a safe stopping point. Purified samples can be stored in the dark at −20 °C. The final elusion volume can be adjusted depending on the format of the array that will be used later on.

3.3 Preparation of the Hybridization Mix and the Blocking Reagent

3.3.1 Variant A (Using Self-Prepared Mixes)

i. Hybridization Mix

1. Measure the volume of DNA needed to make a 50 % Cy3/ 50 % Cy5 mix (=1750 ng of each labeled DNA sample).
2. Add 1/10th volume of sodium acetate, pH 5.2 (3 M).
3. Add 2.5× volumes of ice-cold ethanol (100 % or 96 %; denatured with methanol)
4. Each vial contains:
   (a) 1 volume of Cy3
   (b) 1 volume of Cy5
   (c) 50 μl Cot-1 DNA
   (d) 1 volume of sodium acetate
   (e) 1 volume of ethanol (ice-cold)

ii. Blocking Mixture

For each hybridization set up:

- 16.6 μl COT-1 DNA
- 2.7 μl sodium acetate

- 100 μg salmon sperm DNA
- 73.3 μl ethanol (ice-cold)

iii. Vortexing and Quickspinning the Hybridization Mix and Blocking Solution

1. Wrap in aluminum foil.
2. Place for 30 min at −70 °C for DNA precipitation.
3. Centrifuge for 15 min at 13,000 rpm.
4. Remove the supernatant, add 20 μl hybridization buffer to the blocking mixture, and vortex until the DNA pellet is dissolved.
5. Add 15 μl hybridization buffer to the hybridization mixture and 2 μl of t-RNA.
6. Vortex until the pellet is dissolved.
7. Put all vials in a water bath at 37 °C protected from the light.
8. Vortex and denature the hybridization mixture and the blocking mixture for 10 min at 75 °C.
9. Place for 5 min on ice.
10. Place the hybridization mixture at 37 °C (for 10 min).

iv. Blocking of Unspecific Hybridization

1. Put the array slide on a warm plate (37 °C).
2. Put the blocking solution on the array and add a cover slip.
3. Put the array slides in a humid chamber at 37 °C for 1 h.

3.3.2 Variant B (Using Commercially Available Reagents)

Hybridization mix and blocking reagent can also be purchased ready to use by commercial suppliers. The following protocol should be applied:

1. Pre-warm a water bath at 96 °C.
2. Mix the sample with 1 μg Cot-1 DNA and 10× blocking agent (to a final concentration of 1×).
3. Add equal volume of 2 × hybridization buffer (to a final concentration of 1 ×) (*see* **Note 3**).
4. Vortex and spin down shortly.
5. Denature the mix by incubating for 10 min at 96 °C.
6. Let the DNA renature in an incubator for 30 min–2 h at 37 °C.

3.4 Hybridization

1. Add the hybridization mix on the array and cover the slide.
2. Put the slides in a humid chamber (20 % formamide/2 × SSC) or special hybridization oven.
3. Incubate at 37–65 °C (depending on the type of the array) for at least 16 h under agitation.

3.5 Washing

It is recommended that the slides are washed in batches of maximum 4 slides.

1. Place wash 1 buffer in two glass jars (one for removing the slide cover and another one for the first wash).
2. Carefully remove the slide cover and transfer the slide on a rack into the jar containing the first washing solution.
3. Wash the slide by stirring for 5 min at room temperature.
4. Remove the rack and place it quickly into the jar containing pre-warmed wash 2 buffer.
5. Wash the slide by stirring for 1 min at 37 °C.
6. Remove the rack, and blot and dry the slide by centrifuging for 2 min at 1,200 rpm.
7. Scan the slide immediately or keep in a dark box in a N_2 desiccator.

3.6 Scanning and Analysis

The slides should be scanned with the green laser (~532 nm) and the red laser (~633 nm). The scanner requirements are dependent on the specific microarray format used and the feature extraction and analysis software that will be used during the following steps. We are currently using the Agilent G2565BA Microarray Scanner System. For this type of scanner, it is recommended that higher-resolution arrays are scanned at 2 or 3 μm and lower resolution at 5 μm.

For Illumina SNP arrays, we are using Illumina iScan supplied with the iScan Control Software and for Affymetrix SNP chips the GeneChip® Scanner 3000, controlled by the AGCC software.

For low-density arrays (<32,000 data points), further data analysis can be performed with Excel (Microsoft Inc., Redmond, WA, USA). The following approach is taken for BAC arrays containing 3,500 targets [45]. Spot intensities are corrected for local background, and only spots with signal intensities that are at least 1.5-fold above the background are included in the analysis.

Where useful, further data normalization was achieved by two-dimensional Lowess normalization using Bioconductor software [63]. Following this normalization, the values of the duplicates/triplicates on the array and the color-flip experiments were averaged and a log2 value was calculated. If the signal intensity ratios among replicate spots deviated by more than twice the overall standard deviation of all intensity ratios, the spot was not analyzed any further.

At least 95 % of the spotted clones fulfilled these quality criteria. The experiment was only scored successful if the standard deviation of the log2 of the overall spot intensity ratio was less than 0.096 [45]. This SD value for a combined experiment is typically between 0.035 and 0.06. Two or more flanking targets exceeding a value of the mean ± four times the SD of the log2 of all intensity ratios for that hybridization experiment were considered abnormal, while single targets with hybridization intensity ratios exceeding a value of $\pm[\log2(3/2) - 2SD]$ were also considered abnormal.

While Excel is a cheap analysis platform, it requires some skill to program the necessary functions. In addition, full BAC tiling arrays and oligoarrays contain large numbers of targets that are not easily manipulated in Excel. Several other commercial and noncommercial platforms have been developed (see Tables 1 and 2). Commercial software from array providers is often practically restricted to their own array format. However, data from different sources can be imported into and analyzed in some tools.

3.7 Results from BAC Arrays, Oligoarrays, and SNP Arrays

The results from array CGH are typically presented in plots of intensity ratio (Y axis) against position of the array target (X axis). Figure 2 shows typical array CGH results plotted in this way. In Fig. 2a, b, high-resolution array outputs from, respectively, BlueFuse software (suitable for analyzing BlueGnome/Illumina arrays) and CytoSure software (Agilent) are shown.

For SNP arrays, on top of the aforementioned intensity ration plots, the genotyping information is plotted in the form of B allele frequency (BAF) plots, where the BAF value (Y axis) is plotted against the chromosomal position of the SNP (X axis). In Fig. 2c, d, output from the Genome Viewer tool of GenomeStudio Software (suitable for analyzing Illumina SNP arrays) is shown.

3.8 Important Quality Parameters

3.8.1 Standard Deviation of the Intensity Ratios

The most important parameter in array CGH is the standard deviation (SD) of the intensity ratios at regions with similar copy numbers. The higher the SD, the greater the loss of information. In Fig. 3, the effect of different SD of the Log2 ratios on the technical noise of array experiment is shown. Higher SD will lower the operational resolution of an array. With low SD, single targets deviating from normal can be called copy number variable [45, 64]. With higher SD, several flanking targets should deviate in the same direction to increase the likelihood that there is a copy number difference at a specific genomic locus. Different segmentation algorithms are used to identify copy number variable genomic segments.

3.8.2 Dynamic Range

While it is important to have low SD, the distinction between one, two, and three copies should also be as close to the theoretical values (the $\mathrm{Log2}(1/2) = -1$ and the $\mathrm{Log2}(3/2) = 0.56$) as possible. In reality, we observe a widespread of dynamic ranges, depending mainly on the array platform used, the quality of the targets, and technical variables. With lower dynamic ranges, the

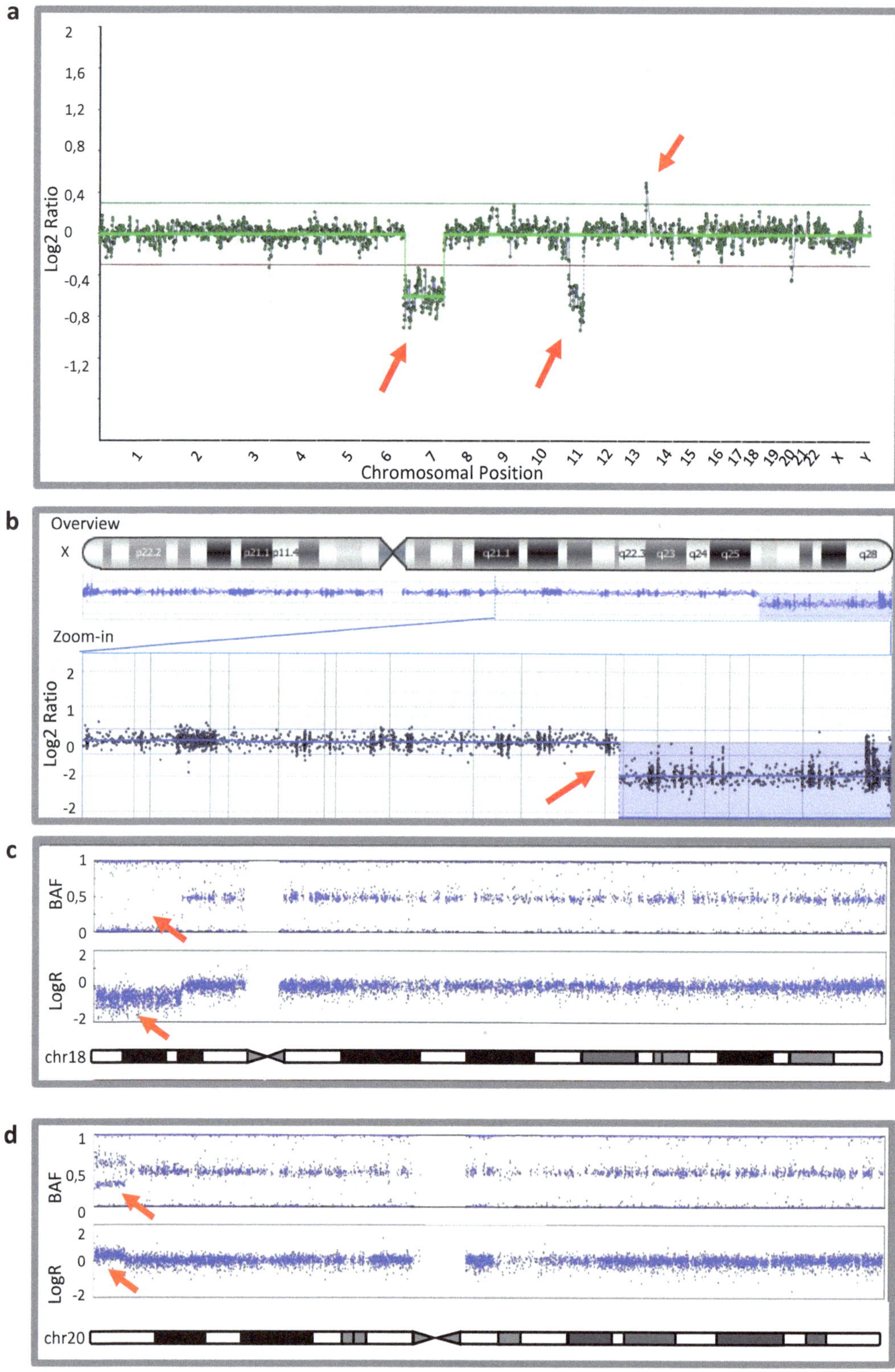
a
2
1,6
1,2
0,8
0,4
0
-0,4
-0,8
-1,2
Log2 Ratio
1 2 3 4 5 6 7 8 9 10 11 12 13 14 15 16 17 18 19 20 21 22 X Y
Chromosomal Position
b
Overview
X
p22.2
p21.1
p11.4
q21.1
q22.3
q23
q24
q25
q28
Zoom-in
Log2 Ratio
2
1
0
-2
-2
c
BAF
1
0,5
0
LogR
2
0
-2
chr18
d
BAF
1
0,5
0
LogR
2
0
-2
chr20

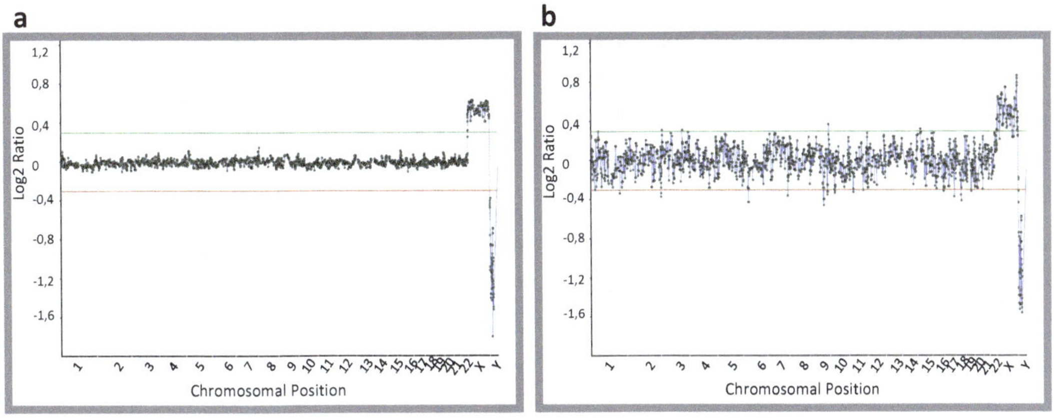

Fig. 3 (**a**) Sex-mismatch array CGH experiment, in which genomic DNA from a female sample labeled in Cy5 is hybridized against a male reference sample labeled in Cy3. (**b**) The experiment is repeated using the same sample, but this time DNA is isolated from a single cell. The effect on the SD is apparent in the image from a single cell, which displays more technical noise leading to an increase in false-positive calls

ability to call a region "variable" decreases. The dynamic range can be reduced by bad hybridization and washing conditions or the saturation of spots. With lower dynamic ranges, it becomes more difficult to discriminate an imbalance from normal variation. Hence, arrays and array protocols should be optimized to get as close as possible to the theoretical values.

3.8.3 Printing and Hybridization Artifacts

Printing or hybridization artifacts are often observed. These artifacts occur for a wide range of reasons, and many are discussed in the troubleshooting section (see Table 3). Clearly, any artifact will reduce the quality of the data (Fig. 4).

3.8.4 Control Experiments

Any user of array CGH technology should ensure the quality of the results obtained using it. To do this, a number of control

Fig. 2 (**a**) Screenshot of BlueFuse Analysis Software (BlueGnome/Illumina). Genome-wide output of a male sample hybridized against a male control on a 24Sure + array from BlueGnome, showing monosomy of chromosome 7 as well as a segmental deletion on the long arm of chromosome 11 and a segmental duplication on the long arm of chromosome 13 (*red arrows*), as a result of an unbalanced translocation t (11;13). (**b**) Screenshot of the CytoSure Analysis Software (Oxford Gene Technology). Chromosome-specific output of a female sample hybridized against a female reference on a custom 8 × 60K oligoarray from Oxford Gene Technology, showing a segmental deletion on the long arm of X-chromosome (*red arrow*). (**c**) and (**d**) Screenshots of GenomeStudio Software (Illumina) using the "Genome Viewer" tool. Chromosome-specific output of a patient sample hybridized on a HumanCytoSNP12-v2.1 bead chip from Illumina, showing a segmental deletion on the short arm of chromosome 18 (**c**) and a segmental duplication on the short arm of chromosome 20 (**d**) (*red arrows*), as a result of an unbalanced translocation t(18;20). Both BAF values (*upper panel*) and the Log2 ratio (*lower panel*) support the presence of the imbalances

Table 3
Troubleshooting; a list of potential problems and how to solve

Symptoms	Potential problem(s)	Potential solution(s)
Low Cy5 signals	Environmental conditions, including ozone, humidity, and temperature	Work in a controlled environment Add antioxidants to hybridization buffers and washing solutions Use ozone-stable fluorescent dyes
High SD	Low DNA quality	Control DNA quality
Low SD	Insufficient suppression of repeat sequences	Control Cot-1 DNA quality
Increased background	Inadequate washing conditions	Adjust the stringency of the washing conditions
Fluorescence signal heterogeneity along the array	Poor laser adjustment Ozone/light degradation Array quality	Normalize by sub-array (block normalization)
Ratio changes artificially: labeling bias	Molecular structure of Cy3 and Cy5	Introduce replicate dye-reversal hybridizations Employ direct labeling using ULS-coupled dyes Employ indirect labeling approaches involving the incorporation of aminoallyl-modified dNTPs, followed by independent direct labeling with reactive fluorochromes
Uneven hybridization	Manual hybridization with drying/leaking	Use an automated hybridization process
Low intensity, high SD	Low incorporation Poor labeling efficiency Poor recovery	Employ stringent QC for DNA quality and labeling Measure the specific activity (typically 1 fluorophore every 30–80 bp)
Low dynamic range	Poor-quality arrays Washing not stringent enough Spot saturation	Check arrays and conditions used
Low SNP call rate	Low-quality DNA Hybridization problems Stringent washing Scanning problems	Control DNA quality Review and adjust wet-lab protocol Repeat scanning of the slide

experiments can be set up [11, 47]. Self-hybridizations permit SD control, and sex-mismatch experiments enable rapid determination of the dynamic range. In addition, each experimenter should determine the operational resolution of the array experiment and the false-positive and false-negative rates of their platform by performing experiments with DNA samples carrying a variety of known CNVs.

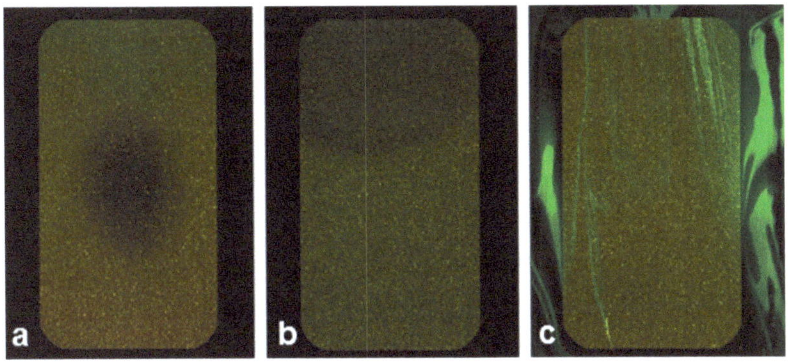

Fig. 4 Troubleshooting: Scanned images of microarrays with hybridization and washing artifacts are displayed. (**a**) and (**b**) Hybridization artifacts resulting from large air bubbles during the hybridization process. Poor hybridization is visible as the *darker regions* which show reduced signal intensities of the underlying targets. (**c**) Typical washing artifacts (seen as *green fluorescence*)

4 Notes

1. Exo-Klenow is sensitive to temperature changes, light, and mixing. Keep always on ice and put back in the freezer as soon as possible. Do not vortex.
2. Alternatively, the samples can be incubated at 37 °C overnight. In this case, the following step (step 14: inactivation of the enzyme) can be omitted.
3. For example, mix 43 μl of purified DNA with 1 μl 1 mg ml^{-1} Cot-1 DNA, 11 μl 10× blocking reagent, and 55 μl 2× hybridization buffer.

Acknowledgments

This work was supported by grants from the University of Leuven (KU Leuven), SymBioSys (PFV/10/016) and GOA/12/015 as well as Belgian Science Policy Office Interuniversity Attraction Poles (BELSPO-IAP) program through the project IAP P7/43-BeMGI.

References

1. Solinas-Toldo S, Lampel S, Stilgenbauer S et al (1997) Matrix-based comparative genomic hybridization: biochips to screen for genomic imbalances. Genes Chromosomes Cancer 20:399–407
2. Pinkel D, Segraves R, Sudar D et al (1998) High resolution analysis of DNA copy number variation using comparative genomic hybridization to microarrays. Nat Genet 20:207–211
3. Cai WW, Mao JH, Chow CW et al (2002) Genome-wide detection of chromosomal imbalances in tumors using BAC microarrays. Nat Biotechnol 20:393–396

4. Albertson DG, Pinkel D (2003) Genomic microarrays in human genetic disease and cancer. Hum Mol Genet 12(Spec No 2): R145–R152
5. Shaw-Smith C, Redon R, Rickman L et al (2004) Microarray based comparative genomic hybridization (array-CGH) detects submicroscopic chromosomal deletions and duplications in patients with learning disability/mental retardation and dysmorphic features. J Med Genet 41:241–248
6. Menten B, Maas N, Thienpont B et al (2006) Emerging patterns of cryptic chromosomal imbalance in patients with idiopathic mental retardation and multiple congenital anomalies: a new series of 140 patients and review of published reports. J Med Genet 43:625–633
7. Wolf M, Mousses S, Hautaniemi S et al (2004) High-resolution analysis of gene copy number alterations in human prostate cancer using CGH on cDNA microarrays: impact of copy number on gene expression. Neoplasia 6:240–247
8. Dhami P, Coffey AJ, Abbs S et al (2005) Exon array CGH: detection of copy-number changes at the resolution of individual exons in the human genome. Am J Hum Genet 76:750–762
9. Brennan C, Zhang Y, Leo C et al (2004) High-resolution global profiling of genomic alterations with long oligonucleotide microarray. Cancer Res 64:4744–4748
10. Vissers LE, de Vries BB, Osoegawa K et al (2003) Array-based comparative genomic hybridization for the genomewide detection of submicroscopic chromosomal abnormalities. Am J Hum Genet 73:1261–1270
11. Vermeesch JR, Fiegler H, de Leeuw N et al (2007) Guidelines for molecular karyotyping in constitutional genetic diagnosis. Eur J Hum Genet 15:1105–1114
12. Dave BJ, Sanger WG (2007) Role of cytogenetics and molecular cytogenetics in the diagnosis of genetic imbalances. Semin Pediatr Neurol 14:2–6
13. Stankiewicz P, Beaudet AL (2007) Use of array CGH in the evaluation of dysmorphology, malformations, developmental delay, and idiopathic mental retardation. Curr Opin Genet Dev 17:182–192
14. Miller DT, Adam MP, Aradhya S et al (2010) Consensus statement: chromosomal microarray is a first-tier clinical diagnostic test for individuals with developmental disabilities or congenital anomalies. Am J Hum Genet 86:749–764
15. Rickman L, Fiegler H, Shaw-Smith C et al (2006) Prenatal detection of unbalanced chromosomal rearrangements by array CGH. J Med Genet 43:353–361
16. Brady PD, Devriendt K, Deprest J et al (2012) Array-based approaches in prenatal diagnosis. Methods Mol Biol 838:151–171
17. Wapner RJ, Martin CL, Levy B et al (2012) Chromosomal microarray versus karyotyping for prenatal diagnosis. N Engl J Med 367:2175–2184
18. Brady PD, Delle Chiaie B et al (2014) A prospective study of the clinical utility of prenatal chromosomal microarray analysis in fetuses with ultrasound abnormalities and an exploration of a framework for reporting unclassified variants and risk factors. Genet Med 16:469–476
19. Shaffer LG, Dabell MP, Fisher AJ et al (2012) Experience with microarray-based comparative genomic hybridization for prenatal diagnosis in over 5000 pregnancies. Prenat Diagn 32:976–985
20. Armengol L, Nevado J, Serra-Juhé C et al (2012) Clinical utility of chromosomal microarray analysis in invasive prenatal diagnosis. Hum Genet 131:513–523
21. Owens B (2012) Genomics: the single life. Nature 491:27–29
22. Harper JC, Harton G (2010) The use of arrays in preimplantation genetic diagnosis and screening. Fertil Steril 94:1173–1177
23. Alfarawati S, Fragouli E, Colls P et al (2011) First births after preimplantation genetic diagnosis of structural chromosome abnormalities using comparative genomic hybridization and microarray analysis. Hum Reprod 26:1560–1574
24. Fiorentino F, Spizzichino L, Bono S et al (2011) PGD for reciprocal and Robertsonian translocations using array comparative genomic hybridization. Hum Reprod 26:1925–1935
25. Iafrate AJ, Feuk L, Rivera MN et al (2004) Detection of large-scale variation in the human genome. Nat Genet 36:949–951
26. Sebat J, Lakshmi B, Troge J et al (2004) Large-scale copy number polymorphism in the human genome. Science 305:525–528
27. Redon R, Ishikawa S, Fitch KR et al (2006) Global variation in copy number in the human genome. Nature 444:444–454
28. Locke DP, Segraves R, Carbone L et al (2003) Large-scale variation among human and great ape genomes determined by array comparative genomic hybridization. Genome Res 13:347–357

29. Wilson GM, Flibotte S, Missirlis PI et al (2006) Identification by full-coverage array CGH of human DNA copy number increases relative to chimpanzee and gorilla. Genome Res 16:173–181
30. Jacobs K, Mertzanidou A, Geens M et al (2014) Low-grade chromosomal mosaicism in human somatic and embryonic stem cell populations. Nat Commun 5:4227
31. Woodfine K, Fiegler H, Beare DM et al (2004) Replication timing of the human genome. Hum Mol Genet 13:191–202
32. Woodfine K, Carter NP, Dunham I et al (2005) Investigating chromosome organization with genomic microarrays. Chromosome Res 13:249–257
33. Vanneste E, Voet T, Le Caignec C et al (2009) Chromosome instability is common in human cleavage-stage embryos. Nat Med 15:577–583
34. Voet T, Vanneste E, Vermeesch JR (2011) The human cleavage stage embryo is a cradle of chromosomal rearrangements. Cytogenet Genome Res 133:160–168
35. Komura D, Shen F, Ishikawa S et al (2006) Genome-wide detection of human copy number variations using high-density DNA oligonucleotide arrays. Genome Res 16:1575–1584
36. Faas BH, van der Burgt I, Kooper AJ et al (2010) Identification of clinically significant, submicroscopic chromosome alterations and UPD in fetuses with ultrasound anomalies using genome-wide 250k SNP array analysis. J Med Genet 47:586–594
37. Treff NR, Su J, Tao X et al (2010) Accurate single cell 24 chromosome aneuploidy screening using whole genome amplification and single nucleotide polymorphism microarrays. Fertil Steril 94:2017–2201
38. Brezina PR, Benner A, Rechitsky S et al (2011) Single-gene testing combined with single nucleotide polymorphism microarray preimplantation genetic diagnosis for aneuploidy: a novel approach in optimizing pregnancy outcome. Fertil Steril 95(1786):e5–e8
39. Treff NR, Northrop LE, Kasabwala K et al (2011) Single nucleotide polymorphism microarray-based concurrent screening of 24-chromosome aneuploidy and unbalanced translocations in preimplantation human embryos. Fertil Steril 95:1606.e1–1612.e2
40. Natesan SA, Bladon AJ, Coskun S et al (2014) Genome-wide karyomapping accurately identifies the inheritance of single-gene defects in human preimplantation embryos in vitro. Genet Med 16:838–845
41. Zamani Esteki M, Dimitriadou M, Mateiu L, Melotte C et al (2015) Concurrent whole-genome haplotyping and copy-number profiling of single cells. Am J Hum Genet 96:894–912
42. Dimitriadou E, Zamani Esteki M et al (2015) Copy number variation analysis by array analysis of single cells following whole genome amplification. Methods Mol Biol 1347:197–219
43. Fiegler H, Carr P, Douglas EJ et al (2003) DNA microarrays for comparative genomic hybridization based on DOP-PCR amplification of BAC and PAC clones. Genes Chromosomes Cancer 36:361–374
44. Fiegler H, Redon R, Carter NP (2007) Construction and use of spotted large-insert clone DNA microarrays for the detection of genomic copy number changes. Nat Protoc 2:577–587
45. Vermeesch JR, Melotte C, Froyen G, Van Vooren S et al (2005) Molecular karyotyping: array CGH quality criteria for constitutional genetic diagnosis. J Histochem Cytochem 53:413–422
46. Nature Protocols. http://www.nature.com/nprot/index.html
47. Fiegler H, Redon R, Andrews D et al (2006) Accurate and reliable high-throughput detection of copy number variation in the human genome. Genome Res 16:1566–1574
48. Le Caignec C, Spits C, Sermon K et al (2006) Single-cell chromosomal imbalances detection by array CGH. Nucleic Acids Res 34:e68
49. Wilson IM, Davies JJ, Weber M et al (2006) Epigenomics: mapping the methylome. Cell Cycle 5:155–158
50. Vissers LE, Gilissen C, Veltman JA (2016) Genetic studies in intellectual disability and related disorders. Nat Rev Genet 17:9–18
51. de Ligt J, Willemsen MH, van Bon BW et al (2012) Diagnostic exome sequencing in persons with severe intellectual disability. N Engl J Med 367:1921–1929
52. Iglesias A, Anyane-Yeboa K, Wynn J et al (2014) The usefulness of whole-exome sequencing in routine clinical practice. Genet Med 16:922–931
53. Carss KJ, Hillman SC, Parthiban V et al (2014) Exome sequencing improves genetic diagnosis of structural fetal abnormalities revealed by ultrasound. Hum Mol Genet 23:3269–3277
54. Drury S, Williams H, Trump N et al (2015) Exome sequencing for prenatal diagnosis of fetuses with sonographic abnormalities. Prenat Diagn 35:1010–1017
55. Fiorentino F, Bono S, Biricik A et al (2014) Application of next-generation sequencing technology for comprehensive aneuploidy screening of blastocysts in clinical

preimplantation genetic screening cycles. Hum Reprod 29:2802–2813

56. Lo YM (2013) Non-invasive prenatal testing using massively parallel sequencing of maternal plasma DNA: from molecular karyotyping to fetal whole-genome sequencing. Reprod Biomed Online 27:593–598
57. Bayindir B, Dehaspe L, Brison N et al (2015) Noninvasive prenatal testing using a novel analysis pipeline to screen for all autosomal fetal aneuploidies improves pregnancy management. Eur J Hum Genet 23:1286–1293
58. Sante T, Vergult S, Volders PJ et al (2014) ViVar: a comprehensive platform for the analysis and visualization of structural genomic variation. PLoS One 9:e113800
59. Lingjaerde OC, Baumbusch LO, Liestøl K et al (2005) CGH-Explorer: a program for analysis of array-CGH data. Bioinformatics 21:821–822
60. Liva S, Hupé P, Neuvial P et al (2006) CAPweb: a bioinformatics CGH array analysis platform. Nucleic Acids Res 34:W477–W481
61. van Houte BP, Binsl TW, Hettling H et al (2009) CGHnormaliter: an iterative strategy to enhance normalization of array CGH data with imbalanced aberrations. BMC Genomics 10:401
62. Hofmann WA, Weigmann A, Tauscher M et al (2009) Analysis of array-CGH data using the R and Bioconductor software suite. Comp Funct Genomics 2009:201325
63. Dudoit S, Gentleman RC, Quackenbush J (2003) Open source software for the analysis of microarray data. Biotechniques Suppl:45–51
64. Vermeesch JR, Brady PD, Sanlaville D et al (2012) Genome-wide arrays: quality criteria and platforms to be used in routine diagnostics. Hum Mutat 33:906–915

Breakpoint Mapping of Balanced Chromosomal Rearrangements Using Array CGH of Microdissection-Derived FISH Probes

Maria Isabel Melaragno and Mariana Moysés-Oliveira

Abstract

The study of de novo balanced chromosome rearrangements associated with phenotypic alterations has provided an important tool for the understanding of pathogenic mechanisms, for they can reveal gene disruptions and their effects in patients. Because balanced rearrangements show no copy number alterations, standard array techniques are inefficient in studying these cases. For this reason, a variation of the array technique, named array painting, is required for determining the chromosome breakpoints. Using this technique, only the rearranged chromosomes, or chromosome segments including the rearranged region, are separated by microdissection, differentially labeled, and hybridized in the array CGH. The chromosome region corresponding to transition point from one hybridization fluorochrome signal to the other reveals the breakpoint. The resolution of the technique depends mainly on the array type and the density of probes as well as on the repetitiveness of the DNA sequence at the breakpoint.

Keywords Array painting, Chromosome microdissection, FISH probes, Array CGH, Balanced rearrangements, Breakpoint determination, Gene disruption

1 Introduction

The precise molecular characterization of structural chromosome rearrangements associated with abnormal phenotype is essential for the identification of genes and genomic segments responsible for the pathogenic effects in patients, for revealing gene functions, and for a better genotype–phenotype correlation. For unbalanced chromosomal abnormalities, the molecular characterization of the rearrangement and genotype–phenotype correlation is usually straightforward (chapter by Thomas Liehr and Anja Weise "Background"). Segments with deletion or duplication are easily detected by chromosome microarray techniques, and in these cases copy number alterations of one or more genes or regulatory elements (dosage effect) are the most common causes of the observed clinical phenotypes. The application of chromosome

Thomas Liehr (ed.), *Fluorescence In Situ Hybridization (FISH)*, Springer Protocols Handbooks,
DOI 10.1007/978-3-662-52959-1_56,

microarray analysis in apparently balanced rearrangements at the cytogenetic level, such as balanced translocations and inversions, is also required to identify the pathogenic cause for phenotypic alterations because microdeletions or microduplications undetectable by karyotype may be revealed (chapter by Eftychia Dimitriadou, Joris Vermeesch "Array CGH").

On the other hand, in cases of proven balanced alterations, in that there is no genomic imbalance, different pathogenic mechanisms have been attributed as causes of the phenotypic alterations, such as gene disruptions of dosage-sensitive genes at the breakpoints [1–3], gene disruption followed by formation of chimeric genes expressing hybrid transcripts [2, 4, 5], disruption of regulatory regions by separation of cis-regulatory elements from the genes they control [6, 7], and also position effects affecting nearby genes [8, 9]. Alternatively, the relation between an abnormal phenotype and a karyotypic alteration must be considered coincidental rather than causal [10].

In order to properly interpret the clinical impact of a balanced chromosomal rearrangement, precise breakpoint mapping is crucial [11]. Many genes have been mapped from patients with balanced rearrangements and phenotypic alterations. An example is the dystrophin gene that was mapped to Xp21 from the study of female patients with Duchenne muscular dystrophy with different balanced X-autosome translocations but all with breakpoints consistently at Xp21 [12]. Balanced rearrangements have also been important to the understanding of the role of genes in development [13]. In addition to revealing pathogenic causes, the breakpoint determination can give insights into the mechanisms underlying the formation of rearrangements [14, 15].

Despite the recognition of this approach as a powerful tool in searching pathogenic genomic regions, the investigation of balanced chromosomal rearrangements in routine diagnosis is generally performed only by low-resolution techniques, such as karyotyping. This occurs because such rearrangements do not result in large gains or losses of genetic material and are undetected by chromosomal microarray-based genome-wide surveys [16]. Furthermore, the breakpoint definition of balanced chromosomal aberrations using cytogenomic and molecular methods in research laboratories, e.g., FISH and long-range PCR followed by Sanger sequencing, are generally time consuming or technically challenging. Therefore, other methodologies, such as array painting, are more appropriate for achieving breakpoint mapping with high-resolution definition [17].

The array painting method [18] is a modification of the array-CGH technique that permits mapping breakpoints in balanced chromosome rearrangements. This method was based on the concept of reverse chromosome painting [19], in which the aberrant chromosome is purified, amplified using degenerate oligonucleotide-primed PCR (DOP-PCR), and used as a probe to hybridize onto normal metaphase spreads using FISH. For array painting, rearranged chromosomes are isolated either by flow sorting cytometry [17, 18] or by microdissection [20, 21], then amplified, differentially labeled, and hybridized onto the DNA microarray. Hybridization will show distinct fluorescence with the different fluorochromes changing their patterns (high or low) at the breakpoints.

The breakpoint resolution achieved by array painting depends mostly on the array type and its density of probes as well as on the repetitiveness of the DNA sequence at the breakpoints. Chromosomal rearrangements with one of the breakpoints at heterochromatic and repetitive regions (e.g., centromeres or the short arm of acrocentric chromosomes) and the other breakpoint at non-repetitive regions are not possible to have their breakpoints mapped with high resolution by standard sequencing-based methods. In these cases, the array painting technique corresponds to a practical and cost-effective approach to define the non-repetitive side of the chromosomal junction.

For rearrangements with both breakpoints in non-repetitive regions, the array painting results allow the design of primers for Sanger sequencing in order to map the breakpoint at the base-pair level and better describe the formation mechanism of the rearrangement (chapter by Jiří Štika and Oldřich Mazal "Sequencing of Microdissection Derived FISH Probes").

2 Material

Standard cell biological and molecular cytogenetic equipment, including standard solutions, are needed; besides following more specialized items and solutions are required.

2.1 Instruments and Equipment

- Sterile, nuclease-free pipette tips
- Vacuum concentrator
- Hybridization oven
- Microarray scanner [here exemplified by DNA Microarray Scanner (Agilent Technologies)]
- Gasket slides and hybridization chamber
- Magnetic stir plates and stir bars
- Slide holder for scanning

2.2 Reagents

- Material for chromosome microdissection, DNA amplification, and reverse chromosome painting (see chapter by Nadezda Kosyakova et al. "FISH-Microdissection").

- DNA labeling kit [here exemplified by Genomic DNA High-Throughput ULS Labeling Kit (Agilent Technologies)].
- DNA purification module [here exemplified by Agilent KREA-pure columns (Agilent Technologies)].
- Array-CGH slides Wash Buffers [here exemplified by Agilent Oligo aCGH Wash Buffer 1 and Agilent Oligo aCGH Wash Buffer 2 (Agilent Technologies)].
- Array-CGH hybridization kit [here exemplified by Agilent Oligo aCGH Hybridization Kit (Agilent Technologies)].
- Human Cot-1 DNA.
- Array-based CGH slides [here exemplified by SurePrint G3 8 × 60k (Agilent Technologies)]. Different microarray formats can be used, such as 1X, 2X, 4X, and 8X slides.

2.3 Software

- Software to scan the microarray slide, extract data, and perform quality control [here exemplified by Agilent Scan Control (Agilent Technologies)]
- Software to extract the data and perform the quality control [here exemplified by Agilent Feature Extraction (Agilent Technologies)]
- Software to determine the breakpoint region [here exemplified by Agilent CytoGenomics (Agilent Technologies)]

3 Methods

The implementation of the array painting technique requires a combination of different methods. The main steps for which are chromosome microdissection, DNA amplification, DNA labeling, hybridization in array-CGH slide, and analysis (Fig. 1). The first steps have been described in detail in specific chapters of this volume. The DNA labeling and array-CGH hybridization protocols are very flexible and vary according to DNA quality, amplicon size, array platform, and slide design. In this chapter, a protocol using 8 × 60k CGH array SurePrint G3 (Agilent Technologies) is described, but other array-CGH slide formats or slides from other companies can be used, following their respective protocols.

3.1 Chromosome Microdissection

1. Under an inverted microscope, identify the rearranged chromosomes in metaphase cells from lymphocyte cultures fixed in slides.
2. Using extended glass microneedles controlled by a micromanipulator, collect the rearranged chromosomes or chromosome regions around breakpoints.

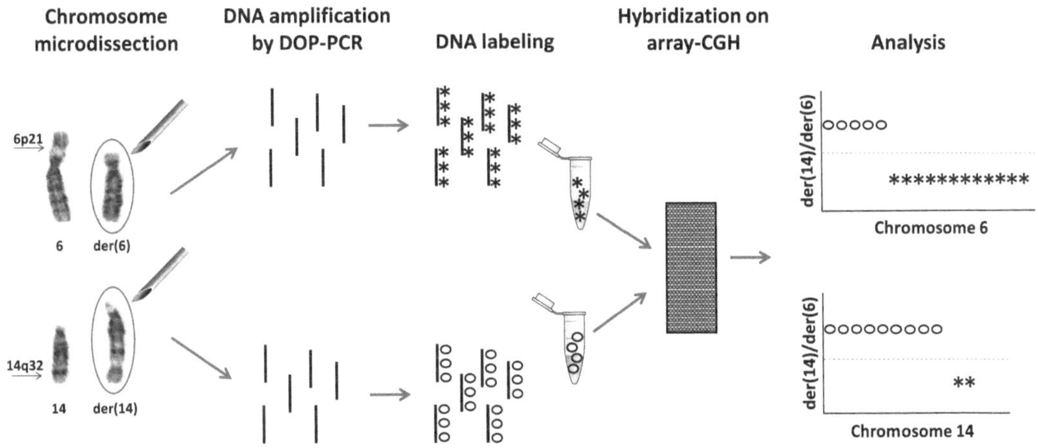

Fig. 1 Diagram of the steps for breakpoint mapping of a balanced translocation t(6;14)(p21;q32) using array CGH of microdissection-derived FISH probes: Chromosome microdissection of the derivative (der) chromosomes using a glass microneedle, DNA amplification by degenerated oligonucleotide-primed polymerase chain reaction (DOP-PCR), differential DNA labeling with Cy3-dCTP or Cy5-dCT, hybridization on an array-CGH slide, and analysis after scanning

3. Collect at least ten copies of each chromosome region and poll them separately in different tubes (e.g., one derivative chromosome or chromosome region polled per tube). For inversions, the region around each breakpoint must be microdissected separately—for details, see chapter by Nadezda Kosyakova et al. "FISH-Microdissection".

3.2 DNA Amplification

From this step onward to Sect. 3.3, the material from each different breakpoint region has to be manipulated separately.

1. Amplify acquired DNA from chromosome microdissection by degenerated oligonucleotide-primed polymerase chain reaction (DOP-PCR).
2. Apply the amplified material on a gel to check the size of the fragments. Fragments ranging from 200 to 500 bp are optimal for the array-CGH protocol—see Sect. 3.6 in chapter by Nadezda Kosyakova et al. "FISH-Microdissection".
3. Reverse chromosome painting may be done optionally for quality control as follows: Dye label an aliquot of DOP-PCR product of each breakpoint region.
4. Use the labeled DNA as FISH probe to hybridize on a normal chromosome spread acc. to chapter by Nadezda Kosyakova et al. "FISH-Microdissection" and chapter by Thomas Liehr "Homemade Locus-Specific FISH Probes: Bacterial Artificial Chromosomes".

3.3 DNA Labeling

DNA labeling and array CGH must be performed by essentially following the protocol of the manufacturer of the kits and slide format of choice. This chapter describes a protocol using 8 × 60k CGH array SurePrint G3 (Agilent Technologies).

1. Label 8 μl of DOP-PCR product from each breakpoint region with distinct fluorescent dyes (e.g., one derivative—der(A)—should be labeled with Cy3-dCTP and the other, der(B), with Cy5-dCT) using Genomic DNA High-Throughput ULS Labeling Kit (Agilent Technologies).
   - In PCR tubes or plates, mix the 8 μl DOP-PCR product with 0.25 μl ULS-Cy3 or ULS-Cy5, 0.75 μl nuclease-free water, and 1 μl 10× labeling solution.
   - In a thermocycler, incubate the mix at 85 °C for 30 min, hold it at 4 °C for 3 min, and keep on ice until purification for dye removal.
2. Purify the labeled DNA using clean-up spin columns (such as Agilent KREApure columns), in order to remove the non-reacted fluorophores.
3. Resuspend the columns with vortex mixer, loosen the cap, and place the columns in collection tubes.
4. Microcentrifuge for 1 min at maximum speed and wash the columns with 300 μl nuclease-free water.
5. Load 10 μl ULS-labeled DNA in the column and spin for 1 min at maximum speed to collect the purified DNA (*see* **Notes 1–5**).

3.4 Preparing DNA for Hybridization

Before hybridization on the array slide, the Cy5-labeled derivative material [e.g., der(A)] and Cy3-labeled derivative material [e.g., der(B)] are combined, repetitive DNA sequences are blocked with Cot-1, and reagents are added to eliminate background noise.

1. Mix the Cy-3 and Cy-5 labeled material (final volume of 20 μl) and adjust the final volume of labeled DNA for the required in the hybridization mix and transfer to a 1.5 ml tube. For the protocol example here described, use the vacuum concentrator to bring 20 μl down to 9 μl.
2. Prepare the DNA for hybridization by adding 2 μl Cot-1 DNA (1 mg ml^{-1}), 0.5 μl Agilent 100× Blocking Agent, and 22.5 μl Agilent 2× HI-RPM hybridization buffer.
3. Incubate the mix at 96 °C for 3 min and immediately incubate at 37 °C for 30 min.
4. Add 11 μl Agilent-CGH block equilibrated to room temperature (RT) (*see* **Note 6**).

3.5 Labeled DNA Hybridization on an Array-CGH Slide

The array-CGH sandwich is assembled, and DNA is hybridized in the microarray slide, here exemplified by the Agilent SurePrint G3 8 × 60k.

1. Load a gasket slide into a hybridization chamber.
2. Apply the total volume of the hybridization mix plus labeled DNA (final volume of 34 μl) on the delimited space for each reaction in the gasket slide.
3. Put the active side of the array-CGH slide Agilent SurePrint G3 8 × 60k (side with the probes printed) down onto the gasket slide.
4. Close the hybridization chamber and hand-tighten firmly and check the mobility of the bubbles.
5. Place the assembled array-CGH sandwich into the hybridization oven and incubate at 65 °C rotating at 20 rpm for 40 h (*see* **Notes 7–9**).

3.6 Array-CGH Slide Washing and Scanning

The array-CGH sandwich must be disassembled; the microarray slide washed and scanned.

1. Prepare three slide-staining dishes:

   Dish 1: filled with Agilent Oligo aCGH Wash Buffer 1 at RT on the bench.

   Dish 2: filled with Agilent Oligo aCGH Wash Buffer 1 at RT, on the magnetic stir plate and with the magnetic stir bar.

   Dish 3: filled with Agilent Oligo aCGH Wash Buffer 2 at 37 °C, on the magnetic stir plate and with the magnetic stir bar.
2. Remove the array slide/gasket slide sandwich from the hybridization chamber, submerge it into dish 1, and separate the two slides.
3. Place the microarray slide on the slide rack, submerge it into dish 2, and stir for 5 min.
4. Transfer the slide rack to dish 3 and stir for 1 min.
5. Place the microarray slide on a slide holder and scan it immediately on a microarray scanner, such as DNA Microarray Scanner (Agilent Technologies), using a software such as Agilent Scan Control.

3.7 Array Painting Analysis

After the TIFF image extraction and hybridization quality control, the breakpoint region is determined analyzing the Cy-3 and Cy-5 patterns of hybridization in the chromosomes involved in the rearrangement.

1. After scanning the microarray with the DNA Microarray scanner (Agilent Technologies), extract the TIFF image from the

scanned slide using feature extraction software, such as Agilent Feature Extraction.

2. Check the QC metrics for signal noise, intensity, and reproducibility generated by Agilent Feature Extraction.
3. Using software for probe and signal visualization, such as Agilent CytoGenomics, calculate the log2 ratios for Cy5/Cy3 intensities and plot this data along the chromosome position.
4. Verify the transition point where the log2 ratios change from high to low ratios (or vice versa) that corresponds with the rearrangement breakpoint (*see* **Notes 10–13**).

4 Notes

1. The commercial kits for array CGH are different depending on the DNA labeling chemistry: enzymatic and direct. Enzymatic labeling is more efficient and results in a better signal quality during analysis. However, some chemical modifications added to the DNA can interfere with the enzymatic labeling reaction and may jeopardize the quality of the hybridization signal. The alternative is to use kits with direct labeling, which work well with any DNA quality and have a lower cost. The reagent and sample volumes used depend on the microarray format.
2. The standard array-CGH protocol generally starts with the fragmentation step using restriction enzymes. Since the DOP-PCR generates amplicons with the appropriate size for Cy3-dCTP and Cy5-dCTP labeling, the fragmentation step can be skipped in the array painting procedure.
3. The reagent and sample volumes as well as the incubation time can be variable depending on the microarray format. The experiments should be carried out according to the manufacturer's instructions.
4. For the study of different chromosome rearrangements involving distinct chromosomes, DNA from more than one patient can be used simultaneously in the same slide. For the 1×1 M slide design, pairs of dissected derivative chromosomes from different patients and different chromosomes can be mixed in a pool before the labeling step. Thus, one array-CGH reaction will define different chromosome breakpoints.
5. The experiments with Cy3 and Cy5-dCTPs must be performed in dark conditions.
6. For incubations at high temperatures, heat blocks are preferred, and holding down the lid with a support is required.

7. The hybridization time can vary depending on the microarray format and kit used for DNA labeling. The experiments should be carried out according to the manufacturer's instructions.
8. The slide design and the probe density at the breakpoint region have a direct impact on the resolution achieved.
9. A finer mapping of chromosomal breakpoints can be obtained by using custom oligonucleotide arrays with a higher number of probes covering the region of interest. The slide customization can be obtained by using specific software and online design applications, such as eArray (Agilent; https://earray.chem.agilent.com/suredesign/), which allows the design of new probes or the selection of predesigned validated probes from databases that focus on the targets of interest.
10. The precision of the chromosome breakpoint determination depends on the resolution of the array that is based on the type of array and its probe density as well as on the DNA sequence of the genomic region where the break occurred. If the break is located in a region with repetitive DNA, such as a heterochromatic region or a segmental duplication, the accuracy is lower compared to breaks located in a single-copy DNA region.
11. In order to confirm the breakpoints identified by array, it is recommended to perform FISH with BAC probes comprising the breakpoints. The FISH hybridization signal must appear divided in both sides of the breakpoint (chapter by Thomas Liehr and Sven Hauke "Interphase FISH in Diagnostics").
12. For the precise determination of the breakpoint at nucleotide level, the sequencing of the junction fragment should be performed using Sanger sequencing. For PCR amplification of the flanking region of the junction of the two fragments, primers around the putative breakpoints should be designed.
13. Breakpoint mapping in patients carrying apparently balanced chromosome rearrangement can be also achieved by next-generation sequencing from microdissected chromosomes [22], from captured fragments of the predicted breakpoint regions [23], or from whole genome sequencing ([24], chapter by Jiří Štika and Oldřich Mazal "Sequencing of Microdissection Derived FISH Probes").

References

1. Menten B, Buysse K, Vermeulen S et al (2007) Report of a female patient with mental retardation and tall stature due to a chromosomal rearrangement disrupting the OPHN1 gene on Xq12. Eur J Med Genet 50:446–454
2. Moysés-Oliveira M, Guilherme RS, Meloni VA et al (2015) X-linked intellectual disability related genes disrupted by balanced X-autosome translocations. Am J Med Genet B 168:669–677
3. Schneider A, Puechberty J, Ng BL et al (2015) Identification of disrupted AUTS2 and EPHA6 genes by array painting in a patient carrying a de novo balanced translocation t

(3;7) with intellectual disability and neurodevelopment disorder. Am J Med Genet A 167:3031–3037
4. Di Gregorio E, Bianchi FT, Schiavi A et al (2013) A de novo X;8 translocation creates a PTK2-THOC2 gene fusion with THOC2 expression knockdown in a patient with psychomotor retardation and congenital cerebellar hypoplasia. J Med Genet 50:543–551
5. Córdova-Fletes C, Domínguez MG, Delint-Ramirez I et al (2015) A de novo t(10;19)(q22.3;q13.33) leads to ZMIZ1/PRR12 reciprocal fusion transcripts in a girl with intellectual disability and neuropsychiatric alterations. Neurogenetics 16:287–298
6. Fonseca AC, Bonaldi A, Bertola DR et al (2013) The clinical impact of chromosomal rearrangements with breakpoints upstream of the SOX9 gene: two novel de novo balanced translocations associated with acampomelic campomelic dysplasia. BMC Med Genet 14:50
7. Crippa M, Bestetti I, Perotti M et al (2014) New case of trichorinophalangeal syndrome-like phenotype with a de novo t(2;8)(p16.1;q23.3) translocation which does not disrupt the TRPS1 gene. BMC Med Genet 15:52
8. Velagaleti GV, Bien-Willner GA, Northup JK et al (2005) Position effects due to chromosome breakpoints that map approximately 900 Kb upstream and approximately 1.3 Mb downstream of SOX9 in two patients with campomelic dysplasia. Am J Hum Genet 76:652–662
9. Moysés-Oliveira M, Guilherme R dos S, Dantas AG et al (2015) Genetic mechanisms leading to primary amenorrhea in balanced X-autosome translocations. Fertil Steril 103:1289–1296
10. Gribble SM, Prigmore E, Burford DC et al (2005) The complex nature of constitutional de novo apparently balanced translocations in patients presenting with abnormal phenotypes. J Med Genet 42:8–16
11. Bugge M, Bruun-Petersen G, Brøndum-Nielsen K et al (2000) Disease associated balanced chromosome rearrangements: a resource for large scale genotype-phenotype delineation in man. J Med Genet 37:858–865
12. Koenig M, Hoffman EP, Bertelson CJ et al (1987) Complete cloning of the Duchenne muscular dystrophy (DMD) cDNA and preliminary genomic organization of the DMD gene in normal and affected individuals. Cell 50:509–517
13. Alkuraya FS, Saadi I, Lund JJ et al (2006) SUMO1 haploinsufficiency leads to cleft lip and palate. Science 313:1751
14. Gu W, Zhang F, Lupski JR (2008) Mechanisms for human genomic rearrangements. Pathogenetics 1:4
15. Le Scouarnec S, Gribble SM (2012) Characterising chromosome rearrangements: recent technical advances in molecular cytogenetics. Heredity 108:75–85
16. Talkowski ME, Ernst C, Heilbut A et al (2011) Next-generation sequencing strategies enable routine detection of balanced chromosome rearrangements for clinical diagnostics and genetic research. Am J Hum Genet 88:469–481
17. Gribble SM, Kalaitzopoulos D, Burford DC et al (2007) Ultra-high resolution array painting facilitates breakpoint sequencing. J Med Genet 44:51–58
18. Fiegler H, Gribble SM, Burford DC et al (2003) Array painting: a method for the rapid analysis of aberrant chromosomes using DNA microarrays. J Med Genet 40:664–670
19. Carter NP, Ferguson-Smith MA, Perryman MT et al (1992) Reverse chromosome painting: a method for the rapid analysis of aberrant chromosomes in clinical cytogenetics. J Med Genet 29:299–307
20. Backx L, Van Esch H, Melotte C et al (2007) Array painting using microdissected chromosomes to map chromosomal breakpoints. Cytogenet Genome Res 116:158–166
21. Obenauf AC, Schwarzbraun T, Auer M et al (2010) Mapping of balanced chromosome translocation breakpoints to the basepair level from microdissected chromosomes. J Cell Mol Med 14:2078–2084
22. Weise A, Timmermann B, Grabherr M et al (2010) High-throughput sequencing of microdissected chromosomal regions. Eur J Hum Genet 18:457–462
23. Sobreira NL, Gnanakkan V, Walsh M et al (2011) Characterization of complex chromosomal rearrangements by targeted capture and next-generation sequencing. Genome Res 21:1720–1727
24. Shluth-Bolard C, Labalme A, Cordier MP et al (2013) Breakpoint mapping by next generation sequencing reveals causative gene disruption in patients carrying apparently balanced chromosome rearrangements with intellectual deficiency and/or congenital malformations. J Med Genet 50:144–150

Sequencing of Microdissection-Derived FISH Probes

Jiří Štika and Oldřich Mazal

Abstract

It is of the greatest importance to assess the effectiveness of an anticancer therapy in acute lymphoblastic and acute myeloid leukemia patients by the quantitative detection of residual leukemic cell populations by minimal residual disease (MRD) monitoring. Regular targets for MRD monitoring are mutations in clinically relevant genes (e.g., *NPM1*, *CEBPα*) and clonal chromosomal abnormalities that generate fusion transcripts (e.g., *AML1-ETO* or *PML-RARα*). However, a significant proportion of patients lack a molecular marker that would be feasible to quantify MRD. The aim of this protocol is to obtain a chromosomal breakpoint sequence that would serve as a marker for PCR assessment of MRD after finding a potential target on a chromosome level (typically a reciprocal translocation). The procedure involves microdissection of the aberrant chromosome, multiple parallel sequencing (MPS), long-range PCR, and Sanger sequencing.

Keywords Molecular cytogenetics, Multiple parallel sequencing (MPS), Leukemia, Molecular marker, Minimal residual disease (MRD), Long-range PCR (LR-PCR)

1 Introduction

The progress of hematological cancer and the effectiveness of an anticancer therapy can be assessed by monitoring the residual population of cancer cells (chapter by Anja Weise and Thomas Liehr "Pre- and Postnatal Diagnostics and Research on Peripheral Blood, Bone Marrow, Chorion, Amniocytes, and Fibroblasts"; chapter by Eyad Alhourani et al. "Tumorcytogenetic Diagnostics and Research on Blood and Bone Marrow Smears or Effusions"). This approach is called minimal residual disease (MRD) monitoring and is typically used in the case of lymphoblastic and myeloid leukemia. There is a well-described set of mutations in clinically relevant genes (e.g., *NPM1*, *CEBPα*) and clonal chromosomal abnormalities that generate fusion transcripts (e.g., *AML1-ETO* or *PML-RARα*) that are routinely used as markers for the MRD monitoring. However, a certain proportion of patients do not carry any known markers for MRD monitoring by real-time PCR. Methods like multicolor fluorescence in situ hybridization (mFISH) could give us information

Thomas Liehr (ed.), *Fluorescence In Situ Hybridization (FISH)*, Springer Protocols Handbooks,
DOI 10.1007/978-3-662-52959-1_57, © Springer-Verlag Berlin Heidelberg 2017

about various chromosomal abnormalities, but without knowledge of the DNA sequence of the chromosomal breakpoint, we cannot prepare a specific polymerase chain reaction (PCR) assay for such a patient. To achieve this goal, we have to combine several methods.

As a first step, we have to screen the patient's chromosomes for any specific abnormalities using mFISH (chapter by Thomas Liehr and Nadezda Kosyakova "Multiplex FISH and Spectral Karyotyping"). Provided such an abnormality is found, we subsequently have this part of the affected chromosome dissected (chapter by Nadezda Kosyakova et al. "FISH-Microdissection"). The next step is to perform array-comparative genomic hybridization (chapter by Maria Isabel Melaragno and Mariana Moysés-Oliveira "Breakpoint Mapping of Balanced Chromosomal Rearrangements Using Array-CGH of Microdissection Derived FISH-Probes") or MPS of the dissected sample. Using multiple parallel sequencing (MPS), a great number of sequencing reads that cover our dissected part of the chromosome are generated. These reads are then mapped to the chromosomal references in order to narrow the location of possible chromosomal breakpoint.

The resolution of this technique is still several kilobases, which is insufficient to find the breakpoint sequence directly. For this reason the next step is performing several long-range PCRs (LR-PCR) with several sets of primers designed for different positions in both chromosomes. Products of LR-PCR are then sequenced using the standard Sanger method until the breakpoint is finally identified. This sequence can now serve as a basis for patient-specific real-time PCR assay for MRD monitoring [1].

The procedure is as follows:

- Microdissect the region around the chromosomal breakpoint, amplify dissected material by degenerate oligonucleotide-primed (DOP) PCR, and verify specificity of amplified DNA by dye labeling an aliquot of DOP-PCR product and subsequent hybridization to control metaphase chromosomes (reverse FISH) (chapter by Nadezda Kosyakova et al. "FISH-Microdissection").
- Perform MPS of amplified DNA fragments, and approximately localize the breakpoint by mapping the obtained reads on the reference sequence of the respective chromosome. Any MPS system can be used. The system used in this protocol is GS Junior (Roche).
- Use the last mapped reads from both chromosomes as docking sites for sets of primers for long-range PCR to amplify the putative breakpoint.
- Identify the exact breakpoint by Sanger sequencing of the purified product of LR-PCR.
- Design real-time PCR assay to monitor MRD.

2 Materials

2.1 DOP-PCR Product Purification

- MinElute PCR Purification Kit (cat. no. 28004, Qiagen)
- 96–100 % molecular biology grade ethanol

2.2 Library Preparation

- GS Rapid Library (RL) Rgt/Adaptor Kit (cat. no. 5619203001, Roche)
- TE buffer (10 mM Tris-Cl, pH 8.0; 1 mM EDTA)

2.3 Library Double AMPure Purification

- Magnetic particle concentrator (MPC) for 1.5 ml tubes
- Agencourt AMPure XP (cat. no. A63880, Beckman)
- 96–100 % molecular biology grade ethanol

2.4 MPS

- Magnetic particle concentrator (MPC) for 1.5 ml tubes
- GS Junior Titanium emPCR Kit (Lib-L) (cat. no. 5996481001, Roche)
- GS Junior Titanium Sequencing Kit (cat. no. 5996554001, Roche)
- GS Junior Titanium PicoTiterPlate Kit (cat. no. 5996619001, Roche)
- 96–100 % molecular biology grade ethanol

2.5 LR-PCR

- Expand Long Range dNTPack (cat. no. 04829034001, Roche)
- Molecular biology grade water

2.6 For LR-PCR Product Purification

- QIAquick Gel Extraction Kit (cat. no. 28704, Qiagen)
- 96–100 % molecular biology grade ethanol
- Molecular biology grade isopropanol
- Scalpel

2.7 Sequencing

- GeneAmp PCR Systems 9700 (*see* **Note 1**)
- Vortex mixer, for example, Vortex 3 (IKA; other possibilities can be found in BigDye XTerminator Purification Kit Protocol)
- Capillary sequencer, for example, Genetic Analyzer 3500 (Applied Biosystems)
- BigDye Terminator v3.1 Cycle Sequencing Kit (Applied Biosystems)
- BigDye XTerminator Purification Kit (Applied Biosystems)
- Molecular biology grade water

3 Methods

3.1 DOP-PCR Product Purification

1. Put approximately 500 ng of the DOP-PCR product into a 1.5 ml tube (*see* **Note 2**).
2. Add 5 volumes of Buffer PB to 1 volume of the DOP-PCR product and mix.
3. Place a MinElute column in a provided 2 ml collection tube in a suitable rack.
4. To bind the DNA, apply the sample to the MinElute column and centrifuge for 1 min exact.
5. Discard the flow-through. Place the MinElute column back into the same tube.
6. To wash, add 750 μl Buffer PE (with 96–100 % ethanol added—see bottle label for volume) to the MinElute column and centrifuge for 1 min.
7. Discard the flow-through, and place the MinElute column back in the same tube.
8. Centrifuge the column for an additional 1 min at the maximum speed.
9. Place the MinElute column in a clean 1.5 ml tube.
10. To elute the DNA, add 16 μl Buffer EB (10 mM Tris-Cl, pH 8.5) to the center of the membrane.
11. Incubate the column for 1 min, and then centrifuge for 1 min (*see* **Note 3**).

3.2 Library Preparation

Prepare the library according to the MPS system you use. Here we use the GS Junior system as an example. Library preparation comprises fragment end repair and adaptor ligation.

3.2.1 Fragment End Repair

1. Prepare the End-Repair Mix containing 2.5 μl RL 10× buffer, 2.5 μl RL ATP, 1 μl RL dNTP, 1 μl RL T4 polymerase, 1 μl RL PNK, and 1 μl RL Taq polymerase. The total volume is 9 μl.
2. Add 16 μl of purified DOP-PCR product.
3. Vortex for 5 s, then briefly centrifuge the tube.
4. Run the End Repair program on a thermocycler, with the heated lid turned on: 25 °C/20 min and 72 °C/20 min and 4 °C/∞.

3.2.2 Adaptor Ligation

1. Add 1 μl of RL Adaptor to the reaction tube.
2. Add 1 μl of RL Ligase to the reaction tube.
3. Vortex for 5 s, then briefly centrifuge the tube.
4. Incubate on the thermocycler at 25 °C/10 min.

3.3 Library Double AMPure Purification

1. Briefly centrifuge the tube.
2. Transfer 25 μl of prepared library to a 1.5 ml tube.
3. Vortex an aliquot of AMPure beads for 20 s or until the beads are completely re-suspended.
4. Add 40 μl of AMPure beads to the tube (final ratio of library to AMPure beads of 1:1.6), and vortex for 5 s.
5. Incubate at room temperature for 10 min. Briefly centrifuge the tube.
6. Place the tube in a magnetic particle collector (MPC), and incubate at room temperature for 5 min.
7. With the tube still on the MPC, carefully remove and discard the supernatant without disturbing the beads.
8. Briefly centrifuge and put the tube back on the MPC. Remove the rest of supernatant completely.
9. Add 25 μl of 1× TE to each tube. Vortex for 5 s or until the pellet is completely re-suspended. Briefly centrifuge the tube.
10. Place the tube in the MPC and incubate at room temperature for 2 min.
11. With the tube still in the MPC, carefully transfer the supernatant (25 μl library) to new 1.5 ml tube.
12. Repeat steps 4–9, then go to step 13.
13. Remove the tube from the MPC, and add 200 μl of 70 % ethanol (freshly prepared) to each tube.
14. Vortex the tube for 5 s. The pellet may not re-suspend completely; this is acceptable. Spin down for 2 s.
15. Place the tube on the MPC and incubate for 1 min.
16. With the tube still on the MPC, carefully remove and discard the supernatant without disturbing the beads.
17. Perform a quick spin and put the tube back on the MPC. Remove the rest of supernatant completely.
18. Repeat steps 13–17. Remove as much of the supernatant as possible.
19. Briefly centrifuge the tube.
20. Place the open tube on a heat block set at 37 °C until the pellet is completely dry (about 2 min). Do not leave the tube on the heat block longer than necessary to avoid overdrying (avoid visible cracks on AMPure bead pellet).
21. Remove the tube from the heat block.
22. Add 10 μl of 1× TE to each tube. Vortex for 5 s or until the pellet is completely re-suspended (spin down if necessary).

23. Place the tube in the MPC, and incubate at room temperature for 2 min.
24. With the tube still in the MPC, carefully transfer the supernatants to a set of fresh screw cap o-ring 1.5 ml tube.
25. Store the purified amplicon pools individually at −20 °C until ready to proceed.
26. Repeat whole double AMPure purification from step 1 once. For this whole part, also *see* **Notes 4** and **5**.

3.4 MPS

Before sequencing on GS Junior, the library must be amplified by emulsion PCR (). For emulsion PCR and sequencing on GS Junior, follow the manufacturer's instructions. Map the obtained reads on the reference sequence of respective chromosomes.

3.5 Long-Range (LR) PCR

1. Use the last mapped reads from both chromosomes as docking sites for sets of primers for long-range PCR to amplify the putative breakpoint.
2. Prepare reaction mixes containing 200 ng DNA, 5 U Expand Long Range Enzyme Mix, 5 % DMSO, 1 × Expand Long Range Buffer with $MgCl_2$, 500 μM each dNTP, and 0.4 μM of each primer in 50 μl final volume.
3. Run PCR—92 °C/8 min. followed by 35 cycles of 92 °C/30 s, 58 °C/30 s, and 68 °C/6 min.
4. Run 10 μl of the PCR product on 0.7 agarose gel to check the quality of the PCR product.

3.6 LR-PCR Product Purification

1. Excise the LR-PCR product from the agarose gel with a clean, sharp scalpel, and transfer to 1.5 ml tube.
2. Add approximately 3 volumes of Buffer QG to 1 volume of the gel.
3. Incubate at 50 °C for 10 min (or until the gel slice has completely dissolved). Vortex the tube every 2–3 min to help dissolve the gel.
4. Add 1 gel volume of isopropanol to the sample and mix.
5. Place a QIAquick spin column in a provided 2 ml collection tube.
6. To bind the DNA, apply the sample to the QIAquick column and centrifuge for 1 min.
7. Add 500 μl Buffer QG to the QIAquick column and centrifuge for 1 min. Discard flow-through and place the QIAquick column back into the same tube.

8. To wash, add 750 µl Buffer PE to the QIAquick column; let the column stand 2–5 min after addition of Buffer PE, and centrifuge for 1 min. Discard flow-through, and place the QIAquick column back into the same tube.
9. Centrifuge the QIAquick column in the provided 2 ml collection tube for 1 min to remove residual wash buffer.
10. Place QIAquick column into a clean 1.5 ml tube.
11. To elute the DNA, add 50 µl Buffer EB (10 mM Tris-Cl, pH 8.5) to the center of the QIAquick membrane, and centrifuge the column for 1 min.

3.7 Sequencing

1. Mix together: 2 µl Ready Reaction Premix, 1 µl BigDye Sequencing Buffer, 1.5 pM appropriate sequencing primer, and appropriate amount of template (usually 1 µl is sufficient to provide a satisfactory signal). Adjust the final volume by molecular biology grade water to 10 µl.
2. Run sequencing reaction on thermal cycler at 96 °C/1 min, followed by 25 cycles of 96 °C/10 s, 50 °C/5 s, and 60 °C/4 min. Use Rapid thermal ramp 1 °C/s.
3. Briefly centrifuge the product.
4. Add 10 µl of XTerminator Solution and 45 µl of SAM Solution.
5. Vortex at 2,000 rpm for 30 min.
6. Centrifuge at 1,000× *g* for 1 min.
7. Run the sequencing analysis on the capillary sequencer according to the manufacturer's protocol.
8. Based on the sequence of the PCR product, a pair of primers and a TaqMan fluorescently labeled hybridization probe can be designed for real-time PCR assay to monitor MRD.

4 Notes

1. Other PCR systems can be also used but may require adjustment of the cycling program.
2. An aliquot of the DOP-PCR product can be quantified on agarose gel using molecular mass standard.
3. Purification of the DOP-PCR product by the MinElute PCR Purification Kit can be preceded by QIAquick Gel Extraction Kit (cat. no. 28704, Qiagen). Since no noticeable improvement was detected, we recommend using only the MinElute PCR Purification Kit.

4. The quality of the library can be checked by agarose gel electrophoresis or Agilent Bioanalyzer. If the adapters are removed successfully, you can omit the second double AMPure purification.
5. Library can be stored 2 weeks at 4 °C or 6 months at −20 °C.

Reference

1. Jancuskova T, Plachy R, Stika J et al (2013) A method to identify new molecular markers for assessing minimal residual disease in acute leukemia patients. Leuk Res 37:1363–1373

INDEX

Thomas Liehr (ed.), *Fluorescence In Situ Hybridization (FISH)*, Springer Protocols Handbooks, DOI 10.1007/978-3-662-52959-1, © Springer-Verlag Berlin Heidelberg 2017

Kolibri 360 GmbH
Ferdinand-Jühlke-Straße 7
99095 Erfurt, Deutschland
produktsicherheit@kolibri360.de